Construction Materials: Engineering Essentials

Construction Materials: Engineering Essentials

Editor: Jonathan Hunter

MURPHY & MOORE
www.murphy-moorepublishing.com

Published by Murphy & Moore Publishing
1 Rockefeller Plaza,
New York City, NY 10020, USA
www.murphy-moorepublishing.com

Construction Materials: Engineering Essentials
Edited by Jonathan Hunter

International Standard Book Number: 978-1-63987-131-5 (Hardback)

Cataloging-in-Publication Data

Construction materials : engineering essentials / edited by Jonathan Hunter.
p. cm.
Includes bibliographical references and index.
ISBN 978-1-63987-131-5
1. Building materials. 2. Materials. 3. Structural engineering. I. Hunter, Jonathan.
TA403.6 .C66 2022
691--dc23

Contents

Preface

This book has been a concerted effort by a group of academicians, researchers and scientists, who have contributed their research works for the realization of the book. This book has materialized in the wake of emerging advancements and innovations in this field. Therefore, the need of the hour was to compile all the required researches and disseminate the knowledge to a broad spectrum of people comprising of students, researchers and specialists of the field.

Construction materials are the category of materials that are used for construction purposes. There are both man-made and natural construction materials. Natural construction materials include clay, rocks, sand, wood twigs, thatch, etc. Man-made construction materials include cement, bricks, tiles, metal concrete, ceramics, etc. Construction materials are divided into specific specialty trades, such as carpentry insulation, plumbing, and roofing work and many more. The two most widely used construction materials are cement and concrete. Cement bonded composites are used for making pre-cast building components. These are made of hydrated cement paste which binds fibres, wood or particles. Portland cement concrete and reinforced concrete are the two most common forms of concrete used in construction. This book contains some path-breaking studies in the field of construction materials. It is a compilation of chapters that discuss the most vital concepts and emerging trends in the field of construction materials. Researchers and students in this field will be greatly assisted by this book.

At the end of the preface, I would like to thank the authors for their brilliant chapters and the publisher for guiding us all-through the making of the book till its final stage. Also, I would like to thank my family for providing the support and encouragement throughout my academic career and research projects.

Editor

1

Carbonation of Water Repellent-Treated Concrete

Peng Zhang,[1,2] Peibing Li,[1] Hong Fan,[1] Huaishuai Shang,[1] Siyao Guo,[1] and Tiejun Zhao[1]

[1]*Center for Durability & Sustainability Studies, Qingdao University of Technology, Qingdao 266033, China*
[2]*Institute of Concrete Structures and Building Materials, Karlsruhe Institute of Technology, Karlsruhe 76131, Germany*

Correspondence should be addressed to Peng Zhang; zhp0221@163.com and Huaishuai Shang; shanghuaishuai@aliyun.com

Academic Editor: Enzo Martinelli

Water repellent treatment has been considered an effective preventive method against water and aggressive ions penetration into concrete and consequently can improve the durability of concrete structures. In reality, many concrete structures are exposed to conditions with high risk of carbonation. In this contribution, one type of ordinary concrete had been prepared and surface impregnated by 400 g/m^2 silane cream and 100 g/m^2 and 400 g/m^2 silane gel. In addition, integral water repellent concrete was produced by adding 2% silane emulsion. Then, the specimens were exposed to accelerated carbonation for 7, 28, and 72 days. The effect of water repellent treatment on carbonation of concrete has been investigated. The results indicate that surface impregnation reduced carbonation depth of concrete under RH 70%, but integral water repellent concrete increased carbonation. Carbonation reaction started behind the hydrophobic layer in the surface-impregnated concrete. The coefficient of carbonation can be described better by a hyperbolic function of time. Treatment by 400 g/m^2 silane gel and silane cream showed better efficiency on reducing carbonation than usage of 100 g/m^2. Coefficient of water capillary suction was decreased significantly by both surface impregnation and integral water repellent treatment. It is an effective method to protect concrete from water penetration into the material.

1. Introduction

Ordinary cement concrete is a cheap and versatile building material which has been applied for nearly 200 years with great success and in most diverse structures. It is hardly possible to imagine the construction of the infrastructure in all industrialized nations today without concrete. But pure cement concrete is rather brittle with a small ultimate strain capacity. This often leads to early crack formation. Moreover, cement concrete is porous, which allows aggressive substances to migrate into the material. This is at the origin of early deterioration of concrete in aggressive environment and of limited protection of embedded steel reinforcement.

Moisture transport in cement-based materials is an important physical process with direct influence on service life and durability. If the surface of concrete is in direct contact with water-soluble compounds, hydration products may be washed out by leaching. In this way, near-surface zones are weakened and become more frost sensitive. If the liquid in contact with the surface is an acid, this deteriorating process is even accelerated. In case water gets in contact with a predried surface of concrete, it will be absorbed quickly and deep into the pore space. Aggressive compounds dissolved in water can be transported deep into the pore space of concrete. Drying of concrete always imposes shrinkage of the dry zones and leads to crack formation. Gases of the environment may be dissolved in the aqueous pore solution of concrete. In this way, acids are formed, for instance, by dissolution of CO_2, which could react rapidly with the hydration products of cement.

It is obvious that all these deterioration mechanisms act from the surface of concrete. Throughout history, a range of protective materials have been applied on the exposed surfaces of structural concrete elements to prevent the ingress of water. Nowadays, water repellent treatment has proved to be effective to prevent concrete from water penetration and consequently can improve the durability of concrete structures [1–8]. Application of hydrophobic treatment with an approved product has even become standard practice on new bridge decks in the Netherlands [9]. For broader information about the studies on water repellent treatment, the proceedings of the Hydrophobe conference series (Hydrophobe I–VIII) can be referred [10–17].

When applying the water repellent agents on actual structures, however, many concrete surfaces are exposed to carbonation condition, and some had been carbonated already before the surface treatment. Heinrichs et al. [18] and Gerdes [19] reported that water repellent treatment alters the drying attitude of concrete. This in turn mainly influences the process of carbonation. They had simulated the transport, storage, and chemical reaction of the main influencing variables for carbonation of treated concrete. It was believed that the carbonation process can be delayed by surface impregnation under proper conditions. The study from Courard and Lucquiaud [20] also showed that the application of a water repellent delayed the propagation of the carbonation front. However, there is some concern that the gradual drying out due to hydrophobic treatment causes deeper carbonation [9]. Besides, very little study has focused on the influence of integral water repellent treatment on the carbonation process of concrete. Therefore, it has to be clarified if the cement-based materials could be protected by water repellent treatments any more when carbonation reaction occurs simultaneously.

In this contribution, one type of ordinary concrete with a water-to-cement ratio of 0.5 has been produced. Two types of water repellent treatment were applied. One was surface impregnation. The surface of the samples was covered by different amounts of silane, in this study, 400 g/m^2 silane cream and 100 g/m^2 or 400 g/m^2 silane gel. Another type was integral water repellent treatment by adding 2% of silane emulsion in fresh concrete during mixing. All surface-impregnated and integral water repellent concrete were exposed to accelerated carbonation for 7, 28, and 72 days. In this way, the effect of water repellent treatment on carbonation of concrete has been investigated. Another emphasis had been concentrated on the water capillary absorption behavior of water repellent-treated concrete after carbonation.

2. Basic Mechanism of Water Repellent Treatment

There are basically two different technologies to make concrete water repellent: (1) surface impregnation and (2) integral water repellent treatment. In the first case, the surface of hardened concrete is impregnated with a reactive agent, and in the second case, the reactive agent is added to the fresh mix. Different organic compounds have been used in the past as water repellent agents. Historically, natural polymers such as oils, fats, and waxes have been used to protect statutes and buildings. Today, in many cases, silanes are applied to make concrete or stones water repellent [15, 16]. The general composition of higher silanes can be written as H-$(SiH_2)_n$-H with $n = 1, 2, 3, \ldots$. Similar to alkanes, silanes with more than three Si atoms can form structural isomers. In the presence of water, silanes react to form siloxanes and finally silicon resin. The resin forms a network, which is firmly fixed on the huge surface of the hydration products of Portland cement. In Figure 1, the replacement of an adsorbed water layer by the silicon resin network is shown schematically.

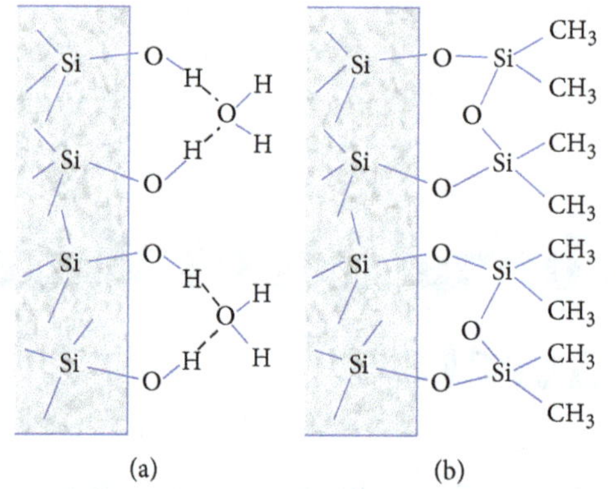

FIGURE 1: Schematic representation of (a) adsorbed water layer on the surface of hardened cement paste and (b) silicon resin fixed on the surface replacing the water layer.

In the case of surface impregnation, liquid silane penetrates into the pore space of concrete by capillary action. In order to achieve a sufficient penetration depth, the contact time between the concrete surface and the liquid silane must be long enough [21–23]. On a horizontal surface, this can be achieved easily, but vertical surfaces have to be sprayed or brushed with liquid silane for several times. This problem can be overcome by applying a so-called cream, a silane emulsion which can be applied once with the necessary thickness or silane gel which is a suspension of fine clay articles in silane. The gel is stiff enough to be applied with the necessary thickness. If integral water repellent concrete is to be produced, silane emulsion can be added to the fresh mix. After hardening and initial drying, the emulsion breaks and silicon resin is formed on the pore walls. In principle, the distribution of silicon resin in integral water repellent concrete should be uniform all over the volume. This is, however, not automatically achieved.

The effect of water repellents is essentially based on their low surface tension. The behavior of water when contacting the surface of a material is governed by the surface tension which can be measured by the contact angle. The intensity of the water repellent property is associated with the contact angle between water and the treated surface. Contact angles of a water droplet of more than 90° represent hydrophobic property while less than 90° hydrophilic property. The organofunctional groups will reduce the critical surface tension of the material surface and thus provide hydrophobicity.

3. Materials and Methods

3.1. Materials and Preparation of Specimens. The composition of the concrete matrix has been the same for all specimens used in this project, which is shown in Table 1. Portland cement 42.5 type II with a compressive strength at 28 days of no less than 42.5 MPa, local crushed granitic aggregates with a maximum diameter of 20 mm and density of 2620 kg/m^3, and river sand with a maximum grain size of

Table 1: Composition of the concrete used in this project.

Type	W/C	Cement, kg/m^3	Sand, kg/m^3	Aggregate, kg/m^3	Water, kg/m^3	Compressive strength at 28 d, MPa
Concrete, *C*	0.5	320	653	1267	160	40.1

Table 2: The type of water repellent agents used in this project and their amounts.

	Type	Component	Amount of use	Note
Ref.	—	—	—	No treatment, reference sample
C400	Silane cream	Alkylalkoxysilane	400 g/m^2	Surface impregnation
G100	Silane gel	Triethoxysilane	100 g/m^2	Surface impregnation
G400	Silane gel	Triethoxysilane	400 g/m^2	Surface impregnation
IW	Silane emulsion	Octyltriethoxysilane	2%	Adding to the fresh mixture

5 mm and density of 2610 kg/m^3 were used. The water-to-cement ratio was 0.5. To improve workability, 0.5% superplasticizer of naphthalene type, related to the mass of cement, had been added to the fresh concrete. In addition, some integral water repellent concrete specimens were prepared with addition of 2% (related to the mass of cement) silane emulsion (octyltriethoxysilane) into the fresh concrete during mixing. It shall be noticed that the compressive strength of integral water repellent concrete at 28 days was reduced by approximately 11%, from 40.1 MPa for reference-untreated concrete to 35.7 MPa.

Prismatic specimens with the dimensions of 100 × 100 × 400 mm^3 and cubic specimens with the edge length of 100 mm were produced. All specimens were compacted in steel forms and cured for one day before demoulding. Then, the specimens were stored in the curing room at a temperature of 20 ± 2°C and relative humidity of 95% until an age of 14 days. After that, all specimens were taken out of the curing room and further stored under laboratory conditions at 20 ± 2°C and 60% RH for another 7 days before applying surface impregnation.

3.2. Water Repellent Treatment. After drying in the lab for 7 days, all the samples were divided into 5 groups. One group was maintained as reference specimens, that is, without water repellent treatment. One group was integral water repellent concrete, as produced above. The other three groups of concrete specimens were treated by different types and amounts of water repellent agents, as listed in Table 2. When one of the moulded surfaces was treated by covering 400 g/m^2 of silane cream, the series was named C400, while the specimens treated by 100 and 400 g/m^2 of silane gel were called G100 and G400, respectively. After surface impregnation, the specimens were exposed to laboratory atmosphere again for another 7 days in order to allow sufficient polymerization of silane. Then, both water repellent-treated and -untreated concrete specimens were ready for accelerated carbonation test.

3.3. Accelerated Carbonation. All the water repellent-treated and -untreated specimens were then submitted to accelerated carbonation for 7, 28, and 72 days. According to the method in Chinese standard [24], the concentration of CO_2 gas had been maintained constant at 20 ± 2%, relative humidity in the carbonation box was about 70%, and the temperature was 20 ± 3°C. Four surfaces except for the treated surface and its opposite surface of concrete specimens had been sealed with wax before placing in the carbonation box. After 7, 28, and 72 days, carbonation depth for each sample was determined by spraying phenolphthalein solution of 1% in ethanol. It was measured every 10 mm along a side length of 100 mm. Except for the two values from the border, the average value of the other nine measurements was the carbonation depth of each sample. For each case, three specimens were measured. Then, the average value of these three samples represented the carbonation depth for each case.

3.4. Water Capillary Absorption. In addition, in order to study the effect of water repellent treatment on water capillary absorption of concrete before and after carbonation, the water capillary absorption test has been performed according to the partial immersion method as explained in the ISO standard [25]. The specimens were cut into two halves first and then were dried in an aerated oven at a temperature of 50°C for 7 days until mass equilibrium. When the specimens cooled down to room temperature, they were put in contact with water. The amount of absorbed water was then measured by weighing the specimens at different time. In this test, it was at 0.5, 1, 2, 4, 8, and 24 hours in the first day and then measured every day until 7 days or so.

4. Results and Discussion

4.1. Carbonation in Water Repellent-Treated and -Untreated Concrete. Carbonation depth in water repellent-treated and -untreated concrete has been measured after exposure to 7, 28, and 72 days of carbonation. The results are shown in Figure 2. It is obvious that surface-impregnated concrete had lower carbonation depth than untreated samples. But the integral water repellent concrete has comparatively higher carbonation depth. The possible reason is that the addition of silane emulsion can retard the hydration process of cement [26]. This also explains why

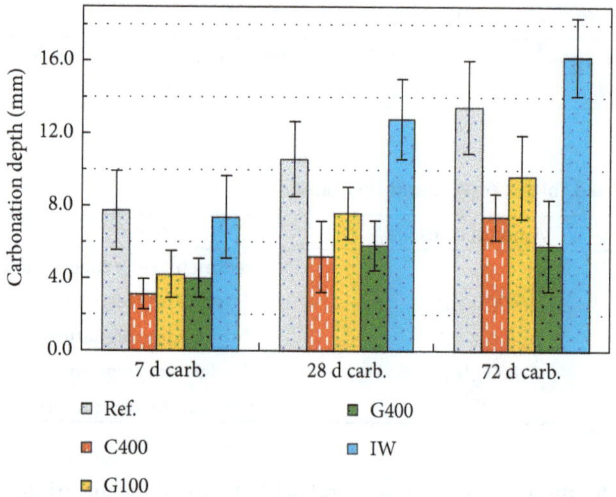

FIGURE 2: Carbonation depth of water repellent-treated and -untreated concrete after 7, 28, and 72 days of accelerated carbonation.

the compressive strength of integral water repellent concrete was reduced slightly. In this case, the penetration of carbon dioxide would be quicker and then cause higher carbonation depth.

For the surface-impregnated concrete, application of 400 g/m^2 silane cream and silane gel reduced approximately one-half of the carbonation depth compared to the reference-untreated samples. For instance, the carbonation depth of untreated concrete was 7.7 mm after 7 days of carbonation. After surface impregnation, carbonation depth of the samples C400 and G400 was 3.1 mm and 4.1 mm, respectively. The efficiency of reducing carbonation action is better than 100 g/m^2 covering usage of silane.

After surface impregnation, the silane penetrated by capillary transport into the concrete. Consequently, the silicon functional group reacts with water and yields a reactive silanol group (hydrolysis stage); further, condensation by cross linking to the hydroxyl group forms polysiloxane as the active water repellent film which is linked to the inorganic substance by way of covalent siloxane bounds.

The rate of carbonation process depends very much on the moisture content of concrete. The neutralization between CO_2 gas and calcium hydrate or C-S-H gel requires the presence of water because carbon dioxide needs to dissolve in water first forming carbonic acid. If the concrete is too dry, CO_2 gas may enter the material rapidly. But it cannot dissolve, and no carbonation occurs. If, on the other hand, it is too wet, CO_2 is not able to enter the pore and no carbonation occurs, either. The optimal conditions for carbonation occur at a relative humidity of 50%–80% [27–29]. In this test, all the treated and untreated specimens were kept in carbonation box under relative humidity of 70%. After 7, 28, and 72 days, the carbonation depth had been measured by spraying the phenolphthalein solution on the fractured surface. The typical results are shown in Figure 3. During the measurements, it has been found that surface-impregnated specimens G400 had three different layers, as shown in Figure 3 (the left photo). The first layer was pink. This means the near surface region was not carbonated as the water content in this layer was very low [30]. Carbonation reaction did not occur in this hydrophobic layer as it is too dry. But CO_2 gas certainly migrated through this layer and reached the inner part where there was necessary water for carbonation reaction. Thus, the second layer without colour indicates that the carbonation process started behind the water repellent layer. However, the carbonation process in this untreated region was delayed because its drying rate was lowered by the hydrophobic layer, and consequently, a high content of water in the pores hindered the carbonation process. Therefore, the carbonation depth in surface-impregnated concrete was reduced in comparison with the reference untreated concrete.

The carbonation mechanism has been studied by many researchers, and various mathematical models have been developed with the purpose of predicting carbonation depth [31–34]. For the sake of simplicity, however, the carbonation kinetics can be described as a first approximation to be a diffusion process of CO_2 through porous material with given moisture content. By applying Fick's law, (1) can be found after integration:

$$x = \sqrt{\frac{2D \cdot c}{d}}\sqrt{t}, \quad (1)$$

where x is the carbonated depth at time t, D is the effective diffusivity of CO_2, c is the CO_2 concentration in the atmosphere, and d is the concentration of the reactive compounds. Diffusivity D is not a constant but a function of numerous parameters. It mainly depends on time (degree of hydration), location (curing effect), and on the moisture content. Equation (1) can be written in the well-known form as

$$x = K\sqrt{t}, \quad (2)$$

where K denotes the coefficient of carbonation. Assuming the coefficient of carbonation rate changes with carbonation time, K can be described by a hyperbolic function, as shown in (3), in which P_1, P_2, and P_3 are parameters fitted from the experimental results. Another exponential model can also be used to describe the decrease of carbonation coefficient, as shown in (4), in which a and b are parameters fitted from the experimental results:

$$K = P_1 - \frac{P_2 t}{P_3 + t}, \quad (3)$$

$$K = a \cdot b \cdot e^{-b\sqrt{t}}. \quad (4)$$

It has been learned already from Figure 2 that with increasing time, the carbonation depth increases. The relationship between carbonation depth x and the square root of time t is shown in Figure 4. According to (3) and (4), the coefficient of carbonation for water repellent-treated and -untreated concrete can be calculated. The results are shown in Figure 5. It can be seen from the results that the carbonation coefficient was decreased significantly by surface impregnation. With the increasing time, finally the coefficient values would reach constant. This proves again that

FIGURE 3: Determination of carbonation depth in surface-impregnated (left: G400) and reference untreated concrete specimens (right: Ref.).

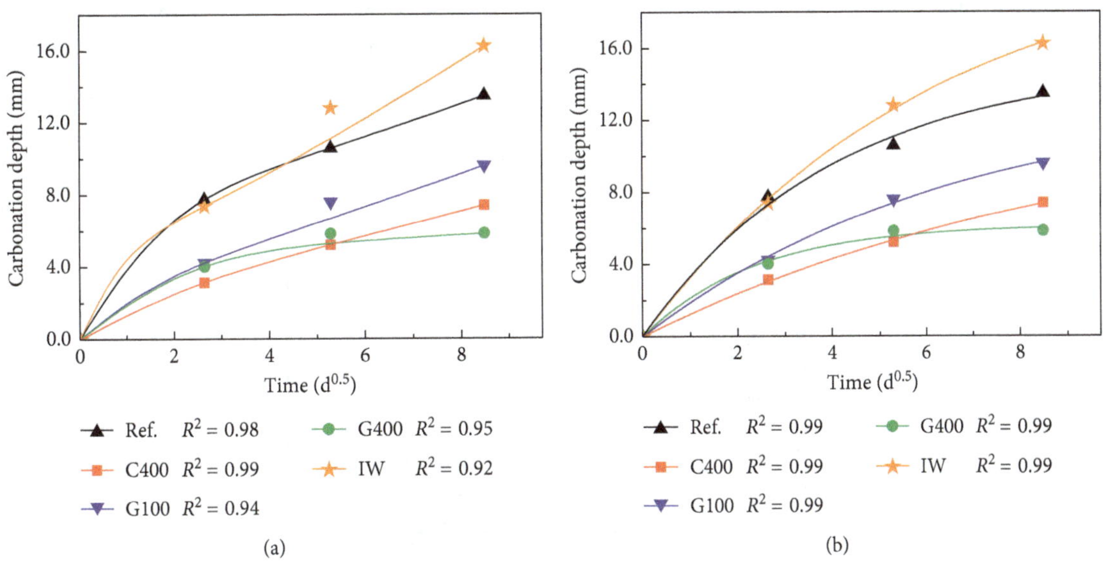

FIGURE 4: Carbonation depth versus square root of carbonation time. (a) Fitted by (3). (b) Fitted by (4).

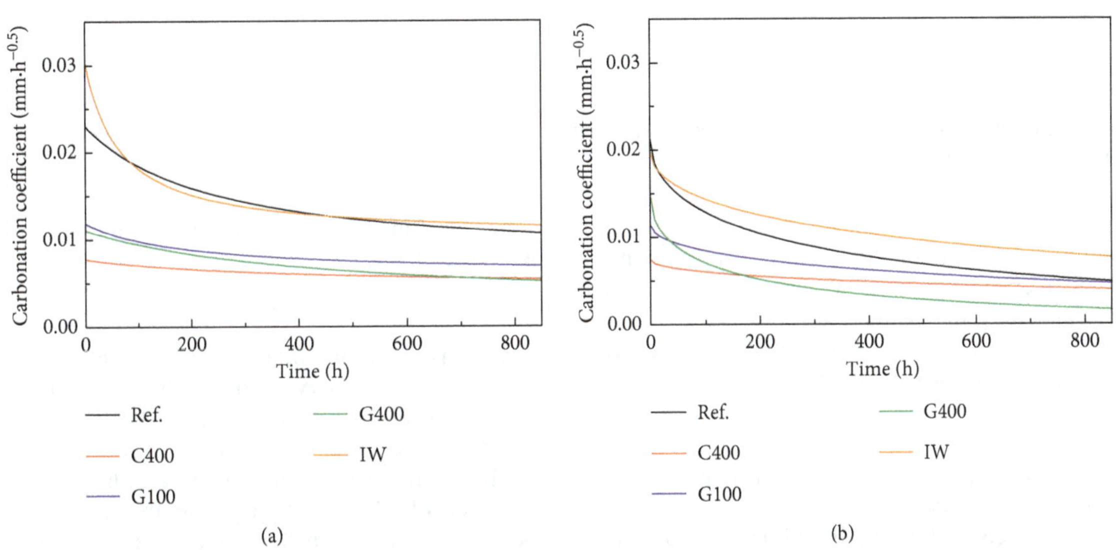

FIGURE 5: Changes of carbonation coefficient with increasing time. (a) Results obtained from (3). (b) Results obtained from (4).

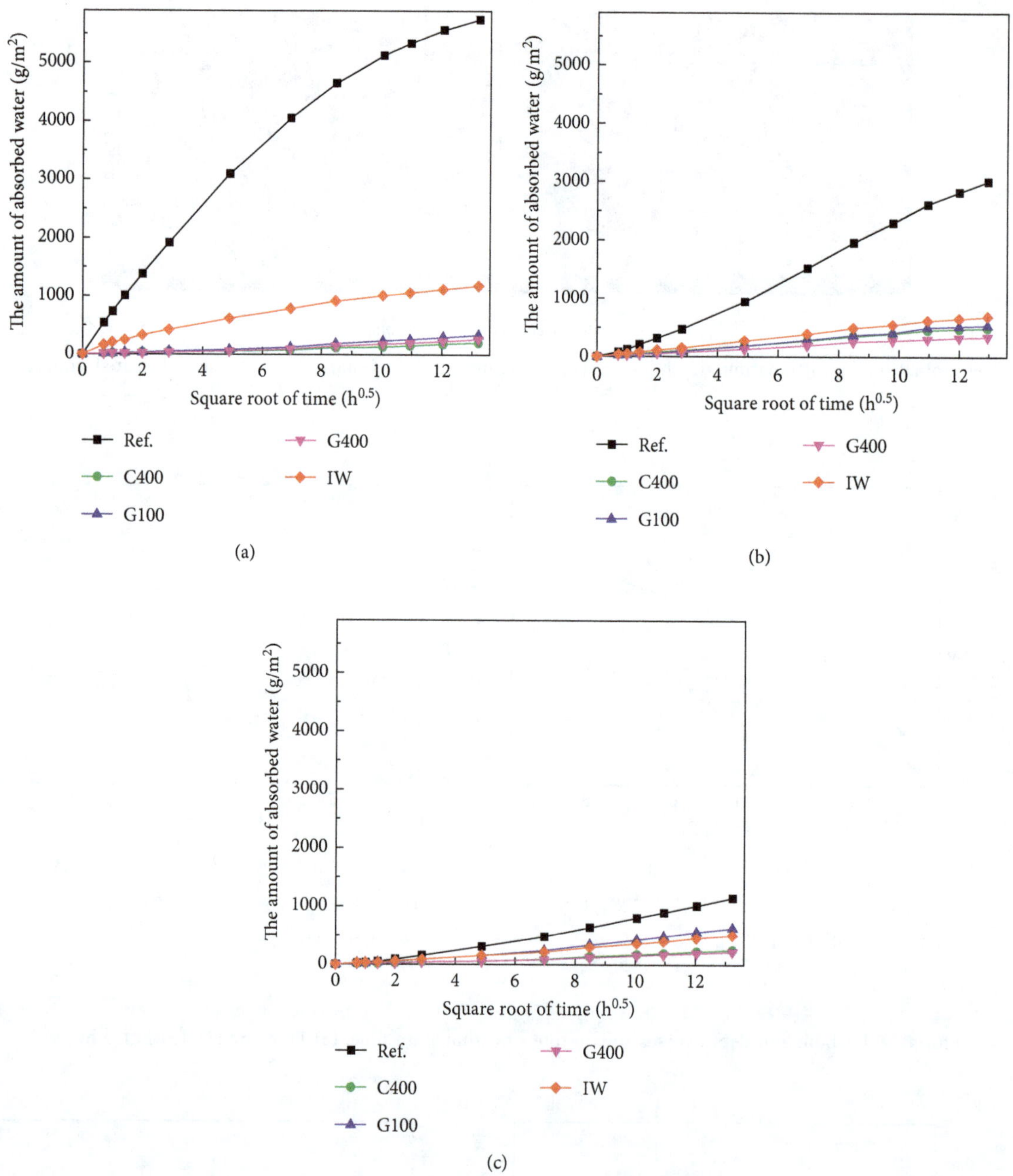

FIGURE 6: Water capillary absorption of water repellent-treated and -untreated concrete before and after 7 and 28 days of carbonation. (a) No carbonation. (b) 7 days of carbonation. (c) 28 days of carbonation.

surface impregnation can slow the carbonation process and consequently reduce carbonation depth. Application of 400 g/m^2 silane cream and 400 g/m^2 silane gel shows better efficiency than 100 g/m^2 silane gel. Comparing the two models, it can be found that the exponential model achieves better agreement with the experiment results.

4.2. Water Capillary Absorption of Treated and Untreated Concrete after Carbonation. If the surface of a porous material is put in contact with a liquid, the liquid will be absorbed by capillary action. Within a limited range of time and as first approximation, the amount of absorbed water as function of duration of contact time can be expressed by

$$\Delta W = A\sqrt{t}, \quad (5)$$

where ΔW is the amount of absorbed water; A stands for the coefficient of capillary suction (g/m^{2}h$^{0.5}$), which depends both on the properties of the absorbed liquid and the porous structure; and t is the absorption time.

If the microstructure of concrete changed due to carbonation of CH or C-S-H gel, or the surface of capillary pores was water repellent treated, the capillary suction would be changed consequently. The coefficient of capillary suction is considered to characterize this change. In this

TABLE 3: Coefficient of capillary absorption (A ($g/m^2h^{0.5}$)) by linear fitting of the results in Figure 6.

Condition/concrete	Ref.	C400	G100	G400	IW
No carbonation	584.4	12.9	19.5	15.1	116.9
After 7 days of carbonation	214.1	40.1	40.7	29.0	56.3
After 28 days of carbonation	71.0	14.6	16.7	13.3	33.0

test, this value for each type of specimen had been obtained experimentally.

The results of capillary suction for all the water repellent-treated and -untreated concrete, which were imposed to accelerated carbonation for 7 and 28 days, are shown in Figure 6. With (5), the coefficient of capillary suction within the first 72 hours can be calculated. The results are compiled in Table 3. The results clearly indicate that both surface impregnation and integral water repellent treatment can reduce considerably the water absorption of concrete.

In the surface-impregnated concrete samples C400, G100, and G400, the silane molecules ($d \approx 1$–3 nm) penetrated into concrete microstructures and made the surfaces of capillary pores hydrophobic. The coefficient of capillary suction was reduced enormously compared to the reference untreated concrete. With respect to the integral water repellent concrete IW, the coefficient of capillary suction also decreased. But the reduction is relatively lower. The origin of this phenomenon is due to the size of silane emulsion. Because the particle size of silane emulsion is approximately 100 nm, nanopores of C-S-H gel are not penetrated. Consequently, the fine pores will not become water repellent by the addition of silane emulsion. If the surface is in contact with water, some water can enter the porous structure of cement-based materials.

It shall be noticed that the carbonation resistance of integral water repellent concrete was lower than the reference untreated concrete mainly because the addition of silane emulsion delayed the hydration process of cement as mentioned previously. But this would not change the situation that the particle of silane emulsion still can break and form a siloxane network on the internal surface of some capillary pores in hardened cement paste. Therefore, the water capillary absorption can be still reduced by internal hydrophobic treatment in comparison with the reference untreated concrete.

After exposure to 7 and 28 days of carbonation, water absorption of concrete was also reduced by surface impregnation and integral water repellent treatment. This implies that water repellent treatment can still protect concrete from water absorption effectively even under carbonation condition. But the carbonation reaction can destroy the hydrophobic film inside the capillary pores and consequently may cause slightly more water absorption. For instance, the coefficient of capillary suction of the surface-impregnated concrete G400 was $15.1\,g/m^2h^{0.5}$ before carbonation. While after 7 days of carbonation, it became $29.0\,g/m^2h^{0.5}$. However, after 28 days of carbonation, water capillary absorption became lower again since the carbonation products filled the pores in the near-surface zone.

5. Conclusions

Based on the results presented herein, the following conclusions can be drawn:

(1) Carbonation depth of concrete can be reduced by one-half by surface impregnation. But for integral water repellent concrete, carbonation depth increased compared to the untreated concrete. One has to be careful while using integral water repellent concrete under severe carbonation atmosphere concerning steel corrosion.

(2) Three layers with different colours can be found when determining the carbonation depth of surface-impregnated concrete by means of phenolphthalein solution. CO_2 migrated through the impregnated layer quickly without carbonation as it is very dry. Carbonation process started behind the hydrophobic layer in this case.

(3) The coefficient of carbonation can be described better by an exponential function than by a hyperbolic function of time. Surface impregnation by silane cream and silane gel with the amount of $400\,g/m^2$ has better efficiency on reducing carbonation than using $100\,g/m^2$ silane gel.

(4) Coefficient of water capillary absorption of concrete can be reduced significantly by both surface impregnation and integral water repellent treatment, even under carbonation condition. It is an effective method to protect concrete from water penetration despite carbonation reaction.

It must be noticed that in this study accelerated carbonation test was performed with relatively high CO_2 concentration as the natural carbonation process was too slow. There might be different trends between the two types of test. This has to be further clarified.

Conflicts of Interest

The authors declare no conflicts of interest.

Acknowledgments

Financial supports of ongoing projects by the National Natural Science Foundation of China (51420105015 and 51778309), the Basic Research Program of China (2015CB655100), the Natural Science Foundation of Shandong Province (ZR2017ZC0737), and the 111 Project are gratefully acknowledged.

References

[1] J. Vries, R. B. Polder, and H. Borsje, "Durability of hydrophobic treatment of concrete," in *Proceedings of Water Repellent Treatment of Building Materials*, pp. 77–90, Aedificatio Publishers, Zurich, Switzerland, September 1998.

[2] P. Zhang, Y. Cong, M. Vogel et al., "Steel reinforcement corrosion in concrete under combined actions: the role of

freeze-thaw cycles, chloride ingress, and surface impregnation," *Construction and Building Materials*, vol. 148, pp. 113–121, 2017.

[3] P. Zhang, P. G. Wang, D. S. Hou, Z. L. Liu, M. Haist, and T. Zhao, "Application of neutron radiography in observing and quantifying the time-dependent moisture distributions in multi-cracked cement-based composites," *Cement and Concrete Composites*, vol. 78, pp. 13–20, 2017.

[4] P. K. Hou, X. Cheng, J. S. Qian, and S. P. Shah, "Effects and mechanisms of surface treatment of hardened cement-based materials with colloidal nanoSiO_2 and its precursor," *Construction and Building Materials*, vol. 53, pp. 66–73, 2014.

[5] C. Schrofl, V. Mechtcherine, A. Kaestner, P. Vontobel, J. Hovind, and E. Lehmann, "Transport of water through strain-hardening cement-based composite (SHCC) applied on top of cracked reinforced concrete slabs with and without hydrophobization of cracks–investigation by neutron radiography," *Construction and Building Materials*, vol. 76, pp. 70–86, 2015.

[6] J. G. Dai, Y. Akira, F. H. Wittmann, H. Yokota, and P. Zhang, "Water repellent surface impregnation for extension of service life of reinforced concrete structures in marine environments: the role of cracks," *Cement and Concrete Composites*, vol. 32, no. 2, pp. 101–109, 2010.

[7] P. Zhang, H. Shang, D. Hou, S. Guo, and T. Zhao, "The effect of water repellent surface impregnation on durability of cement-based materials," *Advances in Materials Science and Engineering*, vol. 2017, Article ID 8260103, 9 pages, 2017.

[8] X. Y. Pan, Z. G. Shi, C. J. Shi, T. Ling, and N. Li, "A review on concrete surface treatment. Part I: types and mechanisms," *Construction and Building Materials*, vol. 132, pp. 578–590, 2017.

[9] R. B. Polder, H. Borsje, and H. de Vries, "Prevention of reinforcement corrosion by hydrophobic treatment of concrete," *Heron*, vol. 46, no. 4, pp. 227–238, 2001.

[10] F. H. Wittmann, T. A. J. M. Siemes, and L. G. W. Verhoef, *Hydrophobe I*, TU Delft, Delft, Netherlands, 1995.

[11] F. H. Wittmann and A. Gerdes, *Hydrophobe II*, ETH Zürich, Zürich, Switzerland, 1998.

[12] K. Littmann and A. E. Charola, *Hydrophobe III*, University of Hannover, Hannover, Germany, 2001.

[13] J. Silfwerbrand, *Hydrophobe IV*, Stockholm, Sweden, 2005.

[14] H. De Clercq and A. E. Charola, *Hydrophobe V*, KIK-IRPA, Brussels, Belgium, 2008.

[15] E. Borrelli and V. Fassina, *Hydrophobe VI*, Rome, Italy, 2011.

[16] J.-M. Mimoso, *Hydrophobe VII*, Lisbon, Portugal, 2014.

[17] J. G. Dai, H. Yokota, and T. J. Zhao, *Hydrophobe VIII*, Hong Kong, 2017.

[18] J. Heinrichs, S. Schmeiser, and A. Gerdes, "Numerical simulation of the influence of water repellent treatment on carbonation of concrete," in *Proceedings of 4th International Conference on Water Repellent Treatment of Building Materials*, pp. 27–44, Aedificatio Publishers, Stockholm, Sweden, 2005.

[19] A. Gerdes, "Transport und chemische Reaktion siliziumorganischer Overindulgent in der Betonrandzone," Building Materials Rep. No. 15, Aedificatio Verlag, Freiburg, Germany, 2001.

[20] L. Courard and V. Lucquiaud, "Influence of hydrophobic treatments applications on the concrete carbonation delay," in *Proceedings of 14th International Conference on Durability of Building Materials and Components*, pp. 403–404, RILEM Publications, Ghent, Belguim, May 2017.

[21] E. De Witte, H. De Clerq, A. De Bruyn, and R. Pien, "Systematic testing of water repellent agents," in *Proceedings of Surface Treatments of Building Materials with Water Repellent Agents*, pp. 1–10, Delft University of Technology, Delft, Netherlands, November 1995.

[22] S. J. Meier and M. F. Bäuml, "Internal impregnation of concrete by means of silanes," *Restoration of Buildings and Monuments*, vol. 12, no. 1, pp. 43–52, 2006.

[23] H. Zhan, F. H. Wittmann, and T. Zhao, "Relation between the silicon resin profiles in water repellent concrete and the effectiveness as a chloride barrier," *Restoration of Buildings and Monuments*, vol. 11, no. 1, pp. 35–45, 2005.

[24] GB/T 50082-2009, *Standard for Test Methods of Long-term Performance and Durability of Ordinary Concrete*, Ministry of Construction, Beijing, China, 2009.

[25] ISO 15148, *Hygrothermal Performance of Building Materials and Products—Determination of Water Absorption Coefficient by Partial Immersion*, ISO, Geneva, Switzerland, 2002.

[26] V. Spaeth and M. P. Delplancke-Ogletree, "Hydration and microstructure development of integral water repellent cement based materials," in *Proceedings of 5th International Conference on Water Repellent Treatment of Building Materials*, pp. 245–254, Aedificatio Publishers, Brussels, Belgium, April 2008.

[27] Y. F. Houst, "The role of moisture in the carbonation of cementitious materials," *Internationale Zeitschrift für Bauinstandsetzen*, vol. 2, no. 1, pp. 49–66, 1996.

[28] D. Russell, P. A. Basheer, G. I. Rankin, and A. E. Long, "Effect of relative humidity and air permeability on prediction of the rate of carbonation of concrete," *Structures and Buildings*, vol. 146, no. 3, pp. 319–326, 2001.

[29] J. H. Cahyadi and T. Uomoto, "Influence of environmental relative humidity on carbonation of concrete," in *Proceedings of Durability of Building Materials and Components*, E&FN Spon, London, UK, September 1993.

[30] P. Zhang, F. H. Wittmann, T. J. Zhao, and E. H. Lehmann, "Neutron imaging of water penetration into cracked steel reinforced concrete," *Physica B-Condensed Matter*, vol. 405, no. 7, pp. 1866–1871, 2010.

[31] Z. Kersner, B. Teply, and D. Novák, "Uncertainty in service life prediction based on carbonation of concrete," in *Proceedings of 7th International Conference on Durability of Building Materials and Components*, pp. 13–20, E & FN Spon, Stockholm, Sweden, May 1996.

[32] V. G. Papadakis, M. N. Fardis, and G. G. Vayenas, "Effect of composition, environmental factors and cement-lime coating on concrete carbonation," *Materials and Structures*, vol. 25, no. 5, pp. 293–304, 1992.

[33] A. Steffens, D. Dinkler, and H. Ahrens, "Modeling carbonation for corrosion risk prediction of concrete structures," *Cement and Concrete Research*, vol. 32, no. 6, pp. 935–941, 2002.

[34] Y. F. Houst and F. H. Wittmann, "Depth profiles of carbonates formed during natural carbonation," *Cement and Concrete Research*, vol. 32, no. 12, pp. 1923–1930, 2002.

2

Experimental Study and Numerical Simulation of Seismic Behavior for RC Columns Subjected to Freeze-Thaw Cycles

Qing Qin, Shansuo Zheng, Lei Li, Liguo Dong, Yixin Zhang, and Sha Ding

Department of Civil Engineering, Xi'an University of Architecture and Technology, Xi'an 710055, China

Correspondence should be addressed to Qing Qin; qinqingjd@163.com

Academic Editor: Yao Luan

Freeze-thaw of concrete is significantly responsible for serious damage to RC buildings, which may result in premature failure with little warning. Therefore, it is necessary to consider the effects of freeze-thaw environment when assessing seismic performance for RC structures. In this study, pseudo-static experiments of four RC columns were conducted in terms of different number of freeze-thaw cycles (FTCs). The results showed that the FTCs had an influence on the bearing capacity, ductility, strength, and stiffness of RC columns. What is more, the simulation results were commonly smaller than the experimental ones when simply assuming that the degrees of freeze-thaw damage for RC components were uniform. Thus, a numerical model considering uneven distribution of freeze-thaw damage was proposed by utilizing the results of Petersen's test for relative dynamic modulus of elasticity (RDME) for different depths of concrete sample and based on Berto's method which was proved to be effective to convert the numbers of FTCs under different freeze-thaw conditions. On the basis of the existing constitutive relations for concrete, four RC columns subjected to different numbers of FTCs were simulated by OpenSees. As a result, the simulation hysteretic curves were in good agreement with the experimental ones.

1. Introduction

In cold region, concrete material may be damaged because of freeze-thaw penetration [1]. The concrete material property and bond-slip between concrete and steel bar can be affected greatly by freeze-thaw cycles, and this kind of damage will be increased with the increase of the numbers of freeze-thaw cycles (FTCs) [1–3]. For lifecycle structural engineering, it is important to know the effect of freeze-thaw cycles on materials and structures. However, most of the relevant researches are mainly focused on the effect on concrete material and bond behavior; very scarce literature on the effect on structural level can be seen.

For the past few years, the experimental study on the material and bond behavior of frost-damaged concrete had more achievements. Hasan et al. (2004) [4] proposed a stress-strain model for concrete damaged by freezing and thawing based on plasticity and fracture of concrete elements which was empirically formulated as a function of plastic tensile strain from experimental data. Shang and Song (2006) [5] tested the biaxial compressive strength of the plain concrete cubes subjected to freeze-thaw. Petersen et al. (2007) [2] studied the freeze-thaw effects on concrete elastic modules and interfacial bond behavior and proposed a bond-slip model using relative dynamic elastic modules of concrete as damage variable. In Petersen et al.'s study, the distribution of the internal damage of concrete was tested. The results showed that the damage distribution is uneven. More recently, Duan et al. (2011) [1] studied the freeze-thaw effects on the stress-strain relationship for plain concrete and confined concrete. In Duan et al.'s study, the number of freeze-thaw cycles and the concrete strength were changed; then a general constitutive model was proposed but the damage of the concrete was assumed to be uniformly distributed. Several types of tests were conducted on frost-damaged concrete by Hanjari et al. (2011) [3]. As a result, the compressive strength, tensile strength, fracture energy, the tensile-crack opening relationship, bond strength, and slip value for frozen-thawed damaged concrete were measured.

While previous research has been primarily concerned with the frozen-thawed deterioration of material and bond properties, relatively little attention has been given to

TABLE 1: The design parameters of specimens.

Specimen number	Concrete strength grade	λ	n	N (kN)	ρ (%)	ρ_v (%)	Freeze-thaw cycles
Z-C1	C50	5	0.3	300.6	3.01	1.18	0
Z-C2	C50	5	0.3	300.6	3.01	1.18	100
Z-C3	C50	5	0.3	300.6	3.01	1.18	200
Z-C4	C50	5	0.3	300.6	3.01	1.18	300

Note: λ is the shear span ratio; n is the axial load ratio with design; N is the axial force; ρ is the ratio of longitudinal reinforcements of columns; ρ_v is the volume-stirrup ratio of columns.

assessing the load-carrying and seismic capacity of frost-damaged RC structural component. With the development of artificial climate simulation, to be simulating the freeze-thaw effects on a 1/2 scaled structural components in laboratory has been become a reality. Xu et al. (2016) [6] investigated the seismic performance of RC columns subjected to freeze-thaw tests which were carried out in Climate Simulation Laboratory. Subsequently, Yang et al. (2016) [7] investigate the influence of FTCs on the seismic performance of squat RC shear wall specimens which were frozen and thawed in same way with Xu et al.'s. Although these test results have shown that FTCs have significant effects on the seismic performances of RC structural components, how to consider the influence of the freeze-thaw effects in a numerical model could not be involved in these literatures. Ueda et al. (2009) [8] used a two-dimensional rigid body spring model which could consider the experimentally observed cracking and plastic deformation caused by frost damage. Hanjari et al. (2013) [9] introduced a methodology to predict the load-carrying capacity of frozen-thawed RC beams using nonlinear finite element analyses; the uneven distribution of frozen-thawed damage in the large beams was not considered, which leads to smaller prediction results compared with test. Hayashida et al. (2014) [10] performed nonlinear finite element analysis using distributed reinforcement and smeared cracking models to evaluate the structural properties of frozen-thawed RC beams; one case was assigned a compressive strength for every element based on ultrasonic propagation velocity results, leading to poor results because of the restriction of numerical models. Berto et al. (2014, 2015) [11–13] extended the coupled environmental-mechanical damage model to model the frozen-thawed damaged RC beams; the effect of FTCs on compressive and tensile strength was considered separately and a new relationship to properly evaluate the equivalent number of FTCs was proposed; this equivalent relationship made it possible to use the constitutive models of frost-damaged concrete with different experimental conditions and sample size in the numerical model of RC structures. Thus, the uneven distribution of frozen-thawed damage in RC structures with different environment conditions should be considered in the numerical model as a further study.

In this paper, the authors mainly focused on the freeze-thaw effects on structural level. Both the experimental and numerical methods were employed. Cyclic loading tests for four 1/2 scale RC columns subjected to different numbers of FTCs were carried out. And then a new concrete constitutive model which considered the uneven distribution of frozen-thawed damage was proposed based on Petersen's experimental data. At last, the available constitutive models of frost-damaged concrete and the model proposed in this paper were used in OpenSees, respectively, to simulate the behavior of RC columns experiencing different numbers of FTCs, and compared with the test results.

2. Experimental Program

2.1. General Information of the Experiment. In RC moment-resisting frame structures under severe ground motions, the points of inflection in the columns are assumed to lie at their midpoints. Therefore, the segment between midpoint and beam-column joint is intercepted as a specimen. Considering the limitation of the dimensions of the environmental chamber used in this study, four 1/2 scale RC columns with invariable axial load ratio were designed for FTCs tests based on the current Chinese code (GB 50011-2010) [14]. Axial and lateral loads were transferred to the columns through the top with 200 mm range which needed more stirrups. The bases are clamped to the laboratory strong floor. Figure 1 shows the plan views and important reinforcement details for the specimens. The values of the primary study variables are summarized in Table 1.

2.2. Timeline of the Test. The specimens were divided into columns and foundation blocks and were cast in sequence due to the limiting dimensions of Climate Simulation Laboratory. The columns were first cast and cured for 4 weeks in a natural environment and then placed in 15–20°C water for a week before being placed into the laboratory. The freeze-thaw testing process would last 4~12 weeks according to the desired levels of frozen-thawed damage. When the setting cycles were finished, the column was recast with foundation block. After the foundation blocks were cured for 4 weeks, the low cyclic loading tests were carried out. The detailed timeline of total experiment was shown in Figure 2.

2.3. Material Properties. The mixing ratio for C50 concrete used in these specimens was 400 kg/m^3 of cement, 980 kg/m^3 of sand, 810 kg/m^3 of fine gravel, and 165 kg/m^3 of water. The real strength of concrete and reinforcement bars was determined via material property tests. Three standard cubic specimens of concrete were prepared and the average cubic

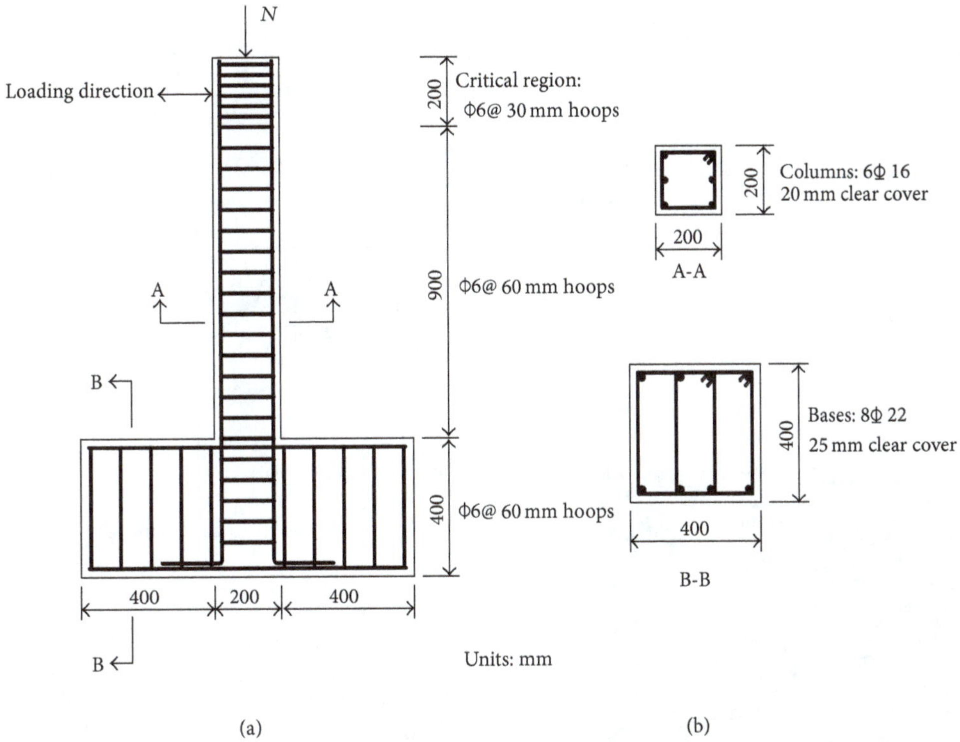

FIGURE 1: Column details including (a) specimen dimensions and (b) section details.

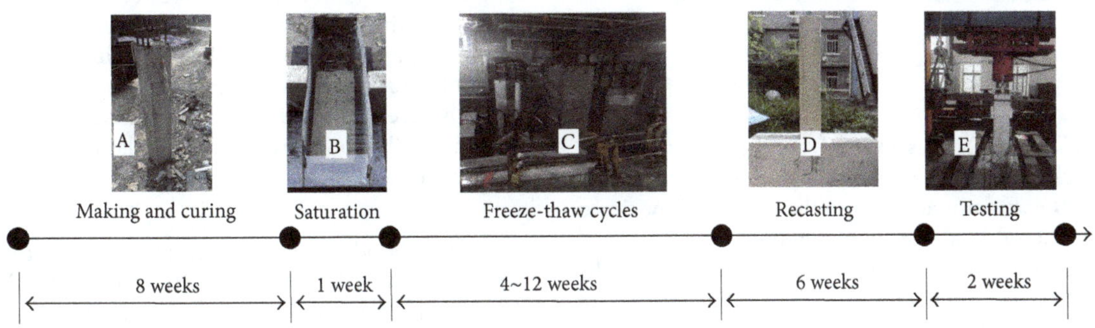

FIGURE 2: Timeline of the experiment.

TABLE 2: Mechanical properties of reinforcement bars.

Diameter (mm)	Yielding strength f_y (MPa)	Ultimate strength f_u (MPa)	Elastic modulus E_s (MPa)
Φ6	270	470	2.1×10^5
Φ16	373	537	2.0×10^5

compressive strength of the concrete was 55.08 MPa. Referring to [15], the axial compressive strength of concrete was 41.86 Mpa. Meanwhile, the longitudinal reinforcements for columns were Grade HRB335, and the stirrups were Grade HPB235. The properties of the reinforcement bars are shown in Table 2.

2.4. Details of Freeze-Thaw Cycles. The freeze-thaw tests for columns were carried out in Climate Simulation Laboratory as shown in Figure 2(C). The detailed design parameters of climate chamber could refer to the literature [7]. According to the Chinese code GB/T50082-2009 [16], the temperature setting of the freeze-thaw cycles was presented in Figure 3. To produce better freeze-thaw effects, 5 spray cycles were applied before each cycle began; each spraying cycle lasted for 3 minutes (one minute of spraying, followed by two minutes without spraying). Each freeze-thaw cycle lasted 5.5 hours, and 100 times FTCs would last 23 days. After 0 times, 100 times, 200 times, and 300 times FTCs, the compressive strength of concrete cylinders which were drilled out in walls was 53.73 Mpa, 49.64 Mpa, 44.25 Mpa, and 28.86 Mpa [7]. The columns and the walls belonged to the same batch FTCs tests.

2.5. Loading Scheme. After the foundation blocks were cured for 4 weeks, the low cyclic loading tests were carried out. The test devices are shown in Figure 4. Firstly, the target

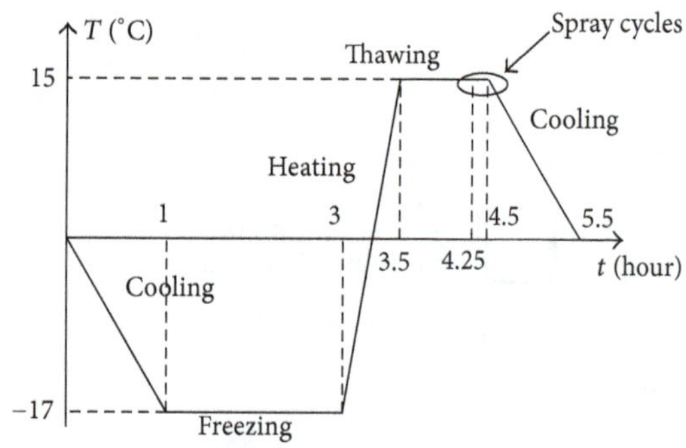

FIGURE 3: Details of temperature change in each cycle.

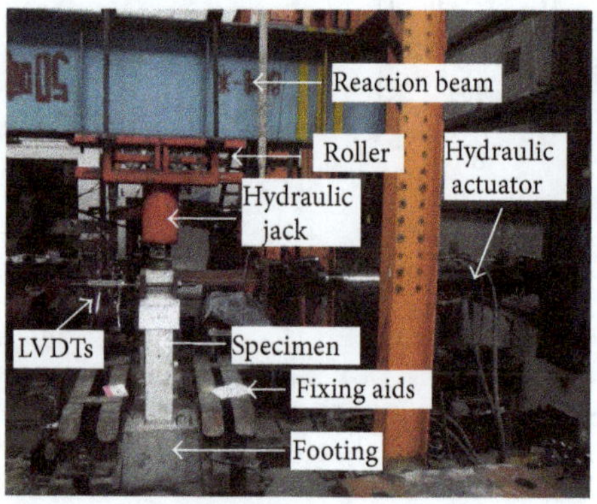

FIGURE 4: Loading system.

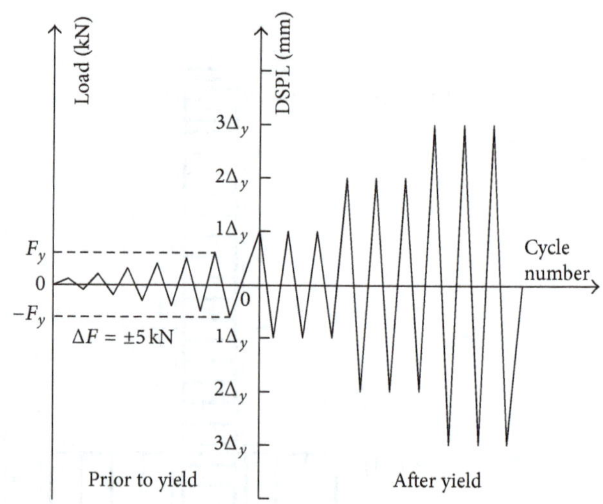

FIGURE 5: Loading protocol.

vertical load 300.6 kN was exerted on the specimens via hydraulic jack, and the axial load ratio with experiment was 0.183. Then, the actuator attached to the top of the specimen through the lateral load transfer assembly applied a lateral load; the vertical load was held constant throughout the testing process. Figure 5 illustrates the loading history of columns. Prior to the yielding of the specimen, the lateral load mode was exerted by the force-controlled load transfer assembly, increasing by an incremental load of ±5 kN. After the yielding of the specimen, three cycles of loading and unloading were conducted for each subsequent displacement level, which increased by multiples of the yield displacement. When the bearing capacity of columns dropped to 85 percent of the peak load, loading ended. The lateral displacements of columns at the loading point were measured by horizontal LVDTs with 100 mm range, and the horizontal load was documented by a pull-pressure sensor.

3. Experimental Results

3.1. Cracks after Freeze-Thaw Cycles. After different freeze-thaw cycles, the characteristics of specimens' surface are shown in Figure 6. It is observed that some tiny cracks appeared on the surface of Z-C2 which was subjected to 100 times FTCs, as shown in Figure 6(a). When the number of FTCs increased, the number of cracks present increased and began to widen and extend. What is more, as the number of FTCs increased to 300, the surface was no longer smooth and had "rise sand" phenomenon, as shown in Figure 6(c). Compared with the surface of frozen-thawed prism concrete specimens tested in literatures [1, 2], the freezing-thawing damage of columns was relatively light. One reason is because the specimens were drier in the climate chamber than when experiencing traditional frozen-thawed method with water. As a result, the ice inside concrete was less so that the stress was not large enough to spall the unconfined concrete which was similar to the real freeze-thaw environment. On the other hand, the size of cross section of columns is larger so that the uneven distribution of frozen-thawed damage in RC columns is more outstanding.

3.2. Failure Mode. Figure 7 illustrates the failure mode of columns. It can be seen that FTCs did not change the destruction form of RC columns where the shear span ratio was equal to 5. All specimens experienced typically bending-shear failures mainly dominated by flexure. Firstly, the flexural horizontal cracks were found at the bottom of the columns that ranged within 50 mm~100 mm. Subsequently, with the lateral load increased, the longitudinal reinforcements yielded; thus the horizontal cracks developed obliquely and some new inclined cracks emerged. Meanwhile, some vertical cracks appeared on the side perpendicular to the load direction at the bottom of columns. And then the concrete cover at the footing of columns began to spall. Furthermore, when the specimens were continued to load, the stirrups yielded and the longitudinal reinforcements were buckling, and the core concrete was crushed in large area so that the columns lost bearing capacity. After freeze-thaw cycles, the longitudinal reinforcements yielded early and the concrete at the footing of columns was crushed prematurely due to the fact that the strength of frozen-thawed concrete decreased gradually. The massive crushed concrete was transformed into granule depending on the degree of freezing-thawing damage, as shown in Figures 7(b)~7(d). In addition, the degree of granulation reduced gradually from the outer to

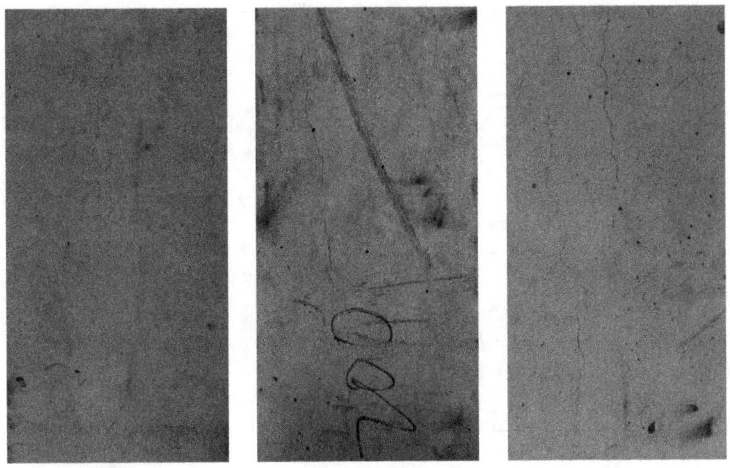

(a) 100 cycles (b) 200 cycles (c) 300 cycles

FIGURE 6: Columns surface after freeze-thaw cycles.

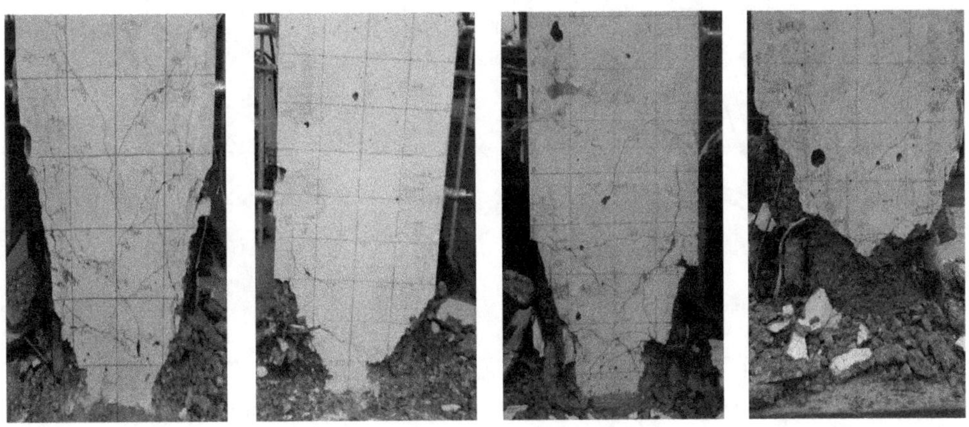

(a) 0 cycles (b) 100 cycles (c) 200 cycles (d) 300 cycles

FIGURE 7: Failure mode of the columns.

the inner part of columns, which illustrated the uneven distribution of freezing-thawing damage.

3.3. Experiment Results

3.3.1. Hysteresis Behavior. The hysteretic curves of the lateral load versus the top displacement for the four specimens are shown in Figure 8. All specimens were destroyed with typical bending-shear failures; the hysteretic characteristics of specimens subjected to different number of FTCs are similar; that is, the intact specimen Z-C1 without FTCs was taken as an example to describe the failure process. At the initiation of loading, the specimen had yet to yield and was in elastic stage, and the curves of loading and unloading were keep in a straight line approximately. When the specimen yielded, the specimen began to access elastic-plastic stage, the loading and unloading stiffness was reduced gradually, and the residual deformation for positive and negative loading direction increased with smaller amplitude prior to the peak state. After the specimen reached the peak, the loading and unloading stiffness decreased sequentially and the residual deformation increased with larger amplitude, and the degradation of strength and stiffness for hysteretic loops was not obvious at the same displacement level.

With the increase of freeze-thaw cycles, the bearing capacity of specimens decreased gradually; however the displacements for different damage states increased slightly, which mainly attributed these results to the decrease of strength and the increase of ductility for concrete material subjected to FTCs [2, 5]. Prior to the peak, the loading and unloading stiffness had not diminished much compared with the intact specimen Z-C1, and the residual deformation had a tendency to increase. After the peak state, not only did the degradation of stiffness become more serious and the residual deformation increase with larger amplitude, but also the degradation of strength and stiffness for hysteretic loops

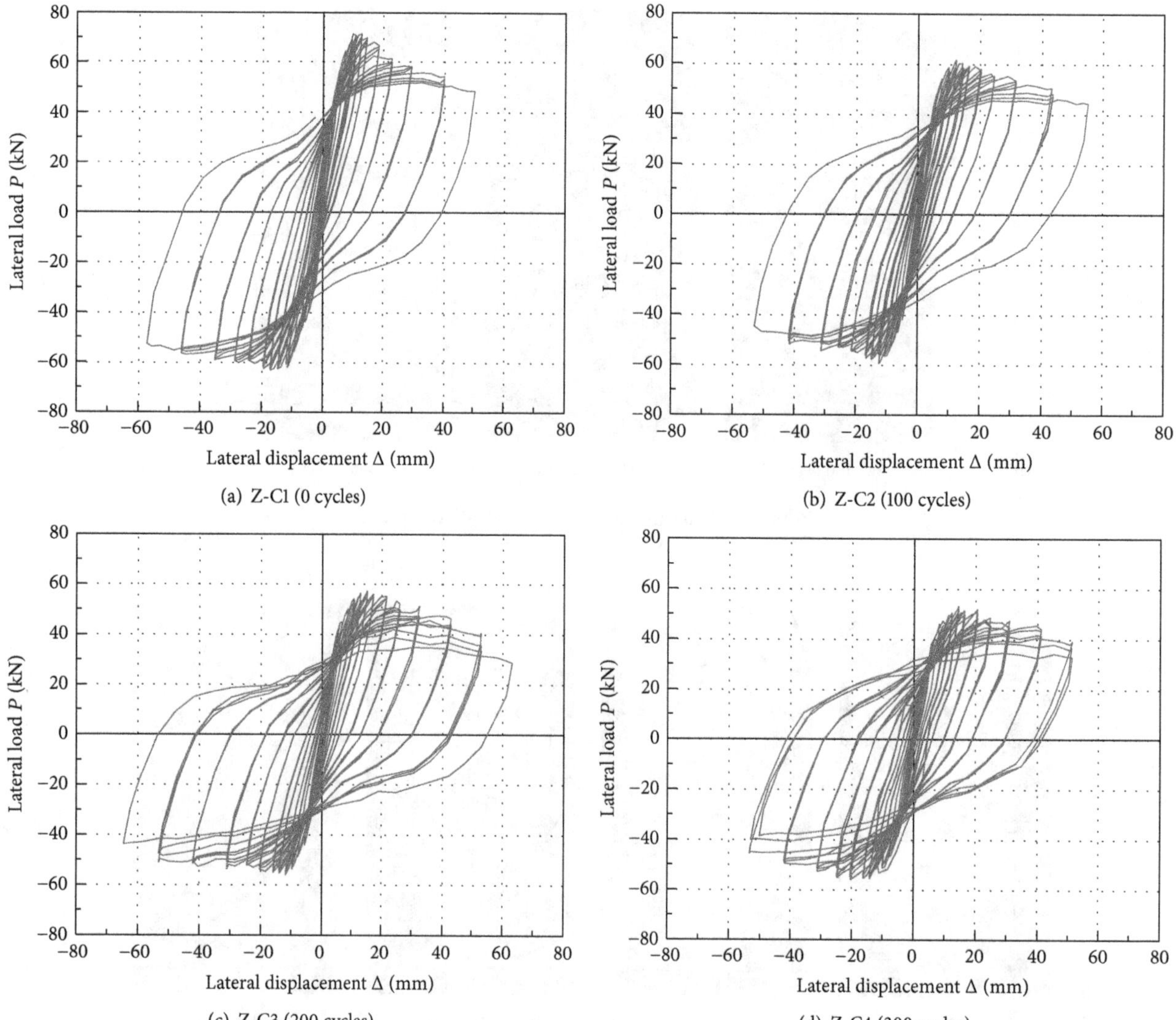

(a) Z-C1 (0 cycles)

(b) Z-C2 (100 cycles)

(c) Z-C3 (200 cycles)

(d) Z-C4 (300 cycles)

FIGURE 8: Hysteric curves of columns.

was increasingly obvious at the same displacement level. The effect of the freeze-thaw cycles on the hysteretic behavior of the specimens is significant.

3.3.2. Characteristic Parameters of Skeleton Curves. Skeleton curves of the columns are plotted in Figure 9. Based on the equivalent elastoplastic energy absorption theory (recommended by Park [17]), the yield load and displacement were calculated. And then, the ultimate state was defined by the point on the descending branch with 85% of the maximum load. What is more, the ductility factor which reflects the deformation capacity of structures was expressed as the ratio of the ultimate displacement and the yield displacement [18]. Because the skeleton curves are asymmetric, the characteristic parameters are calculated as the average of absolute values of pull and push direction, which are presented in Table 3.

Based on Figure 9 and Table 3, some observations and regularity understandings are obtained as follows.

(1) Influence of FTCs on the Load-Carrying Capacity. Note that the yielding and peak load decreased as the number of FTCs increased. After 100 times, 200 times, and 300 times FTCs, the yielding load of frozen-thawed specimens compared with specimen Z-C1 has fallen by 9.72%, 14.90%, and 15.70%, respectively, and the correspondingly peak load has fallen by 9.44%, 15.62%, and 18.68%. These results indicated that the effect of the FTCs on the load-carrying capacity was significant.

(2) Influence of FTCs on the Deformation Capacity. According to the test researches of constitutive relations for freezing-thawing concrete in literatures [1, 3, 5], the compressive strength of concrete was decreased as the number of FTCs increased; however the ductility was increased mainly due

TABLE 3: Force and deformation of the tested columns (unit: mm & kN).

Specimen	Cycles	F_y (kN)	Δ_y (mm)	F_{max} (kN)	Δ_{max} (mm)	F_u (kN) ($0.85F_{max}$)	Δ_u (mm)	Δ_u/Δ_y
Z-C1	0	55.99	7.26	66.92	14.42	56.88	38.45	5.30
Z-C2	100	50.55	7.14	60.60	14.58	51.51	44.03	6.16
Z-C3	200	47.65	7.40	56.47	14.6	48.00	44.48	6.01
Z-C4	300	47.20	8.50	54.42	14.28	46.26	42.64	5.02

Note: F_y and Δ_y are the yield load and displacement; F_{max} and Δ_{max} are the peak load and displacement; F_u and Δ_u are the ultimate load and displacement; Δ_u/Δ_y is the displacement ductility coefficient.

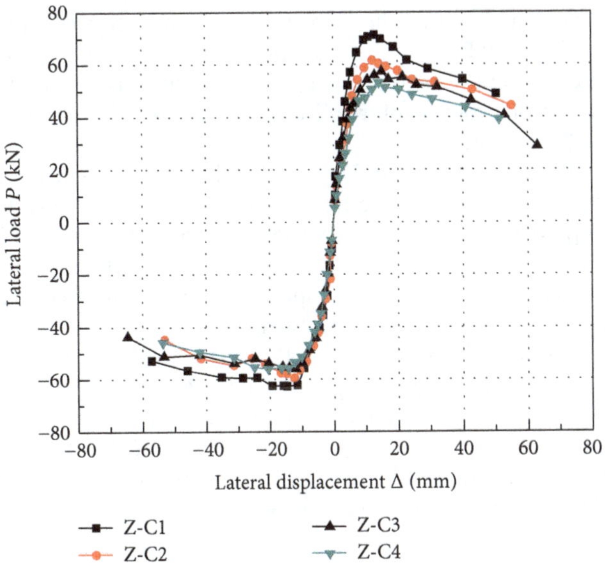

FIGURE 9: Experimental skeleton curves.

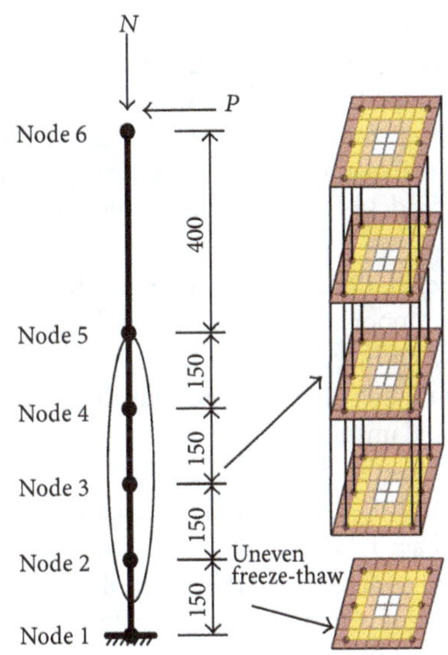

FIGURE 10: Analytical model of column subjected to FTCs.

to the internal microcracks for concrete with different numbers of FTCs. As for the RC specimens designed in this paper, with the increases of the number of FTCs, the internal microcracks in concrete developed continuously, which resulted in the increasing trend for yielding displacement, peak displacement, and ultimate displacement. After 100 times, 200 times, and 300 times FTCs, the yielding displacement of frozen-thawed specimens compared with specimen Z-C1 increased by −1.65%, 1.93%, and 17.08%, the peak displacement increased by 1.11%, 1.25%, and −0.97%, and the ultimate displacement increased by 14.51%, 15.68%, and 10.9%. Because the increases of ultimate displacement are less than the decreases of yield displacement, the ductility factors of freezing-thawing specimens declined, which is consistent with the test results in literature [6].

4. Modeling of the Columns Subjected to Freeze-Thaw Cycles

4.1. General Model. Generally, four element types were commonly used to simulate RC components, which were divided into force-based beam-column element (FB), displacement-based beam-column element (DB), solid finite element (SFE), and beam-column element with concentrated plastic hinges (CPH), respectively [19]. When the RC column was subjected to FTCs, the length of plastic hinge had been changing, as shown in Figure 7. Therefore, the displacement-based beam-column element which is based on the displacement formulation and considers the spread of plasticity along the element was used in this paper. In order to improve the accuracy of model prediction, the optimal number of elements for DB elements should be defined as [19].

When the RC column was modeled in OpenSees, firstly, five elements were divided into by six nodes along the vertical direction of column, because the elastic-plastic deformation was mainly focused on the bottom of RC column, which was divided into four elements with 150 mm. In addition, the top of RC column tended to be more flexible, which was divided into one element with 400 mm. What is more, the footing was assumed to be stiff end, and Node 1 was fixed. And the illustration of analytical model was shown in the left-hand side of Figure 10. Secondly, the number of Gauss integration points of the displacement-based beam-column element was 5. Fiber section model was used to define the element, and the rectangular section was dispatched by 40 fibers × 40 fibers. Finally, each fiber should be defined as a Uniaxial Material which represents uniaxial stress-strain relationships for steel and concrete. As for the longitudinal reinforcement, the steel fiber constitutive law followed Menegotto and Pinto (1973)

[20], combined with the isotropic hardening rules proposed by Filippou et al. (1983) [21]. Longitudinal rebar buckling was not considered in the analysis. According to the material test results, the yield strength and the elastic modules of the reinforcement bars are 363 MPa and 128918 MPa. The strain harden ratio is 1.5%. In allusion to the concrete fibers, concrete cover material was represented through Kent-Park model which followed the constitutive relationship modified by Hisham and Yassin (1994) [22]. And the compression envelope of confined concrete was calibrated using the empirical relations proposed by Mander et al. (1988) [23] for the peak stress and the strain. For unloading and reloading in compression, the Karsan-Jirsa (1969) [24] was used to determine the slope of the curve for confined concrete. In order to consider the ductility of postcrushing concrete, a descending branch was calibrated such that the post-peak slope of the present stress-strain relationship agreed with Roy and Sozen's model [25] for confined concrete after reaching the peak value.

Petersen et al. (2007) [2] pointed out that the damage of FTCs on RC components was uneven; with the depth of freezing and thawing increasing, the degree of internal damage expressed as relative dynamic modulus of elasticity reduced gradually; the damage of uneven distribution over the cross section was illustrated in the right-hand side of Figure 10. Hanjari et al. (2013) [9] modified the concrete constitutive relationship based on the related experimental results for frozen-thawed concrete, and the effect of bond-slip between longitudinal reinforcement and frozen-thawed concrete was considered; in addition, the load-carrying capacity of frozen-thawed RC beams had been predicted using nonlinear finite element analyses. For smaller cross section of RC beams, the predicted curves of load-deformation were in good agreement with test results. Moreover, for larger cross section of RC beams, because the uneven distribution of frozen-thawed damage in the large beams was not considered, the effect of freezing-thawing damage on concrete had been amplified, which leads to smaller prediction results compared with test. Therefore, a modified model had been proposed to consider the uneven distribution of freeze-thaw in view of the concrete constitutive relationship applied to the above process of modeling RC columns, which was referenced to the measurements for relative dynamic modulus of elasticity (RDME) at various distances from the exposed surface in literature [2], as described in Section 4.3.

4.2. The Equivalent Number of FTC, N_{eq}. When the freezing-thawing tests for concrete specimens (prismoids, cylinders) were carried out, various freezing-thawing devices and test specifications were adopted in different literature [1, 3, 5], which led to the inconsistent frozen-thawed deterioration conditions, for example, freezing-thawing velocity, moisture content, and sample size. As a result, the observed degree of frozen-thawed damage was varied for specimens subjected to the same FTCs in different literatures, and this limits the mutual verification and statistical analysis for test data in different literature. Berto et al. (2015) [13] used the equivalent number of FTCs (N_{eq}) according to ASTM C666 [26] "procedure A" which led to the same strength reduction to consider the above different FTC conditions. The relation for evaluation of N_{eq} was proposed:

$$N_{\text{eq}} = \chi N^{\gamma}, \tag{1}$$

where N was the actual number of FTCs, while χ and γ were parameters depending on the experimental conditions. The actual number N of FTCs in Duan et al. tests, Shang and Song tests, and Hanjari et al. tests had been, respectively, transformed into the equivalent number N_{eq} through (1) by Berto et al. (2015), and the proposed FTD model had been validated with better applicability. Therefore, the above transformational method was used to transform the actual number N of FTCs in the above freeze-thaw tests for columns and Petersen's tests, and then the modified model considering uneven distribution of freeze-thaw was proposed to simulate the RC columns.

Petersen et al. (2007) [2] used the same specification with Hanjari et al. experimental tests (RILEM TC176-IDC [27]) to conduct freeze-thaw tests of concrete specimens. This standard was rather different than recommendations of ASTM C666 in terms of specimen dimension, freezing-thawing rate, moisture content, and so on. Berto et al. (2015) investigated the influence of these experimental tests on the evaluation of χ and γ based on a sensitivity analysis. The Ordinary Least Squares approach was used to obtain optimization parameters ($\chi = 11.71$ and $\gamma = 0.87$) that best fit experimental results with small error based on observed RDME. The transformed N_{eq} was written as

$$N_{\text{eq}} = 11.71 \left(N_{\text{Hanjari}}\right)^{0.87}. \tag{2}$$

Because Petersen et al.'s freezing-thawing tests were carried out on concrete samples immersed in water with one side, on the contrary, Hanjari's samples were immersed in water with all sides. When the sample reached the same degree of freeze-thaw damage, Petersen's test experienced more cycles than Hanjari's test. So we assumed that $N_{\text{Petersen}} = \eta N_{\text{Hanjari}}$, where $\eta = 2.5$.

As for the FTCs tests of RC columns mentioned above, the target temperatures were performed by GB/T50082-2009 [16], which is similar to the ASTM C666 proc. A. However, the freezing-thawing method described in Section 2.4 was different from the ASTM C666 because the specimens were not frozen-thawed in water; therefore a relatively lighter freeze-thaw condition was provided in this paper, and the section size of RC columns did not meet the standard requirements. For these reasons, the equivalent number N_{eq} of FTCs needed to be estimated. To this aim the parameters χ and γ were estimated considering the degradation of strength experimentally observed on cylinder specimens drilled out from squat RC walls which belonged to the same batch FTCs tests in climate chamber. In order to consider the degradation of concrete compressive strength with the deterioration of environment, according to the literature [13], the environmental damage parameter may be written as

$$d_{\text{env}} = 1 - \frac{f_c^d}{f_c}, \tag{3}$$

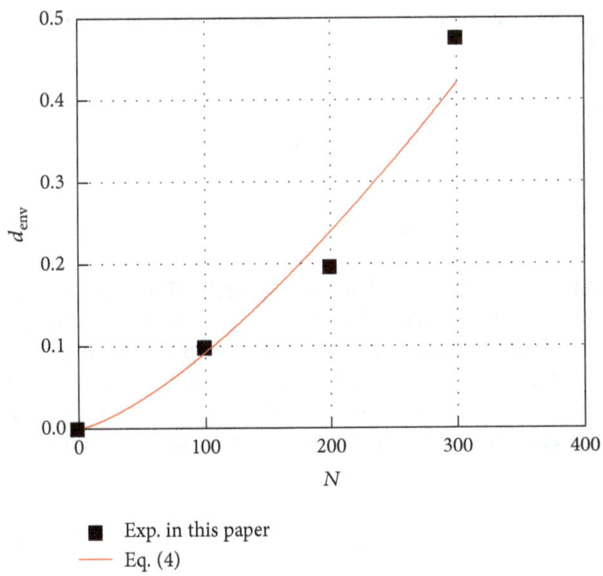

FIGURE 11: Relation between d_{env} and $N_{exp.}$.

where f_c^d and f_c are, respectively, the values of compressive peak stress in degraded and sound conditions. The corresponding values of d_{env} for 0, 100, 200, and 300 FTCs are plotted in Figure 11 with rectangular block and evaluated by regression analysis shown in Figure 11 with red line. The results of the regression analyses are summarized as follows:

$$d_{env} = 1.565 \times 10^{-4} N_{exp.}^{1.384}, \quad R^2 = 0.87. \tag{4}$$

According to the literature [11], the frozen-thawed deterioration condition with ASTM C666 proc. A was considered, and the environmental damage parameter d_{env} was calculated by

$$d_{env} = (0.0114 - 0.00022 f_C) N_{eq}, \tag{5}$$

where f_C [MPa] is the cylindrical compressive strength. In this paper, the compressive strength of cube concrete was 55.08 MPa, and the corresponding cylinder compressive strength was 43.49 MPa calculated by [15]. In principle the degradation of concrete strength was the same under different freezing-thawing conditions; the actual number $N_{exp.}$ in the above FTCs tests about RC columns was transformed by

$$N_{eq} = 8.542 \times 10^{-2} N_{exp.}^{1.384}. \tag{6}$$

Based on (2) and (6), the corresponding numbers of FTCs with different frozen-thawed conditions were calculated as listed in Table 4.

4.3. Proposed Model. Petersen et al. (2007) [2] studied the effects of freezing-thawing depth on relative dynamic modulus of elasticity, the test results were shown in Figure 13 with labeled lines, and the internal damage was more pronounced and progresses into the specimen with an increasing number of FTCs. According to the characteristics of the labeled lines in Figure 13, we assumed that (1) the freezing-thawing environment around the specimens' surface was identical and the corresponding damage degree of freeze-thaw at the same distance was identical as well; (2) the relative dynamic modulus of elasticity (RDME) was held constant when the number of FTCs did not exceed the critical value N_i' where the corresponding freeze-thaw depth was equal to x_i; and (3) the RDME was decreased linearly when the number of FTCs exceeded its critical value N_i'. In this paper, the calculation model of RDME considering the uneven freeze-thaw damage was proposed based on the above assumption, as follows:

TABLE 4: The corresponding numbers of FTC with different frozen-thawed conditions.

$N_{exp.}$	N_{eq} (ASTM C666 proc. A)	$N_{Petersen}$ (RILEM TC176-IDC)
100	50.07	13.27
200	130.67	40.03
300	229.03	76.25

$$\text{RDME} = \begin{cases} 1, & \text{if } N \le N_i', \\ 1 - m_i \left(N - N_i'\right) \ge 5\%, & \text{if } N > N_i', \end{cases} \tag{7}$$

where N was the actual number of FTCs taken from Table 4 in Petersen's condition; N_i' was the corresponding number of FTCs at the x_i freeze-thaw depth when the RDME began to decline, known as the critical value mentioned above, as shown in Figure 13; and m_i was the corresponding slope for the linear decline lines shown in Figure 13 at the x_i freeze-thaw depth. As for the freeze-thaw depth x_i, it is the minimum distance from exposed surface with both directions to the frozen-thawed concrete fiber, as shown in Figure 12. Based on the observation from the distribution of the internal damage over the specimen tested by Petersen which was shown in Figure 13 with labeled lines, the critical value N_i' increased with the freeze-thaw depth deepening, and the corresponding slope m_i of linear decline line decreased continuously. Therefore, the linear regression analysis was used to confirm N_i' and m_i which were written as (8), respectively. The calculated results were shown in Figure 13 with dashed lines, which was in good agreement with the test results.

$$N_i' = 0.9x_i + 3, \quad m_i = \frac{(1.3 - 0.008x_i)}{100}. \tag{8}$$

According to the British Code of Practice for Structural Use of Concrete CP110: 1972 [28], the relationship between the dynamic modulus of elasticity of concrete E_{dyn} and its strength f_C was written as

$$E_{dyn} = 22 + 2.8\sqrt{f_C}, \tag{9}$$

where E_{dyn} [GPa] was the dynamic modulus of elasticity of concrete without FTC and f_C [MPa] was the corresponding cylindrical compressive strength which was equal

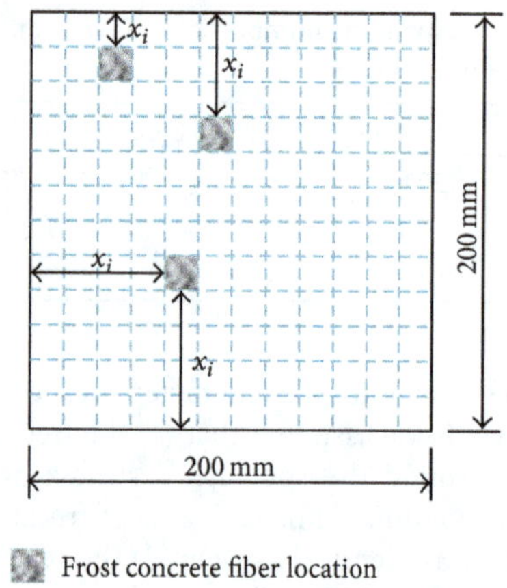

FIGURE 12: The diagram for freeze-thaw depth x_i with RC columns.

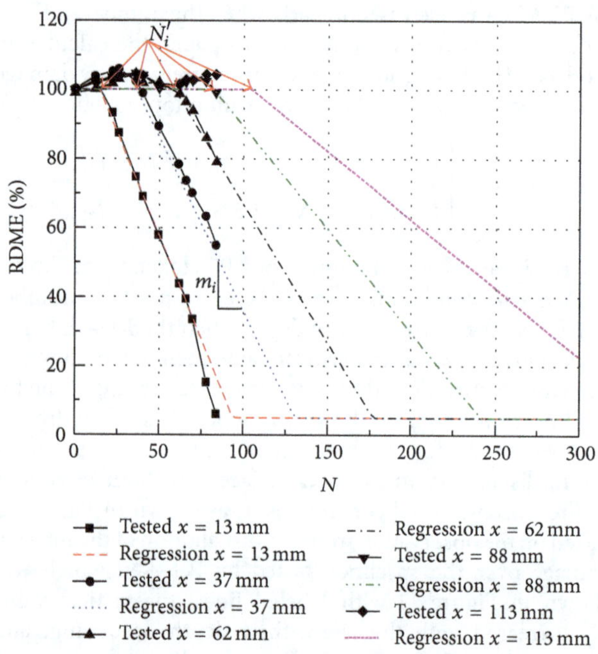

FIGURE 13: The comparison diagram between calculation and test results for RDME with various freeze-thaw depths.

to 43.49 MPa. Thus E_{dyn} for undamaged concrete was 40.465 GPa calculated by (9). According to (7) considering the uneven distribution of freeze-thaw damage, the damaged dynamic elastic modulus $E^d_{\text{dyn},i}$ [GPa] for frozen-thawed concrete fiber with various freeze-thaw depth was written as follows:

$$E^d_{\text{dyn},i} = \text{RDME} \cdot E_{\text{dyn}}. \tag{10}$$

Hanjari et al. (2013) [9] had studied the regression analysis for relationships between the damaged dynamic elastic modulus and the cylindrical compressive strength for frozen-thawed concrete which were tested by Suzuki et al. (2007), Hanjari et al. (2009a), Fagerlund et al. (1994), and Shang et al. (2006). The compressive strength $f^d_{C,i}$ [MPa] of frozen-thawed concrete fiber at various freeze-thaw depths was derived from Hanjari et al.'s (2013) regression model, as follows:

$$f^d_{C,i} = 30.4 \cdot \sqrt[3.3]{\frac{E^d_{\text{dyn},i}}{52.8 - E^d_{\text{dyn},i}}}. \tag{11}$$

In allusion to the unconfined concrete subjected FTCs, the compressive strength of fiber at different freeze-thaw depth was directly calculated by (11). However, in view of confined concrete, the stirrup constraint coefficient k_i for different frozen-thawed depth was introduced here to consider the confinement effect on freeze-thaw confined concrete. The compressive strength $f^d_{CC,i}$ [MPa] for confined concrete fiber at different freeze-thaw depth was calculated by

$$f^d_{CC,i} = k_i f^d_{C,i}, \tag{12}$$

where k_i was calculated for freeze-thaw concrete with different depth by Mander's method.

According to Petersen et al. [2] proposed equation derived from regression analysis of test results, the static modulus of elasticity $E^d_{c,i}$ [GPa] for frozen-thawed concrete at different freeze-thaw depth was calculated by

$$E^d_{c,i} = (0.9\text{RDME} - 0.06) \times E_c, \tag{13}$$

where E_c was the modulus of elasticity for undamaged concrete which was equal to $5000\sqrt{f_C}$.

Based on the results calculated by (12) and (13), the compressive strength of frozen-thawed unconfined and confined concrete fiber and the corresponding static modulus of elasticity at different freeze-thaw depth were determined. The peak strain $\varepsilon_{0,i}$ for frozen-thawed unconfined concrete fiber was calculated by Kent-Park's constitutive relationship, as follows:

$$\varepsilon_{0,i} = \frac{2 f_{C,i}}{E^d_{C,i}}. \tag{14}$$

The peak strain $\varepsilon_{c0,i}$ for frozen-thawed confined concrete fiber was calculated by Mander's constitutive relationship, as follows:

$$\varepsilon_{c0,i} = \varepsilon_{0,i} \times [1 + 5(k_i - 1)]. \tag{15}$$

The ultimate strain $\varepsilon_{cu,i}$ for frozen-thawed unconfined concrete fiber was easily taken as $2\varepsilon_{0,i}$, and the ultimate strain $\varepsilon_{ccu,i}$ for frozen-thawed confined concrete was taken as 0.02 which is consistent with Mander's constitutive relationship. The input constitutive parameters for frozen-thawed concrete fibers with various freeze-thaw depths were calculated by looping statement in OpenSees script. At this point, the numerical model considering the uneven freeze-thaw for RC columns had been built. The bond-slip for steel bars was not taken into account in this paper, which needs to be thought about in future studies.

Table 5: Input parameters of constitutive model for frozen-thawed unconfined concrete (*Concrete02* in OpenSees).

Specimen	Cycles	Depth (mm)	$f^d_{C,i}$ (Mpa)	$\varepsilon_{0,i}$	$\sigma^d_{cu,i}$ (Mpa)	$\varepsilon_{cu,i}$
Z-C1	0	0–20	43.57	0.0018	0	0.036
Z-C2	100	0–10	39.92	0.0031	0	0.0062
		10–20	43.57	0.0018	0	0.0036
Z-C3	200	0–10	28.69	0.0037	0	0.0074
		10–20	32.45	0.0033	0	0.0067
Z-C4	300	0–10	15.75	0.0158	0	0.0316
		10–20	20.94	0.0062	0	0.0124

Note: $\sigma^d_{cu,i}$ is the ultimate stress of frozen-thawed concrete.

Table 6: Input parameters of constitutive model for frozen-thawed confined concrete (*Concrete04* in OpenSees).

Specimen	Cycles	Depth (mm)	$f^d_{CC,i}$ (Mpa)	$\varepsilon_{c0,i}$	$\varepsilon_{ccu,i}$	$E^d_{c,i}$
Z-C1	0	20–100	46.41	0.0038	0.02	32973.5
Z-C2	100	20–100	46.41	0.0038	0.02	32973.5
Z-C3	200	20–30	38.74	0.0046	0.02	22954.6
		30–40	43.31	0.0045	0.02	26023.8
		40–100	46.41	0.0038	0.02	32973.5
Z-C4	300	20–30	26.43	0.0065	0.02	11131.0
		30–40	30.10	0.0055	0.02	15060.1
		40–50	33.60	0.0049	0.02	18561.9
		50–60	37.07	0.0047	0.02	21636.4
		60–70	40.59	0.0046	0.02	24283.5
		70–80	44.14	0.0045	0.02	26503.3
		80–100	46.41	0.0038	0.02	32973.5

4.4. Validation of Proposed Model. Four RC columns subjected to FTCs with 0 cycles, 100 cycles, 200 cycles, and 300 cycles were simulated, respectively, by using the above-mentioned models considering the uneven distribution of frozen-thawed damage. The most important input parameters for frozen-thawed unconfined and confined concrete with *d* were listed in Tables 5 and 6. The comparison between tests and simulations for hysteretic response is shown in Figure 14. It can be seen that the simulation results of frozen columns with 100 cycles, 200 cycles, and 300 cycles are in good agreement with the experiment. However, the proposed model did not consider the bond-slip phenomenon between longitudinal steel bars and freeze-thawed concrete; thus the strength degradation after peak point was inconsistent with the tests, so it is necessary to establish the bond-slip constitutive relations for frozen-thawed concrete with steel bars in the future research. As for the test result of intact column with 0 FTCs, the positive bearing capacity was larger than the negative one due to the test error with misalignment of the centerlines for hydraulic jack and column before the cyclic loading imposed. When the column with 0 FTCs was simulated, this misalignment above was not considered; therefore, the simulative positive bearing capacity was lower than the experimental value; on the contrary, the simulative negative initial stiffness and negative bearing capacity were larger than experimental values. To sum up, the proposed model in this paper has a good applicability, which can simulate the freezing-thawing process of RC members reasonably, and has better accuracy than that simulated by using conventional methods assuming even distribution of frozen-thawed damage.

5. Conclusions

An experimental study was undertaken to investigate the influence of FTCs on the seismic behavior of RC columns. The effects of the numbers of FTCs on the failure modes, hysteretic response, and skeleton curves of the specimens were analysed. Meanwhile, a methodology considering uneven distribution of frozen-thawed damage to analyse the hysteretic behavior of frozen-thawed RC columns was introduced. The results of the simulation were compared with the above experiments for RC columns. The following conclusions can be drawn based on this study:

(1) FTCs did not change the destruction form of RC columns and all specimens experienced typically bending-shear failures mainly dominated by flexure. In addition, the degree of granulation reduced gradually from the outer to the inner part of columns, which revealed the uneven distribution of freezing-thawing damage.

(2) The hysteretic characteristics of specimens subjected to different number of FTCs are similar. With the

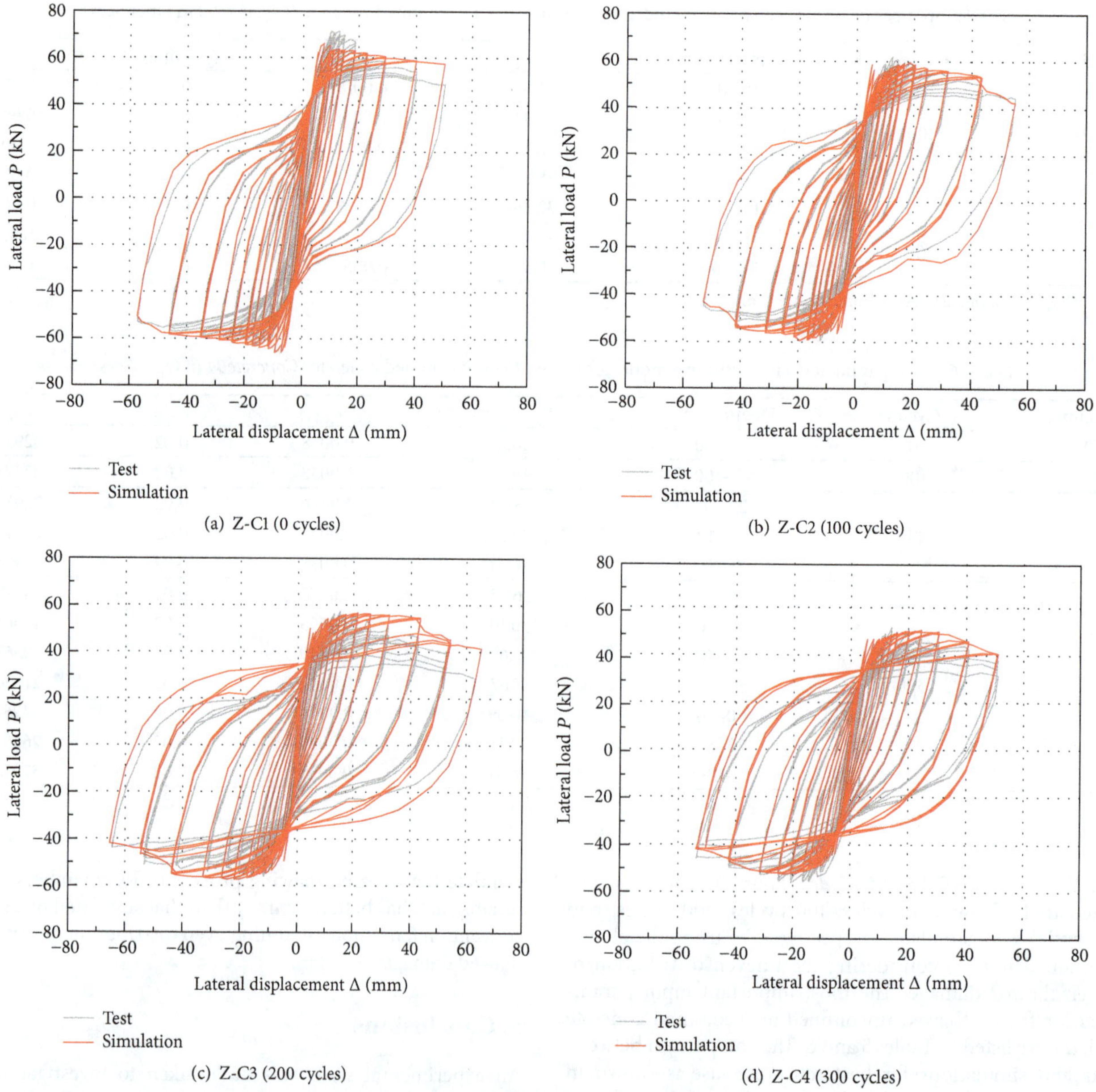

FIGURE 14: Comparison between hysteretic responses of experiment and modeling.

increase of freeze-thaw cycles, the bearing capacity of specimens decreased gradually; however the displacements for different damage states increased slightly, and the corresponding ductility factors declined gradually.

(3) The numbers of FTCs with different freeze-thaw conditions could be transformed mutually by utilizing Berto's method. As a result, the 100 cycles, 200 cycles, and 300 cycles of freeze-thaw tests in this paper were equivalent to 50 cycles, 130 cycles, and 229 cycles, respectively, under the ASTM C666 proc. A condition and were equivalent to 13 cycles, 40 cycles, and 76 cycles under the RILEM TC176-IDC condition.

(4) The calculation model of RDME considering the uneven freeze-thaw damage was proposed by linear regression analysis for Petersen's test data based on some assumptions. And then the compressive strength of frozen-thawed confined/unconfined concrete and the corresponding static modulus of elasticity with various freeze-thaw depths were calculated. On the basis of the existing constitutive relations for concrete, the simulation hysteretic curves calculated by OpenSees were in good agreement with the experimental ones.

The proposed methodology did not consider the bond-slip phenomenon between longitudinal steel bars and freeze-thawed concrete; it is necessary to establish the bond-slip constitutive relations in future research.

Competing Interests

The authors declare that they have no competing interests.

Acknowledgments

This study was supported by the National Key Technology R&D Program under Grant no. 2013BAJ08B00, the Natural Science Foundation of China under Grant nos. 50978218 and 51108376, and the Natural Science Foundation of Shaanxi Provincial Department of Education under Grant nos. 2016KJXX-93 and 2014JQ7231.

References

[1] A. Duan, W. L. Jin, and J. R. Qian, "Effect of freeze-thaw cycles on the stress-strain curves of unconfined and confined concrete," *Materials and Structures*, vol. 44, no. 7, pp. 1309–1324, 2011.

[2] L. Petersen, L. Lohaus, and M. A. Polak, "Influence of freezing-and-thawing damage on behavior of reinforced concrete elements," *ACI Materials Journal*, vol. 104, no. 4, pp. 369–378, 2007.

[3] K. Z. Hanjari, P. Utgenannt, and K. Lundgren, "Experimental study of the material and bond properties of frost-damaged concrete," *Cement and Concrete Research*, vol. 41, no. 3, pp. 244–254, 2011.

[4] M. Hasan, H. Okuyama, Y. Sato, and T. Ueda, "Stress-strain model of concrete damaged by freezing and thawing cycles," *Journal of Advanced Concrete Technology*, vol. 2, no. 1, pp. 89–99, 2004.

[5] H. S. Shang and Y. P. Song, "Experimental study of strength and deformation of plain concrete under biaxial compression after freezing and thawing cycles," *Cement and Concrete Research*, vol. 36, no. 10, pp. 1857–1864, 2006.

[6] S. Xu, A. Li, Z. Ji, and Y. Wang, "Seismic performance of reinforced concrete columns after freeze-thaw cycles," *Construction and Building Materials*, vol. 102, pp. 861–871, 2016.

[7] W. Yang, S.-S. Zheng, D.-Y. Zhang, L.-F. Sun, and C.-L. Gan, "Seismic behaviors of squat reinforced concrete shear walls under freeze-thaw cycles: a pilot experimental study," *Engineering Structures*, vol. 124, pp. 49–63, 2016.

[8] T. Ueda, M. Hasan, K. Nagai, Y. Sato, and L. Wang, "Mesoscale simulation of influence of frost damage on mechanical properties of concrete," *Journal of Materials in Civil Engineering*, vol. 21, no. 6, pp. 244–252, 2009.

[9] K. Z. Hanjari, P. Kettil, and K. Lundgren, "Modelling the structural behaviour of frost-damaged reinforced concrete structures," *Structure and Infrastructure Engineering*, vol. 9, no. 5, pp. 416–431, 2013.

[10] H. Hayashida, Y. Sato, and T. Ueda, "Evaluation of structural properties of the freeze-thaw-damaged RC beam members by nonlinear finite-element analysis," in *Proceedings of the 4th International Conference on the Durability of Concrete Structures (ICDCS '14)*, pp. 45–51, Purdue University, West Lafayette, Ind, USA, July 2014.

[11] L. Berto, A. Saetta, D. A. Talledo, and R. Vitaliani, "Structural analysis of frost damaged constructions by means of a coupled environmental-mechanical damage model," in *Proceedings of the Joint 11th World Congress on Computational Mechanics (WCCM '14)*, pp. 904–915, Barcelona, Spain, July 2014.

[12] L. Berto, A. Saetta, R. Scotta, and D. Talledo, "A coupled damage model for RC structures: proposal for a frost deterioration model and enhancement of mixed tension domain," *Construction and Building Materials*, vol. 65, pp. 310–320, 2014.

[13] L. Berto, A. Saetta, and D. Talledo, "Constitutive model of concrete damaged by freeze-thaw action for evaluation of structural performance of RC elements," *Construction and Building Materials*, vol. 98, pp. 559–569, 2015.

[14] GB 50011-2010, "Code for seismic design of buildings," China Architecture and Building Press, Beijing, China, 2010.

[15] GB, "Standard for test method of concrete structures," GB/T 50152-2012, Chinese Standards, Beijing, China, 2012.

[16] Chinese Standard, *Standard for Test Methods of Long-term Performance and Durability of Ordinary Concrete (GB/T50082-2009)*, China Architecture and Building Press, Beijing, China, 2009.

[17] R. Park, "Evaluation of ductility of structures and structural assemblages from laboratory testing," *Bulletin of the New Zealand National Society for Earthquake Engineering*, vol. 22, no. 3, pp. 155–166, 1989.

[18] C. K. Gulec and A. S. Whittaker, "Performance-based assessment and design of squat reinforced concrete shear walls," MCEER-09-0010, 2009.

[19] V. Terzic, J. Matthew, I. Jose, and A. Stephen, "Concrete Column Blind Prediction Contest 2010: Outcomes and Observations," PEER Report, 2015.

[20] M. Menegotto and P. E. Pinto, "Method of analysis for cyclically loaded reinforced concrete plane frames including changes in geometry and non-elastic behavior of elements under combined normal force and bending," *Preliminary Report IABSE*, vol. 13, 1973.

[21] F. C. Filippou, E. P. Popov, and V. V. Bertero, "Effects of bond deterioration on hysteretic behavior of reinforced concrete joints," Report EERC 83-19, Earthquake Engineering Research Center, University of California, Berkeley, Calif, USA, 1983.

[22] M. Hisham and M. Yassin, *Nonlinear analysis of prestressed concrete structures under monotonic and cycling loads [Ph.D. dissertation]*, University of California, Berkeley, Calif, USA, 1994.

[23] J. B. Mander, M. J. Priestley, and R. Park, "Theoretical stress-strain model for confined concrete," *Journal of Structural Engineering*, vol. 114, no. 8, pp. 1804–1826, 1988.

[24] I. D. Karsan and J. O. Jirsa, "Behavior of concrete under compressive loading," *Journal of Structural Division ASCE*, vol. 95, no. ST12, 1969.

[25] H. E. H. Roy and M. A. Sozen, "Ductility of concrete," in *Flexural Mechanics of Reinforced Concrete*, ACI Special Publication SPI2, pp. 213–235, ACI, Miami, Fla, USA, 1964.

[26] American Standard ASTM C666, "Standard test method for resistance of concrete to rapid freezing and thawing," 2008.

[27] M. J. Setzer, P. Heine, S. Kasparek et al., "Test methods of frost resistance of concrete: GIF-Test: capillary suction, internal damage and freeze thaw test—reference method and alternative methods A and B," *Materials and Structures*, vol. 37, no. 274, pp. 743–753, 2004.

[28] CP110, "Code of practice for the structural use of concrete," British Standards Institution, London, UK, 1972.

3

Investigation on Durability Performance in Early Aged High-Performance Concrete Containing GGBFS and FA

Seung-Yup Jang,[1] Subbiah Karthick,[2] and Seung-Jun Kwon[3]

[1] *Korea Railroad Research Institute, Suwon, Republic of Korea*
[2] *Innovative Durable Building and Infrastructure Research Center, Hanyang University, 55 Hanyangdaehak-ro, Sangrok-gu, Ansan-si, Gyeonggi-do, Republic of Korea*
[3] *Department of Civil and Environmental Engineering, Hannam University, 133 Ojeong-ding, Daejeon 306-791, Republic of Korea*

Correspondence should be addressed to Seung-Jun Kwon; jjuni98@hannam.ac.kr

Academic Editor: Xiao-Jian Gao

The significance of concrete durability increases since RC (Reinforced Concrete) structures undergo degradation due to aggressive environmental conditions, which affects structural safety and serviceability. Steel corrosion is the major cause for the unexpected failure of RC structures. The main cause for the corrosion initiation is the ingress of chloride ions prevailing in the environment. Hence quantitative evaluation of chloride diffusion becomes very important to obtain a chloride diffusion coefficient and resistance to chloride ion intrusion. In the present investigation, 15 mix proportions with 3 water-to-binder ratios (0.37, 0.42, and 0.47) and 3 replacement ratios (0, 30, and 50%) were prepared for HPC (high-performance concrete) with fly-ash and ground granulated blast furnace slag. Chloride diffusion coefficient was measured under nonstationary condition. In order to evaluate the microstructure characteristics, porosity through MIP was also measured. The results of compressive strength, chloride diffusion, and porosity are compared with electrical charges. This paper deals with the results of the concrete samples exposed for only 2 months, but it is a part of the total test plan for 100 years. From the work, time-dependent diffusion coefficients in HPC and the key parameters for durability design are proposed.

1. Introduction

Reinforced concrete (RC) structures are an economical and versatile construction material in civil infrastructure such as bridges, buildings, and nuclear reactors, and involve major construction to the tune of millions of dollars [1, 2]. However, the long term durability and service life of RC structures are one of the major problems faced by the construction industry for the past few decades [3]. The durability of reinforced concrete is largely affected by the migration of aggressive ions (chloride and sulphate) via capillary absorption and hydrostatic pressure of cementitious matrix. The ions reach the reinforced rebar and destroy the passive film [4, 5] subsequently corroding the steel. Chloride induced corrosion of RC structure has become a major problem worldwide [6–8], especially in buildings, bridges, parking decks, tunnels, and other buildings exposed to seawater or deicing salts. As a result of this deterioration of the RC structure, the repair costs nowadays constitute a major part of spending on infrastructure. Further, corrosion of steel does not only damage the RC structure but also causes safety concerns.

Corrosion of steel in the RC structure can be mitigated by adopting various preventive measures, namely, cathodic protection, using corrosion inhibitors, coating to steel rebar, coating to concrete, using blended cement, and realkalization of concrete [9–14]. The use of corrosion inhibitors is the more appropriate method, and its maintenance cost is high. One alternative way to prevent the corrosion of high-performance concrete (HPC) by improving impermeability, resistance to chloride ion diffusion [15], and abrasion resistance can be achieved through the partial replacement of cement with industrial byproducts (supplementary cementing materials (SCMs)) such as fly-ash (FA) [16], ground granulated blast furnace slag (GGBFS) [17], silica fume, rice husk ash [18], and micro silica.

TABLE 1: Chemical composition and physical properties of OPC, GGBFS, and FA.

Types	Items								
	Chemical composition (mass %)							Physical properties	
	SiO_2	Al_2O_3	Fe_2O_3	CaO	MgO	SO_3	Lg.loss	Specific gravity (g/cm^3)	Blaine (cm^2/g)
OPC	21.96	5.27	3.44	63.41	2.13	1.96	0.79	3.16	3,214
GGBFS	32.74	13.23	0.41	44.14	5.62	1.84	0.2	2.89	4,340
FA	55.66	27.76	7.04	2.70	1.14	0.49	4.3	2.19	3,621

In South Korea, the annual production amount of FA was 8.5 million tons in coal-fired power plants and GGBFS (produced by POSCO and Hyundai Steel) was about 12 million tons/year as of 2011 [19]. Compared to the Ordinary Portland Cement (OPC), it is very low regarding emission of environmental load. Hence, it has been mostly used as a component of low carbon emission concrete [20]. Replacing OPC with GGBFS and FA in high-performance concrete (HPC) and self-compacting concretes (SCC) are becoming increasingly common in civil engineering structures [21]. Utilizing FA and GGBFS for the production of HPC and SCC not only reduces the total material cost of the construction industry but also results in considerable benefits to the environment [22]. In addition, the use of FA and GGBFS in the SCC has a unique and distinctive effect on the properties of the HPC. Further, the replacement of OPC with GGBFS and FA in SCC had a lower resistance to carbonation than the pure OPC-blended SCC. This effect appeared to be more pronounced with an increase in the replacement level of FA and GGBFS [23]. Moreover, utilizing FA and GGBFS in SCC is more effective in resisting the chloride ion and sulphate ion migration and reducing the capillary pores. The chloride ion diffusion coefficient of the SCC mixtures with FA and GGBFS was lower than the control SCC [24–26]. In addition, GGBFS and FA have pozzolanic activity which is attributed to the presence of SiO_2 and Al_2O_3. It reacts with calcium hydroxide during cement hydration to form additional calcium silicate hydrate (CSH) and calcium aluminate hydrate (CAH), which are effective in forming denser matrix leading to higher strength and better durability [27–29]. Furthermore, according to Yuan et al., GGBFS and FA increase chloride binding due to the high content of aluminate hydrates and hence chloride migration was reduced in HPC [30]. The research field on the evaluation of chloride diffusion in concrete is growing with consideration for diffusion, permeation [31], and binding capacity of chloride ions [32, 33]. Recently, numerical techniques covering chloride diffusion in partially saturated condition [34], chloride behavior in concrete with early-age cracking [35], and micro structures formation modeling in high-performance concrete [36, 37] were proposed based on behavior in early-age concrete considering hydration and micro pore structure.

In this work, 4000 specimens for 15 mix proportions with various water-binder (w/b) material ratios (0.37, 0.42, and 0.47) and various replacement percentages of GGBFS and FA (0%, 30, and 50%) are prepared and cured in tap water. As a part of the work, half of the samples analysis is going on at KCL (Korea Conformity Laboratory, Korea) by exposing to tidal, atmosphere, and submerged conditions to know long term durability for 100 years. The apparent diffusion coefficient and porosity will be measured by using RCPT (Q) and MIP to know long term durability for 100 years. The samples will be measured at different intervals of time period: 0.6, 1, 2, 3, 4, 5, 7, 10, 20, 40, 80, and 100 years. The compressive strength, electrical charges, and chloride diffusion coefficient in nonsteady state and steady state conditions were measured. In order to evaluate the microstructure characteristics, porosity through MIP was also measured. From this work, time-dependent diffusion coefficients in HPC and key parameters for durability design are proposed. The results showed that partial replacement of OPC with GGBFS and FA contributed considerable improvement to various properties of HPC.

2. Experimental Program

2.1. Materials Used. Ordinary Portland Cement (OPC) of Type 1 with specific gravity 3.16 was used. Chemical compositions of OPC, FA, and GGBFS are given in Table 1. River sand was used as fine aggregate, passing through 2.36 mm sieve, falling under zone and with specific gravity of 2.60. Coarse aggregates used were crushed angular aggregates of normal size 19 mm with specific gravity 2.62.

2.2. Concrete Mix Proportion. In this study, a total of 15 HPC mixtures were prepared; OPC partial replacement with GGBFS and FA in three replacement ratios like 0, 30, and 50% were considered for HPC mixtures. Three water-to-binder material (w/b) ratios (0.37, 0.42, and 0.47) were used. The details of the mixing proportions of the HPC are shown in Table 2.

2.3. Test Program

2.3.1. Compressive Strength. The compressive strength of the various concrete mixers was calculated according to ASTM C39/C39M [39] using cylindrical specimens of 100 mm in diameter and 200 mm in height, cast with different percentage of FA and GGBFS (HPC). After curing in room condition for 24 hours, the specimens were demoulded and immersed in water for curing at 25°C. The compressive strength was measured after 28 and 49 days of curing, the concrete cylinders were tested in the compression-testing machine, with 100 T capacities at the rate of loading 140 kN/min. The ultimate load at which the cube failed was taken.

TABLE 2: Mixing proportion of HPC.

System	w/b ratio (%)	Water (kg/m^3)	Cement (kg/m^3)	Fly-ash (kg/m^3)	GGBFS (kg/m^3)	Coarse aggregates (kg/m^3)	Fine aggregates (kg/m^3)
O100	0.37	168	454	0.0	0.0	952	767
	0.42	168	400	0.0	0.0	976	787
	0.47	168	357	0.0	0.0	960	838
O70S30	0.37	168	318	0.0	136	946	762
	0.42	168	280	0.0	120	972	783
	0.47	168	250	0.0	107	956	835
O50S50	0.37	168	227	0.0	227	943	760
	0.42	168	200	0.0	200	969	780
	0.47	168	179	0.0	179	853	832
O70F30	0.37	168	318	136	0.0	952	745
	0.42	168	280	120	0.0	953	768
	0.47	168	250	107	0.0	939	820
O50F50	0.37	168	227	227	0.0	950	662
	0.42	168	200	200	0.0	982	685
	0.47	168	179	179	0.0	1008	703

FIGURE 1: The photograph image of RCPT experiment setup.

2.3.2. Rapid Chloride Ion Penetration Test (RCPT). The rapid chloride ion permeability test (RCPT) was conducted in accordance with ASTM C1202-10 [38] using a concrete disc of size 100 mm diameter and 50 mm thickness. After 28 days of curing, the concrete specimens were subjected to RCPT test by impressing a voltage of 60 V between two containers filled with 3% NaCl solution and 0.3 N NaOH solutions, as shown in Figure 1. Electrical current was measured every 30 minutes for up to 6 hours. The amount of electrical current passing through the specimen was measured and the total charge passed (in coulombs) was used as an indicator of the resistance of the concrete to chloride ion penetration. The total charge passed through the concrete specimens was calculated using the following formula [38]:

$$Q = 900\left(I_0 + 2I_{30} + 2I_{60} + \cdots + 2I_{300} + 2I_{330} + I_{360}\right), \quad (1)$$

where Q is charge passed (coulombs), I_0 is current (amperes) immediately after voltage is applied, and It is current (amperes) at t min after the voltage is applied.

2.3.3. Chloride Diffusion. Chloride diffusion coefficient method is the extension of the RCPT test. The chloride diffusion coefficient values were calculated by the two conditions such as steady state condition based on the results from ASTM [38, 40] and nonsteady state condition from Tang's method [40, 41]. In RCPT test, the time duration is 6 hours where it is under nonsteady state condition. The diffusion cell and experimental set up is provided in ASTMC 1202 [38] and the calculation of the diffusion coefficient is performed by an electrical method proposed by previous researches [40, 41]. Silver nitrate solution (0.1 N, $AgNO_3$) is used as an indicator [42, 43]. In this test, chloride diffusion coefficient in nonsteady state conditions and steady state condition (effective diffusion) was calculated using (2a) and (2b) and (3), respectively.

$$D_{\text{cpd}} = \frac{RTL}{zFU} \cdot \frac{X_d - \alpha\sqrt{X_d}}{t}, \quad (2a)$$

$$\alpha = 2\sqrt{\frac{RTL}{zFU}} \cdot \text{erf}^{-1}\left[1 - \frac{2C_d}{C_0}\right], \quad (2b)$$

$$D_{\text{eff}} = 0.0103\left(Q_0\right)^{0.84} E - 8, \quad (3)$$

where D_{cpd} is diffusion coefficient in nonsteady state and steady state condition from RCPT (m^2/s), R is universal gas constant (8.314 J/mol K), T is absolute temperature (K), L is thickness of specimen (m), z is ionic valence (=1.0), F is Faraday constant (=96,500 J/V mol), U is applied potential (V), t is test duration time (s), C_d is the chloride concentration at which the color changes when using a colorimetric method to measure X_d based on the reference [40, 42], C_0 is chloride concentration in the upstream solution (mol/l), α is an experimental constant, and erf^{-1} is the inverse function of the error function. According to the previous research [40], $\text{erf}^{-1}(1 - (2C_d/C_0))$ is calculated as 0.764 and zFU/RTL

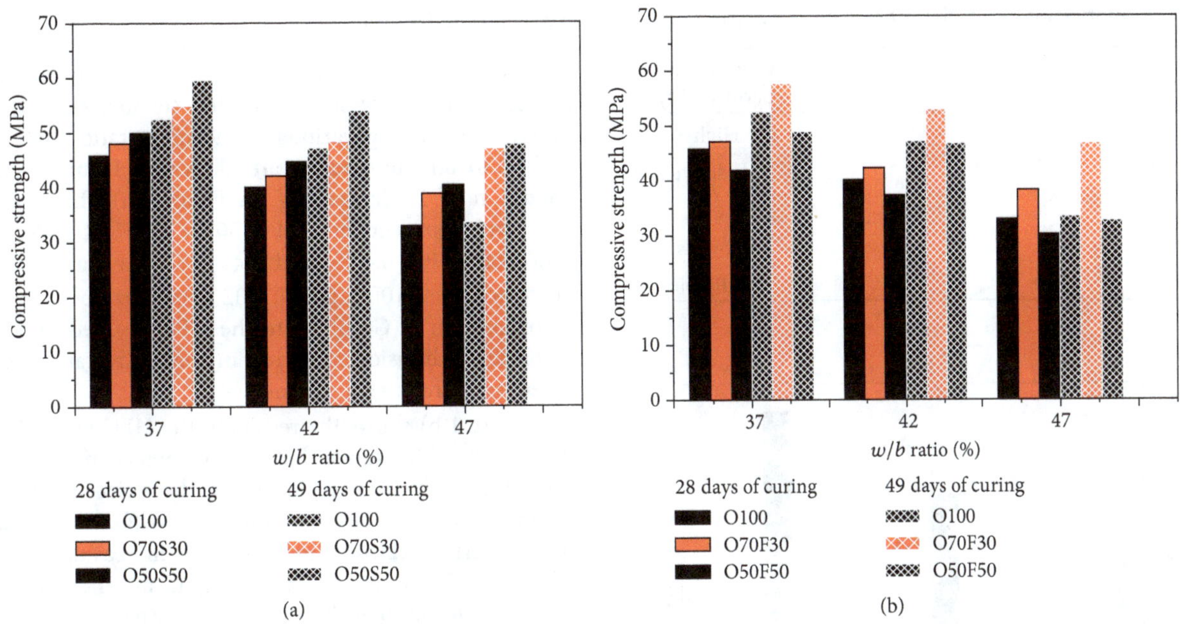

FIGURE 2: Compressive strength of HPC with (a) GGBFS and (b) FA at 28- and 49-day curing period.

is 23,600 (m^{-1}) in this study. D_{eff} means effective diffusion coefficient in RCPT test.

2.4. Porosity Measurement. The porosity of concrete specimens at the age of 28 days was also investigated through the mercury intrusion porosimetry (MIP) test (ASTM D 4404) [44]. MIP (Micromeritics, Autopore IV 9520, USA) has been one of the most widely used methods to analyze the pore structure of HPC samples. When preparing samples for MIP test, aggregates were avoided from sampling and about 1 cm^3 volumes (2.4~3.6 g) were placed in the quanta chrome porosity analyzer and a sample was used for each measurement. The intruded volume could be read to an accuracy of $\pm 0.001\ cm^3$.

3. Results and Discussion

3.1. Compressive Strength. The compressive strength of HPC with different percentages of GGBFS and FA at the age of 28 and 49 days as per KS F 2405 is shown in Figure 2. It was observed from the results that, at the age of 28 days and above, there has been an increase in compressive strength up to 30–50% replacement level of GGBFS in 0.37 of w/b ratio. The increase in 28 days strength of HPC mixes is due to the improvement in the effectiveness of the mineral admixture. The pozzolanic action of GGBFS reacts with OPC, which yields early strength at lower w/b ratios. The similar results are also reported by Tripathi et al. [45]. They reported that, at the age of 28 days at lower w/b ratio, the compressive strength of concrete with ISF slag was higher than control mix concrete even at 60% replacement level. Moreover, the compressive strength values of GGBFS concrete at 49-day curing period was higher when compared to all the other HPC mixes, where the mix proportions of OPC : GGBFS were 70 : 30 and 50 : 50 at w/b of 0.37.

On the other hand, another system of compressive strength of HPC with different percentage of FA at the age of 28 and 49 days has increased up to 30% replacement level of FA in 0.37 of w/b ratio and decreasing thereafter (50% replacement of FA). An increase in the compressive strength of HPC by up to 30% in replacement FA when compared with the control mix may be due to the pozzolanic action and packing effect of FA particles. FA contains more of silica and alumina; it reacts with calcium hydroxide to form C-A-S-H and C-S-H, which contributes to the higher strength of 30% of FA in HPC. Further, by replacing FA by up to 50%, strength has been reduced due to the weak bonding between cement paste and fly-ash particles and insufficient alkali from the reduced OPC amount.

In addition, the compressive strength was slightly reduced by increasing w/b (0.42 and 0.47) ratio in HPC mix with GGBFS and FA replacement level. The increasing w/b ratio dilutes the cement paste and creates more water-filled pore space between the grains, that is, less nuclei for the hydrates in each volume unit. Hydrates have to grow larger and larger to cover the spatial gap (the water) between them and to interact and to develop strength—either physically (interlocked growth) or chemically (e.g., van-der-Waals attraction).

3.2. Rapid Chloride Ion Penetration Test (RCPT). Rapid chloride permeability test was conducted to investigate the performance of HPC against chloride ingress. The total charge passed through the concrete matrix is lower and it means the resistance to chloride penetration is higher. Table 3 shows the classification of concrete for chloride ion penetrability based on total charge passed ASTM C1202 [38].

Figure 3 shows the RCPT test results at the end of 28 days for the HPC with various percentages of GGBFS and FA. Figure 3 illustrates that the cement replacement with

TABLE 3: Chloride ion penetrability based on charge passed [38].

Charge passed (c)	Chloride ion permeability
>4,000	High
2,000–4,000	Moderate
1,000–2,000	Low
100–1,000	Very low
<100	Negligible

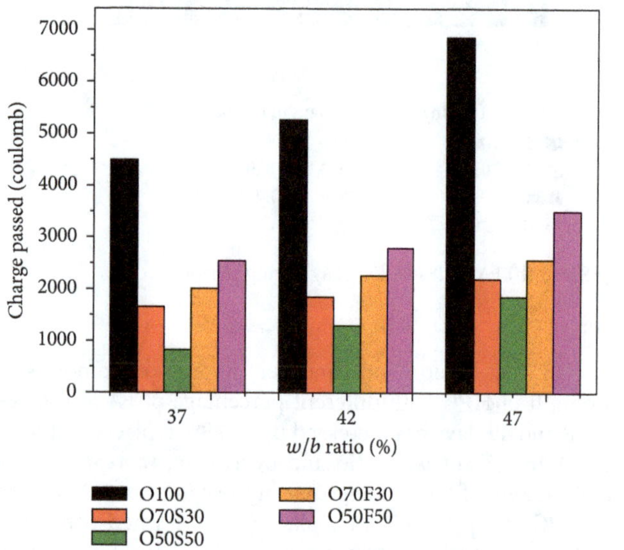

FIGURE 3: RCPT results of HPC with GGBFS and FA.

GGBFS in HPC (0.37% of w/b) had lesser coulomb values when compared to control mix. For example, the 30% and 50% replacement level of GGGFS in 0.37% w/b ratio of HPC charge passed coulomb values are 1659.6 and 829.8 coulomb. This may be due to the reaction of pozzolanic materials like GGBFS reacting with $Ca(OH)_2$ to form C-S-H gel. The C-S-H gel considerably reduces pores between fine aggregate and coarse aggregate, so charge passed values may be significantly reduced through the GGBFS replaced HPC. At the same time with increasing w/b (0.42 and 0.47%) ratio, the charge passed coulomb values slightly increase. It may be due to the more dilute cement paste which creates more water-filled pore space between the grains.

Figure 3 shows the charge passed results in cement replacement with FA in HPC. From the figure, it is observed that 30% of FA in HPC has lower values when compared to the control mix. For example, the charge passed result in the system with 30% replacement and 0.37 of w/b ratio is 2012.4 coulomb. Further, the replacement of FA up to 50% charge passed coulomb values (2541 coulomb) significantly increases when compared to 30% of FA in HPC. This result indicates that the presence of 50% GGBFS and 30% FA in HPC can be more efficient in preventing chloride ion migration.

3.3. *Chloride Diffusion*

3.3.1. Nonsteady State Condition. The chloride diffusion coefficient at nonsteady state condition (6 hours during test) in HPC containing various replacement ratios (0%, 30%, and 50%) and w/b ratios are shown in Figure 4. In the condition of 28-day curing. In Figure 4(a), the results in HPC with replacement level of 50% GGBFS at various w/b ratios of 0.37%, 0.42%, and 0.47 at 28 days are 6.4640 × 10^{-12}, 6.5330 × 10^{-12}, and 7.1102 × 10^{-12}, respectively. The incorporation of GGBFS into the HPC resulted in a lower chloride ion diffusion coefficient in nonsteady state condition when compared to control mix.

Figure 4(b) shows the results in the HPC with FA from 0% to 30%, which shows that the diffusion coefficients in the HPC drastically decrease. Regarding the replacement level of FA up to 50%, the chloride diffusion coefficient in the FA-blended HPC begins to increase with increasing w/b ratio of HPC with FA and it has also significant increases when compared to lower w/b ratio of HPC. Zhao et al. [21] also reported that, at the age of 28 days at lower w/b ratio, the chloride diffusion coefficient of concrete with FA was lower than control mix even at 30% replacement level. Further, they also reported that increasing the replacement level FA up to 50% of FA the chloride ion diffusion coefficient also increased.

3.3.2. Steady State Condition (Effective Diffusion). The chloride diffusion coefficient at steady state condition in HPC with the replacement of GGBFS and FA in different w/b ratio systems is shown in Figure 5. The chloride diffusion coefficient also decreases with the replacement level of 30% and 50% GGBFS in HPC when compared to the control mix. For example, control mix and 50% of GGBFS in HPC (0.37 w/b) are 1.2053 × 10^{-11} m^2/sec and 0.2916 × 10^{-11} m^2/sec, respectively. The reduction of chloride diffusion is 75.8% when compared with the control mix. This diffusion coefficient data confirms the better performance of 50% GGBFS in HPC.

On the other hand it was observed that 30% of FA in HPC has lower chloride diffusion coefficient when compared to the control mix. For example, control mix and 30% of FA in HPC (0.37 w/b) are 1.2053 × 10^{-11} m^2/sec and 0.6137 × 10^{-11} m^2/sec, respectively. The reduction of chloride diffusion is 49.1% when compared to control mix. At the same time, the chloride diffusion coefficients of HPC with the replacement of GGBFS and FA (lower w/b ratio) at steady state condition significantly decrease when compared to nonsteady state condition. It may be due to absorption of chloride ion and formation of bound chloride (Friedel salt) [46]. Moreover, the HPC with GGBFS exhibits lower chloride diffusion coefficient values than HPC with FA. It may depend on the aluminate content, where aluminate in GGBFS and FA forms AFm phases which react with chloride and also produces the calcium chloroaluminate hydrate and Friedel's salt [47]. Moreover, at a given replacement level, the HPC samples with GGBFS had a lower chloride ion diffusion coefficient than those with FA.

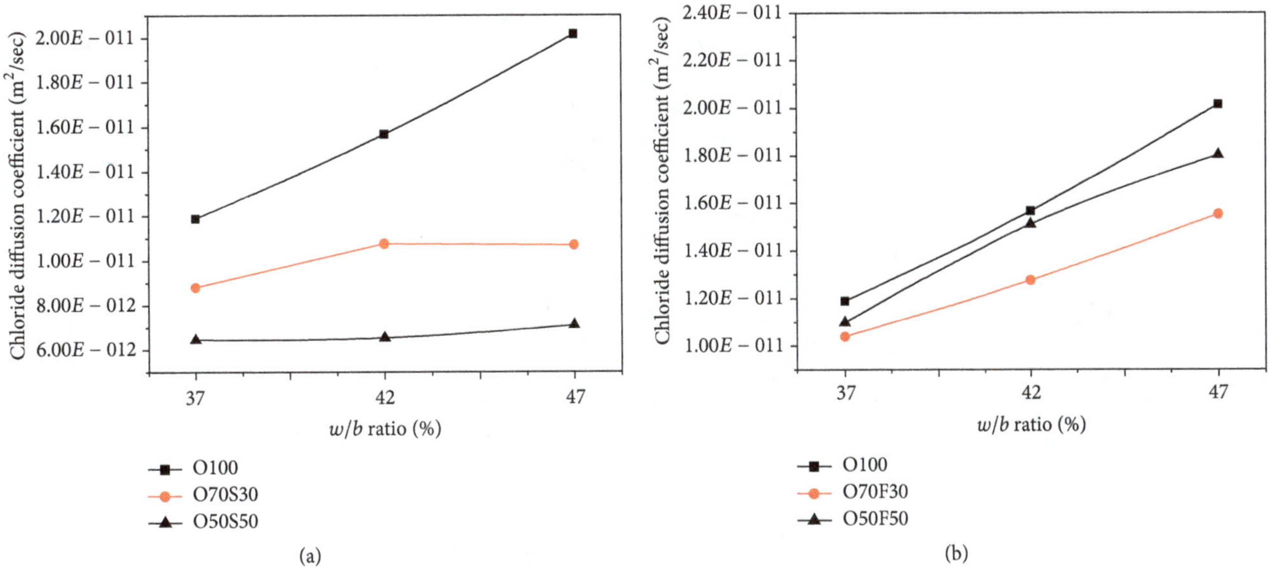

Figure 4: Chloride diffusion coefficient of HPC with (a) GGBFS and (b) FA at nonsteady state condition.

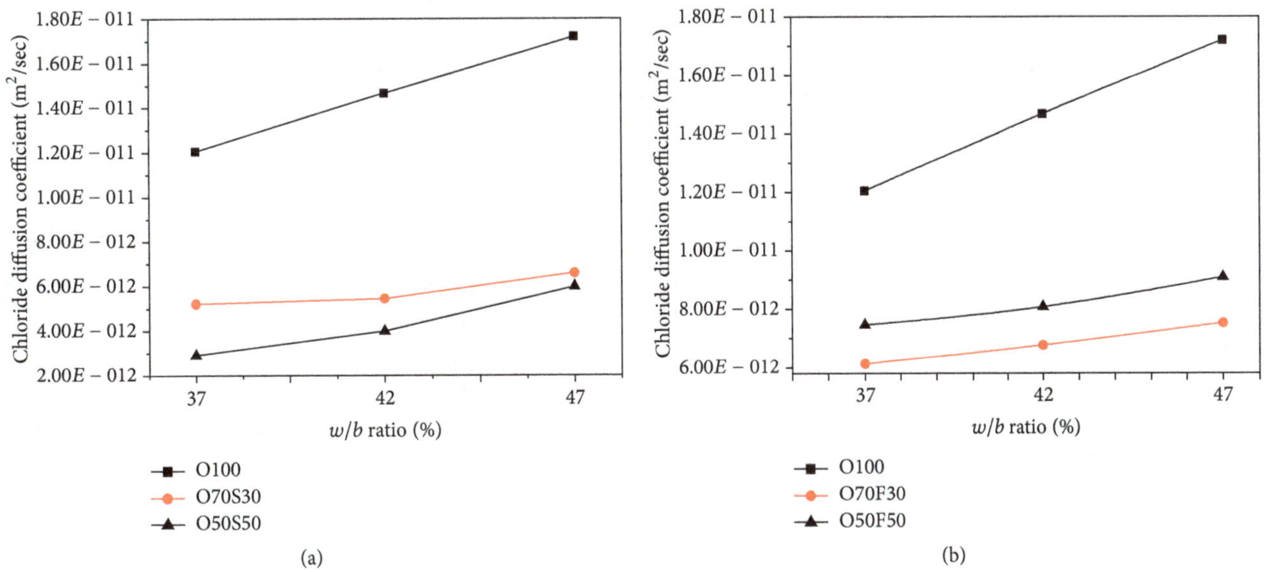

Figure 5: Chloride diffusion coefficient of HPC with (a) GGBFS and (b) FA at steady state condition.

3.4. Porosity Measurement. Figure 6 shows the total porosity measurement for GGBFS (a) and FA (b) replaced HPC by MIP test at 28 days of water curing. As shown in Figure 6(a), an increase in the percentage of GGBFS with HPC reduces the average pore size diameter by 8.53 and 7.33% for 30 and 50% GGBFS replacement, respectively. Further increase in the w/b ratio (0.42 and 0.47) in 30% and 50% GGBFS on porosity at 28 days of curing significantly increases. For example, at various 0.37, 0.42, and 0.47%w/b ratios of 50% GGBFS replacement of HPC, the porosity results are 7.33, 8.95, and 10.38%. As shown in Figure 6(b), the 30% of FA with HPC, porosity is 7.29% at 0.37%w/b ratio. Increasing the percentage of FA and w/b ratio, the porosity slightly increases. At a given replacement level, the HPC samples with GGBFS had a lower pore diameter than those with FA.

3.5. Correlation between Test Results. Figures 7–9 present the relationship between compressive strength (CS), chloride diffusion coefficient (CDC), and porosity (P) against charge passed coulombs (CPC). Table 4 shows the numerical representation to determine the correlation between these properties of HPC with GGBFS and FA.

Figures 7(a) and 7(b) show the increase in CS attributed to the decrease of the CPC for all mixes of HPC. However, CS decreases with an increase in w/b ratio for all mixes of HPC. Also, it meagerly increases by replacing GGBFS and FA of HPC (Figures 7(a) and 7(b)). In addition, a good correlation is observed between CPC and CS. The same CS value is obtained in all mixes of HPC that showed different values of CPC. For example, at CS of 40 MPa in all mixes of HPC, the CPC values are around 5266.8 coulombs for control mix and

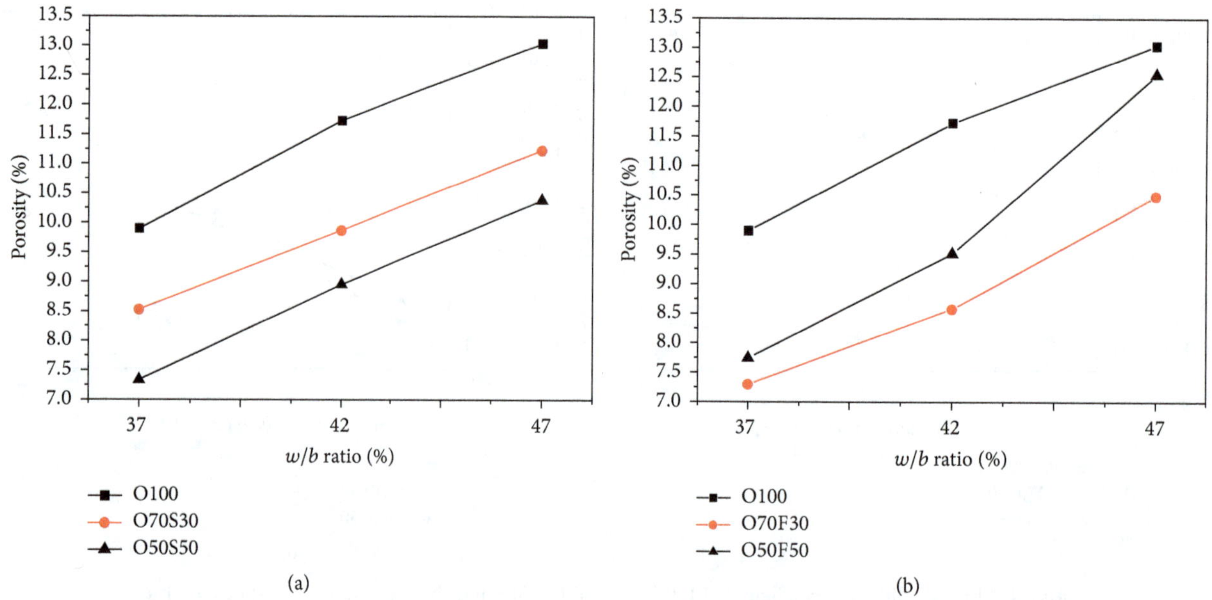

FIGURE 6: Porosity of HPC with (a) GGBFS and (b) FA.

TABLE 4: Result of compressive strength, charge passed coulomb, chloride diffusion coefficient, and porosity of concrete.

System	w/b ratio (%)	Cement (kg)	*CPC (coulomb)	*CS (MPa)	*CDC (m^2/sec) $\times 10^{-11}$	Porosity (%)
O100	0.37	104.4	4494.6	45.85	1.2053	9.89
	0.42	92.0	5266.8	40.09	1.4670	11.72
	0.47	82.1	6863.4	32.9	1.7200	13.03
O70S30	0.37	73.1	1659.6	48.09	0.5219	8.53
	0.42	64.4	1847.8	42.02	0.5452	9.86
	0.47	57.5	2194.2	38.69	0.6599	11.22
O50S50	0.37	52.2	829.8	49.94	0.2916	7.33
	0.42	46.0	1300	44.67	0.3993	8.95
	0.47	41.2	1854.8	40.36	0.5989	10.38
O70F30	0.37	73.1	2012.4	47.17	0.6137	7.29
	0.42	64.4	2262.6	42.33	0.6772	8.57
	0.47	57.5	2566.8	38.2	0.7529	10.49
O50F50	0.37	52.2	2541.6	41.92	0.7466	7.74
	0.42	46.0	2791.8	37.33	0.8079	9.51
	0.47	40.9	3509.4	30.2	0.9083	12.54

*CPC: charge passed coulomb; *CS: compressive strength; *CDC: chloride diffusion coefficient.

1847 coulombs and 1854 coulombs for 30 and 50% GGBFS of HPC, respectively. 2262 coulombs and 2541 coulombs are for 30 and 50% FA of HPC. This is due to the microstructure of concrete and secondary formation of calcium silicate hydrate by using various percentages of GGBFS and FA of HPC in different w/b ratio. Similar correlations between CPC and CDC are shown in Figures 8(a) and 8(b). Moreover, Figures 9(a) and 9(b) illustrated a good correlation between CPC and porosity of HPC. However, the same porosity values are obtained in all mixes of HPC that showed different values of CPC. For example, with 10% of porosity in all mixes of HPC, the CPC values obtained around 4494.6 coulombs for control mix and 1847 coulombs and 1854 coulombs for 30 and 50% GGBFS of HPC. 2566 coulombs and 2791 coulombs are for 30 and 50% FA of HPC. This may be due to the difference in the microstructure of concrete prepared using different type of pozzolanic materials [48] (GGBFS, FA) and mix design. It causes the reduction of the pores, resulting in a dense microstructure and hence reduces the charge passed coulomb on concrete [49, 50].

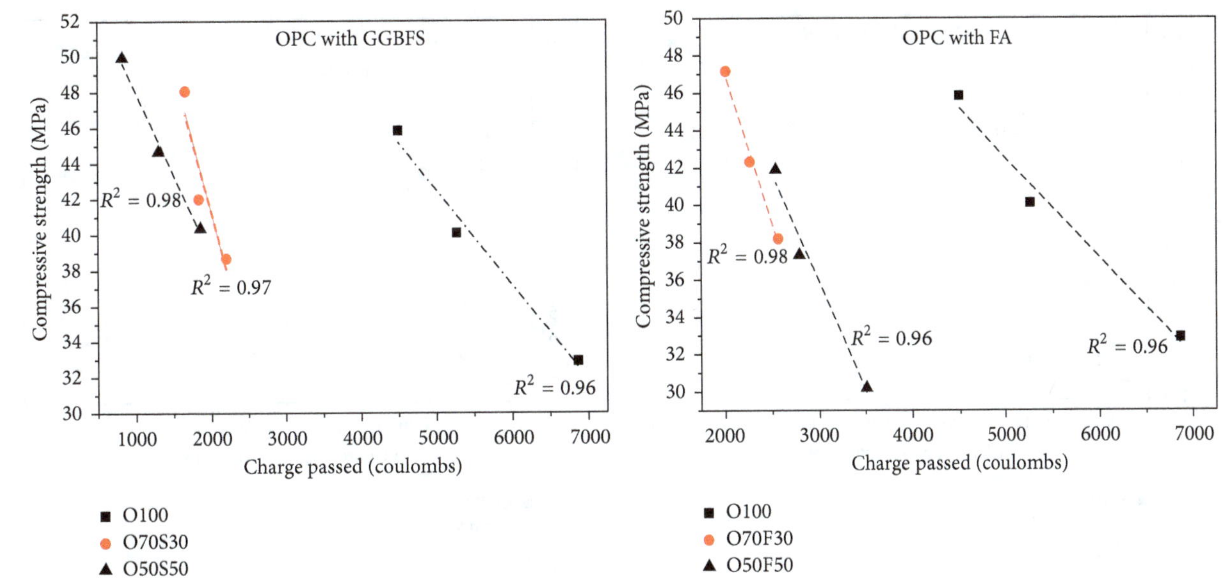

(a) (b)

Figure 7: Correlation graph between charge passed coulomb versus compressive strength: (a) GGBFS in HPC and (b) FA in HPC.

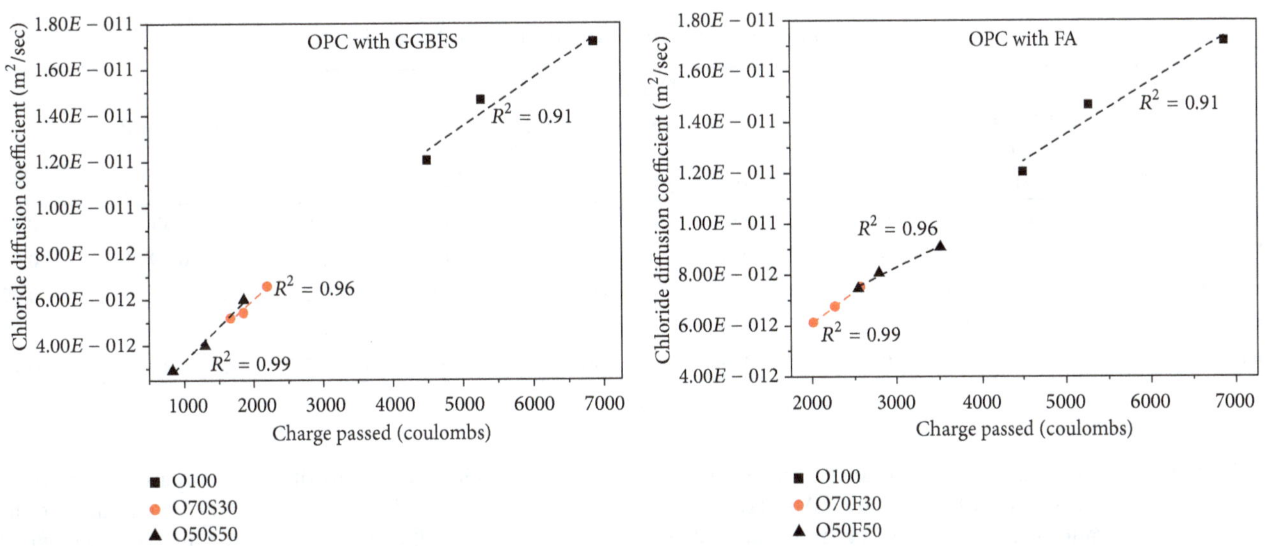

(a) (b)

Figure 8: Correlation graph between charge passed coulomb versus chloride diffusion coefficient: (a) GGBFS in HPC and (b) FA in HPC.

The correlation equation can be expressed by the following single formula as reported elsewhere [51]:

$$\mathrm{DI} = \frac{a}{(f_{\mathrm{CPC}})^b}, \tag{4}$$

where DI is the durability index and "a" and "b" are the experimental constants. The f_{CPC} charge passed coulomb. The constants "a" and "b" were obtained through the regression analysis of the data in Table 4. The best-fit values of constants a and b and the coefficient of determination (R^2) are summarized in Table 5.

R^2 over 0.85 indicates an excellent correlation between the fitted parameters [52]. Therefore, the data in Table 5 indicate a valid agreement between the CPC and CS of the replacement with GGBFS and FA of HPC. Furthermore, the data in Table 5 also indicates that an excellent fit correlation between CPC is plotted against CS and CDC of all types of mix concrete. However, in the case of control mix of concrete, the degree of the bit between the CPC and porosity of concrete is on the lower side (R^2 < 0.83) but reasonably correlated.

4. Conclusion

The conclusions drawn from this work are as follows.

(1) The 0.37 w/b ratio of HPC containing 50% GGBFS and 30% FA yields the highest compressive strength values at 28 and 49 days of curing.

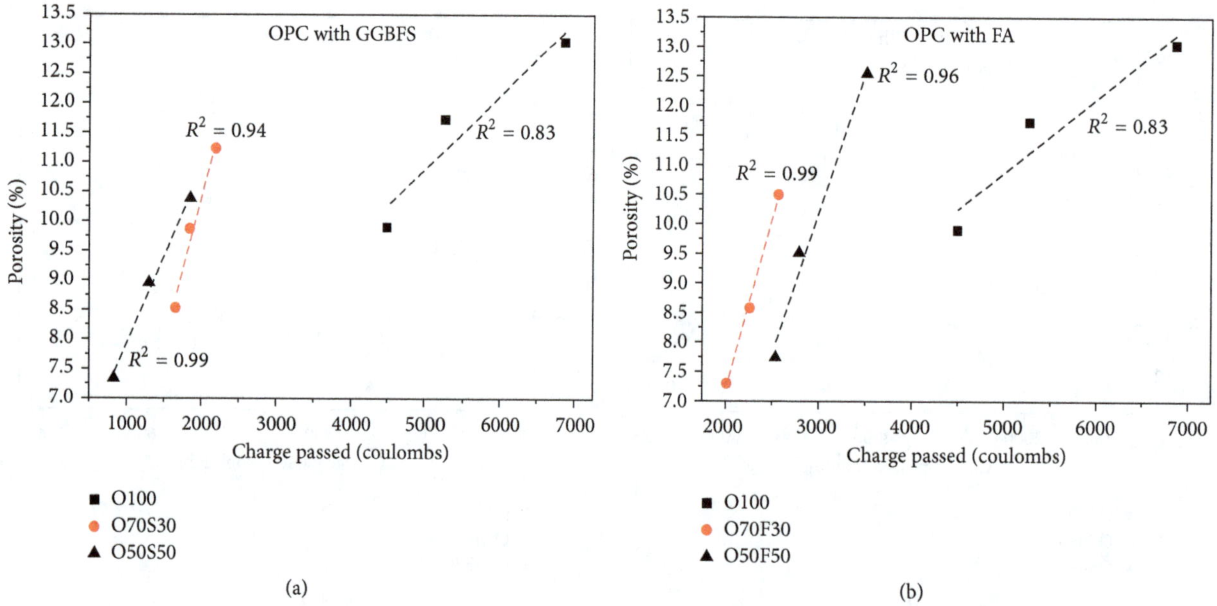

FIGURE 9: Correlation graph between charge passed coulomb versus porosity: (a) GGBFS in HPC and (b) FA in HPC.

TABLE 5: Constants "a" and "b" and regressions coefficients R^2.

System	Correlation: CPC versus CS a	b $\times 10^{-2}$	R^2	CPC versus CDC a 10^{-12}	b 10^{-15}	R^2	CPC versus P a	b $\times 10^{-2}$	R^2
O100	69	0.532	0.96	3.09	2.08	0.91	4.62	0.13	0.83
O70S30	76	1.72	0.97	0.68	2.67	0.96	0.56	0.49	0.94
O50S50	57	0.93	0.98	0.29	3.02	0.99	0.49	0.29	0.99
O70F30	79	1.6	0.98	1.09	2.51	0.99	4.40	0.58	0.99
O50F50	71	1.16	0.96	3.46	1.61	0.96	4.21	0.48	0.96

(2) Better chloride diffusion resistance properties are observed for 50% GGBFS, and 30% FA in HPC with 0.37 w/b ratio of concrete showing minimum value. The effective porosity values are decreased as follows: 9.89% for HPC, 7.30% for 30% FA in HPC, and 7.33% for 50% GGBFS in HPC. Similarly, chloride diffusion coefficient also follows the same trend: 1.2053×10^{-11} for HPC, 0.6137×10^{-11} for 30% FA in HPC, and 0.2916×10^{-11} for 50% GGBFS in HPC. These results conclude the excellent durability performance in HPC containing FA and GGBFS.

(3) Among 15 mixes of concrete, 30% FA and 50% GGBFS in HPC with 0.37 w/b ratio show the least pores, minimum current flow, and less chloride diffusion coefficient. This study indicates that better durability performance for chloride environment is obtained in HPC containing GGBFS and FA compared to a control mix.

(4) Through regression analysis, several durability performances such as compressive strength, diffusion coefficient, and porosity are compared with charge passed from RCPT results. They have high determination coefficient over 0.85 for all the cases, which indicates that electrical charge from RCPT can be another index for both durability and structural performance.

Conflicts of Interest

The authors declare that there are no conflicts of interest regarding the publication of this paper.

Acknowledgments

This research was supported by Basic Science Research Program through the National Research Foundation of Korea (NRF) funded by the Ministry of Science, ICT & Future Planning (no. 2015R1A5A1037548) and Basic Science Research Program through the National Research Foundation of Korea (NRF) funded by the Ministry of Education (NRF-2016R1D1A1A09919224).

References

[1] B. Lomborg, *The Skeptical Environmentalist: Measuring the Real State of the World*, Cambridge University Press, Cambridge, UK, 2001.

[2] X. Shi, N. Xie, K. Fortune, and J. Gong, "Durability of steel reinforced concrete in chloride environments: an overview," *Construction and Building Materials*, vol. 30, pp. 125–138, 2012.

[3] G. Horrigmoe, "Future needs in concrete repair technology," in *Proceedings of the Concrete Technology for a Sustainable Development in the 21st Century*, E&FN Spon, London, UK, 2000.

[4] L. Bertolini, B. Elsener, P. Pedeferri, and R. Polder, *Corrosion of Steel in Concrete: Prevention, Diagnosis, Repair*, Wiley-VCH, Weinheim, Germany, 2004.

[5] C. L. Page, "Nature and properties of concrete in relation to reinforcement corrosion," in *Proceedings of the Corrosion of Steel in Concrete*, Aachen, Germany, 1992.

[6] T. H. Wee, A. K. Suryavanshi, and S. S. Tin, "Evaluation of rapid chloride permeability test (RCPT) results for concrete containing mineral admixtures," *ACI Structural Journal*, vol. 97, no. 2, pp. 221–232, 2000.

[7] U. Angst, B. Elsener, A. Jamali, and B. Adey, "Concrete cover cracking owing to reinforcement corrosion—theoretical considerations and practical experience," *Materials and Corrosion*, vol. 63, no. 12, pp. 1069–1077, 2012.

[8] L. Samples and J. Ramirez, "Methods of corrosion protection and durability of concrete bridge decks reinforced with epoxy-coated bars phase I," Tech. Rep. FHWA/IN/JTRP-98/15, Purdue University, West Lafayette, Ind, USA, 1999.

[9] C. Christodoulou, G. Glass, J. Webb, S. Austin, and C. Goodier, "Assessing the long term benefits of impressed current cathodic protection," *Corrosion Science*, vol. 52, no. 8, pp. 2671–2679, 2010.

[10] V. Saraswathy and H.-W. Song, "Improving the durability of concrete by using inhibitors," *Building and Environment*, vol. 42, no. 1, pp. 464–472, 2007.

[11] V. Saraswathy and H.-W. Song, "Corrosion performance of rice husk ash blended concrete," *Construction and Building Materials*, vol. 21, no. 8, pp. 1779–1784, 2007.

[12] S. Muralidharan, V. Saraswathy, S. P. M. Nima, and N. Palaniswamy, "Evaluation of a composite corrosion inhibiting admixtures and its performance in Portland pozzolana cement," *Materials Chemistry and Physics*, vol. 86, no. 2-3, pp. 298–306, 2004.

[13] H.-W. Song and V. Saraswathy, "Analysis of corrosion resistance behavior of inhibitors in concrete using electrochemical techniques," *Metals and Materials International*, vol. 12, no. 4, pp. 323–329, 2006.

[14] S. P. Karthick, A. Madhavamayandi, S. Muralidharan, and V. Saraswathy, "Electrochemical process to improve the durability of concrete structures," *Journal of Building Engineering*, vol. 7, pp. 273–280, 2016.

[15] S. Bhaskar, B. Ravindra Gettu, H. Bharatkumar, and M. Neelamegam, "Strength, bond and durability related properties of concretes with mineral admixtures," *Indian Concrete Journal*, vol. 86, no. 2, pp. 9–16, 2012.

[16] X.-Y. Wang and K.-B. Park, "Analysis of compressive strength development of concrete containing high volume fly ash," *Construction and Building Materials*, vol. 98, pp. 810–819, 2015.

[17] B. S. Divsholi, T. Y. D. Lim, and S. Teng, "Durability properties and microstructure of ground granulated blast furnace slag cement concrete," *International Journal of Concrete Structures and Materials*, vol. 8, no. 2, pp. 157–164, 2014.

[18] P. Kathirvel, V. Saraswathy, S. P. Karthik, and A. S. S. Sekar, "Strength and durability properties of quaternary cement concrete made with fly ash, rice husk ash and limestone powder," *Arabian Journal for Science and Engineering*, vol. 38, no. 3, pp. 589–598, 2013.

[19] T. H. Kim, S. H. Tae, C. U. Chae, and W. Y. Choi, "The environmental impact and cost analysis of concrete mixing blast furnace slag containing titanium gypsum and sludge in South Korea," *Sustainability*, vol. 8, no. 6, pp. 502–520, 2016.

[20] Technical Report, "Korea Ground Granulated Furnace Slag Association, Seoul, South Korea," 2015, http://www.kslag.or.kr/site/.

[21] H. Zhao, W. Sun, X. Wu, and B. Gao, "The properties of the self-compacting concrete with fly ash and ground granulated blast furnace slag mineral admixtures," *Journal of Cleaner Production*, vol. 95, pp. 66–74, 2015.

[22] M. Ondova, N. Stevulova, and E. Zelenakova, "Energy savings and environmental benefits of fly ash utilization as partial cement replacement in the process of pavement building," *Chemical Engineering Transactions*, vol. 25, pp. 297–302, 2011.

[23] P. Sulapha, F. S. Wong, H. T. Wee, and S. Swaddi wudhipong, "Carbonation of concrete containing mineral admixtures," *Journal of Materials in Civil Engineering*, vol. 15, no. 2, pp. 135–143, 2003.

[24] E. Güneyisi, M. Gesoglu, and E. Özbay, "Evaluating and forecasting the initial and final setting times of self-compacting concretes containing mineral admixtures by neural network," *Materials and Structures*, vol. 42, no. 4, pp. 469–484, 2009.

[25] M. Uysal, K. Yilmaz, and M. Ipek, "The effect of mineral admixtures on mechanical properties, chloride ion permeability and impermeability of self-compacting concrete," *Construction and Building Materials*, vol. 27, no. 1, pp. 263–270, 2012.

[26] H. Siad, A. H. Mesbah, H. Khelafi, K. S. Bernard, and M. Mouli, "Effect of mineral admixture on resistance to sulphuric and hydrochloric acid attacks in self-compacting concrete," *Canadian Journal of Civil Engineering*, vol. 37, no. 3, pp. 441–449, 2010.

[27] S. C. K. Bendapudi, "Contribution of fly ash to the properties of mortar and concrete," *International Journal of Earth Science Engineering*, vol. 4, no. 6, pp. 1017–1023, 2011.

[28] L. J. Malvar and L. R. Lenke, "Efficiency of fly ash in mitigating alkali-silica reaction based on chemical composition," *ACI Materials Journal*, vol. 103, no. 5, pp. 319–326, 2006.

[29] F. U. A. Shaikh and S. W. M. Supit, "Compressive strength and durability properties of high volume fly ash (HVFA) concretes containing ultrafine fly ash (UFFA)," *Construction and Building Materials*, vol. 82, pp. 192–205, 2015.

[30] Q. Yuan, C. Shi, G. De Schutter, K. Audenaert, and D. Deng, "Chloride binding of cement-based materials subjected to external chloride environment—a review," *Construction and Building Materials*, vol. 23, no. 1, pp. 1–13, 2009.

[31] M. Nagesh and B. Bishwajit, "Modeling of chloride diffusion in concrete and determination of diffusion coefficients," *ACI Materials Journal*, vol. 95, no. 2, pp. 113–120, 1998.

[32] L. Tang, *Chloride transport in concrete: measurement and prediction [Ph.D. thesis]*, Department of Building Materials, Chalmers University of Technology, Gothenburg, Sweden, 1996.

[33] T. Luping and L.-O. Nilsson, "Chloride binding capacity and binding isotherms of OPC pastes and mortars," *Cement and Concrete Research*, vol. 23, no. 2, pp. 247–253, 1993.

[34] K. Maekawa, T. Ishida, and T. Kishi, "Multi-scale modeling of concrete performance," *Journal of Advanced Concrete Technology*, vol. 1, no. 2, pp. 91–126, 2003.

[35] H. W. Song, S. J. Back, C. H. Lee, and S. J. Kwon, "Service life prediction of concrete structures under marine environment considering coupled deterioration," *Restoration of Buildings and Monuments*, vol. 12, no. 2, pp. 265–284, 2006.

[36] H.-W. Song, J.-C. Jang, V. Saraswathy, and K.-J. Byun, "An estimation of the diffusivity of silica fume concrete," *Building and Environment*, vol. 42, no. 3, pp. 1358–1367, 2007.

[37] T. Ishida and K. Maekawa, "Modeling of durability performance of cementitious materials and structures based on thermo-hygro physics," in *Proceedings of the 2nd International Workshop on Life Prediction and Aging Management of Concrete Structures (RILEM PRO 29)*, pp. 39–49, Paris, France, 2003.

[38] ASTM, "Standard test method for electrical indication of concrete's ability to resist chloride ion penetration," ASTM-C 1202, 2012.

[39] ASTM-C39, "Standard Test Method for Compressive Strength of Cylindrical Concrete Specimens," 2012.

[40] N. S. Berke and M. C. Hicks, "The life cycle of reinforced concrete decks and marine piles using laboratory diffusion and corrosion data," in *Corrosion Forms and Control of Infrastructure*, V. Chaker, Ed., ASTM STP-1137, pp. 207–231, American Society for Testing and Materials, Philadelphia, Pa, USA, 1992.

[41] L. Tang and L. O. Nilsson, "Rapid determination of the chloride diffusivity in concrete by applying an electrical field," *ACI Materials Journal*, vol. 89, no. 1, pp. 49–53, 1992.

[42] N. Otsuki, S. Nagataki, and K. Nakashita, "Evaluation of $AgNO_3$ solution spray method for measurement of chloride penetration into hardened cementitious matrix materials," *ACI Materials Journal*, vol. 89, no. 6, pp. 587–592, 1992.

[43] H.-W. Song and S.-J. Kwon, "Evaluation of chloride penetration in high performance concrete using neural network algorithm and micro pore structure," *Cement and Concrete Research*, vol. 39, no. 9, pp. 814–824, 2009.

[44] ASTM, "Standard test for determination of pore volume distribution of soil and rock by MIP," ASTM-D 4404, 2004.

[45] B. Tripathi, A. Misra, and S. Chaudhary, "Strength and abrasion characteristics of ISF slag concrete," *Journal of Materials in Civil Engineering*, vol. 25, no. 11, pp. 1611–1618, 2013.

[46] G.-F. Peng, N.-Q. Feng, and Q.-M. Song, "Influence of chloride-ion adsorption agent on chloride ions in concrete and mortar," *Materials*, vol. 7, no. 5, pp. 3415–3426, 2014.

[47] H. Justnes, "A review of chloride binding in cementitious systems," *Nordic Concrete Research*, vol. 21, pp. 48–63, 1998.

[48] M. Maslehuddin, C. L. Page, and Rasheeduzzafar, "Temperature effect on the pore solution chemistry in contaminated cements," *Magazine of Concrete Research*, vol. 49, no. 178, pp. 135–145, 1997.

[49] O. S. B. Al-Amoudi, M. Maslehuddin, and M. A. Bader, "Characteristics of silica fume concrete and its impacts on concrete in the Arabian Gulf," *Concrete*, vol. 35, no. 2, pp. 45–50, 2001.

[50] O. S. B. Amoudi, M. Rasheeduzzafar, Maslehuddin, and A. I. Al-Mana, "Prediction of longterm corrosion resistance of plain and blended cement concretes," *ACI Materials Journal*, vol. 90, no. 6, pp. 564–570, 1993.

[51] O. S. B. Al-Amoudi, W. A. Al-Kutti, S. Ahmad, and M. Maslehuddin, "Correlation between compressive strength and certain durability indices of plain and blended cement concretes," *Cement and Concrete Composites*, vol. 31, no. 9, pp. 672–676, 2009.

[52] D. C. Montgomery and E. A. Peck, *Introduction to Linear Regression Analysis*, John Wiley & Sons, New York, NY, USA, 1982.

4

Long-Term Properties of Cement-Based Composites Incorporating Natural Zeolite as a Feature of Progressive Building Material

Alena Sičáková, Matej Špak, Mária Kozlovská, and Marek Kováč

Faculty of Civil Engineering, Technical University of Košice, Vysokoškolská 4, 042 00 Košice, Slovakia

Correspondence should be addressed to Matej Špak; matej.spak@tuke.sk

Academic Editor: Prinya Chindaprasirt

This study is aimed at the utilization of natural zeolite as a prospective type of Supplementary Cementitious Material in connection with the innovations of construction solutions through the materials. The influence of zeolite on several properties of cement-based composites is studied. Basic characteristics of input powdery materials as well as the technological parameters of zeolite tested on mortar containing 50% of cement replacement are presented. The technical parameters of concrete containing 8, 13, and 25% of cement replacement by zeolite are presented as well. The paper is valuable due to the three-year testing period. The density of the hardened concrete was found to be decreasing with increasing zeolite content. While no significant differences in compressive strength were found after 28 days, differences between individual samples were clear after the long-term period. The presence of zeolite influenced the compressive strength positively in the case of the 8% and 13% replacement, while the higher proportion of natural zeolite (25%) caused lower compressive strength after 28 days but was similar to the reference concrete after three years. A major increase in strength was detected for all samples in the long-term period. It confirms the long-term potential of zeolite for strength improvement.

1. Introduction

The ordinary Portland cement is one of the most utilized binder materials for the production of building materials. However, its production has negative environmental impact due to the requirement of high amount of energy, production of greenhouse gases, and so on. Utilization of supplementary cementitious materials (SCMs) for the production of cement-based composites can help to significantly minimalize these adverse influences. Additionally, different SCMs bring several types of improvement to composites properties, for example, better workability or durability. Moreover, the costs of concrete can decrease. The incorporation of SCM is a feature of the current advanced concrete and is connected with the modern approaches to concrete production. Thus, it can be successfully involved in the modern methods of construction. Natural zeolite is one of the best-applicable SCMs, mainly due to its pozzolanic activity.

The quality and durability of concrete are strongly dependent on the kind, properties, and dosage of the basic components (cement, water, and aggregates) and the additives and admixtures. The alternative materials play an important role in current concrete technology such as SCMs. Due to the improvement of mechanical properties and durability of concrete, the trend of utilization of SCMs has increased [1–3]. SCMs improve the microstructure of concrete by minimizing its porosity and improving the resistance against the aggressive attack due to their chemical nature [2, 4–7]. In the standard EN 206 [8], three kinds of Type II additions—fly ash, blast furnace slag, and silica fume—are mentioned, and the conditions of their application to the concrete are specified.

The other kinds of addition can be used in specific cases of concrete structure exposure. Both zeolite and silica fume are recommended if a chemical attack occurs. The standard [8] defines the kind of addition in accordance with both the nature and intensity of chemical environment, but the recommended dosage is not specified in the standard. Therefore, the practical production of concrete can be uncertain. There are a number of studies investigating zeolite as cement supplement (pozzolanic additions) [9–17]. Some results of these studies are ambiguous, as shown below. There is a lack of information on the influence of specific dosage of additions on the long-term properties of the concrete and information about other relevant properties, such as length deformation and water absorption capacity.

Natural zeolites are hydrated aluminosilicates that have high amounts of reactive SiO_2 and Al_2O_3. Their reactivity is mainly attributed to their highly porous structure, which is related to the large external surface area giving interaction of zeolite with lime [11, 12], ability of ion exchange, and meta-stability, which supports the dissolution of zeolitic crystals and precipitation of hydrated calcium silicates and aluminates during interaction of OH^- ions available in the saturated lime solution [13, 14].

Researchers have found that the zeolites of different modifications act as pozzolanic additions in concretes; during cement hydration, CSH and CAH gel phases increase, which also increases the resistance of Portland cement compositions to acids and sulphate corrosion and increases its durability as well [11].

Regarding the early-age strength, the references are different. According to Yun-Sheng et al., zeolite added at 15% to the cement mix increases the early-age strength of concrete [15]. Turkish researchers carried out tests on concrete specimens containing 5, 10, 20, and 40% of zeolite. Compressive strength was measured after one, two, seven, and 28 days of curing. The results showed that the concrete specimens with zeolite additive had lower compressive strength compared to the control specimen after 24 hours of curing. The same trend was observed after two and seven days of curing. After 28 days of curing, the compressive strength of specimens containing 5% of zeolite increased by 6.8% compared to the control specimens; the compressive strength of the next specimens increased by 15.9%, 22.3%, and 4.1% for specimens containing 10%, 20%, and 40% of zeolite, respectively [16].

According to Ramezanianpour et al. [17], the addition of natural zeolite delays the strength development during the first seven days, after which the concretes containing 10% of natural zeolite provide almost similar compressive strength compared to that of the reference concretes. The strength of concrete mixtures containing 15% of natural zeolite, however, is marginally lower than that of the reference concrete. The pozzolanic activity of natural zeolite is significantly high, as most of the pozzolanic reactions occurred between seven and 28 days. The use of natural zeolite leads to considerable reductions in water permeability and capillary absorption for each of the selected water-to-cementitious material ratios. The highest improvements through the use of natural zeolite were observed in the rapid chloride permeability and electrical resistivity tests. The electrical resistivity of concrete samples containing natural zeolite was two to four times better than those of the reference samples. These improvements were more significant for concretes with higher water-to-cementitious materials ratios (w/cm). The electrical resistivity and rapid chloride penetrability of mixtures with 10 and 15% natural zeolite and w/cm of 0.50 was better than that of the reference concrete with w/cm of 0.35, indicating that using 10–15% of natural zeolite was more effective in improving permeability than reduction in w/cm from 0.50 to 0.35. The depth of carbonation increases with the use of natural zeolite. While the reduction of chloride penetration enhances the concrete resistivity against corrosion, the increased carbonation results in the opposite performance. Study of the SEM images shows significant improvements in porosity of the studied pastes through the use of natural zeolite. In addition, studying the images taken from transition zone reveals that natural zeolite enhances the structure of the transition zone favourably. The calcium hydroxide content is considerably reduced by the use of natural zeolite as an SCM, which results in the production of secondary C-S-H.

According to Vejmelková et al. [18], both the bulk and the matrix density of cement-based composites decrease with increase in the zeolite supplementary. Poon et al. [19] also state that 15% replacement of zeolite results in lower porosity, while a higher replacement level (25%) increases the porosity at all studied ages.

The drying shrinkage of products made with Portland-pozzolan cements is dependent on the hydration products and water demand of the mixtures. Although tests should be conducted to determine the drying shrinkage of Portland cement and natural pozzolans combinations, there is a scarcity of studies in this field [20]. Jana [21] investigates the effects of using 10, 20, and 30% of zeolite as an SCM. He observes that drying shrinkage of zeolite mixtures at 10 and 20% of Portland cement replacement levels is similar or slightly higher than that of the control mixture, whereas by using 30% of zeolite, the drying shrinkage is about 20% higher than the control mixture. Moreover, Kasai et al. [22] examined drying shrinkage for the mortars blended with clinoptilolite and mordenite. Both clinoptilolite and mordenite blended mortars experienced higher shrinkage than the control one.

It is reported that although natural zeolite reduces the slump of concrete, it can prevent bleeding and segregation. Regarding hardened concrete, natural zeolite increases compressive strength due to its pozzolanic property. Moreover, it enhances the durability of conventional concrete by reducing concrete permeability and, mainly, by improving resistance to alkali-aggregate reaction [23]. According to Najimi et al. [24], natural zeolite can be properly used as an SCM in normally consolidated concrete, considering the environmental protection and sustainable development. Chan and Ji [25] report that the pozzolanic reactivity of natural zeolite is between that of silica fume and fly ash.

This paper is aimed at the utilization of zeolite as a perspective kind of SCM and its influence on the long-term (up to three years) properties of concrete, since there is a lack of information about the long-term properties. Most reports are oriented towards the 28-day properties, or eventually up to one year, because of the challenges of long-term research. However, the pozzolanic additions need a longer time to develop their properties in the construction due to the slower progress of pozzolanic hydration processes. The short-term values of properties could be misleading.

TABLE 1: Chemical composition and particle size characteristics of binders.

Binder	SiO_2	CaO	Fe_2O_3 [% wt.]	Al_2O_3	MgO	CaO/SiO_2 [—]	$d(0.5)$ [μm]	$d(0.9)$ [μm]
OPC (CEM I 42.5 N)	20.34	64.1	2.99	2.97	9.03	3.150	26.68	67.87
Natural zeolite	78.75	3.50	1.73	11.6	1.16	0.040	20.68	91.97
Silica fume	97.00	0.70	1.50	1.10	—	0.007	0.226	8.647

Comprehensive results for zeolite, including basic characteristics (chemical composition and particle size distribution), technological parameters (mixing water demand, initial setting time, activity index, and relative linear deformation), and technical parameters of concrete incorporating zeolite (density, compressive strength, and total water absorption) are presented in the paper. For comparison, addition of silica fume is presented as well.

2. Materials and Methods

The main scope of the experiment is to observe the long-time technical parameters (density, compressive strength, and water absorption) of zeolite-based concrete and to assess the possibility of the practical utilization of natural zeolite for the production of progressive building material.

2.1. Materials Characteristics. Three types of mineral binders are used in the experiment. Ordinary Portland cement (OPC) class CEM I 42.5 N came from Turňa nad Bodvou, Slovakia, natural zeolite (NZ) came from Lehôtka pod Brehmi, Slovakia, and silica fume (SF) came from Istebné, Slovakia.

The chemical composition of ordinary Portland cement, natural zeolite, and silica fume was determined through X-ray fluorescence analysis (XRF). SPECTRO iQ II (Ametek, Germany) with silicon drift detector (SDD) with resolution of 145 eV at 10,000 pulses was used for the analysis. The primary beam was polarized using Bragg crystal and Highly Ordered Pyrolytic Graphite (HOPG) target. The samples were measured at 300 and 180 s at voltage of 25 kV and 50 kV and current of 0.5 and 1.0 mA in a helium atmosphere by using the standardized method of fundamental parameters for cements or liquids. The chemical compositions of binder solids are listed in Table 1.

The particle size distribution of binder solids was determined using a laser granulometric analyser (Mastersizer 2000; Malvern Instruments Ltd, UK). The parameters of grain size distribution are characterized by $d(0.5)$ and $d(0.9)$ medians, which are also listed also in Table 1. The curves of particle size distribution of binders are shown in Figures 1–3, respectively.

The differences in the particle size of binding materials are evident and in accordance with their character. Results also inform indirectly about the specific surface area in terms of mutual comparison: the smaller the particle size, the larger the specific surface area.

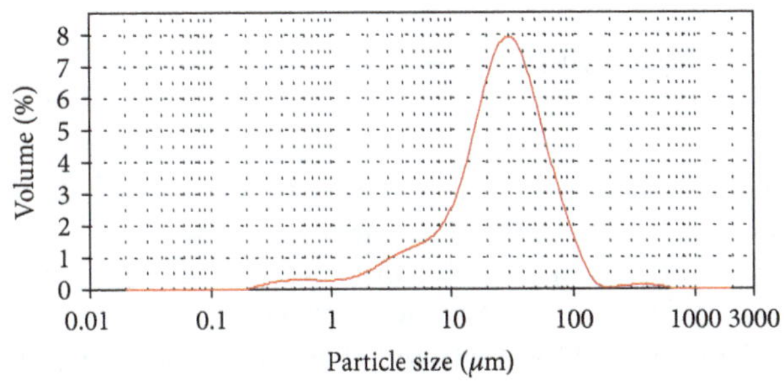

FIGURE 1: Particle size distribution of OPC.

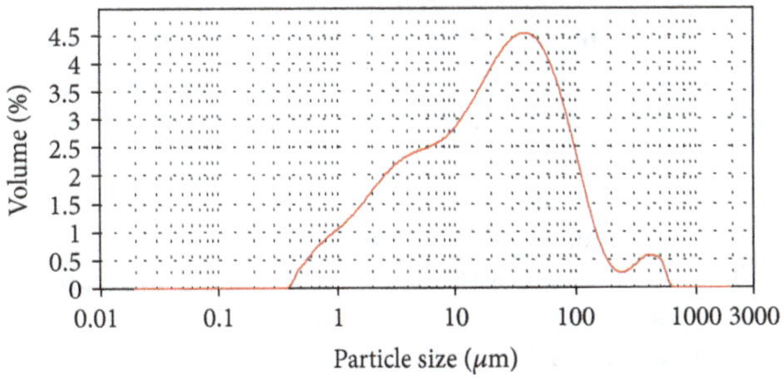

FIGURE 2: Particle size distribution of natural zeolite.

2.2. Technological Characterization of Natural Zeolite. Technological characterization (water demand for standard consistency, initial setting time, activity index, and relative linear deformation) of natural zeolite was investigated. Both the water demand and initial setting time were tested on pastes composed of 50% of OPC and 50% of NZ and pastes consisting of OPC only, while the water-to-binder ratio was 0.5. Both the strength activity index and relative linear deformation were tested on mortars consisted of 1350 g of normalized silica sand, 500 g of binder, and 225 g of water, while two samples with different binder composition were tested. The control mixture contained 100% of OPC and research mixture contained 50% of OPC and 50% of NZ.

The water demand for standard consistency (the consistency that will permit the distance between plunger of Vicat apparatus and base-plate of 4–8 mm) and the initial setting time (the elapsed time, measured from zero to the time at which the distance between the needle and the base-plate is 3–9 mm) were measured using standard methods for testing cements (according to [26]).

The activity index was found in accordance with [27]. Here, the activity index refers to the ratio of the average compressive strength of SCM-based mortar and the reference cement-based mortar at the designated ages. Compressive strength was tested in accordance with [28].

The deformation changes of tested mortars were measured through active linear deformation [mm], using the

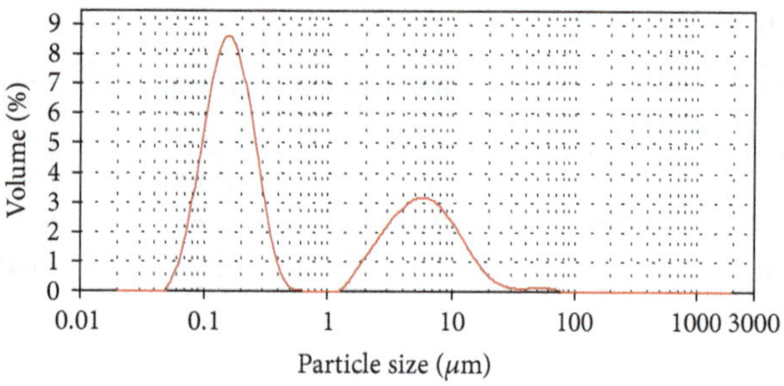

FIGURE 3: Particle size distribution of silica fume.

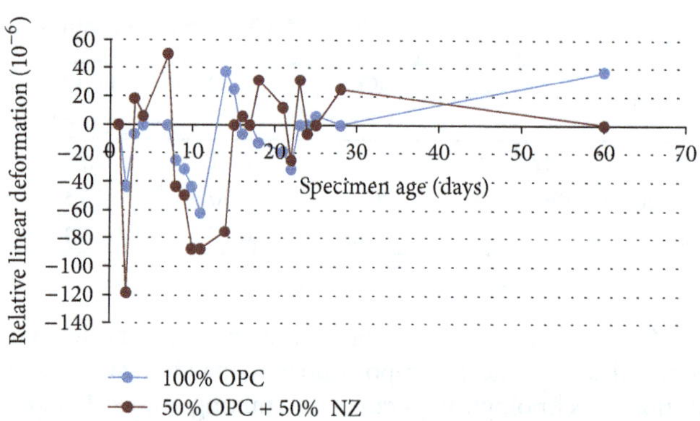

FIGURE 4: Relative linear deformation of the 100% OPC and 50% OPC + 50% NZ mortar.

length comparator, while three samples were measured for each recipe. The relative linear deformation was then calculated as the ratio of the active linear deformation [mm] to the actual length of the sample [mm]. Specimens of dimensions 40 × 40 × 160 mm were measured during 60 days at regular intervals. During this time, samples were kept in standard wet conditions (20°C temperature and water curing).

2.3. Physical-Mechanical Properties of Natural Zeolite-Based Concrete. The mineral binders (OPC, NZ, and SF) were used for concrete preparation. Natural normal-weight fine and coarse aggregates of fractions 0/4, 4/8, and 8/16 (NFA-0/4, NCA-4/8, NCA-8/16) were used. The superplasticizer based on polycarboxyl ether (PCE) was used to obtain optimal consistence together with tap water.

The experiment was focused on testing the concrete mixture designed by the acceptance of the standard recommendations for the composition for aggressive exposure. The recipe of the reference concrete was designed to achieve the compressive strength corresponding to C 35/45 strength class (according to [8]). Water-to-binder ratio (w/b) was 0.45 and cement content was 360 kg per 1 m^3 of fresh concrete. Natural zeolite was applied in varying replacement percentages to cement (8%, 13%, and 25% by weight). For directly showing binder modification by well-known silica fume, it was decided to use the combination of 8% of zeolite and 8% of SF. The materials and compositions of the tested concretes are given in Table 2.

The samples of cubes shape with dimensions 150 × 150 × 150 mm were prepared with standard methods. Specimens were released after one day and consequently cured in water under laboratory conditions until the tests execution. The standard tests (density, compressive strength, and water absorption) were executed after 28 days and the compressive strength after 365, 730, and 1095 days (one, two, and three years) according to [29–31].

3. Results and Discussion

3.1. Technological Characterization of Natural Zeolite. The water demand for standard consistency, initial setting time, and activity index of mortars are listed in Table 3. The consistent results comparing [17, 32, 33] have been found, in fact, to determine the expansion of water demand (probably due to a larger specific surface area) and to increase the setting time increasing of zeolite-blended mortars. The presented results for the zeolite paste also refer to the higher need for water to obtain the standard consistency. The initial setting time is practically three times longer than that of cement. Activity index after seven days (20.4) and 28 days (48.5%) shows low increase in strength at an early age and high cement replacement (50%).

Deformation changes in tested mortars are given through 60-day development of relative linear deformation, as shown in Figure 4.

Results of drying shrinkage presented worldwide are often not consistent because of nonuniform methods, apparatus, and curing conditions; these results are discussed in both positive and negative ways [21, 24, 34]. Here, the method of relative linear deformation is given for the direct comparison of two kinds of mortars cured under the same conditions, while the effect of zeolite is clearly visible. The relationship between the linear deformation changes and the risk of cracking is generally known.

The linear deformations of both samples are quite active up to 60 days, while the linear deformation of zeolite samples is somewhat larger than that of the cement-only samples; the values ranges of NZ and OPC based samples are −120 to +50 and −62 to +37, respectively. Jana [21] reports similar results when using 30% of zeolite as SCM; he obtained about 20% higher drying shrinkage than that of control mixture without zeolite.

3.2. Physical-Mechanical Properties of Natural Zeolite-Based Concrete

3.2.1. Density. The density of hardened concrete is referred as one of the durability-related properties, due to the porosity and permeability context. Concrete porosity is inversely proportional to density. Usually, the concrete permeability increases with an increase in porosity and decreases with an increase in density. The relation between permeability and porosity depends on the pore system properties. Lower concrete porosity leads to reduction in permeability and consequently in better durability. A very highly impermeable concrete reduces or eliminates the ingress of water and other aggressive chemicals and gases. This leads to improved concrete durability due to avoided expansive reactions, which can occur in the presence of these aggressive agents [35].

TABLE 2: The recipes of concrete mixtures. Components proportion was calculated for 1 m^3 of fresh concrete.

Mixture	OPC $[kg \cdot m^{-3}]$	NZ $[kg \cdot m^{-3}]$	NZ [%]	SF $[kg \cdot m^{-3}]$	SF [%]	NFA-0/4 $[kg \cdot m^{-3}]$	NCA-4/8 $[kg \cdot m^{-3}]$	NCA-8/16 $[kg \cdot m^{-3}]$	PCE $[kg \cdot m^{-3}]$	w/b* [—]
V0	360	—	—	—	—	825	235	740	0.8	0.45
VZ1	330	29	8	—	—	825	235	740	0.8	0.45
VZ2	313	47	13	—	—	825	235	740	0.8	0.45
VZ3	270	90	25	—	—	825	235	740	0.8	0.45
VZ-SF	302	29	8	29	8	825	235	740	0.8	0.45

*Using the k-value = 2 for silica fume in accordance with EN 206.

TABLE 3: Technological parameters of zeolite comparing to cement.

Binder	Water demand [%]	Initial setting time [min.]	7-day activity index [%]	28-day activity index [%]
OPC	100.00	110	100	100
50% OPC + 50% NZ	126.62	320	20.4	48.5

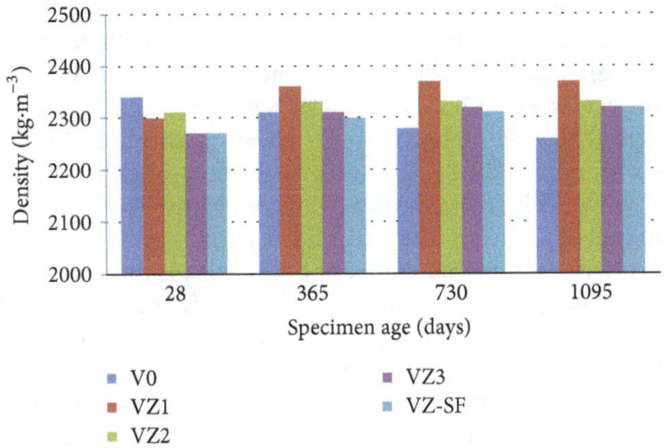

FIGURE 5: Density of hardened concretes after different time of hardening.

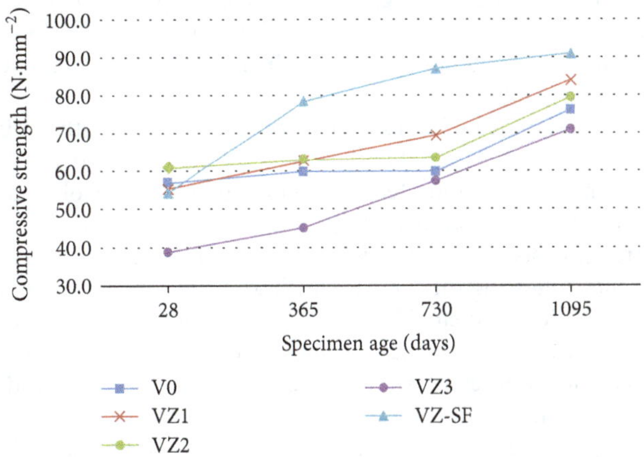

FIGURE 6: Time dependence of compressive strength of hardened concretes.

The results of density measurement are summarized in Figure 5. Values at 28 days and three years of setting and hardening ranged from 2270 to 2340 $kg \cdot m^{-3}$ and from 2260 to 2370 $kg \cdot m^{-3}$, respectively. The density of all concretes increased during the time, excluding the concrete V0. The highest growth of density was found between measurements after 28 and 365 days—that is, during the first year of hardening. The density increased during the later stage only gradually. Concrete V0 obtained the highest density of all after 28 days but the lowest at the end of the experiment. Concrete VZ1 obtained highest density after one year and also in later stages. The amount of zeolite proportionally affected the density; thus, the higher dosage of zeolite caused the lower concrete density. This is in contrast to some research [36, 37], but, there, the zeolite content was up to 10%. According to Vejmelková et al. [18], both the bulk and matrix densities of the cement-based composites decrease with increase in the zeolite supplementary. Also, Poon et al. [19] observe that 15% replacement of zeolite resulted in lower porosity (affecting density), but a higher replacement amount (25%) increased the porosity at all the studied ages. Based on those comments and our results, the decrease in density can be attributed to the increase in zeolite amount.

The long-term results (one-, two-, and three-year) of all concretes incorporating the natural zeolite are higher than that of the OPC sample (V0), unlike the 28-day values. In this case, the results are lower and do not show the clear dependence on the amount of zeolite. This is clearly visible just after the long-term period. The presence of zeolite manifests itself differently over a longer period of setting and hardening of concrete.

Silica fume in combination with zeolite (concrete VZ-SF) decreased the density significantly compared to the reference concrete (V0) after 28 days. On the other hand, the density of VZ-SF was about 2.6% higher than V0 after three years.

3.2.2. Compressive Strength. The compressive strength was measured on 150 mm cubes after 28, 365, 730, and 1095 days of hardening. The time-development of compressive strength is shown in Figure 6. Generally, compressive strength of all concretes increased during the time, as also reported in previous research [17, 24]. The values after 28 days of setting and hardening ranged from 40 to 60 MPa, while values after three years ranged from 70 to 90 MPa. The initial reference strength (after 28 days) was almost the same for all concretes except VZ3. In the case of VZ3, the strength development was slower because of the high dosage of zeolite in the binder (25%). The early strength of concrete is commonly supplied by the hydration of CaO and silicates from cement clinker, while the pozzolanic activity of zeolite later influences the strength. The CaO/SiO_2 ratio of the zeolite used was under 0.5 (Table 1), which refers to its pozzolanic nature. It was in accordance with results from Ramezanianpour et al. and others [17, 36]. The amount of zeolite in total binder influenced the compressive strength, especially after two and three years of hardening. Concretes VZ1 and VZ2 with 8% and 13% of zeolite supplementary achieved about 16.4% and 6.2% higher compressive strength compared to the reference concrete after two years and 10.4% and 4.5% after three years, respectively. Concrete VZ3 with 25% zeolite supplementary obtained about 3.8% and 6.7% lower compressive strength compared to sample V0 after two and three years, respectively. According to Valipour et al. [38], the optimal content of zeolite to obtain highest compressive strength is 10%.

The significant increase in the compressive strength of all concretes was observed in the long term (three-year period). It confirms the long-term potential of natural zeolite to improve the mechanical properties. The positive change in strength was observed between 28-day and 3-year measurements, following the increase of about 33%, 53%, 30%, 83%, and 67% for concretes V0, VZ1, VZ2, VZ3, and VZ-SF, respectively. The concrete VZ3 achieved the most significant increase in compressive strength during the monitored period.

The silica fume significantly influenced the compressive strength of concrete. While the strength of VZ-SF sample after 28 days was similar to that of the other samples, the long-term strength values after forwarding periods were much higher. It is also evident that the strength development shows convex behaviour for all VZ concretes and concave behaviour for only VZ-SF concrete. Thus, the compressive strength of SF concrete increased faster than zeolite-based ones, but the strength of zeolite samples grew faster at the end of the monitored period.

The activity index was calculated as a relation between compressive strength of the related concrete and the reference concrete (VZ-/V0) after three years of hardening. This value characterizes the improvement or deterioration of zeolite-based concrete compared to the control one. It was 1.1, 1.0, 0.9, and 1.2 for samples VZ1, VZ2, VZ3, and VZ-SF, respectively. As expected, the higher zeolite content in binder caused lower compressive strength against the reference concrete.

With the exception of VZ-3, the long-term results (one-, two-, and three-year) of all concretes incorporating the natural zeolite are higher than that of the OPC sample (V0), except 28-day values. In this case, the results are close to each other and do not show the clear dependence on the amount of zeolite. This is clearly visible just after the long-term period.

3.2.3. Water Absorption. The water absorption of hardened concretes is summarized in Figure 7. Values after 28 days of

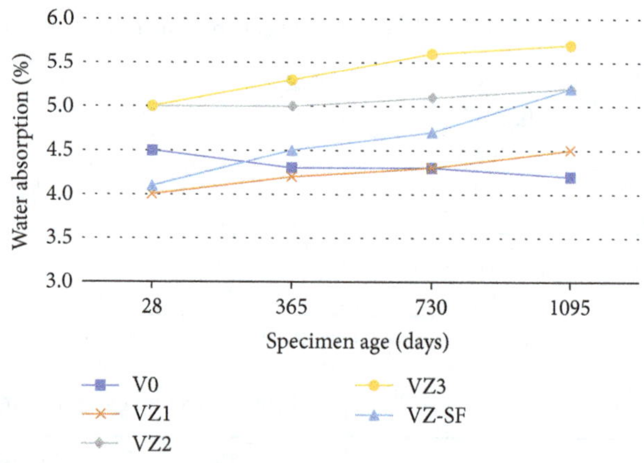

FIGURE 7: Time dependence of water absorption of hardened concretes.

setting and hardening ranged from 4.0 to 5.0%, while values after three years ranged from 4.2 to 5.7%. The long-term changes in water absorption of zeolite-based concretes were not significant. The relative differences in 28-day and three-year values were 12.5, 4.0, and 14.0% for VZ1, VZ2, and VZ3, respectively. The highest increase in water absorption (26.8%) was found for VZ-SF. Contrary to the control OPC sample (V0), values increase during the three-year period while the amount of zeolite in concrete influences the water absorption negatively. The higher content of zeolite leads to increase in water absorption. It is connected with the water absorption of natural zeolite, which is higher than that of cement [34]. Poon et al. [19] report that the porosity increases due to the higher replacement amount (25%) of zeolite at all studied ages; this may also be attributed to the higher absorption ability. It was recognized that the water absorption grew together with density. This result is unexpected, because several research works [17, 24] obtained adverse results.

The long-term results (one-, two-, and three-year) of all concretes including the natural zeolite are higher than that of the OPC sample (V0), unlike 28-day values. In this case, VZ1 and VZ-SF obtained lower values. The presence of zeolite manifests itself differently over a longer period of setting and hardening.

4. Conclusion

In this paper, the technological properties of natural zeolite and the long-term mechanical properties, density, and water absorption of concrete based on natural zeolite as SCM were evaluated. The results can be summarized as follows:

(i) At high cement replacement (50%) by the zeolite, the zeolite paste led to a higher demand for mixing water to obtain the standard consistency than that of OPC paste. The initial setting time was practically 2.9 times longer. Strength activity index after 28 days was 48.5%; this low increase in strength at an early age reflects the slow hydration process.

(ii) The long-term density of all zeolite-based concretes was higher than that of OPC concrete, unlike the

28-days values. Although the increase in zeolite amount caused the decrease in density, samples incorporating 25% zeolite achieved higher density than that of OPC in all later ages.

(iii) A major increase in strength was observed for all samples in the long term (three-year period). An increase in the zeolite amount caused a decrease in compressive strength, but the replacement of 8% and 13% of cement by natural zeolite brought about an improvement in long-term compressive strength compared to the only-OPC-based concrete. A higher replacement (25%) caused a decrease in strength after 28 days of setting and hardening; however, significant positive change (83%) was observed during the three-year period. This confirms the long-term potential of zeolite to improve the strength. Silica fume, in combination with natural zeolite, significantly improved the strength.

(iv) Long-term changes in water absorption of zeolite-based concretes were not significant, up to 14.0% in relative difference (26.8% for VZ-SF). Except for the V0 sample, values were increasing during the three-year period, while the higher amount of zeolite in concrete negatively influenced the water absorption.

(v) Generally, the long-term properties of zeolite-based concrete are different from that of 28-day properties, while both the density and compressive strength are influenced positively. The differences in concretes composition are manifested after a longer time, when the results are ordered more clearly and show a better sequence dependence on the sample's composition.

The results demonstrate the long-term potential of SCMs for the improvement of the concrete properties, which are important for the long-term role of concrete in the structure.

Conflicts of Interest

The authors declare that they have no conflicts of interest.

Acknowledgments

This article has been carried out within the project ITMS "Center of Excellent Integrated Research of Progressive Building Constructions, Materials and Technologies" no. 26220120037 and also presents a partial research result of project VEGA–1/0677/14 "Research of Construction Efficiency Improvement through MMC Technologies."

References

[1] K. Obla, C. Lobo, and L. Lemay, "Specifying Concrete for Durability".

[2] P. C. Aïtcin, *High-Performance Concrete*, Taylor & Francis, Abingdon, UK, 1998.

[3] P. Tikalsky, B. Mather, and J. Olek, *Concrete Durability*, Transp. New Millenium - State Art Futur. Dir., 2000.

[4] M. Najimi, J. Sobhani, and A. Pourkhorshidi, "Durability of copper slag contained concrete exposed to sulfate attack," *Construction and Building Materials*, vol. 25, no. 4, pp. 1895–1905, 2011.

[5] S. Levy and P. Helene, "Durability of concrete mixed with fine recycled aggregates," *Exacta*, vol. 5, no. 1, 2007.

[6] K. Byfors, "Influence of silica fume and flyash on chloride diffusion and pH values in cement paste," *Cement and Concrete Research*, vol. 17, no. 1, pp. 115–130, 1987.

[7] V. G. Papadakis, "Effect of supplementary cementing materials on concrete resistance against carbonation and chloride ingress," *Cement and Concrete Research*, vol. 30, no. 2, pp. 291–299, 2000.

[8] European Committee for Standardization, "(20025). EN 206-1 "Concrete. Part 1:Specification, performance, production and conformity". Brussels".

[9] D. Caputo, B. Liguori, and C. Colella, "Some advances in understanding the pozzolanic activity of zeolites: The effect of zeolite structure," *Cement and Concrete Composites*, vol. 30, no. 5, pp. 455–462, 2008.

[10] T. Perraki, E. Kontori, S. Tsivilis, and G. Kakali, "The effect of zeolite on the properties and hydration of blended cements," *Cement and Concrete Composites*, vol. 32, no. 2, pp. 128–133, 2010.

[11] I. Janotka and L. Krajči, "Utilization of natural zeolite in Portland pozzolan cement of increased sulfate resistance," in *Proceedings of the 5th CANMET/ACI Conf. Durab. Concr.*, pp. 223–238, 2000.

[12] T. Perraki, G. Kakali, and F. Kontoleon, "The effect of natural zeolites on the early hydration of Portland cement," *Microporous and Mesoporous Materials*, vol. 61, no. 1-3, pp. 205–212, 2003.

[13] I. Janotka, A. Ray, and S. C. Mojumdar, "Acid and sulfate resistance of Portland cement - natural zeolite mortar," in *Proceedings of 8th CANMET/ACI International Conference on Fly Ash, Silica Fume, Slag and Natural Pozzolans in Concrete, Las Vegas*, pp. 639–652, 2004.

[14] R. Sersale and G. Frigione, "Natural zeolites as constituents of blended cements," *Studies in Surface Science and Catalysis*, vol. 24, pp. 523–530, 1985.

[15] T. Yun-Sheng, C. Huang, and K. Hsu, "The pozzolanic activity of a calcined waste FCC catalyst and its effect on the compressive strength of cementitious materials," *Cement and Concrete Research*, vol. 35, no. 4, pp. 782–787, 2005.

[16] B. Yılmaz, A. Uçar, B. Öteyaka, and V. Uz, "Properties of zeolitic tuff (clinoptilolite) blended portland cement," *Building and Environment*, vol. 42, no. 11, pp. 3808–3815, 2007.

[17] A. A. Ramezanianpour, R. Mousavi, M. Kalhori, J. Sobhani, and M. Najimi, "Micro and macro level properties of natural zeolite contained concretes," *Construction and Building Materials*, vol. 101, pp. 347–358, 2015.

[18] E. Vejmelková, D. Koňáková, T. Kulovaná et al., "Engineering properties of concrete containing natural zeolite as supplementary cementitious material: Strength, toughness, durability, and hygrothermal performance," *Cement and Concrete Composites*, vol. 55, pp. 259–267, 2015.

[19] C. S. Poon, L. Lam, S. C. Kou, and Z. S. Lin, "A study on the hydration rate of natural zeolite blended cement pastes," *Construction and Building Materials*, vol. 13, no. 8, pp. 427–432, 1999.

[20] V. J. Inglezakis and A. A. Zorpas, *Handbook of Natural Zeolites*, Cyprus Open University, 2012.

[21] D. Jana, "A new look to an old pozzolan, clinoptilolite A promising pozzolan in concrete," in *Proceedings of the 29th ICMA Conference on Cement Microscopy*, pp. 168–206, Quebec City, West Chester, 2007.

[22] Y. Kasai, K. Tobinai, E. Asakura, and N. Feng, "Comparative study of natural zeolites and other inorganic admixtures in terms of characterization and properties of mortars," in *Proceedings of the 9th Canmet/ACI Int. Conf. Fly Ash, Silica Fume, Slag Nat. Pozzolans Concr.*, V. D. Malhotra, Ed., pp. 615–634, 1992.

[23] N. Feng and H. Jia, "Prevention of AAR in cement concrete with natural zeolite," *Concr. Cem. Prod*, vol. 2, 1995.

[24] M. Najimi, J. Sobhani, B. Ahmadi, and M. Shekarchi, "An experimental study on durability properties of concrete containing zeolite as a highly reactive natural pozzolan," *Construction and Building Materials*, vol. 35, pp. 1023–1033, 2012.

[25] S. Y. N. Chan and X. Ji, "Comparative study of the initial surface absorption and chloride diffusion of high performance zeolite, silica fume and PFA concretes," *Cement and Concrete Composites*, vol. 21, no. 4, pp. 293–300, 1999.

[26] European Committee for Standardization, *EN 196 -3 "Methods of Testing Cement - Part: Determination of Setting Times and Soundness*, 2005, Brussels.

[27] European Committee for Standardization, *EN 196 - 1, Methods of testing cement - Part 1: Determination of strength*, Brussels, 2005.

[28] European Committee for Standardization, *EN 15167-1, Ground granulated blast furnace slag for use in concrete, mortar and grout. Part 1: Definitions, specifications and conformity criteria*, Brussels, 2007.

[29] European Committee for Standardization, *EN 12390-7, Testing Hardened Concrete, part 7: Density of hardened concrete*, Brussels, 2011.

[30] European Committee for Standardization, *EN 12390-3, Testing Hardened Concrete, part 3: Compressive Strength of Test Specimens*, Brussels, 2010.

[31] *STN 731316 Determination of moisture, water absorption and capillarity of concrete*, Slovak Office of Standards, Metrology and Testing, Bratislava, 1989.

[32] S. Özen, *Pozzolanic activity of natural zeolites: mineralogical, chemical and physical characterization and examination of hydration products [Thesis]*, Middle East Technical University, 2013.

[33] F. Canpolat, K. Yılmaz, M. Köse, M. Sümer, and M. Yurdusev, "Use of zeolite, coal bottom ash and fly ash as replacement materials in cement production," *Cement and Concrete Research*, vol. 34, no. 5, pp. 731–735, 2004.

[34] T. Markiv, K. Sobol, M. Franus, and W. Franus, "Mechanical and durability properties of concretes incorporating natural zeolite," *Archives of Civil and Mechanical Engineering*, vol. 16, no. 4, pp. 554–562, 2016.

[35] T. R. Naik, "Concrete durability as influenced by density and/or porosity," Report No. CBU-1997-27, Center for By-Products Utilization, 1997.

[36] D. Nagrockiene and G. Girskas, "Research into the properties of concrete modified with natural zeolite addition," *Construction and Building Materials*, vol. 113, pp. 964–969, 2016.

[37] D. Nagrockienė, G. Girskas, and G. Skripkiūnas, "Properties of concrete modified with mineral additives," *Construction and Building Materials*, vol. 135, pp. 37–42, 2017.

[38] M. Valipour, M. Yekkalar, M. Shekarchi, and S. Panahi, "Environmental assessment of green concrete containing natural zeolite on the global warming index in marine environments," *Journal of Cleaner Production*, vol. 65, pp. 418–423, 2014.

5

Water Absorption Properties of Cement Pastes: Experimental and Modelling Inspections

Ludovica Casnedi, Ombretta Cocco, Paola Meloni, and Giorgio Pia

Dipartimento di Ingegneria Meccanica, Chimica e dei Materiali, Università degli Studi di Cagliari, Piazza d'Armi, 09123 Cagliari, Italy

Correspondence should be addressed to Giorgio Pia; giorgio.pia@dimcm.unica.it

Academic Editor: Renal Backov

An intermingled fractal units' model is shown in order to simulate pore microstructures as pore fraction and pore size distribution. This model is aimed at predicting capillary water absorption coefficient and sorptivity values in cement pastes. The results obtained are in good agreement with the experimental ones. For validating this model, a comparison with other procedures has been shown. It is possible to establish that the newly proposed method matches better with the experimental results. That is probably due to the fact that pore size distribution has been considered as a whole. Moreover, even though the proposed model is based on fractal base units, it is able to simulate and predict different properties as well as nonfractal porous microstructure.

1. Introduction

Porous building materials may suffer from weathering process caused by physical, chemical, and biological factors [1, 2]. The influence of these deterioration processes depends on the *topoclimatic* environmental conditions, the material type, its preservation state, and location on the building [3, 4].

The circulation of water within microstructures of porous materials causes them to deteriorate considerably more rapidly, consequently compromising their behaviour [5–7]. Their thermal (insulation) performance, mechanical properties, and degradation kinetics are therefore proportional to the amount of water that they may contain.

For this reason, the study of the capillary suction properties of construction materials remains a matter much debated [8–13]. Its importance is connected to the analysis of their performance when installed as well as their durability [14–16]. In order to preserve these materials, it is necessary, in particular when it comes to historical building, to apply inspection methods, which not only assess the conservation state of the microstructure of materials but also help in the future evolution of a number of characteristics related to porosity for traditional or advanced engineering applications [17–21].

In this regard, a large number of studies have dealt with these properties using several modelling procedures [9, 12, 22]. Over the last few decades, a new type of geometrical analysis has been developed which utilises figures with fractional dimensions, known as *fractals*. This term is derived from the Latin word *fractus* (broken/fractured). The importance of this approach is due to its usefulness in describing the analysis of complex systems. According to Mandelbrot, there are a large number of fractals in nature and these forms may be used for describing some common aspects such as vegetables' growth, mountains and rivers geomorphology, geometrical organisation of human organs, and in our case, materials' microstructures [23]. A fractal (deterministic fractal) is a geometric figure whose each and every part is similar to the whole and is repeated on different scales (self-similarity and autosimilarity). Many fractals are constructed using the iteration procedure. For example, the well-known Menger sponge is created by splitting each of the edge of the primary solid cube into three equal parts, creating 27 subcubes. Seven of these subcubes will be removed: the subcube in the middle of each face and the subcube in the centre of the cube. For the later iterations, which can be infinite, the same subdivision and removal process are repeated on the remaining solid subcubes.

In microstructure modelling terms, the development of a fractal, as the Menger sponge, is dimensionally limited to the measurements of the edge of the larger and smaller cubes. While the number of iteration increases, geometrical structure is fine, highly intricate, and detailed at all scales. When it comes to porosity, fractal geometry displays a well-known series of "sponges" that, regardless of being originally very simple, recall the pore size distribution or the morphology of their outlines (e.g., the Menger sponge or its corresponding bidimensional model known as the Sierpinski carpet).

The development into space of the scaling procedures (self-similarity, applied on no less than two/three orders of magnitude) allows the process of the porosimetric data obtained by mercury intrusion porosimetry or gas absorption. These applications can be found in the literature in several different cases and include traditional as well as the study of advanced ceramic materials, namely, sandstones and soils as well as cement materials [24–27].

Quite recently, it has also been possible to connect the fractal geometry to a number of relations enabling the effective and rational (not empirical) evaluation of physical quantities which are technologically relevant [5, 28, 29].

For example, Winslow [30] described the X-ray scattering technique for measuring the dimension of a fractal surface to demonstrate that the surface of hydrated cement paste is fractal in character and has a large fractal dimension. A model to describe nucleation and growth for the hydration of tricalcium silicate has been presented by Livingston [31]. Lange et al. [32] explored several image analysis techniques that provide insight into the nature of pore structure as observed in backscattered electron images of polished sections. Diamond and Bonen [33] have shown that pore systems of concrete may present fractal characteristics. In the study by Arandigoyen and Alvarez [34], the microstructure of blended mortars is studied taking into account porosity and pore size distributions. Surface fractal dimension has been calculated and correlated with the percentage of cement/lime. It has been shown that the system with higher percentage of cement has higher surface fractal dimension values. Comegna et al. [35] used "Tyler and Wheatcraft fractal model" for determining soil-water retention curves. Meanwhile, in Xu and Dong [36], the soil-water characteristics, hydraulic conductivity, and soil-water diffusivity of unsaturated soils are derived and expressed by only two parameters, the fractal dimension and the air-entry value, which can be evaluated from the fractal model for the pore size distribution. In [37], Shi et al. presented a series of fractal models developed to investigate the effect of gas diffusion layer's wettability on liquid water and gas permeation. Zheng and Yu [38] proposed a fractal analytical expression in order to predict gas permeability flow through dual-porosity media. The model predictions are in good agreement with the experimental data. Furthermore, the fractal geometry has been used to correlate mechanical behaviour and thermal conductivity with pore size distribution [29, 39–44].

This paper is set out to highlight that, through appropriate fractal modelling techniques applied to the microstructure derived from experimental tests requiring a fairly small amount of material, such as mercury intrusion techniques, it is possible to predict the water content that such microstructures absorb as a function of the exposed surface and the time of exposition: sorptivity coefficient.

Moreover, fractal model data are compared with other models from [9, 12].

The purpose of this communication, in addition to illustrating how experimental data on capillary suction correspond very closely to those obtained from the intermingled fractal unit (IFU) model, is to put forward the idea of developing a new model which is able to predict the properties which are connected to the microstructure in any porous material [5, 42, 44]. In a former article, it was noted how the IFU and the fractal procedure are able to get some values of thermal conductivity and mechanical properties which are very close to those obtained experimentally [29, 43, 44]. For this reason, this approach could be a very useful tool to design cement pastes and concrete materials with peculiar physical properties.

2. IFU Model and Analytical Expression for Calculating Sorptivity Coefficient

The proposed model is based on the Sierpinski carpets as base units. This fractal figure is obtained by commencing from a square with sides which are divided according to a factor (F) of 3, resulting into a figure consisting of 9 subsquares, from which two are removed. The remaining 7 subsquares (s) are treated in the same manner; that is, their sides are divided by the length scale factor (F) of 3, thus generating a further 9 subsquares, from which 2 are removed (Figure 1(a)) and 7 are involved in the iteration process. This procedure may continue *ad infinitum*, but for experimental cases, the number of iterations (i) depends on the porosimetric range (Figure 1(a)). Fractal dimension of the figure under consideration is calculated as

$$D_f = \frac{\log(s)}{\log(F)}. \tag{1}$$

Different Sierpinski carpets can be obtained by changing the number of the removed squares or the number of iterating and noniterating squares. An example of this configuration is shown in Figure 1(b). It is possible to note that there are 5 iterating squares and 2 *solid forever* squares at second iteration. This configuration is repeated for the following iterations.

The extruded Sierpinski carpet has a pore size distribution in which the hypothetical mercury volume intruded is proportional to pore size. The maximum adsorption corresponds to maximum pore ray. Porous materials can have a more complex pore size distribution in which maximum pores coincide with a generic pore radius or there may be more peaks corresponding to the pore size of the radius range [45]. In both cases, these pore size distributions are not typically fractal.

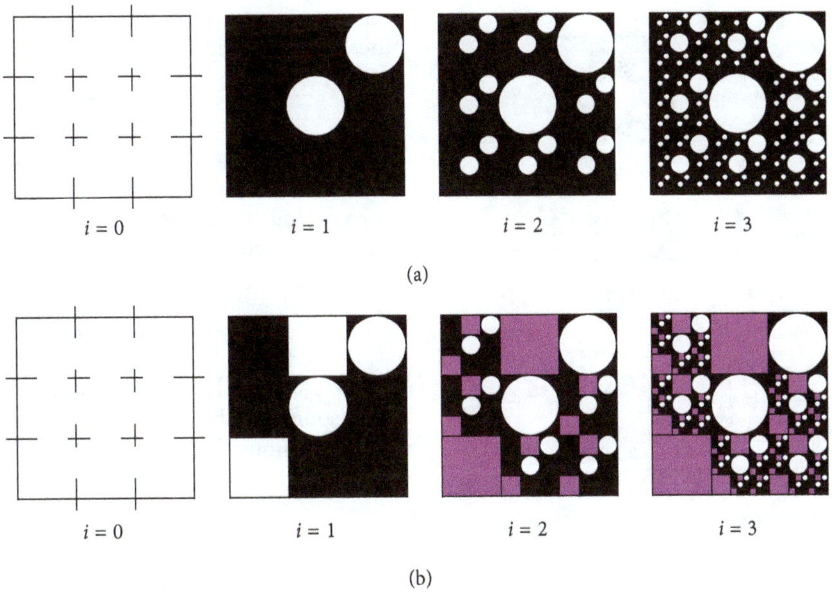

FIGURE 1: Sierpinski carpet construction: (a) geometric construction of the Sierpinski carpet and its iteration process. The fractal dimension D_f is 1.77 and the iterations are 3. (b) Fractal scaling ($i = 0, 1, 2,$ and 3) for the Sierpinski carpet with two pores at first iteration. Starting from second iteration, two squares remain *solid forever* and five squares generate new iteration.

It has recently been shown that even real nonfractal structures can be effectively modelled by laying together different fractal units. This modelling procedure, called intermingled fractal unit (IFU) model, is able to reproduce porous microstructure in terms of agreement with experimental porosimetric data: pore volume fraction (ε_{exp}) and pore size distribution [28, 44] (Figure 2).

Considering the value of ε, obtained from the MIP tests, in a simple case in which only two different types of the Sierpinski carpet have been utilised as base units, the number n_B of B base units, for each A base unit, is calculated by the equation

$$n_B = \frac{\left(A_{Ap} - \varepsilon_{exp} \cdot A_A\right)}{\left(\varepsilon_{exp} \cdot A_B - A_{Bp}\right)}, \tag{2}$$

where A_A, A_B, A_{Ap}, and A_{Bp} represent the total areas and the total pore areas of units A and B (Figure 2).

The base units have different maximum pore sizes, whereas they could have different numbers of removed squares as well as fractal dimension, iterations, pore volume fraction, etc.

In order to gain control on the pore fraction value, it could be necessary to introduce sufficient parts of IFU-*filled surface*. This part of the model is not involved in any pore generation process and is porosity free. So, whenever useful, it is possible to keep porosity constant by varying the pore size distribution. For the same reason, as shown in Figure 1(b), some squares derived from the Sierpinski carpet cannot be part of the iteration process, thus remaining *solid forever* (no porosity). In order to obtain a tridimensional model, the Sierpinski carpets and filled surface are extruded connecting in series more IFU cross sections. The mercury intrusion porosimetric (MIP) technique, from which, in this work, experimental data are acquired, considers separate and distinct capillaries. For this reason, in the proposed model, no connections in any directions are considered. However, capillary rise is perturbed by pore tortuous paths decreasing fluid rate.

At the same time, the obtained IFU arrangement must be considered from analytical point of view. In this sense, for reproducing porous microstructures, it is sufficient to know the numerical characteristics of the base units and the filled surface. Geometrical representations, as reported in Figure 2, are oversimplified scheme (e.g., n_B is a very large value, i.e., $\approx 10^9$, which results into nonfit for the illustration purpose).

Overall, all these model characteristics make it particularly performing and sensitive with regard to the most different pore size distributions experimentally verifiable.

During the water capillary absorption process, the sorptivity value is represented by the slope of the rectilinear section of the curve between water absorption per unit area versus square root of time. Each class of pores, therefore, absorbs at its own speed as a function of the radius size. In order to calculate the sorptivity so that it can be compared with experimental values, it is necessary to study the flow of fluid which passes through every base units of IFUs.

In a porous model with the characteristics expressed above, the number of pores for a given iteration, i, is given by the following formula:

$$N_{\text{pores}(i=n)} = F^{i \cdot D_f}\left(F^{2-D_f} - 1\right). \tag{3}$$

If $i = 2$, $F = 3$, and $D_f = 1.77$, then $N_{\text{pores}(i=n)} = 3^{2 \cdot 1.77}\left(3^{2-1.77} - 1\right) = 14$, as shown in Figure 1(a), indeed the second iteration generates 14 new pores.

The flow of a fluid for an individual capillary with a section calculated as that of the circumference inscribed in

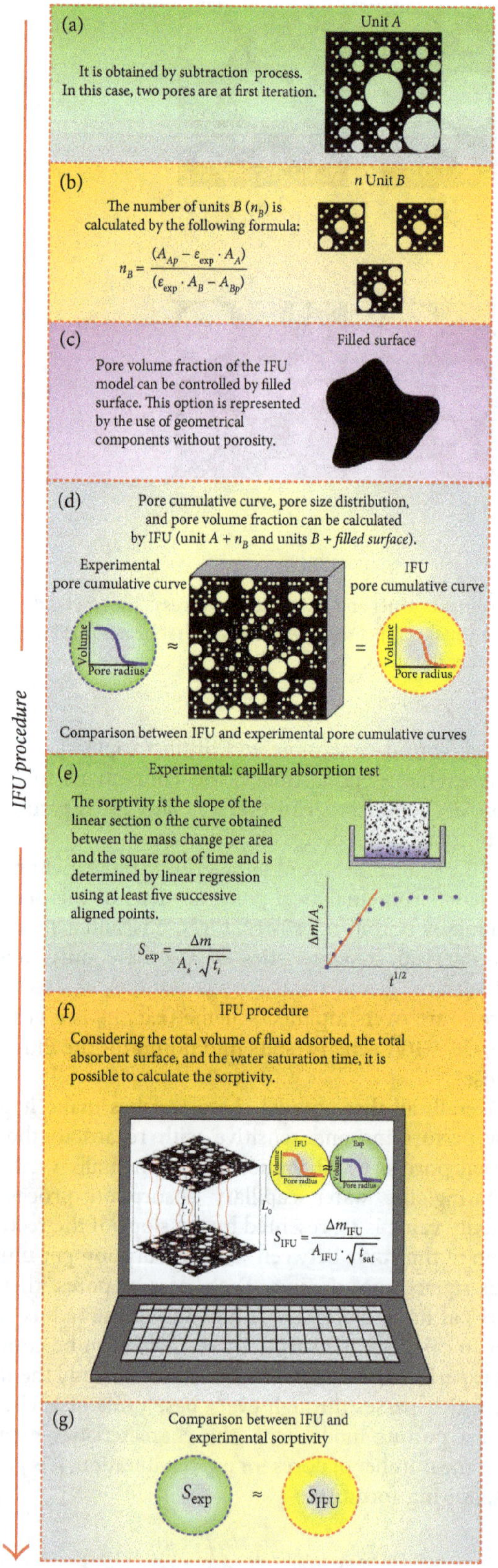

FIGURE 2: Modelling procedure to calculate the sorptivity value. (a) to (c) IFU construction intermingling different types and numbers of the Sierpinski carpets and solid surface; (d) comparison between experimental and IFU pore cumulative curves; (e) and (f) calculation of sorptivity applying the fractal expression; (g) comparison between experimental and model sorptivity data.

the square subtracted from the given iteration i is given by the Hagen-Poiseuille equation:

$$q(\lambda) = \frac{\pi}{128} \frac{\Delta p}{L_t(\lambda)} \frac{\lambda^4}{\mu}, \tag{4}$$

where Δp is the pressure loss, L_t is the capillary length considering tortuosity, λ is the pore diameter, and μ is the dynamic viscosity.

The tortuous path of pores (L_t) is expressed by

$$L_t(\lambda) = L_0^{D_t} \lambda^{1-D_t}, \tag{5}$$

where L_0 is the length of rectilinear capillary and its value corresponds to the dimension (height) of the sample and D_T is the tortuous fractal dimension:

$$D_T = 1 + \frac{\ln \tau}{\ln(L_0/\lambda)}. \tag{6}$$

τ_{class} is the tortuosity of the specific pore class, calculated as

$$\tau_{\text{class}} = 1 + a \ln \frac{1}{\varepsilon_{\text{class}}}, \tag{7}$$

where $\varepsilon_{\text{class}}$ is the pore class volume fraction and a is a constant that depends on pore shape (in this case, its value is 0.63) [46].

Moreover, a generic diameter may be expressed by the following relation:

$$\lambda = \frac{\lambda_{\max}}{F^{(i-1)}}. \tag{8}$$

Simple algebra shows that the Hagen-Poiseuille equation can be rewritten as

$$q_{\text{pore}}(\lambda) = \frac{\pi}{128} \frac{\Delta p}{L_t} \frac{1}{\mu} \frac{\lambda_{\max}^4}{F^{4(i-1)}}. \tag{9}$$

The flow Q_{class} for a generic class of pores is the result of $q_{\text{pore}} \cdot N$, thus,

$$Q_{\text{class}} = \left(\frac{\pi}{128} \frac{\Delta p}{L_t} \frac{1}{\mu} \frac{(\lambda_{\max})^4}{F^{4(i-1)}} \right) \cdot \left(F^{i \cdot D_f} \left(F^{2-D_f} - 1 \right) \right). \tag{10}$$

The average speed of water rising A_p as total porous surface will be given by

$$v = \frac{Q}{A_p}, \tag{11}$$

where Q is the total flow and A_p is the total porous surface.

In this manner, the time that the fluid takes to saturate each single pore could be obtained by the following equation:

$$t_{\text{sat}} = L_0 \left\{ \frac{1}{A_p} \left(\frac{\pi}{128} \frac{\Delta p}{L_t(\lambda)} \frac{1}{\mu} \frac{(\lambda_{\max})^4}{F^{4(i-1)}} \right) \cdot \left[F^{i \cdot D_f} \left(F^{2-D_f} - 1 \right) \right] \right\}^{-1}. \tag{12}$$

At time $t_{\lambda\text{sat}}$ of the most significant class of pores for capillary absorption, it is possible to calculate all the heights

TABLE 1: IFU input data to models A-1 and A-2. The fractal dimension (D_f), the maximum (R_{max}) and minimum rays (R_{min}), the number of iterations, the number of solid forever squares for each base unit (the Sierpinski carpet), the number of B base units (n_B), the filled surface (FS) used to build the IFU, and comparison between experimental (ε_{exp}) and predicted pore volume fractions (ε_{IFU}).

		A-1	A-2
Unit A	D_f	1.89	1.89
	n	1	1
	R_{max} (μm)	1.50	40
	Iteration	1	8
	R_{min} (μm)	0.5	0.06
	Solid forever	0	1
Unit B	D_f	1.77	1.77
	n	31	1991
	R_{max} (μm)	0.3	1.48
	Iteration	4	5
	R_{min} (μm)	0.004	0.06
	Solid forever	0	0
	Filled surface (μm^2)	0	$1.36{\cdot}10^4$
	$\varepsilon_{exp}/\varepsilon_{calc}$	0.36/0.36	0.65/0.65

TABLE 2: IFU input data to models B-1, B-2, and B-3. The fractal dimension (D_f), the maximum (R_{max}) and minimum rays (R_{min}), the number of iterations, the number of solid forever squares for each base unit (the Sierpinski carpet), the number of B base units (n_B), the filled surface (FS) used to build the IFU, and comparison between experimental (ε_{exp}) and predicted pore volume fractions (ε_{IFU}).

		B-1	B-2	B-3
Unit A	D_f	1.63	1.89	1.77
	N	1	1	1
	R_{max} (μm)	2.88	7.39	5.87
	Iteration	6	7	6
	R_{min} (μm)	0.004	0.003	0.008
	Solid forever	2	0	4
Unit B	D_f	1.63	1.89	1.77
	N	$6.77{\cdot}10^8$	$1.62{\cdot}10^7$	$7.79{\cdot}10^5$
	R_{max} (μm)	0.04	0.03	0.07
	Iteration	2	2	2
	R_{min} (μm)	0.004	0.003	0.008
	Solid forever	4	4	1
	Filled surface (μm^2)	$8.70{\cdot}10^7$	$5.29{\cdot}10^5$	$4.08{\cdot}10^4$
	$\varepsilon_{exp}/\varepsilon_{calc}$	0.11/0.11	0.19/0.19	0.37/0.37

reached by the fluid on the remaining pores of different radii (eventually, for the pores higher than the representative radius, at $t_{\lambda sat}$, they are considered filled of water).

Consequently, the volume of fluid absorbed could be calculated by

$$V_{tot} = \sum_{1}^{n} A_p \cdot h_{class}. \quad (13)$$

Considering the mass of the total volume of absorbed water, $\Delta m_{IFU} = V_{tot} \cdot \rho$, in which ρ is the density of water and the total absorbent surface (A_{IFU}) sorptivity is equal to

TABLE 3: IFU input data to models C-1, C-2, C-3, and C-4. The fractal dimension (D_f), the maximum (R_{max}) and minimum rays (R_{min}), the number of iterations, the number of solid forever squares for each base unit (the Sierpinski carpet), the number of B base units (n_B), the filled surface (FS) used to build the IFU, and comparison between experimental (ε_{exp}) and predicted pore volume fractions (ε_{IFU}).

		C-1	C-2	C-3	C-4
Unit A	D_f	1.89	1.89	1.89	1.89
	N	1	1	1	1
	R_{max} (μm)	54.66	36	45	60.66
	Iteration	10	9	10	10
	R_{min} (μm)	0.003	0.002	0.002	0.002
	Solid forever	0	0	0	1
Unit B	D_f	1.63	1.63	1.77	1.77
	N	$1.13{\cdot}10^9$	$4.43{\cdot}10^7$	$41.75{\cdot}10^8$	$1.49{\cdot}10^8$
	R_{max} (μm)	0.008	0.016	0.021	0.028
	Iteration	2	2	3	3
	R_{min} (μm)	0.003	0.002	0.002	0.002
	Solid forever	4	3	3	3
	Filled surface (μm^2)	$4.41{\cdot}10^6$	$5.77{\cdot}10^5$	$3.71{\cdot}10^6$	$1.82{\cdot}10^6$
	$\varepsilon_{exp}/\varepsilon_{calc}$	0.17/0.17	0.23/0.23	0.21/0.21	0.27/0.27

$$S_{IFU} = \frac{\Delta m_{IFU}}{A_{tot} \cdot \sqrt{t_{\lambda sat}}}. \quad (14)$$

In brief, the use of the IFU consists of (Figure 2 shows an outline about IFU sorptivity procedure) (a) cumulative MIP curve data analysis and verification of the fractality of the microstructures [44, 47], (b) model processing (based on the Sierpinski carpets with different geometric size and scaling iterations) in order to reproduce the experimental pore size distribution and the pore volume fraction, (c) applying the fractal expression of capillary adsorption for each unit and considering the various contributions, and (d) comparison between modelling predictions and experimental and/or other model soptivity data [9, 12].

3. Materials

3.1. First Data Set (A). The microstructures analysed are systems with a cementitious CPA CEM I 52.5 (EN 196-1) and argillaceous matrix consisting exclusively of kaolin (with an absolute density, measured using a water pycnometer, of 2.65 g/cm^3) [48, 49]. The wood aggregates (*Picea* wood) have sizes between 3 and 8 mm: A-1 (clay cement matrix) and A-2 (clay cement matrix + 30% wood aggregates). Samples are placed in a storage room for 28 days.

3.2. Second Data Set (B). A commercial Portland cement CEM II 42.5 to make up pastes with water-cement rations (w/c; by weight) of 0.25, 0.34, and 0.43 is used (henceforward referred to as B-1, B-2, and B-3) [50]. The test specimens (25 mm cubes) were aged for 28 days (7 days under water and 21 days in air).

Table 4: IFU input data to models D-1, D-2, D-3, D-4, D-5, and D-6. The fractal dimension (D_f), the maximum (R_{max}) and minimum rays (R_{min}), the number of iterations, the number of solid forever squares for each base unit (the Sierpinski carpet), the number of B base units (n_B), the filled surface (FS) used to build the IFU, and comparison between experimental (ε_{exp}) and predicted pore volume fractions (ε_{IFU}).

		D-1	D-2	D-3	D-4	D-5	D-6
Unit A	D_f	1.89	1.89	1.77	1.63	1.77	1.89
	N	1	1	1	1	1	1
	R_{max} (μm)	2.25	7	19	18	34	34
	Iteration	7	8	9	9	9	9
	R_{min} (μm)	0.002	0.002	0.002	0.002	0.002	0.002
	Solid forever	0	5	4	4	6	0
Unit B	D_f	1.89	1.89	1.63	1.63	1.46	1.46
	N	$5.05{\cdot}10^4$	$1.38{\cdot}10^5$	$9.78{\cdot}10^5$	$9.32{\cdot}10^5$	$7.39{\cdot}10^5$	$3.19{\cdot}10^5$
	R_{max} (μm)	0.083	0.086	0.078	0.074	0.14	0.14
	Iteration	4	4	4	4	4	4
	R_{min} (μm)	0.002	0.002	0.002	0.002	0.002	0.002
	Solid forever	2	1	2	2	3	3
	Filled surface (μm^2)	$4.10{\cdot}10^8$	$7.89{\cdot}10^8$	$1.26{\cdot}10^5$	$4.13{\cdot}10^4$	$4.67{\cdot}10^4$	0
	$\varepsilon_{exp}/\varepsilon_{calc}$	0.22/0.22	0.29/0.29	0.37/0.37	0.48/0.48	48.7/48.7	0.60/0.60

3.3. Third Data Set (C). Four cement-based mortars, hereafter referred to as C-1, C-2, C-3, and C-4, are considered [51]. All the composites were produced on the basis of the Portland cement CEM I 42.5 R, natural quartz sand with maximum grain size of 2 mm (from Bielinek in Poland), and tap water in different quantities. The weight ratio of sand and binder (cement, silica fume) was s : b = 3.1 : 1.0. Curing was performed for 28 days in air (lab conditions).

3.4. Fourth Data Set (D). Sixty cylindrical specimens of 4 cm (diameter of 3.4 cm) were elaborated blending the lime with the cement in order to obtain six different compositions (% lime: 0, 20, 40, 60, 80, and 100%, resp., called D-1, D-2, D-3, D-4, D-5, and D-6) with 10 specimens of each composition [52]. Different amounts of mixing water were employed in order to obtain a similar consistency in all the pastes. The specimens, arranged in a vertical position, were cured in ambient laboratory conditions (RH 60 ± 10% and 20 ± 5°C) for 2 years. Other characteristics are illustrated in the reference.

For all systems, capillary absorption and mercury intrusion porosimetry (MIP) tests to measure sorptivity values and porosity (pore size distribution and pore size range) have been conducted [48–52].

Capillary tests are performed placing regular specimens in a container with a dry bedding layer (minimum thickness of 5 mm). During absorption, water must be added in order to maintain the water-level constant. Samples are weighed at regular interval, which depends on the speed of water absorption. When the weight of the samples remains constant (±1%) for 24 hours, the test is ended.

The water absorbed per unit of exposed surface Q_i (mg/cm^2) at time t_i (s) is estimated by using $Q_i = [(m_i - m_0)/A_s]$, in which m_i (mg) is the mass of the sample at time t_i, m_0 (mg) is the mass of the dry specimen, and A_s (cm^2) is the area of the sample face in contact with the water layer.

The capillary water absorption coefficient (S_{exp}), also called *sorptivity*, is defined as the slope of the linear section of the curve obtained by reporting in a plot the mass variation per area (Q_i) versus the square root of time ($t_i^{1/2}$) and was determined by linear regression using at least 5 successive aligned points.

Mercury intrusion porosimetry (MIP) is carried out by using the Micromeritrics Autopore porosimeter. For these tests, one fragment (dimension ≈ 1 cm^3) is selected by each sample for which sorptivity has been estimated by capillary absorption test.

4. IFU Application and Comparison with Experimental and Model Findings

The comparison between experimental findings and modelling predictions can be used to validate the IFU modelling approach and tests its capabilities and limitations. In particular, Tables 1–4 show IFU input data to set the model in order to reproduce experimental porosimetric cumulative curves. They report the fractal dimension (D_f), the maximum (R_{max}) and minimum rays (R_{min}), the number (i) of iterations (R_{max}, R_{min}, and i strictly depends on experimental porometric data), the number of *solid forever* squares for each base unit (the Sierpinki carpet), the number of B base units (n_B), the filled surface (FS) used to build the IFU, and a comparison between experimental (ε_{exp}) and predicted (ε_{IFU}) porosities expressed as a fraction.

Figure 3 reports a comparison between experimental (from MIP tests [50, 51]) and modelling pore cumulative curves. The following graphs are given as an example for B-2, B-3; C-2, C-4; and D-3, D-4 systems.

It is possible to see that IFU modelling reproductions are in good agreement with experimental findings.

Once the simulation has been successfully performed, the IFU input data (from Tables 1–4) can be used to apply (14) for calculating the sorptivity values (S_{IFU}). The S_{IFU} has been compared in Table 5 with the corresponding experimental measurements (S_{exp}) [48–52].

Moreover, in order to make a comparison complete, Table 5 also contains the sorptivity calculated by adopting the models reported in [9], $S_{[9]}$, and in [12], $S_{[12]}$,

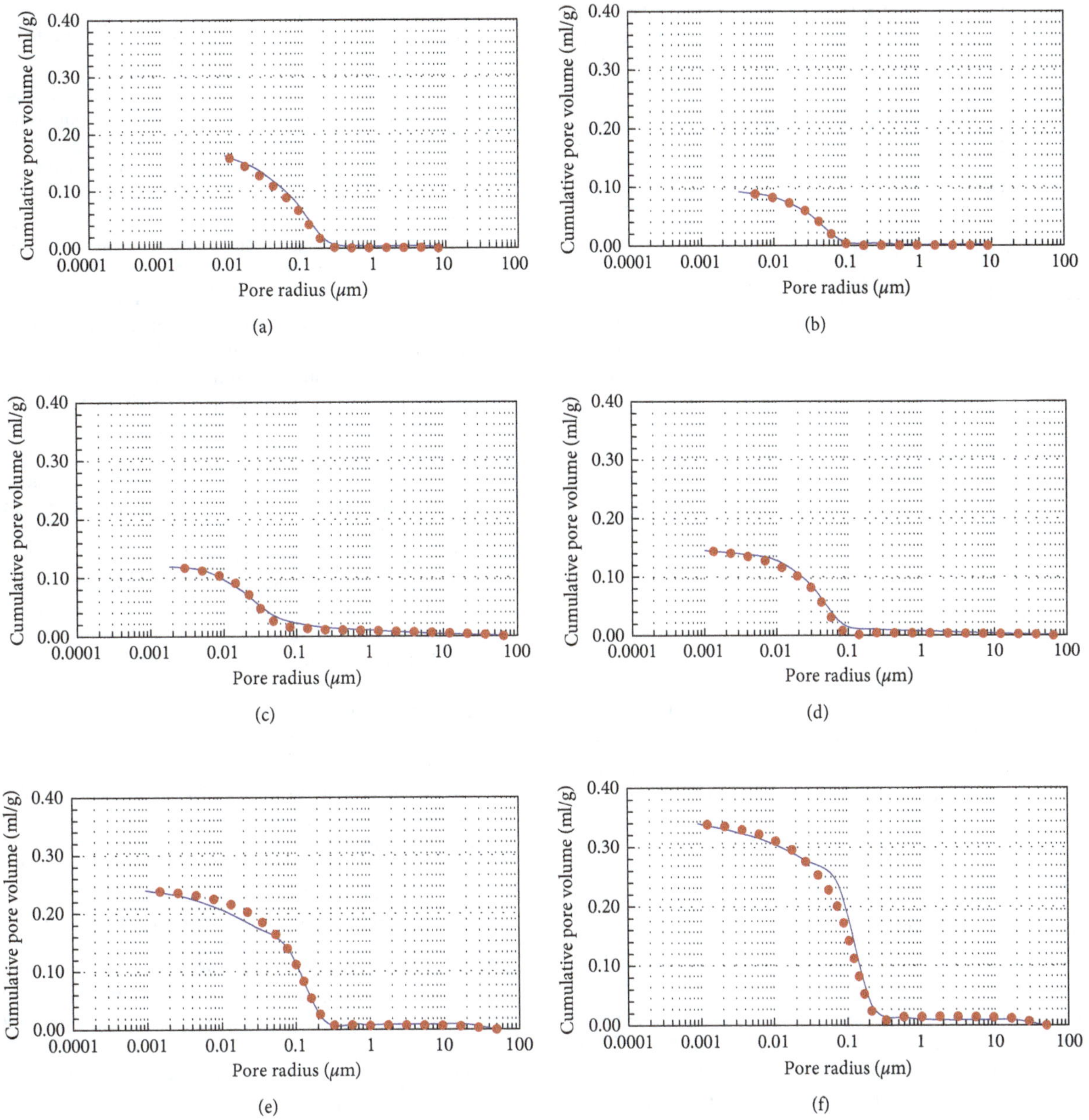

FIGURE 3: Comparison between experimental (continuous line) and IFU (dot line) pore cumulative curves: B-2 (a), B-3 (b), C-2 (c), C-4 (d), D-3 (e), and D-4 (f).

$$S_{[9]} = \rho\left(\frac{\gamma}{\mu}\right)^{0.5}\frac{\varepsilon_{\exp}}{\tau}r_0^{0.5}\left(\cos\frac{\theta}{2}\right)^{0.5},$$

$$S_{[12]} = \varepsilon\rho\sqrt{\frac{\varepsilon_{\exp} r\gamma\cos\theta}{5\mu}}, \qquad (15)$$

where ρ, γ, μ, ε, r_0, τ, and θ represent, respectively, the density, the surface tension, the viscosity of the liquid, the experimental open porosity, the median pore size, the pore tortuosity of the solid, and the liquid-brick contact angle [9]. These models predict sorptivity values with two different analytical procedures. It is possible to note that $S_{[9]}$ and $S_{[12]}$ predictions are definitely much smaller or higher than the experimental findings. The greater difference obtained by applying the latter two procedures [9, 12] is presumably owing to the fact that they only take into consideration the mean pore radius, while the IFU procedure considers the whole pore size distribution. Indeed, pore size distributions can be very different from each other, and they can determine diverse properties and performances of porous materials [5, 29, 42, 43]. Nevertheless, these pore size distributions can have the same average pore radius.

Under these circumstances, since, for some materials, average pore radius can be representative of pore size distribution [9, 12], for others, it is necessary to consider entirely their pore size distribution.

Overall, the IFU modelling procedure proposed has proven to be an important tool to better understand the correlation between microstructures and suction properties

TABLE 5: Comparison among the sorptivity values obtained experimentally (S_{exp}), calculated on the basis of the IFU model (S_{IFU}) and the formulas proposed by Raimondo ($S_{[9]}$) and Scherer ($S_{[12]}$).

Sample	S_{exp} (mg cm^{-2} s$^{-0.5}$)	S_{IFU} (mg cm^{-2} s$^{-0.5}$)	$S_{[9]}$ (mg cm^{-2} s$^{-0.5}$)	$S_{[12]}$ (mg cm^{-2} s$^{-0.5}$)
A-1	5.59	4.34	10.42	37.78
A-2	8.83	6.19	10.75	345.94
B-1	0.42	0.97	5.03	2.68
B-2	0.80	1.51	10.97	5.31
B-3	0.92	1.60	24.46	23.11
C-1	5.71	6.44	39.90	2.47
C-2	2.29	4.93	42.80	5.25
C-3	3.28	4.84	31.35	5.68
C-4	3.81	2.92	63.12	5.12
D-1	2.20	3.84	5.32	6.72
D-2	7.20	9.35	11.94	9.44
D-3	16.90	19.94	10.28	23.39
D-4	35.00	32.20	21.20	33.51
D-5	51.00	50.68	79.40	114.69
D-6	59.00	59.53	91.77	66.45

of porous materials. It allows verifying the data achieved from experimental tests that are often difficult to implement and/or interpret. The IFU model has full potentiality to become a work instrument to plan the mix design of cement pastes.

5. Conclusions

An intermingled fractal units' model has been developed in order to reproduce the pore size distribution of cement. By applying an analytical-fractal procedure, it is possible to predict the sorptivity coefficient. The agreement between experimental findings and modelling predictions indicates that the IFU model developed is able to suitably account for the fundamental mechanisms of water absorption in porous building materials. Moreover, other two analytical procedures to have sorptivity values have been applied, but it is possibly established that the newly proposed method better matches with the experimental results. This good agreement achieved is probably due to the IFU capability to consider pore size distribution entirely and not only an average pore. In fact, the IFU model based on fractal units is able to reproduce every pore size distribution even if they are not fractal.

Conflicts of Interest

The authors declare that they have no conflicts of interest.

References

[1] C. Moses, D. Robinson, and J. Barlow, "Methods for measuring rock surface weathering and erosion: a critical review," *Earth-Science Reviews*, vol. 135, pp. 141–161, 2014.

[2] C. Cardell, F. Delalieux, K. Roumpopoulos, A. Moropoulou, F. Auger, and R. Van Grieken, "Salt-induced decay in calcareous stone monuments and buildings in a marine environment in SW France," *Construction and Building Materials*, vol. 17, no. 3, pp. 165–179, 2003.

[3] F. Sandrolini, E. Franzoni, E. Sassoni, and P. P. Diotallevi, "The contribution of urban-scale environmental monitoring to materials diagnostics: a study on the Cathedral of Modena (Italy)," *Journal of Cultural Heritage*, vol. 12, no. 4, pp. 441–450, 2011.

[4] O. Rudic, D. Rajnovic, D. Cjepa, S. Vucetic, and J. Ranogajec, "Investigation of the durability of porous mineral substrates with newly designed TiO_2-LDH coating," *Ceramics International*, vol. 41, no. 8, pp. 9779–9792, 2015.

[5] G. Pia and U. Sanna, "An intermingled fractal units model and method to predict permeability in porous rock," *International Journal of Engineering Science*, vol. 75, pp. 31–39, 2014.

[6] C. Atzeni, G. Pia, and U. Sanna, "A geometrical fractal model for the porosity and permeability of hydraulic cement pastes," *Construction and Building Materials*, vol. 24, no. 10, pp. 1843–1847, 2010.

[7] E. Franzoni, "Rising damp removal from historical masonries: a still open challenge," *Construction and Building Materials*, vol. 54, pp. 123–136, 2014.

[8] J. Cai, X. Hu, D. C. Standnes, and L. You, "An analytical model for spontaneous imbibition in fractal porous media including gravity," *Colloids and Surfaces A: Physicochemical and Engineering Aspects*, vol. 414, pp. 228–233, 2012.

[9] M. Raimondo, M. Dondi, D. Gardini, G. Guarini, and F. Mazzanti, "Predicting the initial rate of water absorption in clay bricks," *Construction and Building Materials*, vol. 23, no. 7, pp. 2623–2630, 2009.

[10] D. Park, S. Park, Y. Seo, and T. Noguchi, "Water absorption and constraint stress analysis of polymer-modified cement mortar used as a patch repair material," *Construction and Building Materials*, vol. 28, no. 1, pp. 819–830, 2011.

[11] M. Rabehi, B. Mezghiche, and S. Guettala, "Correlation between initial absorption of the cover concrete, the compressive strength and carbonation depth," *Construction and Building Materials*, vol. 45, pp. 123–129, 2013.

[12] G. W. Scherer and G. S. Wheeler, "Silicate consolidants for stone," *Key Engineering Materials*, vol. 391, pp. 1–25, 2009.

[13] Z. Wu, H. S. Wong, and N. R. Buenfeld, "Influence of drying-induced microcracking and related size effects on mass transport properties of concrete," *Cement and Concrete Research*, vol. 68, pp. 35–48, 2015.

[14] E. Franzoni, E. Sassoni, G. W. Scherer, and S. Naidu, "Artificial weathering of stone by heating," *Journal of Cultural Heritage*, vol. 14, no. 3, pp. e85–e93, 2013.

[15] M. Á. García-del-Cura, D. Benavente, J. Martínez-Martínez, and N. Cueto, "Sedimentary structures and physical properties of travertine and carbonate tufa building stone," *Construction and Building Materials*, vol. 28, no. 1, pp. 456–467, 2012.

[16] P. J. McGlinn, F. C. de Beer, L. P. Aldridge et al., "Appraisal of a cementitious material for waste disposal: neutron imaging studies of pore structure and sorptivity," *Cement and Concrete Research*, vol. 40, no. 8, pp. 1320–1326, 2010.

[17] M. Y. J. Liu, U. J. Alengaram, M. Z. Jumaat, and K. H. Mo, "Evaluation of thermal conductivity, mechanical and transport properties of lightweight aggregate foamed geopolymer concrete," *Energy and Buildings*, vol. 72, pp. 238–245, 2014.

[18] U. A. Dogan and M. H. Ozkul, "The effect of cement type on long-term transport properties of self-compacting concretes," *Construction and Building Materials*, vol. 96, pp. 641–647, 2015.

[19] Z. Zhang, "Comparisons of various absorbent effects on carbon dioxide capture in membrane gas absorption (MGA)

process," *Journal of Natural Gas Science and Engineering*, vol. 31, pp. 589–595, 2016.

[20] M. Rezakazemi, I. Heydari, and Z. Zhang, "Hybrid systems: combining membrane and absorption technologies leads to more efficient acid gases (CO_2 and H_2S) removal from natural gas," *Journal of CO_2 Utilization*, vol. 18, pp. 362–369, 2017.

[21] Z. Zhang, Y. Yan, L. Zhang et al., "Theoretical study on CO_2 absorption from biogas by membrane contactors: effect of operating parameters," *Industrial and Engineering Chemistry Research*, vol. 53, no. 36, pp. 14075–14083, 2014.

[22] A. Stazi, M. D'Orazio, and E. Quagliarini, "In-life prediction of hygrometric behaviour of buildings materials: an application of fractal geometry to the determination of adsorption and suction properties," *Building and Environment*, vol. 37, no. 7, pp. 733–739, 2002.

[23] B. B. Mandelbrot, *The Fractal Geometry of Nature*, Freeman, New York, NY, USA, 2004.

[24] S. Sasanian and T. A. Newson, "Use of mercury intrusion porosimetry for microstructural investigation of reconstituted clays at high water contents," *Engineering Geology*, vol. 158, pp. 15–22, 2013.

[25] L. Korat, V. Ducman, A. Legat, and B. Mirtič, "Characterisation of the pore-forming process in lightweight aggregate based on silica sludge by means of X-ray micro-tomography (micro-CT) and mercury intrusion porosimetry (MIP)," *Ceramics International*, vol. 39, no. 6, pp. 6997–7005, 2013.

[26] J. Zhou, G. Ye, and K. van Breugel, "Characterization of pore structure in cement-based materials using pressurization–depressurization cycling mercury intrusion porosimetry (PDC-MIP)," *Cement and Concrete Research*, vol. 40, no. 7, pp. 1120–1128, 2010.

[27] R. M. Novais, M. P. Seabra, and J. A. Labrincha, "Ceramic tiles with controlled porosity and low thermal conductivity by using pore-forming agents," *Ceramics International*, vol. 40, no. 8, pp. 11637–11648, 2014.

[28] C. Atzeni, G. Pia, and U. Sanna, "Fractal modelling of medium–high porosity SiC ceramics," *Journal of the European Ceramic Society*, vol. 28, no. 14, pp. 2809–2814, 2008.

[29] G. Pia, L. Casnedi, M. Ionta, and U. Sanna, "On the elastic deformation properties of porous ceramic materials obtained by pore-forming agent method," *Ceramics International*, vol. 41, no. 9, pp. 11097–11105, 2015.

[30] D. N. Winslow, "The fractal nature of the surface of cement paste," *Cement and Concrete Research*, vol. 15, no. 5, pp. 817–824, 1985.

[31] R. A. Livingston, "Fractal nucleation and growth model for the hydration of tricalcium silicate," *Cement and Concrete Research*, vol. 30, no. 12, pp. 1853–1860, 2000.

[32] D. A. Lange, H. M. Jennings, and S. P. Shah, "Image analysis techniques for characterization of pore structure of cement-based materials," *Cement and Concrete Research*, vol. 24, no. 5, pp. 841–853, 1994.

[33] S. Diamond and D. Bonen, "Microstructure of hardened cement paste. A new interpretation," *Journal of the American Ceramic Society*, vol. 76, no. 12, pp. 2993–2999, 1993.

[34] M. Arandigoyen and J. I. Alvarez, "Pore structure and mechanical properties of cement–lime mortars," *Cement and Concrete Research*, vol. 37, no. 5, pp. 767–775, 2007.

[35] V. Comegna, P. Damiani, and A. Sommella, "Use of a fractal model for determining soil water retention curves," *Geoderma*, vol. 85, no. 4, pp. 307–323, 1998.

[36] Y. F. Xu and P. Dong, "Fractal approach to hydraulic properties in unsaturated porous media," *Chaos, Solitons and Fractals*, vol. 19, no. 2, pp. 327–337, 2004.

[37] Y. Shi, J. Xiao, M. Pan, and R. Yuan, "A fractal permeability model for the gas diffusion layer of PEM fuel cells," *Journal of Power Sources*, vol. 160, no. 1, pp. 277–283, 2006.

[38] Q. Zheng and B. Yu, "A fractal permeability model for gas flow through dual-porosity media," *Journal of Applied Physics*, vol. 111, no. 2, p. 024316, 2012.

[39] C. Atzeni, G. Pia, U. Sanna, and N. Spanu, "Surface wear resistance of chemically or thermally stabilized earth-based materials," *Materials and Structures*, vol. 41, no. 4, pp. 751–758, 2007.

[40] X. Huai, W. Wang, and Z. Li, "Analysis of the effective thermal conductivity of fractal porous media," *Applied Thermal Engineering*, vol. 27, no. 17-18, pp. 2815–2821, 2007.

[41] J. Cai and X. Huai, "Study on fluid–solid coupling heat transfer in fractal porous medium by lattice Boltzmann method," *Applied Thermal Engineering*, vol. 30, no. 6-7, pp. 715–723, 2010.

[42] G. Pia and U. Sanna, "Intermingled fractal units model and electrical equivalence fractal approach for prediction of thermal conductivity of porous materials," *Applied Thermal Engineering*, vol. 61, no. 2, pp. 186–192, 2013.

[43] G. Pia, L. Casnedi, and U. Sanna, "Porous ceramic materials by pore-forming agent method: an intermingled fractal units analysis and procedure to predict thermal conductivity," *Ceramics International*, vol. 41, no. 5, pp. 6350–6357, 2015.

[44] G. Pia and U. Sanna, "A geometrical fractal model for the porosity and thermal conductivity of insulating concrete," *Construction and Building Materials*, vol. 44, pp. 551–556, 2013.

[45] G. Pia and U. Sanna, "An intermingled fractal units model to evaluate pore size distribution influence on thermal conductivity values in porous materials," *Applied Thermal Engineering*, vol. 65, no. 1-2, pp. 330–336, 2014.

[46] J. Comiti and M. Renaud, "A new model for determining mean structure parameters of fixed beds from pressure drop measurements: application to beds packed with parallelepipedal particles," *Chemical Engineering Science*, vol. 44, no. 7, pp. 1539–1545, 1989.

[47] P. Pfeifer and D. Avnir, "Chemistry in noninteger dimensions between two and three. I. Fractal theory of heterogeneous surfaces," *Journal of Chemical Physics*, vol. 79, no. 7, pp. 3558–3565, 1983.

[48] A. Bouguerra, H. Sallée, F. de Barquin, R. Dheilly, and M. Quéneudec, "Isothermal moisture properties of wood-cementitious composites," *Cement and Concrete Research*, vol. 29, no. 3, pp. 339–347, 1999.

[49] A. Bouguerra, A. Ledhem, F. de Barquin, R. M. Dheilly, and M. Quéneudec, "Effect of microstructure on the mechanical and thermal properties of lightweight concrete prepared from clay, cement, and wood aggregates," *Cement and Concrete Research*, vol. 28, no. 8, pp. 1179–1190, 1998.

[50] C. Atzeni, L. Massidda, and U. Sanna, "Densifying of cement pastes by chemical deposition of silica in the pores," in *Proceedings of International Congress on the Chemistry of Cement*, New Delhi, India, 1992.

[51] H. Garbalińska and A. Wygocka, "Microstructure modification of cement mortars: effect on capillarity and frost-resistance," *Construction and Building Materials*, vol. 51, pp. 258–266, 2014.

[52] M. Arandigoyen and J. I. Alvarez, "Blended pastes of cement and lime: pore structure and capillary porosity," *Applied Surface Science*, vol. 252, no. 23, pp. 8077–8085, 2006.

6

Polyfunctional Modifiers for Bitumen and Bituminous Materials with High Performance

Alim Feizrakhmanovich Kemalov, Ruslan Alimovich Kemalov, Ilmira Maratovna Abdrafikova, Pavel Sagitovich Fakhretdinov, and Dinar Zinnurovich Valiev

Department of High-Viscosity Oil and Natural Bitumen, Kazan (Volga Region) Federal University, 18 Kremlyovskaya Street, Kazan 420008, Russia

Correspondence should be addressed to Alim Feizrakhmanovich Kemalov; alim.kemalov@mail.ru

Academic Editor: Hainian Wang

Over the last decade increase in capacity and the intensity of vehicular traffic has increased manifoldly, including heavy trucks, super singles, and higher tire pressures, resulting in significant increase of dynamic loads on the road surface which in turn lead to high quality requirements for bitumen in order to avoid premature wear and failure of asphalt concrete pavements. One of the possibilities to increase the quality of bitumen is to use special additives and modifiers that can provide a high adhesion to mineral filler and inhibit the aging and degradation processes in the asphalt coating. To achieve this, in the present study composite modifiers based on bisimidazolines derivatives were synthesized. The developed polyfunctional modifier (PFM) of complex action provides enhanced thermal stability, significantly improves the adhesion between bitumen binder and aggregates, and also improves the physical-mechanical properties of the asphalt concrete. Based on the test results it is recommended to use the synthesized samples of the PFM additive with complex action in asphalt mixtures for road paving.

1. Introduction

Asphalt concrete (AC) coatings include two main components: bitumen and aggregates (crushed stone). Bitumen performs the function of a substance that binds particles of mineral material to form an asphalt concrete coating. Being nonpolar, bitumen has high waterproofing properties. It is known that high-resinous and low-paraffinic petroleum is the most suitable for the production of road and construction bitumen. However, due to the limited supply of this type of petroleum almost any oil residues are used. As a result, the quality of bitumen is very unsure at the moment and that leads to poor quality of bituminous materials resulting in poor quality asphalt concrete pavements.

At the same time, most of the used aggregates have a greater affinity for water than they do for bitumen binder [1–3]. Polycyclic aromatic compounds, resins, and heteroatomic compounds contained in the bitumen do not provide desired adhesion to the mineral part of asphalt concrete. Oxygenated compounds formed during the oxidation of vacuum residues also have little impact on the adhesive properties of bitumen [4–7].

Poor adhesion between the bitumen binder (BB) and aggregate may lead to stripping (delamination of the binder from the aggregate surface) in the presence of water which leads to pavement failure such as raveling, potholes formation, and rutting [2, 3, 6–8]. The situation is aggravated by the continuous increase in capacity and traffic volumes, including heavy truck axle loads, super singles, and high tire pressure that leads to a significant increase of dynamic loads on the pavement. Therefore, to obtain high quality pavements, key factor is in ensuring high adhesion between bitumen and the mineral component of pavement. For this reason adhesion additives are used; most frequently these are cationic surfactants [3, 7–11]. Classical cationic surfactants are diphilic structures in which nonpolar part is a long chain hydrophobic alkyl and the polar part (so-called "head") is a positively charged ammonium center, or nitrogen atom capable of being protonated in acidic mediums. This fundamental property, diphilicity, is the reason why surfactants are lined up at the interface "bitumen-mineral material" [5–9]. Herewith, the positively charged (hydrophilic) groups are fixed on polar surface of the mineral material and the hydrophobic hydrocarbon chains are fixed in the bitumen. Thus, the adhesive additive acts as bridge between bitumen binder and aggregates that is schematically presented in Figure 1.

Nowadays imidazolines and their derivatives (Ia–Iд) as well as bisimidazolines (IIa, IIб) [9, 11, 12] that exhibit the ability to increase the bitumen adhesion to mineral material are most widely used as additives for bitumen; the mechanism of their action is shown schematically on Figure 2. Adhesion additive "Amidan" [12], which presents the technical product, containing a mixture of imidazoline (Ia–Iд) and bisimidazoline (IIa, IIб) derivatives, can be taken as example.

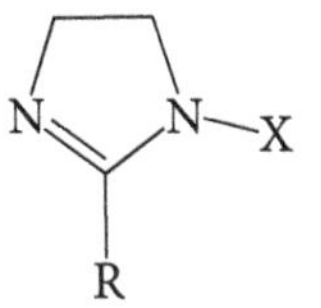

(I)

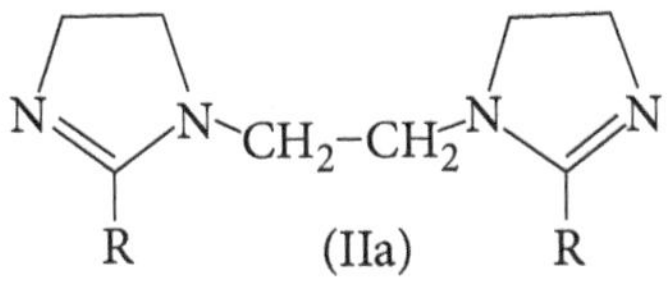

(IIa)

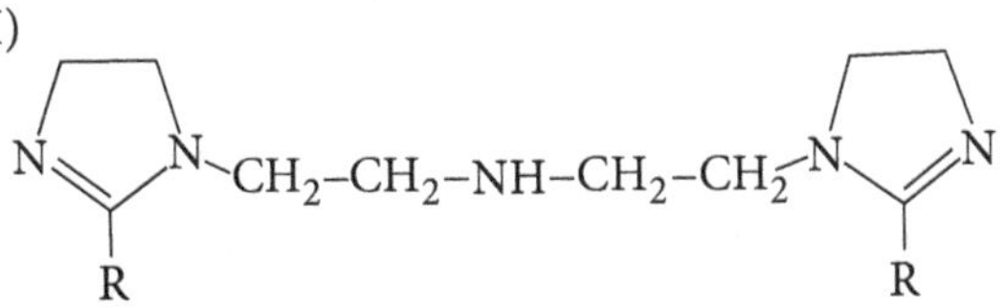

(IIб)

where R = alkyl or alkenyl (in all the structures (I-II)); X are the possible substituents in (I) with the following structures (Ia–Iд):

$X = H$ (Ia),

$X = CH_2{-}CH_2{-}NH_2$ (Iб),

$X = CH_2{-}CH_2{-}NH{-}C(=O){-}R$ (Iв),

$X = CH_2{-}CH_2{-}NH{-}CH_2{-}CH_2{-}NH_2$ (Iг),

$X = CH_2{-}CH_2{-}NH{-}CH_2{-}CH_2{-}NH{-}C(=O){-}R$ (Iд)

Composite modifiers such as sulfur, rubber (polybutadiene, natural rubber, butyl rubber, chloroprene, etc.), organic-manganese compounds, thermoplastic polymers (polyethylene, polypropylene, polystyrene, ethylene-vinyl acetate (EVA)), thermoplastic rubbers (polyurethane, olefin copolymers, and block copolymers of styrene-butadiene-styrene (SBS)) are also used in order to improve physical-chemical properties of BB as well as the physical-mechanical properties of AC [13–15].

The effectiveness and quality of additives are largely related to their thermal stability, and this is due to the following reasons (most relevant for Russia):

(1) Road bitumen is produced in bulk by liquid-phase oxidation of vacuum residue, the optimal process temperature is 250°C, and it is more viable to inject the additive to the oxidized bitumen directly at its outlet from the oxidation column, to obtain bitumen with improved adhesion properties.

(2) The hot-mix asphalt concrete is the most used in road construction, and the compounding of bitumen with mineral material is carried out at relatively high temperatures (160–170°C).

It should be emphasized that a disadvantage of modern adhesive modifiers for bitumen is that they demonstrate a low thermal resistance (up to 120–175°C), while for some systems there is a rapid loss of adhesion at 150°C. On the other hand their use often leads to deterioration of physical-mechanical properties of the corresponding AC [16].

Thus, the aim was to develop the new modifiers with increased thermal stability (up to 220–240°C) that would be capable of increasing the adhesion between bitumen and aggregates and improving elasticity of the binder, that is, the polyfunctional modifiers (PFM).

2. Materials and Methods

According to the concept of nonclassic cationic surfactants [17, 18], they have hydrophobic hydrocarbon chains which are fragmented (broken) by polar heteroatomic groups. This structure ensures their physical and chemical properties and adsorption behavior, different from the classic surfactants. Based on this concept high-performance chemicals were obtained that were useful for the production, treatment, and transportation of petroleum as well as for the production of corrosion inhibitors, emulsifiers, demulsifiers, the biologically active substances, and so on [17–19]. We used the same concept to develop desired nitrogen-containing compounds.

2.1. Modifiers Synthesis. In accordance with the concept of nonclassical surfactants, we had assumed that it is necessary to increase polarity of the molecule of constructed imidazolines by increasing the number of polar imidazoline centers from one to two in order to create compounds with needed properties, as well as to introduce nonclassical spacer into the molecule, wherein the hydrophobicity of an alkyl chain is broken (fragmented) by one or more alkanoylamide fragments:

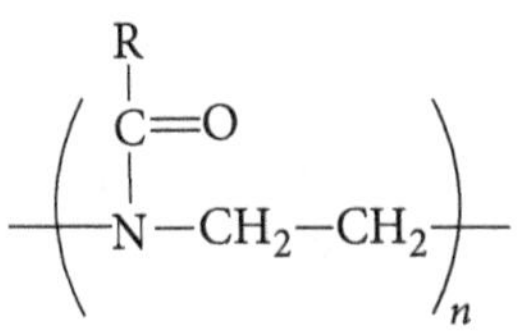

On one hand, this structure of the spacer provides an increase in the molecule polarity that is responsible for interaction between bitumen and hydrophilic surface of the aggregates, because extra polar amide groups appear along with two imidazoline cycles. On the

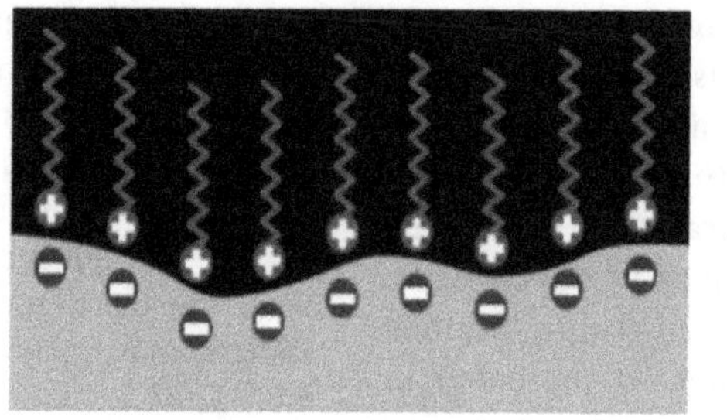

FIGURE 1: Cationic surfactants as a bridge between the bitumen and aggregates.

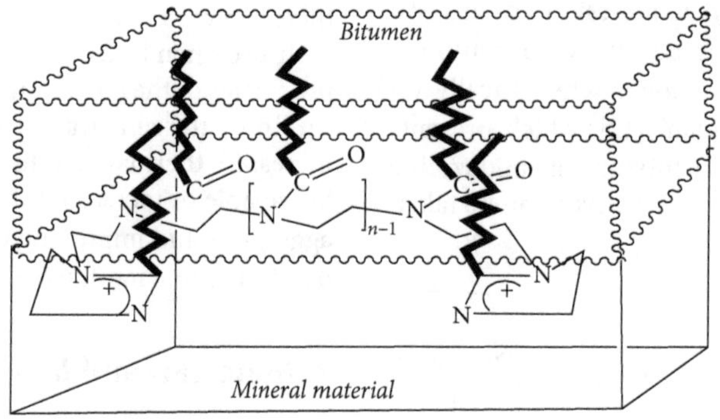

FIGURE 2: Location of bisimidazoline molecules on the interface "bitumen-mineral material."

other hand, the use of alkanoylamide fragments allows introducing additional hydrophobic long chain alkyl groups which increase the possibility of adhesive to fix in the bitumen media. This enhances the adhesive properties of the designed compounds. As a result, we obtained N-acylated bisimidazolines with (ethylene-N-alkanoylamide) spacers, namely, 1-(2[1]-alkylimidazolinyl-1[1])-2-[(2[2]-alkylimidazolinyl-1[2])-poly(ethylene-N-alkanoylamide)]ethanes, containing two imidazoline cycles, connected by spacers:

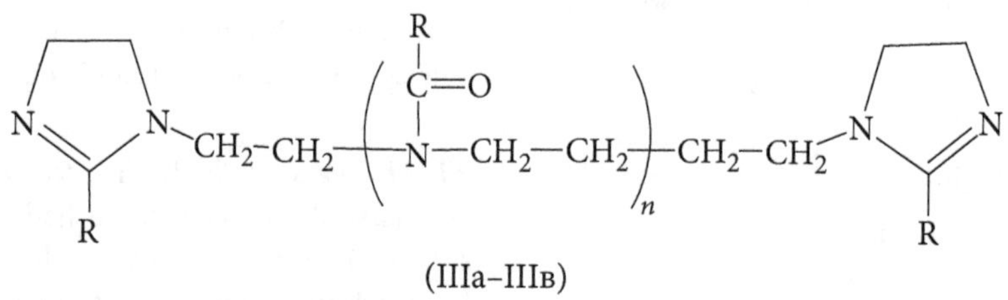

(IIIa–IIIв)

where R = CH_3-$(CH_2)_7$-CH=CH-$(CH_2)_7$-; n = 1 (IIIa), n = 2 (IIIб), and n = 3 (IIIв).

The authors assumed that this construction of the additive would lead to increasing of bitumen adhesion to the mineral component of pavement. Furthermore, the conversion of the secondary amino groups into alkanoylamides should also increase significantly the thermal stability of designed compounds due to the fact that tertiary amides are more stable than the starting amine-compounds.

The desired spacers were obtained with the average yield (the ratio of the mass of the obtained product to the mass of the starting material) of up to 90%, and are viscous and pasty brown liquids, the physicochemical characteristics of which are given in Table 1.

The starting amino-compound and the resulting compounds were analyzed by the conventional method of *potentiometric titration* of a mixture of primary, secondary, and tertiary amines with the definition of the total basicity B_T, the basicity of secondary and tertiary amines $B_{2,3}$, and the basicity of tertiary amines B_3. Then the joint basicity of tertiary amines and amides $B_{3+amide}$ was determined by potentiometric titration of the sample weight of test substance in acetic anhydride (as solvent) with the use of 0.1 N solution of $HClO_4$ in glacial acetic acid. The basicity of amides B_{amide} was determined by the difference between the basicity $B_{3+amide}$ and the basicity B_3 [20]. The results of obtained compounds analysis are shown in Table 1.

The FTIR spectra of the additives and the original bitumen are presented in Figure 3.

TABLE 1: Physicochemical characteristics of N-acylated bisimidazolines derivatives (IIIa–IIIв).

Compound	Element analysis, found/calculated		Basicity of amides, ml 0.1 N $HClO_4$/g, found/calculated
	C, %	N, %	
(IIIa)	78.57/78.64	7.29/7.40	10.49/10.56
(IIIб)	78.41/78.51	6.63/6.70	15.85/15.94
(IIIв)	78.32/78.43	6.19/6.28	19.18/19.21

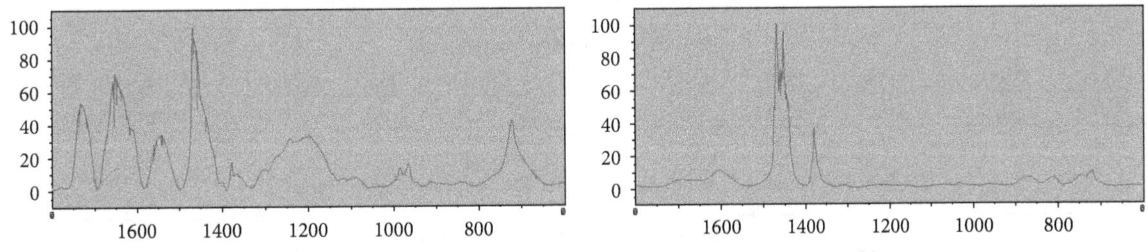

FIGURE 3: IR spectra of the additives (a) and the original bitumen BND 60/90 (TAIF-NK) (b).

Absorption bands with the following wave numbers were observed in the IR spectra of the obtained compounds (Figure 3(a)): υC=N = 1600–1630 cm^{-1}, indicating the presence of imidazoline cycles; υC=O = 1740–1745 cm^{-1}, υ–N(CO)R = 1660 cm^{-1}, and υCN = 1260–1300 cm^{-1} can be related to alkanoylamide structures; $\delta(CH_2)_x = 720$ cm^{-1} – corresponds to pendular oscillation of alkylene structure in a chain [21]. Wide absorption bands at 1600 cm^{-1} in FTIR spectra of the original bitumen (Figure 3(b)) stand for the υCH bond in aromatic ring; sharp absorption bands with high intensity at 1440 and 1460 cm^{-1} correspond to stretch vibration of $-CH_2-$ structure fragment in long chains, confirming high content of paraffin waxes; absorption band at 1375 cm^{-1} is related to methyl groups.

2.2. Bitumen Modification and Performance Properties Analysis. Investigation of the effect of the synthesized compounds on the adhesion between bitumen and the mineral material was carried out by the use of example of oxidized bitumen of 60/90 penetration grade (BND 60/90 (the mark of bitumen in Russia means "petroleum bitumen for roads")) that was produced in TAIF-NK PSC, with the following group chemical composition, wt.%: 22.62 of saturates (including 10.2 % of paraffin waxes), 15.14 of aromatics; 40.26 of resins; 21.98 of asphaltenes.

The bitumen modification process was carried out as follows:

(1) The BND was heated in a metal vessel to the temperature of 80°C with the use of stirring mechanism, kept under thermostatic control for 20 min to obtain homogeneous diluted substance with homogeneous distribution of components;

(2) The BND was gradually heated to the temperature of 120–130°C, at which it was kept for 20 min;

(3) Obtained PFM substances were injected into the heated bitumen and stirred for 20–30 min to ensure homogenous dispersion of the PFM in bitumen media.

The *adhesion* between bitumen and aggregates (mineral material) testing was conducted in accordance with GOST (GOST, Russian state standard) 12801-98 [22], the essence of which is to determine the ability of the viscous bitumen that is precoated on the surface of mineral material to persist on it while being exposed by water. The quality of the adhesion assessment was rated according to Table 2.

The ability of bitumen to coat aggregates was evaluated based on *contact angle measurement*. Original BND 60/90 and modified by PFM in amount of 0.6, 0.8, 1.0, and 1.2 wt.% were taken. The contact angle of wetting was determined on two surfaces: on glass which is the standard of the polar surface, as well as mineral rocks of various genesis: (dolomitized limestone, consisting of 90% dolomite and 10% calcite) from "Bianca gravel factory," chemical compositions of which are presented in Tables 3-4. Pervouralsk aggregates were considered to be of "acidic" nature; Bianca – is of "basic" nature.

Measurements were carried out using cathetometer KM-8 by direct measurement of the base and height of the drop with subsequent calculation [23]. The cosine of the contact angle of the drop is calculated by the formula

$$\cos\theta = \frac{r^2 - h^2}{r^2 + h^2}, \quad (1)$$

where $r = L/2$ is the droplet radius, L is the droplet base, and h is the droplet height. With the help of a cathetometer, the contact angles of the taken aggregates wetting by bitumen

TABLE 2: Evaluation of grip (adhesion).

Feature of the bitumen film on the surface of crushed stone	Evaluation of grip
Binder film is completely retained on the surface, local decreasing of the thickness of a film can be observed	Excellent (five points)
Binder film is completely retained on the surface but is partially separated from the sharp corners and edges	Good (four points)
Binder film is retained on over 50% of crushed stone surface	Satisfactory (three points)
Binder film is retained on less than 50% of crushed stone surface. On the exposed surface individual bitumen droplets are observed	Poor (two points)

TABLE 3: Chemical composition of the "Pervouralsk mine management" gravel.

Content in terms of dry matter, wt.%												
SiO_2	TiO_2	Al_2O_3	Fe_2O_3	FeO	CaO	MgO	MnO	K_2O	Na_2O	V_2O_5	P_2O_5	S
22,1	3,27	11,7	27,2	16,9	8,9	9,0	0,23	0,2	0,14	0,4	0,01	0,03

TABLE 4: Chemical composition of the "Bianca gravel factory" gravel.

Content in terms of dry matter, wt.%							
SiO_2	Fe_2O_3	CaO	MgO	Al_2O_3	$SO_{3\ общ}$	$CaCO_3$	$MgCO_3$
3,3	0,7	20,9	11,9	0,8	0,3	37,2	24,9

binder samples were determined. The drop of bitumen (of the same size and mass, as far as possible) was laid onto the surface of a substrate, at a bitumen temperature of 130°C. As the drop of bitumen cools instantly as soon as it reaches the cold substrate surface, the samples of the stone materials were heated to 40°C, which practically did not change the surface properties of the rock, but retained the liquid state of the bitumen for a sufficient time. The equilibrium value of the wetting contact angle was determined on the basis of the kinetic curves. The wetting angle was measured twice: immediately after the dropping of the bitumen and after 1 minute, necessary to achieve equilibrium wetting angle. The latest value was taken into account.

It is known that adhesion of bitumen to mineral materials is the work required to separate the bitumen layer from these materials. In order to determine the *work of adhesion* (a quantitative measure) between aggregates and bitumen, we measured the surface tension of bitumen in the presence of the PFM. Surface tension at the interface between binder and air phases (σ_{BA}) is the most important factor determining the intensity of wetting of a mineral substrate by bitumen. Surface tension measurements were carried out at a temperature of 25°C using EasyDyne S K20S tensiometer (KRUSS) by the Du Nui ring method. Bitumen samples were dissolved in toluene at a mass ratio of toluene : bitumen equal to 1 : 1.5.

Based on the obtained values of θ and σ, the wetting energy or the adhesive stress was calculated.

$$W_e = \sigma \times \cos\theta. \quad (2)$$

The work of adhesion was calculated using Young-Dupré equation:

$$W_a = \sigma \times (1 + \cos\theta). \quad (3)$$

The work of cohesion was calculated to assess the interaction strength between the bitumen components. The work of cohesion is determined by the expenditure of energy on the reversible rupture of the body over a section equal to unit of area. Since, as a result of a rupture, a surface with two-unit area forms, the work of cohesion is as follows:

$$W_k = 2\sigma. \quad (4)$$

To characterize the wetting process, the spreading coefficient S is used which is taken as the difference between the work of adhesion W_a and the work of cohesion W_k. For the complete wetting $S \to 0$.

$$S = W_a - W_k = \sigma_l \times (\cos\theta - 1). \quad (5)$$

The relationship between the work of adhesion and cohesion can be expressed through a relative value, a relative adhesion work of liquid, Z_a:

$$Z_a = \frac{W_a}{W_k}. \quad (6)$$

If W_a and W_k are expressed in terms of the quantities defined in (3) and (4), then

$$Z_a = \frac{1 + \cos\theta}{2}. \quad (7)$$

To study the *dynamic viscosity* (according to ASTM D4402) of modified bitumen samples, dynamic tests were carried out using Rheotest RN 4.1. We obtained the dependence of dynamic viscosity (mPa·s) from temperature (°C) at a constant shear rate (100 s^{-1}). The measurements were carried out in the following way: first, the measuring system "cone-plate" was heated to 125°C, on which bitumen samples were laid (original bitumen and the bitumen modified by 0.6, 0.8, 1.0, and 1.2 wt.% of PFM, resp.); then the measurements were taken along with cooling of the samples. In order to study the thixotropic properties of the samples, immediately

TABLE 5: Physical-chemical characteristics of the PFM.

Number	The name of indicators	The meaning	Testing method
(1)	Appearance	The viscous honey-like mass	Visually
(2)	Colour	From light yellow to dark brown	Visually
(3)	Smell	Weak, characteristic	Organoleptic
(4)	Density at 20°C, g/cm^3	0.9538	GOST 3900-85
(5)	Kinematic viscosity at 100°C mm^2/s	139.6	GOST 33-2000

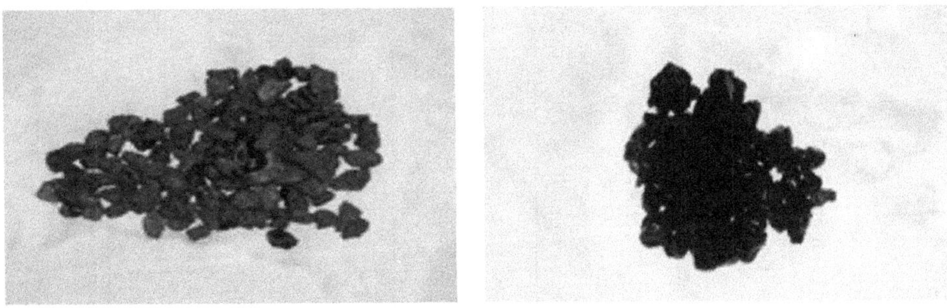

FIGURE 4: The results of the impact of adhesion additive on the adhesion between bitumen and aggregates: (a) the original bitumen 60/90 without additive; (b) bitumen with 1 wt.% of PFM.

after direct measurements, the viscosity was measured in the opposite direction, with a decrease in the shear rate from 1000 to 1 s^{-1} at temperatures of 80–120°C.

The properties of binders such as *softening temperature* (EN 1427, ASTM D36, AASHTO T53), *brittleness temperature* according to Fraas (i.e., the Fraass Breaking Point, EN 12593-2013), *penetration* (EN 1426, ASTM D5, AASHTO T49), *ductility* (EN 13398/13587/13589, ASTM D 113, AASHTO T 51), and the *change of the bitumen softening temperature after heating* were determined according to GOST 22245-90 "Viscous petroleum road bitumen" [24]. First, bitumen samples are heated in the drying oven at a temperature of 163°C for 5 hours in accordance with GOST 18180-72 [25], after which the softening temperature is determined. Density at 20°C and kinematic viscosity at 100°C were determined in accordance with GOST 3900-85 [26] and GOST 33-2000 [27], respectively.

The thermal stability test of modified bitumen was carried out at temperatures of 163, 180, and 200°C and a heating duration of 25 hours (max) with the use of stone materials of basic and acidic nature. The temperature of 163°C is specified by GOST 18180-72 [25] and ASTM D 2872 [28]. The temperatures of 180 and 200°C were chosen upon the recommendations of road constructing organizations. The modified bitumen, that was under a thermal exposure, was sampled every 5 hours, and then the adhesion test was conducted according to the above-mentioned technique. Bitumen binder can be stored at bitumen storage, where it is constantly heated to keep fluidness, for maximum 24 h before it is directed to an asphalt concrete mixing vessel. For that reason the maximum duration of heating (25 h) was chosen that exceeds possible storage period.

Pilot-scale tests were also conducted as well as the technical inspection of asphalt concrete mixtures; the results of the latest were compared with the GOST 9128-2009 [29] requirements.

3. Results and Discussion

3.1. Comparative Study of the Obtained Additives and Their Effect on Bitumen Properties. The features and advantages of the PFM compared to adhesive "Amidan" are the following: high adhesion (at high temperatures as well), thermal stability at 250–270°C within 4 hours (compared to the values of 120–175°C), high flash point, and high processability, the possibility of dosing directly into the bitumen that is outflow of the oxidizing column with a temperature up to 280°C. PFM includes both polymer and adhesive characteristics, which affects the production costs. Despite the fact that additive is doped in small amounts (up to 0.8–1.0 wt.%), they also affect the physical-chemical properties of the final AC. The specially designed spacer's structure enables starting of secondary polymerization processes taking place between asphaltenic and resinous compounds in bitumen media.

Physical-chemical characteristics of the synthesized PFM are presented in Table 5.

Research of the polyfunctional modifier influence on the adhesion between bitumen and the aggregates (mineral material) was carried on the above-mentioned sample of 60/90 bitumen of OAO TAIF-NK production, with the characteristics shown in Table 6. As it can be seen from the data in Table 6, proposed compounds added at amounts of 0.5–1.0% have improved adhesion between bitumen and the mineral filler surface, corresponding to five points in accordance with GOST 12801-98.

Pictures taken after the performed adhesion test are presented on Figure 4. When a mineral material is coated

Table 6: Results of the quality tests of bitumen modified with the use of synthesized compounds.

Compound	Dosage, %	Softening temperature, °C	Penetration, 0,1 mm		Ductility, cm		Adhesion, points	Brittleness temperature by Fraas, °C	The change of bitumen softening point after heating (Δt_{soft}), °C
			25°C	0°C	25°C	0°C			
Original bitumen	—	49.7	67	22	84	5.5	2	−23	5.0
	0.1	49.5	67	22	84	5.5	3	−23	—
	0.25	49.3	70	23	84	5.5	4	−24	3.0
Compound (IIIa) ($n = 1$)	0.5	49	72	24	84	5.5	5	−26	2.1
	1.0	48.8	75	25	85	5.6	5	−27	1.0
	1.5	48.5	78	26	86	5.7	5	−27	—
	0.1	49.5	67	22	84	5.5	3	−23	—
	0.25	49.3	67	22	84	5.5	4	−24	3.2
Compound (IIIб) ($n = 2$)	0.5	49	68	23	84	5.5	5	−26	2.0
	1.0	48.8	69	23	85	5.6	5	−27	1.0
	1.5	48.5	70	24	86	5.7	5	−27	—
	0.1	49.6	67	22	84	5.5	3	−23	—
	0.25	49.5	67	22	84	5.5	4	−24	3.5
Compound (IIIв) ($n = 3$)	0.5	49.3	68	23	84	5.5	5	−26	2.3
	1.0	49.0	69	23	85	5.6	5	−27	1.0
	1.5	48.6	70	24	86	5.7	5	−27	—
	0.1	45	120				3		—
	0.25	45	120				4		7.0
"Amidan"	0.5	45	120				5		6.5
	1.0	46	115				5		6.0
	1.5	46	115				5		5.0

Table 7: Surface tension of the bitumen samples.

Bitumen samples	Original	PFM content, wt.%			
		0.6	0.8	1.0	1.2
Average value, σ_l, mN/m	33.04	31.38	31.34	31.46	31.42

with the original bitumen (i.e., without modifying additives), the binder film is persistent on the area of less than 50% of stones' surface, corresponding to 2 points. Addition of 1 wt.% of modifier leads to complete coverage of the mineral material surface, corresponding to 5 points.

The use of PFM improves the thermal stability of bitumen (Table 6). Thus, bitumen with PFM additive amounts of 0.9–1.0 wt.% has low Δt_{soft}, equal to 1°C, while for the bitumen with "Amidan" Δt_{soft} equals 6°C. In addition, there has been improvement in the penetration, which indicates the bitumen binder improved plasticity.

The test results indicate the fact that the increase in the content of adhesive additives in the range from 0 to 1.5 wt.% leads to increase of penetration values (at 25°C), namely, from 67 to 78 (0.1 mm), respectively (Table 6).

The results of adhesive additive "Amidan" testing (Table 6) indicate deterioration of the bitumen properties. Though the adhesion is equal for both adhesives, "Amidan" acts as a diluting agent, leading to increase in the penetration value. The latest would lead to fractionation (i.e., delamination) of the binder in the process of its application. The changes of "Amidan"-modified bitumen softening temperatures after heating are too high, do not meet the standard requirements, and mean low stability of the binder. For that reason no further tests were carried out for bitumen containing "Amidan."

Regarding the brittleness temperature, determined according to Fraas (the Fraass Breaking Point), it indicates the lower temperature at which the bitumen loses viscoplastic properties and, as a consequence, becomes brittle. It is known that the lower the brittle point, the wider the temperature range in which the bitumen is in the viscoplastic state and, therefore, the better its physical-mechanical characteristics. In this case, the additive doping allows reducing the brittle point and, consequently, providing expansion of ductility temperature range. Apart from the above-mentioned advantages, it should be noted that the compounds are superior to "Amidan" in adhesion at high temperatures. Thus, the modified bitumen with additives amount of 0.9–1.0 wt.% shows high adhesion to sand (5 points) after thermal exposure at temperatures of 220–240°C for 12 hours. It should be noted that the bitumen with the same amounts "Amidan" exhibits much lower thermal stability, providing similar adhesion only at temperatures above 170°C.

3.2. Contact Angle Measurements and Adhesion Work Calculation. Figure 5 shows the dependence of the glass surface wetting contact angle on the content of the additive. It is established that bitumen does not wet the glass because $\cos\theta$ has negative values. Doping the proposed additive into bitumen results in a slight increase in wetting ability, an extremum is observed with an additive content of 0.8%.

Figure 6 shows the dependence of the aggregates wetting contact angle on the content of the additive. It can be seen that the presence of an additive significantly changes the wetting of the mineral substrate by the modified bitumen; a wetting inversion is observed; that is, $\cos\theta$ increases with increasing content of the additive and acquires positive values. We note that, in the case of the surface of mineral materials, an extremum is also observed with an additive content of 0.8% by weight.

Surface tension measurements results are presented in Table 7.

The data in Table 7 indicates that the surface tension decreases with the addition of the additive. This agrees with the data of [30] that the surface tension at the bitumen-solid interface decreases with increasing of surfactants content. Surface tension in combination with adhesion properties provides information of the strength of the bitumen adherence to mineral material.

The calculated values of the above-mentioned characteristics of the "aggregates-organic binder" system surface properties, that is, wetting energy, adhesion work, cohesion work, spreading coefficient, and relative adhesion work, are presented in Table 8.

The analysis of the obtained data shows that the PFM, being adsorbed at the interface, promotes better wetting of the aggregates surface with bitumen. In this case, the contact angle θ decreases, the spreading coefficient and relative adhesion work increase reaching a peak when an additive content is 0.8 wt.%, and cohesion work is reduced.

The value of Z_a approaches 1. Consequently, the adhesion forces between bitumen and mineral materials approximate in magnitude to the binding forces of the bitumen molecules. This contributes to the formation of a material with a homogeneous defect-free structure.

3.3. Component Composition and Dispersity of the Bitumen. Information on the component composition can play an important role in studying the changes occurring as a result of bitumen modification (Table 9). Thus, when introducing the developed additive into the initial bitumen in an amount of about 1% by weight, mainly the content of saturates in the direction of decrease and asphaltenes in the direction of increase is changing. In this case, the relative and absolute changes of the remaining groups of components are insignificant.

The sequence of hydrocarbon group transitions in this case can be delineated as follows [31]:

$$\text{Aromatics (PCA)} \longrightarrow \text{aromatics (MCA)} \longleftarrow \text{saturates} \longrightarrow \text{asphaltenes}, \quad (8)$$

TABLE 8: Physical-chemical characteristics of the "aggregates-organic binder" system surface properties.

Bitumen samples	Diorite					Carbonate				
	Wetting energy, W_e, mN/m	Adhesion work, W_a, mN/m	Cohesion work, W_k, mN/m	Spreading coefficient S	Relative adhesion work, Z_a	Wetting energy, W_e, mN/m	Adhesion work, W_a, mN/m	Cohesion work, W_k, mN/m	Spreading coefficient S	Relative adhesion work, Z_a
0*	−3.97	29.08	66.08	−37.01	0.44	−3.30	29.74	66.08	−36.34	0.45
0.6	1.88	33.26	62.76	−29.50	0.53	3.14	34.52	62.76	−28.24	0.55
0.8	6.89	38.23	62.68	−24.45	0.61	5.33	36.67	62.68	−26.01	0.59
1.0	5.35	36.81	62.92	−26.11	0.59	2.83	34.29	62.92	−28.63	0.55
1.2	4.09	35.50	62.84	−27.34	0.57	2.51	33.93	62.84	−28.91	0.54

*Additive content, wt.%.

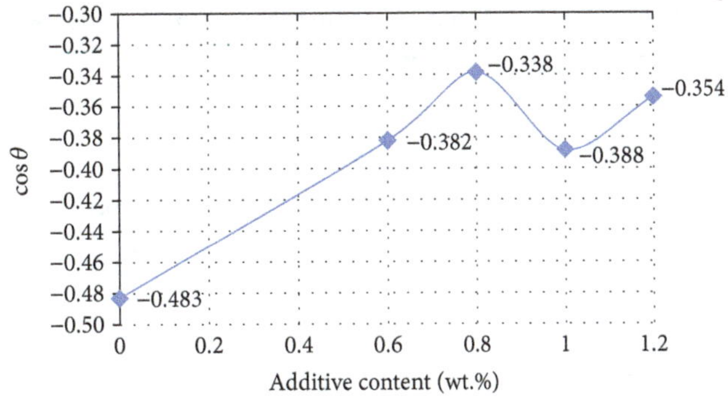

FIGURE 5: Dependence of the glass surface wetting contact angle on the content of the PFM.

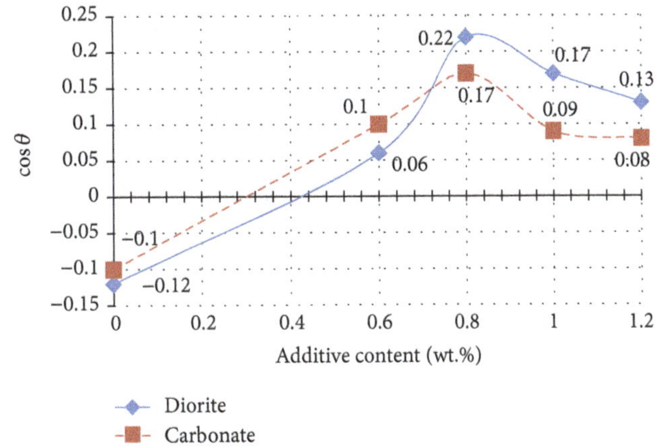

FIGURE 6: Dependence of the substrates surface wetting contact angle on the content of the PFM.

where PCA is polycyclic aromatics and MCA is monocyclic aromatics.

One of the most important characteristics of oil disperse system (ODS) is stability, since it affects the ability of the system to resist internal processes of interaction of particles with each other, which can lead to changes in the particle size of the dispersed phase and maintain an equilibrium distribution of dispersed phase particles in the medium [32]. Therefore, the radius of the particles of the dispersed phase can reach only certain limiting sizes, because the process of stratification can begin because of the difference in the densities of the phase and the medium. The sizes of dispersed particles can change under the influence of external factors (like temperature fluctuations, the action of shear deformations of the medium, the action of various additives, etc.) [32]. In this connection, it was necessary to examine the change in sizes of dispersed particles of bitumen affected by PFM doping. The results obtained are presented in Figure 7.

As can be seen from the results of measurements, in the medium of toluene-heptane mixture, regardless of the ratio of solvents, the smallest particle diameter corresponds to the additive in bitumen composition equal to 0.8% by weight. This is due to the fact that the additive provides the destruction of aggregate combinations with the formation of smaller structures together with their solvation. This may indicate that the bitumen system with this PFM content is in a state that can be conditionally called an "active state" (the core is minimal and the thickness of the solvate shell is maximal) [32]. Apparently, with a further increase in the dosage of the additive, a phase inversion, that is, enlargement of particles, takes place.

Thus, the PFM doping into bitumen and its heating at 120–130°C alters the behavior of some hydrocarbon groups in comparison with the behavior of these groups in the absence of an additive. Based on the results of the group chemical composition, it can be concluded that the additive is distributed mainly between monocyclic aromatic hydrocarbons and asphaltenes. As a result of the study of the particle size, it has been established that the system is in the most "active state" with an additive content of PFM in an amount of 0.8% by weight.

3.4. Rheological Properties of Modified Bitumen. The rheological properties of the ODS essentially depend on their composition and primarily on the content of the main

TABLE 9: Group chemical composition of the original and modified bitumen.

Components, wt.%	Bitumen samples				
	Original	PFM content, wt.%			
		0.6	0.8	1.0	1.2
Saturates (S)	22.62	18.45	18.63	14.18	14.2
*Aromatics (MCA)**	*6.88*	*8.74*	*9.77*	*12.65*	*12.07*
*Aromatics (BCA + PCA)**	*8.26*	*7.98*	*7.66*	*6.85*	*7.92*
Aromatics (A)	15.14	16.72	17.43	19.5	19.99
Resins (R)	40.26	39.47	39.07	40.65	39.92
Asphaltenes (A)	21.98	25.36	24.87	25.67	25.89

*MCA: monocyclic aromatics, BCA + PCA bicyclic and polycyclic aromatics.

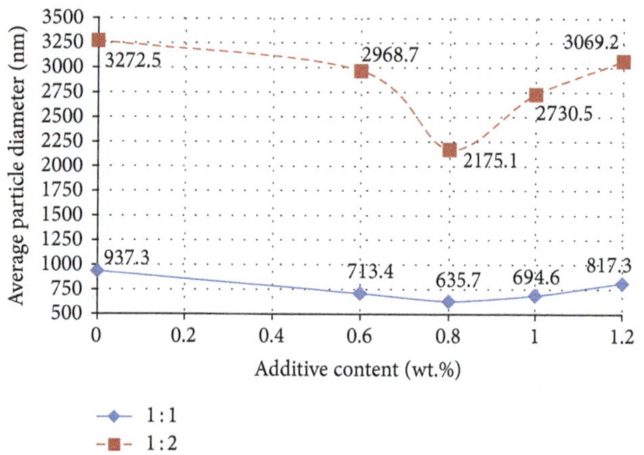

FIGURE 7: Dependence of the average diameter of dispersed particles from the content of the additive in toluene : heptane medium.

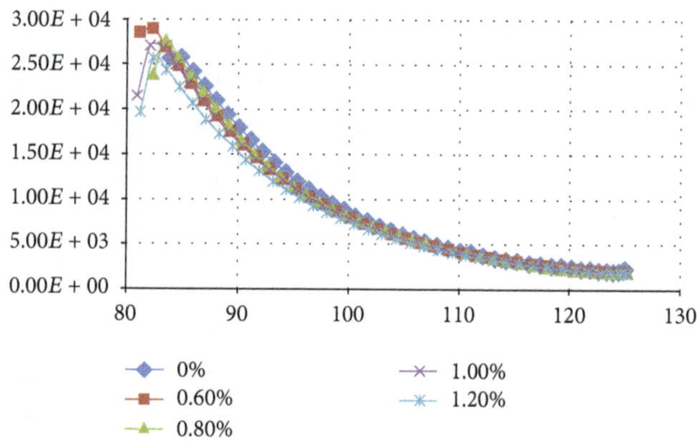

FIGURE 8: Dependence of viscosity on temperature.

structure-forming components, which include the most high-molecular components, resinous-asphaltene substances (RASs), as well as paraffins at temperatures below their crystallization temperature, which can significantly influence on the properties of bitumen [31–33].

Viscosity increases in proportion to the volume of particles of the dispersed phase and the degree of their aggregation [31–33]. Aggregation of particles leads to an increase in apparent viscosity, since aggregates reduce the effective cross-section of the flow channel and create additional inhibition of the displacement of the fluid layers [33]. Destruction of the structure or deformation of solvate shells causes an anomaly of viscosity.

Effective regulators of viscosity and the ultimate shear stress of asphaltene-containing ODS are surfactants that exert peptizing action, blocking the adhesion planes of particles and decreasing the strength of contact between the layers of the dispersion medium through which the particles of the disperse phase contact [33].

The viscosity of bitumen depends not only on the composition of bitumen and its temperature, but also on the duration and nature of the action of the load. The graph of viscosity change with temperature is shown in Figure 8. As can be seen, when all the samples are cooled, the viscosity increases equally, but in the initial bitumen, the fracture, that is, destruction of the structure, begins at a higher temperature (85°C) than in the modified ones. This temperature corresponds to the highest viscosity of the original unmodified bitumen and is equal to $2.58 \cdot 10^4$ mPa·s.

Figure 9 shows the thixotropy curves of bitumen samples at different temperatures in semilogarithmic coordinates. It should be noted that, at temperatures of 80–90°C, viscosity increase is observed in samples of modified bitumen. The additive can promote formation of spatial structures due to the formation of complex structural unit (CSU), conjugated with the available particles of the dispersed phase [32]. In this case, the properties of the binder will be determined by this new spatial conjugate structure [32].

We assume that the additive provides destruction due to peptization of aggregate combinations of asphaltenes with the formation of highly disperse RASs; that is, asphaltenes disperse into smaller structures. This conforms to data on the average particle diameter (Section 3.3), in which the smallest particle diameter corresponds to the dosage of the additive in the bitumen of 0.8 wt.%. As a result, the amount of particles increases and thereby the viscosity increases due to the larger contact surface. Summarizing the foregoing, we should note that the additive has a structuring effect on the properties of road bitumen that is expressed in the formation of a spatial dispersed structure, conjugated with asphaltene complexes of bitumen [31]. The structuring effect is manifested with a certain concentration and only in the presence of a spatial

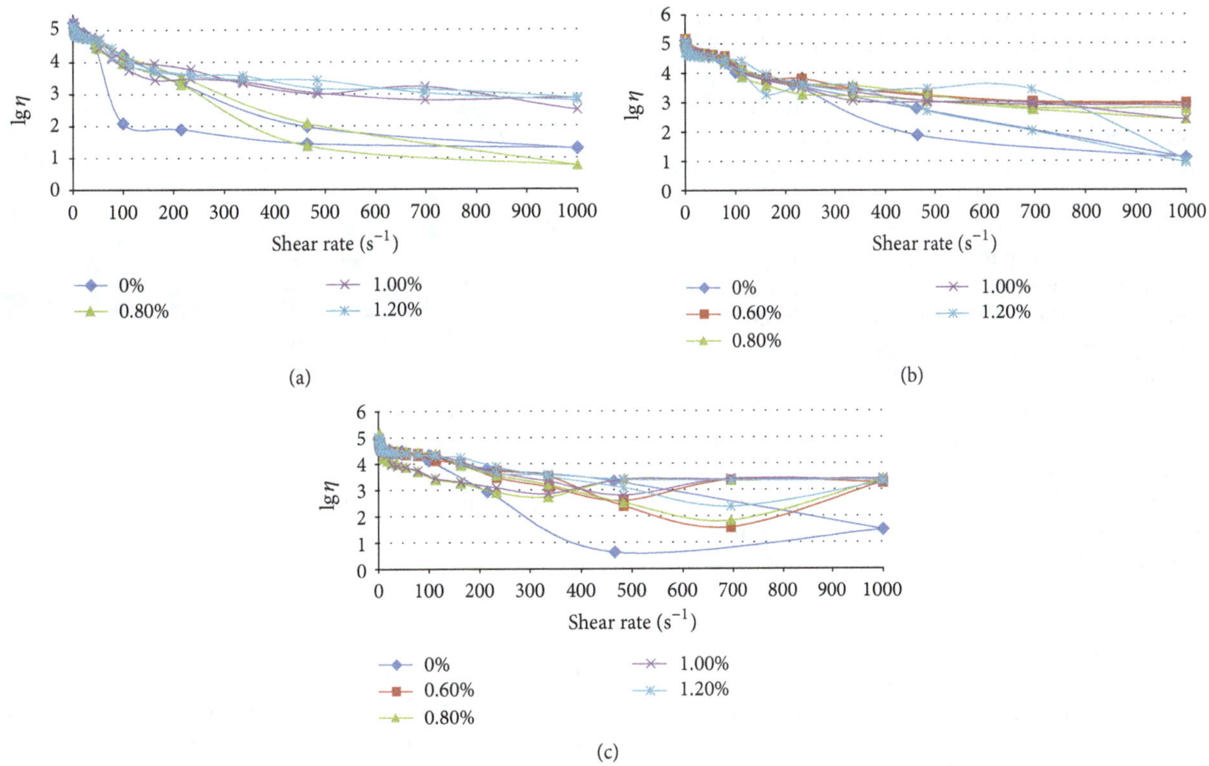

FIGURE 9: Curves of thixotropy of samples at temperatures, °C: (a) 80; (b) 85; (c) 90.

TABLE 10: Results of laboratory tests of oxidized bitumen 60/90.

Number	Indicators	GOST 22245-90 requirements	Original	Bitumen samples Adhezoline content, wt.% 0.6	0.8	1.0	1.2
(1)	Penetration, 0.1 mm:	Not below					
	at 25°C	61–90	73	75	81	86	78
	at 0°C	20	25	30	30	25	23
(2)	Softening temperature (T_{soft}), °C	Not below 47	51.4	49.8	49.6	50	49
(3)	Ductility, cm	Not less than					
	at 25°C	55	89	128	116	110	120
	at 0°C	3.5	4.5	4.9	5.0	4.5	4.9
(4)	Adhesion to crushed stone, points						
	(1) Pervouralsk	Is not regulated	2	4	4	5	5
	(2) Sangalyk		2	4	4	5	5
	(3) Bianca		3	5	5	5	5
	(4) Valegin Bor		2	4	5	5	5
(5)	Penetration index	From −1 to +1	0.1	−0.2	−0,1	0,2	−0,3
(6)	Brittleness temperature, °C	Not higher than minus 15	Minus 22.9	Minus 23.7	Minus 25.7	Minus 29.1	Minus 28.8

coagulation structure in the bitumen [31]. It is likely that this specific concentration in the bitumen is 0.8 wt.%, which we called the "active state" (Section 3.3).

3.5. The Effect of PFM on Bitumen Performance Properties. The results of laboratory tests carried out on bitumen modified using 0.6, 0.8, 1.0, and 1.2 wt.% of additive are presented in Table 10 in comparison with the GOST 22245-90 requirements.

The interrelation between the changes in the quality indicators of bitumen given in Table 10 can be explained on the basis of the specificity of the action of the synthesized reagent and its subsequent polymerization activity when the group component composition and the particle size of the dispersed phase change. Thus, according to T_{soft} value changes, the concentration of PFM of 0.6 wt.% is characterized by an increase in the structure formation (T_{soft} = 128°C) and an increase in the mechanical strength of the ODS; with an increase in the concentration of PFM up to to 0.8–1.0 wt.% we observe a sharp drop in the value of softening point, to 116°C and 110°C, respectively. Apparently, this indicates the disaggregation of resinous-asphaltene particles (RAS) in the dispersion medium of the modified bitumen by the PFM.

The indicator "penetration index" (PI) is considered as a reflection of the colloidal structure of bitumen. Taking into account the fact that with the change of penetration of bitumen the ductility indicator passes through the maximum value, the penetration index is limited from −1 to +1. It is known that the PI indicator accurately characterizes the elastic properties of bitumen and their dependence on temperature [15, 31]. For the studied samples, the PI indicator changes from +0.1 (for the original unmodified bitumen) to the values of −0.2; −0.1; +0.2 for the concentrations (wt.%) of PFM equal to 0.6; 0.8; 1.0, respectively. Thus, the increase in the content of PFM up to 1.0 wt.% increases the elasticity of modified bitumen. At the same time, an important trend is observed: the further increase in the concentration of additive up to 1.2 wt.% leads to a sharp decrease in PI to −0.3, which is directly proportional to the penetration indicator. This indicates the activation of the structure formation of the disperse system of modified bitumen.

The use of bitumen as film-forming base in asphalt materials is justified primarily by the fact that the rate of diffusion of water into the bitumen is $0.83 \cdot 10^{-15}$–$1.66 \cdot 10^{-15}$ kg·m/(s·N) [5]; this causes low solubility of water in the bitumen. As mentioned above, the degree of adhesion between the bitumen and the stone material depends on the chemical nature of the components and, therefore, the variety of types of bitumen and the stone material [4, 16].

The process of adhesive bond forming is usually divided into two stages. The first is the so-called transport stage, during which adhesive molecules move to the surface of the substrate (body on which the adhesive is applied) and the functional groups of the adhesive interact, orientating in the interphase layer. This results in intimate contact of molecules and/or associates of the adhesive with the substrate. An increase in temperature and pressure, as well as conversion of one of the phases (typically an adhesive) to the liquid state by dissolving or melting, contributes to the flow of the first stage of the adhesion process. To achieve a more intimate contact between the adhesive and the substrate we can apply more thorough cleaning of the interacting surfaces. The adhesion second step consists in direct intermolecular interaction of the adhesive and substrate, which may be caused by the different forces, from van der Waals to chemical forces. Intermolecular interactions of contacting phases complete the adhesion process, and that corresponds to the minimum of interfacial (surface) energy [16].

We have developed a polyfunctional modifier of complex action that is a nonclassical cationic surfactant, as mentioned above. Based on the foregoing conceptualization, it is possible to assume that the efficiency of PFM is due to the presence of discontinuities in hydrophobic hydrocarbon chains caused by presence of polar heteroatomic moieties which provide the required hydrophilic-lipophilic balance (HLB) as well as

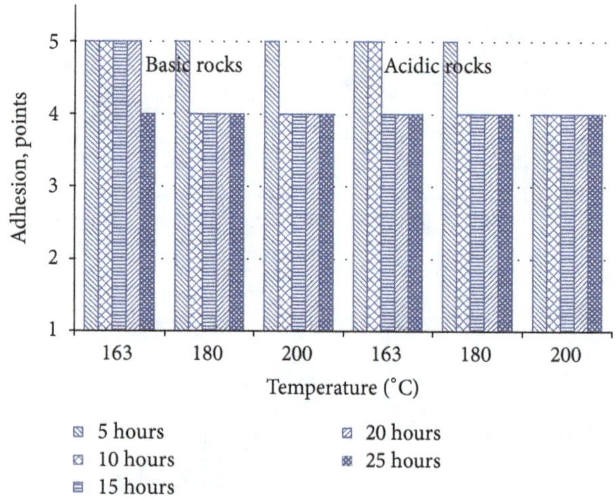

FIGURE 10: Thermal stability of bitumen modified with the use of PFM.

the optimal location of these polar groups and hydrophobic hydrocarbon radicals in the adhesive molecule. In addition, in this case the length of the nonpolar hydrocarbon chains and their location has a great influence. All this allows us to provide the most effective interaction of PFM with the surface of the dispersed phase and the volume of the dispersion medium.

According to the existing technology of preparation of asphalt mixtures used in the asphalt plants, asphalt undergoes high temperatures. The essential drawback of adhesive additives currently used in road construction is their thermal instability (up to 140–150°C).

The results of research for the thermal stability of bitumen modified by PFM (as determined above, the optimal amount of 0.8 wt.% was doped) are presented in Figure 10.

As can be seen from the data presented in Figure 10, the temperature control at 163°C for 20 hours provides the adhesion degree evaluated with 5 points, which corresponds to the control sample No. (1). Increasing the heating duration at the same temperature for up to 25 hours leads to some reduction in adhesion. An influence of temperature of 180°C and 200°C for 10 hours and more leads to a decrease in adhesion down to 4 points. It should be noted that the bitumen without additive has shown insufficient adhesion both with basic and acidic rocks, namely, 3 points after thermal exposure for 5 and 10 hours and 2 points after 15–25 hours thermal exposure (at all the temperatures).

Thus, the synthesized polyfunctional modifier significantly improves the adhesion of the bitumen to the surface of the mineral material and pavements with enhanced thermal stability.

We had conducted pilot-scale tests with the road constructing organization ZAO Trest Kamdorstroy. The results of technical inspection have shown that asphalt concrete mixtures obtained with the use of modified bitumen meet the GOST 22245-90 specification quality requirements.

Field tests carried out to assess the physical and mechanical characteristics of asphalt concrete mixtures based on bitumen BND 60/90 produced by JSC "TAIF-NK," modified using 1 wt.% of PFM, taken from the coating after two years and 9 months of operation of the roadway, have shown that these asphalt mixes fully comply with GOST 9128-2009 "Mixtures of asphalt concrete road, airfield and asphalt concrete."

Thus, the results obtained give grounds to judge the increase in the between-repair time of the road surface, and, as a consequence, its durability as a result of the use of PFM-modified bitumen.

4. Conclusion

The experimental data indicate a high capacity of N-acylated derivatives of bisimidazolines to improve the adhesion of bitumen to mineral materials, as well as the plasticity of bitumen, while providing increased (up to 220–240°C) thermal stability. The polyfunctional modifier can be used as complex additives for bitumen used in road and civil construction, regardless of the nature of used aggregates (basic or acidic) and regardless of the change in composition of the original bitumen.

Thus, it was found that the presence of an adhesive additive in the bitumen significantly changes the wetting of the mineral substrate with modified bitumen. The wetting inversion is observed; $\cos\theta$ increases with increasing content of the additive; it acquires positive values; the work of adhesion increases, reaching a peak with an additive content of 0.8% by weight; the work of cohesion and the surface tension between the liquid and solid phases decrease.

It has been established that the smallest particle diameter corresponds to the dosage of the additive in the bitumen of 0.8% by weight, so that it can be concluded that the additive has a structuring effect on the properties of road bitumen.

The use of PFM leads to improved performance properties enabled by lowering the binder sensitivity to temperature changes and load time susceptibility as well as providing high adhesion between the binder and aggregates at elevated temperatures in combination with high elasticity at low temperatures.

Finally, it should be noted that, from January 2015 to January 2016, the experiments were conducted to evaluate the adhesive properties of bitumen in various weather conditions. So, on January 14, 2015, a previously prepared sample of bitumen modified by PFM in the amount of 1 wt.% was placed outside and kept under different weather conditions for one year, after which the sample of the bitumen was monthly examined for its adhesion properties. Over a year, outside temperatures changed from minus 25°C in winter to plus 35°C in the summer. The obtained data on the evaluation of adhesion proved to be quite satisfactory. Bitumen has not lost its adhesive properties during the whole period of storage in the open air at various temperatures, which can also indirectly confirm the increased durability of this kind of bitumen and asphalt concrete on its basis.

Conflicts of Interest

The authors declare that they have no conflicts of interest.

Acknowledgments

The work is performed according to the Russian Government Program of Competitive Growth of Kazan Federal University. This work was funded by the subsidy allocated to Kazan Federal University for the state assignment in the sphere of scientific activities (10.7636.2017/7.8).

References

[1] R. Hunter, *The Shell Bitumen Handbook*, ICE Publishing, London, UK, 6th edition, 2014.

[2] L. M. Gokhman, "Bitumens, bitumen-polymer binders, asphalt concrete, polymeric asphalt concrete, study guide" (Russian), p. 117, ZAO "ECON-INFORM", 2008.

[3] S. Manolis, "Engineering properties of asphalt cement binders and their relation to pavement performance," Royal Military College of Canada, Coco Paving Inc., 2014, http://www.asphalt-institute.org/wp-content/uploads/Thickness_Mix/2014-pptbyManolis-EngineeringPropertiesofACBindersandRelationtoPavement-Performance.pdf.

[4] A. F. Burger, M. F. C. van de Ven, J. Muller, and K. J. Jenkins, "Rheology of polymer modified bitumen: A comparative study of three binders and three Binder/filler systems," in *Proceedings of the 20th South African Transportation Conference "Meeting the Transport Challenges in Southern Africa"*, 2001.

[5] A. F. Kemalov and R. A. Kemalov, "Study of disperse polymer systems for producing high-quality polymeric-bituminous materials," *Chemistry and Technology of Fuels and Oils*, vol. 48, no. 5, pp. 339–343, 2012.

[6] X. Lu, Y. Talon, and P. Redelius, "Aging of bituminous binders – laboratory tests and field data," in *Proceedings of the E&E Congress*, Copenhagen, Denmark, 2008.

[7] S. Cui, B. R. K. Blackman, A. J. Kinloch, and A. C. Taylor, "Durability of asphalt mixtures: Effect of aggregate type and adhesion promoters," *International Journal of Adhesion and Adhesives*, vol. 54, pp. 100–111, 2014.

[8] A. Chomicz-Kowalska, W. Gardziejczyk, and M. M. Iwański, "Moisture resistance and compactibility of asphalt concrete produced in half-warm mix asphalt technology with foamed bitumen," *Construction and Building Materials*, vol. 126, pp. 108–118, 2016.

[9] A. A. Abramzon, V. V. Bocharov, G. M. Gayevoy et al., *Surfactants*, Khimiya, Leningrad, Russia, 1979, (Russian).

[10] D. N. Rubingh and P. M. Holland, *Cationic Surfactants: Physical Chemistry*, vol. 37 of *Surfactant Science*, Marcel Dekker Inc., New York, NY, USA, 1990.

[11] J.-Z. Ai, X.-P. Guo, J.-E. Qu, and Z.-Y. Chen, "Adsorption behavior of imidazoline amide on the surface of galvanic electrode," *Acta Physico - Chimica Sinica*, vol. 21, no. 10, pp. 1096–1101, 2005.

[12] V. V. Krut', A. B. Solomentsev, V. P. Kolodezny et al., "Adhezion additive of polyfunctional action," pat. 2130954 Russian Federation: IPC C08L95/00, C08K5/3415/ applicant and patent holder OAO Oreldorstroy - No. 97100884/04.

[13] A. F. Kemalov, R. A. Kemalov, and I. M. Abdrafikova, "The use of plant oils for gaining the colored plastoconcrete," *Biosciences Biotechnology Research Asia*, vol. 11, no. 3, pp. 1699–1706, 2014.

[14] S. Kishchynskyi, V. Nagaychuk, and A. Bezuglyi, "Improving Quality and Durability of Bitumen and Asphalt Concrete by Modification Using Recycled Polyethylene Based Polymer Composition," *Procedia Engineering*, vol. 143, pp. 119–127, 2016.

[15] I. I. Mukhamatdinov, *Bitumen binders modified with the use of cationic-active adhesion additive [Ph.D. Thesis]*, Kazan Federal University, 2015.

[16] NPCS Board of Consultants & Engineers, *The Complete Technology Book on Industrial Adhesives*, Asia Pacific Business Press Inc., Delhi, India, 2008.

[17] P. S. Fakhretdinov, "The concept of non-classic cationic surfactants," in *Proceedings of the XVIII Mendeleev Congress on General and Applied Chemistry*, vol. 1, p. 483, Moscow, Russia, 2007.

[18] P. S. Fakhretdinov and G. V. Romanov, "The concept of non-classic surfactants is the basis for the creation of new highly effective reagents for the petroleum industry," in *Proceedings of the the 2nd Russian Scientific-Practical Conference "Practical Aspects of oilfield chemicals"*, p. 104, Ufa, Russia, 2012.

[19] A. F. Kemalov and R. A. Kemalov, "Practical Aspects of Development of Universal Emulsifiers for Aqueous Bituminous Emulsions," *World Applied Sciences Journal*, vol. 23, no. 6, pp. 858–862, 2013.

[20] A. P. Kreshkov, L. N. Bykova, and N. A. Kazaryan, *Acid-Base Titrations in Non-Aqueous Solutions*, Khimya, Moscow, Russia, 1967.

[21] B. N. Tarasevich, *IR Spectra of the Main Classes of Organic Compounds. Reference Materials*, M.V. Lomonosov Moscow State University, Moscow, Russia, 2012.

[22] GOST 12801-98, "Materials based on organic binders for road and airfield construction. Test methods," 1999.

[23] A. W. Neumann and R. J. Good, "Techniques of measuring contact angles," in *Surface and Colloid Science*, R. J. Good and R. R. Stromberg, Eds., vol. 11, pp. 31–91, Springer, New York, NY, USA, 1979.

[24] GOST 22245-90, "Viscous petroleum road bitumens. Specifications," 1991.

[25] GOST 18180-72, "Petroleum bitumens. Method for determination of mass change after heating," 1973.

[26] GOST 3900-85, "Petroleum and petroleum products. Methods for determination of density," 1987.

[27] GOST 33-2000, "Petroleum products. Transparent and opaque liquids. Determination of kinematic viscosity and calculation of dynamic viscosity," 2002.

[28] ASTM D 2872, "Standard test method for effect of heat and air on moving film of asphalt (Rolling thin-film oven test)," Tech. Rep.

[29] GOST 9128-2009, "Asphalt concrete mixtures for roads and airdromes, and asphalt concrete. Technical specifications," 2011.

[30] R. B. Gun, *Petroleum Bitumen*, Chemistry, Moscow, Russia, 1973.

[31] S. M. Petrov, *Modifiers of multifunctional action for obtaining oxidized road bitumen with improved properties [Ph.D. Thesis]*, Kazan State Technological University, Kazan, Russia, 2009.

[32] B. P. Tumanyan, *Scientific and Applied Aspects of the Theory of Oil Dispersed Systems*, Tekhnika, Moscow, Russian, 2000.

[33] G. I. Fuchs, *Viscosity and Plasticity of Petroleum Products*, Institute for Computer Research, Moscow-Izhevsk, 2003.

7

Modeling of Reinforced Concrete Beams Exposed to Fire by Using a Spectral Approach

Rui Sun,[1] **Bo Xie**,[1] **Ricardo Perera**,[2] **and Yongjun Pan**[3]

[1]*Key Laboratory of New Technology for Construction of Cities in Mountain Area, Ministry of Education, School of Civil Engineering, Chongqing University, Chongqing 400045, China*
[2]*Department of Mechanical Engineering, Technical University of Madrid, José Gutiérrez Abascal 2, 28006 Madrid, Spain*
[3]*School of Automotive Engineering, Chongqing University, Chongqing 400044, China*

Correspondence should be addressed to Rui Sun; ruisun@cqu.edu.cn

Academic Editor: Patrice Berthod

This paper presents a numerical procedure for predicting the mechanical behavior of reinforced concrete (RC) beams exposed to fire. Thermal behavior is captured using a two-dimensional finite element (FE) model in an environment of elevated temperatures, while a one-dimensional spectral numerical model is formulated to simulate the mechanical response of the beam under increasing loads. Both models are integrated. Previous thermal results obtained from the FE model are provided for the proposed numerical model to calculate the responses of the RC beam for different levels of fire exposure. Few elements are deployed in this study due to the simplicity and efficiency of the proposed numerical model. Consequently, the computational burden of the numerical simulation for predicting the structural performances of RC beams exposed to fire is released comparing with conventional FE model. The numerical results are consistent with the test data, which demonstrate the model's capability of efficiently simulating the thermomechanical behaviors of RC beams due to its simplicity and accuracy.

1. Introduction

Behavior of RC buildings exposed to fire during their service lifetime is one of the crucial topics that need to be addressed. Degradation of the mechanical properties as well as the residual resistance of RC members is essential for the structural performance under fire [1]. Design codes [2–4] have been developed providing reliable guidelines that can conveniently be employed for the analysis and design of RC structures under fire. They propose simplified expressions empirically obtained from experimental data by means of a standard fire test [5].

As design codes are usually derived from empirical approaches and based on limited experimental results, they are conservative enough from a design point of view. Additionally, they focus on the global behavior of RC members, without addressing possible local phenomena. The constitutive model of concrete, boundary conditions, geometrical configuration, and concrete cracking, which have a significant influence on the performance of RC structures, is not suitably addressed. To understand the mechanical response of materials exposed to fire, experimental tests, in which the residual properties of materials are usually tested after cooling down for the convenience of measurement, are performed [6, 7]. The results directly illustrate the reduction of mechanical strength of concrete and reinforced steel under high temperatures. However, although the strength capacity of steel can become recovered, the same is not true for concrete. The interfacial bond strength between steel and confined concrete under elevated temperatures reduces even faster than steel according to experimental and numerical studies [8–10], whereas satisfactory accuracy may be achieved using perfect bond assumption for the prediction of global responses of RC structures in an environment of fire or elevated temperatures.

Destructive fire tests have been performed [11–14] to experimentally obtain the residual load-carrying capacity of RC members, in which RC beams are subjected to standard fire tests and loaded to failure as soon as they cool down to room temperature. Many factors influence the degradation of the structural strength or stiffness of RC beams at elevated temperatures, such as support conditions, load level, geometrical properties, and fire scenarios. Thus, the residual strengths of RC members exposed to fire that are experimentally obtained vary over an extensive range due to these uncertainties.

The difficulty of measurement and the cost of experimental tests promote the development of numerical models, which can be conducted as an economical choice for evaluating the residual resistance of RC beams in a nondestructive manner. Simplified cross-sectional analysis has been developed by researchers [15–18]; reduction factors are introduced to strength equations at room temperature to evaluate the residual capacities of RC beams with increasing temperature. However, this methodology may cause the underestimation of the bearing capacities of RC beams since the strain hardening of the reinforcement is not considered. As the most extensively applied numerical method, finite element model is constructed to study the degradation of the structural strengths and residual capacities of RC members exposed to fire. A thermomechanical model that employs a procedure of heat transfer and mechanical response analysis in sequence has been already proposed [19–23]; the temperature-dependent constitutive relationships are utilized to represent the reduction of material properties of concrete and reinforcement at elevated temperatures. The results demonstrate that FE models are appropriate for predicting the mechanical response of RC members that are exposed to fire and increasing loads.

Although satisfactory results can be achieved by performing complex FE analyses, the computational costs would undoubtedly increase since a specific constitutive model would be required for each different level of temperature. To overcome this drawback, a two-step simplified approach is presented in this study to predict the structural performance of RC beams exposed to fire. The first step corresponds to a typical thermal analysis performed with a two-dimensional FE model. With this analysis, temperature distribution on the beam is obtained under fire conditions. Subsequently, a one-dimensional spectral element model is developed to perform the mechanical analysis of the RC beam exposed to fire. For it, the beam cross section is divided into several segments with the purpose of including the temperature effects on the material strength throughout the entire cross section. In this way, the mechanical behavior and the degradation of the load-carrying capacity of RC beams after fire exposure can be efficiently evaluated in a nonexpensive way due to the simplicity of the proposed spectral model. The accuracy of the proposed approach is validated by comparing its predictions with experimental tests and more complex FEM analyses published previously [23].

2. Spectral Element Model for RC Beams

Although more complex FE models [16, 24–26] have been established to study the mechanical behavior of RC structures at elevated temperatures, one of the main objectives of this study is to develop a simplified numerical approach able to predict the global response of RC beams exposed to fire and simultaneously releasing the computational burden. A one-dimensional spectral element model will be implemented in this study to analyze the mechanical behaviors of RC beams; it will be briefly presented since it has been detailed in [27, 28], to simulate RC beams that are externally bonded with fiber-reinforced polymer strips and interfacial debonding damage. First, the interfacial slip between the steel reinforcement and confined concrete is not considered, which indicates that full composite action is assumed to study the global mechanical responses of RC beams in a simplified way. The coordinates and kinematic of the transverse section of RC beam are shown in Figure 1, and the axial and transverse displacements of a beam can be expressed using first-order shear deformation theory:

$$u_C(x,z,t) = u_0(x,t) - z\phi(x,t), \tag{1}$$

$$w(x,z,t) = w(x,t), \tag{2}$$

where u_0 and u_C are the axial displacement at midplane and a certain point with a depth of z from midplane, respectively, and w and ϕ denote the transverse displacement and rotation of the cross section about the y-axis. The stress of the concrete and steel bars can be formulated according to the perfect bond assumption as follows:

$$\begin{aligned} \sigma_C &= E_C\varepsilon_C, & \varepsilon_C &= u_{0,x} - z_C\phi_{,x}, \\ \sigma_B &= E_B\varepsilon_B, & \varepsilon_B &= u_{0,x} - z_B\phi_{,x}, \\ \tau_C &= G_C\gamma_C, & \gamma_C &= w_{,x} - \phi, \end{aligned} \tag{3}$$

where σ and τ represent the axial stress and shear stress; the subscripts C and B represent the concrete and steel bars, respectively; ε_C and ε_B are the longitudinal strain of concrete and the longitudinal strain of the bars derived from (1) with the coordinates z_C and z_B, respectively; and the elastic modulus of concrete and steel reinforcement E_C and E_B and shear modulus of concrete G_C are highly dependent on the stress-strain constitutive relationship, which would change significantly with the variation of temperature.

Using Hamilton's principle, the differential equation of motion, which is related to the variables u_0, w, and ϕ, can be derived based on the following expression:

$$\begin{aligned} \delta u_0 &: I_0\ddot{u}_0 - I_1\ddot{\phi} - A_{11}u_{0,xx} + B_{11}\phi_{,xx} = 0, \\ \delta w &: I_0\ddot{w} - A_{22}w_{,xx} + A_{22}\phi_{,x} = 0, \\ \delta\phi &: I_2\ddot{\phi} - I_1\ddot{u}_0 + B_{11}u_{0,xx} - D_{11}\phi_{,xx} - A_{22}w_{,x} + A_{22}\phi = 0. \end{aligned} \tag{4}$$

The boundary force equations are expressed as

$$\begin{aligned} N &= A_{11}u_{0,x} - B_{11}\phi_{,x}, \\ V &= A_{22}w_{,x} - A_{22}\phi, \\ M &= -B_{11}u_{0,x} + D_{11}\phi_{,x}, \end{aligned} \tag{5}$$

where N, V, and M are the boundary forces associated with u_0, w, and ϕ, respectively; $(\ )_{,x}$ and $(\cdot)$ represent the differentiation with respect to length and time; A_{11}, B_{11}, D_{11}, and A_{22} are the stiffness coefficients defined by the material properties; and I_0, I_1, and I_2 are related to the inertial items.

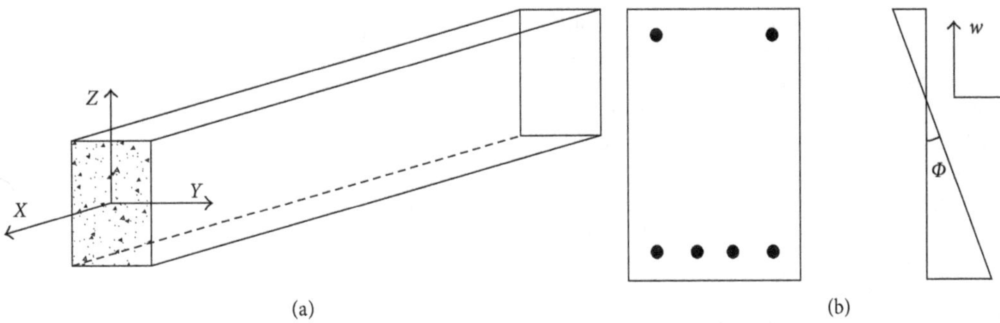

FIGURE 1: Coordinates and kinematic of the transverse section of RC beam.

Unlike the conventional FE model, one of the main differences of the proposed spectral element method (SEM) is the implementation of a fast Fourier transform (FFT), which is capable of transferring the displacement field from the time domain to the frequency domain. The three variables are formulated as

$$\{u_0, w, \phi\} == \sum_{n=1}^{N}\sum_{m=1}^{M}\{\widehat{u}_{mn}^{*}\}e^{-jk_{mn}x}e^{-j\omega_n t}, \quad (6)$$

where ω_n denotes circular frequency at the nth sampling point, and k_{mn} is the mth wave number related to ω_n in FFT. $\{\widehat{u}_{mn}^{*}\} = (\widehat{u}_0, \widehat{w}, \widehat{\phi})_{mn}$ represents the vector of wave coefficients related to the mth mode of the wave at the frequency point ω_n. By introducing (6), governing equations (4) turn into the following formula:

$$[W]\{\widehat{u}^{*}\} = 0, \quad (7)$$

where W is a matrix with 3×3 dimensions in function of frequency and wave number as follows:

$$[W] = \begin{pmatrix} A_{11}k^2 & 0 & -B_{11}k^2 + I_1\omega_n^2 \\ 0 & A_{22}k^2 - I_0\omega_n^2 & -jA_{22}k \\ -B_{11}k^2 + I_1\omega_n^2 & jA_{22}k & D_{11}k^2 - I_2\omega_n^2 + A_{22} \end{pmatrix}. \quad (8)$$

For each frequency point, the wave number k_{mn} can be obtained by solving a 6th order equation when W equals zero in (8), and the six related eigenvectors $R_i (i = 1, \ldots, 6)$ can be used to formulated the general solution of nodal displacement as

$$\begin{aligned}\{u(x)\} &= \begin{pmatrix} \hat{u}_0(x) \\ \hat{w}(x) \\ \hat{\phi}(x) \end{pmatrix} \\ &= [\{R_1\}\cdots\{R_6\}]\ \underset{i=1,\ldots,6}{\mathrm{diag}}\left(e^{-jk_i x}\right)\{A\} \\ &= [R][D(x)]\{A\}. \end{aligned} \quad (9)$$

Furthermore, the displacements of two nodes in an element with the length of L can be expressed as

$$\begin{Bmatrix} u_1 \\ u_2 \end{Bmatrix} = \begin{bmatrix} R_1 \\ R_2 \end{bmatrix}\begin{bmatrix} D(0) \\ D(L) \end{bmatrix}\{A\} = [T_1]\{A\}, \quad (10)$$

where $\{A\}$ is the vector of constants which depends on the associated boundary conditions. Substituting (10) into (9), the solutions of nodal displacements can be obtained as

$$\{u(x)\} = [R][D(x)][T_1]^{-1}\begin{Bmatrix} u_1 \\ u_2 \end{Bmatrix} = [N]\begin{Bmatrix} u_1 \\ u_2 \end{Bmatrix}, \quad (11)$$

where $[N]$ is the matrix of shape functions of the proposed spectral model. By combining boundary force equations (5), the dynamic stiffness matrix (DSM) $[K]$ is obtained to calculate the displacement in frequency field:

$$\begin{Bmatrix} \{f_1\} \\ \{f_2\} \end{Bmatrix} = \begin{bmatrix} F(0) \\ F(L) \end{bmatrix}\{A\} = [T_2][T_1]^{-1}\begin{Bmatrix} u_1 \\ u_2 \end{Bmatrix} = [K]_{6\times 6}\begin{Bmatrix} u_1 \\ u_2 \end{Bmatrix}. \quad (12)$$

Because the governing equations and boundary conditions were previously derived in the frequency domain, the DSM in (12), which is frequency-dependent, can be established. Compared with the conventional FE model, the DSM may be considered as a combination of a mass matrix and a stiffness matrix using exact dynamic shape functions, which are obtained according to the governing differential equations. Considering the variation in the DSM, which depends on frequency, one spectral element sufficiently represents a structure as long as its material and geometrical properties remain uniform throughout the entire element, which indicates that the mesh refinement in a high-frequency range is no longer necessary leading to significant reduction of the degree of freedom and computational costs. Therefore, the mechanical behaviors of a structure can be captured in a fast way with a very small number of elements due to its simplicity and efficiency. The static behaviors of RC beams can be directly calculated by making frequency value in DSM tend to zero to prevent numerical problems in this study. Details of the proposed model can be checked in [27, 28]. The temperature-dependent material constitutive model is discussed in Section 4.

3. Transient Thermal Analysis of RC Beams

In this study, a thermomechanical analysis of RC beams is performed in two steps: firstly, a thermal analysis and secondly, a mechanical analysis. Temperature distributions obtained in the first step will be taken as a basis for a subsequent mechanical analysis. To simplify the thermal analysis, the air temperature has been assumed spatially homogeneous, and the longitudinal thermal transfer along the length of beam has been disregarded. Furthermore, in this study, concrete is treated as homogeneous material which is represented by a stress-strain relationship. According to this assumption, the thermal distribution would be uniform along the length of the RC beam, since the nonuniform distribution of components of concrete (e.g., cement and aggregates) is neglected in a macroscopic perspective. Hence, a 2D transient thermal analysis throughout a typical cross section is representative for predicting the temperature distribution, which varies along the length of the RC beam similarly, and the thermal problem is reduced to a 2D problem as in [29, 30].

When a structure is exposed to fire, convection, radiation, and conduction phenomena must be considered. While convection and radiation involve heat flux through the surface of the structure, conduction involves heat transfer through the body of structure. The three phenomena should be appropriately considered in a transient thermal analysis, in which heat conduction is governed using partial differential equations [1] as

$$\frac{\partial}{\partial x}\left(k\frac{\partial T}{\partial x}\right)+\frac{\partial}{\partial y}\left(k\frac{\partial T}{\partial y}\right)+\frac{\partial}{\partial z}\left(k\frac{\partial T}{\partial z}\right)+Q-\rho c\frac{\partial T}{\partial t}=0, \tag{13}$$

where T is the temperature distribution with respect to xyz coordinate, ρ is the material density, and k and c represent the thermal conductivity and specific heat, respectively, which are highly temperature-dependent. The specific volumetric heat source Q equals zero since no inherent heart source exists in this case. The initial temperature is needed to solve the differential equation, which can be identified as the ambient temperature at the beginning of the fire test. The effect of heat convection and radiation can be considered using the following surface boundary condition:

$$-k\frac{\partial T}{\partial n}=h_c(T-T_f)+\sigma\varepsilon_m\varepsilon_f\left[(T-T_z)^4-(T_f-T_z)^4\right], \tag{14}$$

where n represents the normal direction of the outside surface, h_c is the coefficient of convective heat transfer, which is expressed as $25\,\mathrm{W/(m^2{\cdot}K)}$, $\sigma = 5.67\times10^{-8}\cdot\mathrm{W/(m^2{\cdot}K^4)}$ denotes the Stefan–Boltzmann constant, ε_m and ε_f are the surface emissivity of a structure and the surface emissivity of fire, respectively, and T_f and T_z represent the temperature measured at the fire-exposed surface of a structure and the absolute zero temperature, respectively.

In this study, a 2D FE model of a typical cross section of an RC beam will be established using commercial FEM software, and the ambient temperature of a furnace during the fire test will be directly applied to the surface of an RC beam as thermal loads according to ISO 834 guidelines [5] to perform the transient thermal analysis. Considering that the steel reinforcement only takes a small part in the cross section of the beam, the temperature differences between concrete and steel bars can be ignored [31]. Therefore, the reinforcement will be disregarded in the transient thermal analysis for simplicity. The temperature distribution will be

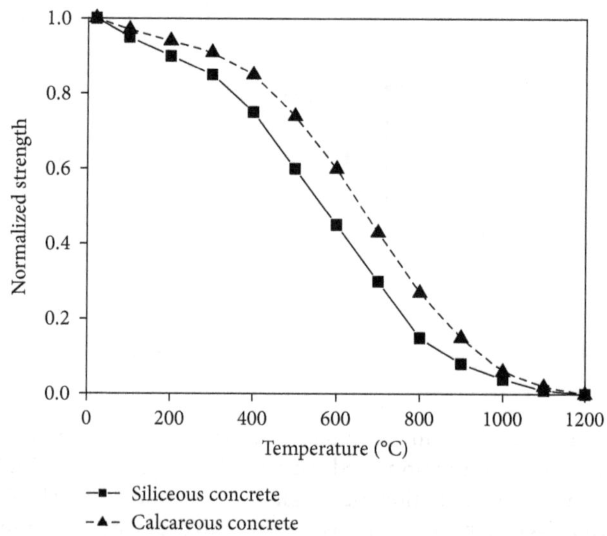

FIGURE 2: Normalized compressive strength of concrete at elevated temperatures.

employed to predict the structural behaviors of the RC beam in the next step.

4. Temperature-Dependent Material Properties

The suitable specification of the material properties highly dependent on the temperature is essential in the numerical simulation due to their influence on the thermal and mechanical responses of RC beams exposed to fire. The relationships provided in the European code [2] will be implemented in this study. The density of concrete does not significantly change at elevated temperatures (less than 5%) [32, 33], and it usually has a minor influence on thermal responses. Thus, a constant value of $2300\,\mathrm{kg/m^3}$ will be assumed in this study. The conductivity k and specific heat c of concrete in (13) can be expressed as functions of temperature according to European code [2]:

$$\begin{aligned} k &= 1.6-0.16\frac{T}{120}+0.008\left(\frac{T}{120}\right)^2, \\ c &= 900+80\frac{T}{120}-4\left(\frac{T}{120}\right)^2. \end{aligned} \tag{15}$$

Since the steel reinforcement has a minor influence on the heat conduction analysis, its thermal properties are not considered.

The variation in the mechanical properties of concrete and reinforcement has been specified according to many experimental tests and design guidelines. The Eurocode model [2] is deployed to describe the uniaxial compressive constitutive model of concrete at evaluated temperatures. The compressive strength f_c and the related strain ε_{c1} are expressed as a function of temperature since these critical parameters define the stress-strain relationship of concrete for a given temperature. As shown in Figure 2, a higher temperature would undoubtedly cause a reduction of compressive strength of concrete, and its corresponding strain would increase as illustrated in [2]. The linear stress-strain relationship of compressive concrete is

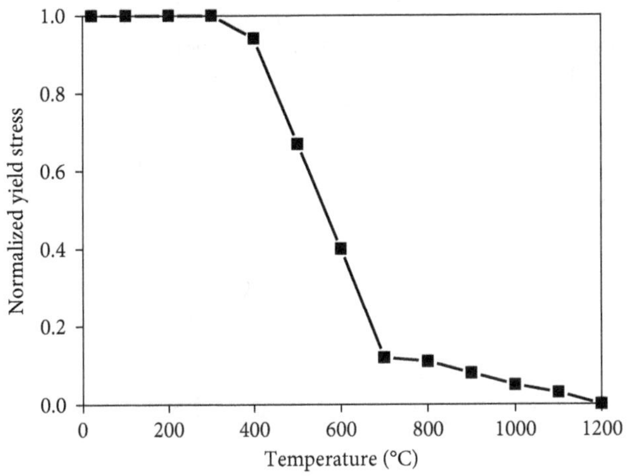

FIGURE 3: Normalized yield stress of steel at elevated temperatures.

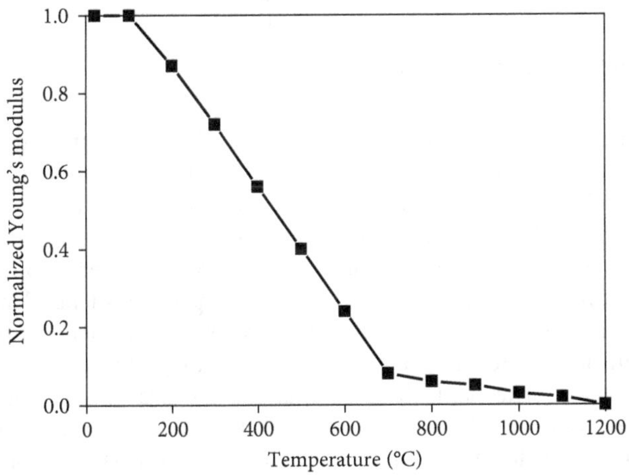

FIGURE 4: Normalized Young's modulus of steel at elevated temperatures.

assumed until 0.33 f_c is attained. Then, a nonlinear increase in compressive stress is adopted with the development of concrete strain to represent the strain-hardening curve to the compressive strength f_c, which can be expressed as

$$\sigma_c = \frac{3\eta}{2 + (\eta)^3}, \quad \eta = \frac{\varepsilon_c}{\varepsilon_{c1}}, \tag{16}$$

where σ_c and ε_c denote the concrete stress and the related strain, respectively. A linear descending branch is used to represent the postpeak softening behavior of concrete based on the ultimate strain in compression, in which empirical values are recommended in [2]. A bilinear stress-strain relationship is used to represent the concrete behavior in tension with a linear ascending up to the tensile strength f_{ct}. The value of f_{ct} decreases to zero with the increase in temperature up to 600°C as presented in [2], and some modifications are introduced to avoid the numerical problem of simulation [34]. The linearly descending branch is defined based on the ultimate strain in tension ε_{ct} whose value does not depend on temperature and fall between 0.002 and 0.004 according to Terro [35].

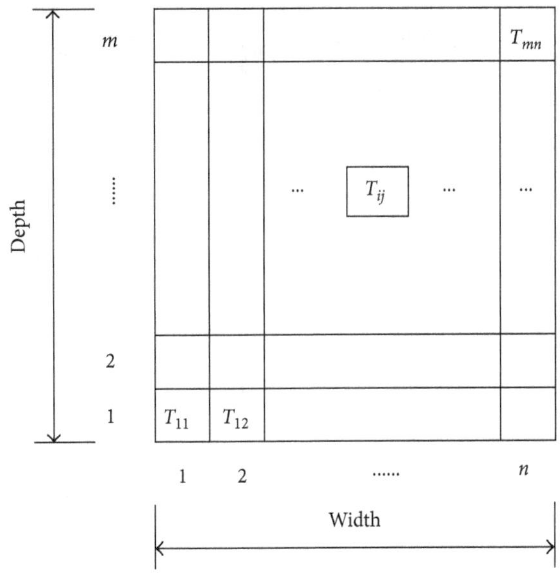

FIGURE 5: Discretization of cross section for structural analysis.

Figures 3 and 4 illustrate the reduction of yield stress and Young's modulus of steel at elevated temperatures, respectively. The mechanical properties of steel reinforcement are defined by Eurocode 2 [2], in which a set of empirical formulas were concluded to represent the temperature reduction of constitutive relationships.

Besides, one of the critical points is the modeling of bond-slip behavior between steel reinforcement and concrete. In this study, a perfect bond performance is assumed, which represents that the average strain in the reinforcement equals to that of the concrete at the same location. This assumption is able to represent compressive section of RC beams accurately. While in the tensile section, the appearance of cracks might cause weakness of bond performance leading to a slipping of reinforcement. However, since a smeared crack model is employed in the proposed spectral approach, in which one element would be enough to model an entire segment as long as the geometrical and material properties remain to be constant, and the cracking process is simulated by specifying the reduction of concrete strength using constitutive equations. According to this, the average strain of steel and concrete at the same location is approximately the same since no stress concentration is presented by using the smeared crack model. Thus, a perfect bond behavior between concrete and reinforcement is deemed to be adequate in this study for simplicity.

5. Framework to Evaluate the Fire Performances of RC Beams

In the first stage, the temperature distribution along the time is computed by performing a transient thermal analysis on a 2D FE model of the beam cross section since the longitudinal heat transfer is disregarded. The heat transfer coefficients presented in Section 3 will be used.

As long as the temperature distribution has been obtained at one thermal load step, the spectral element model

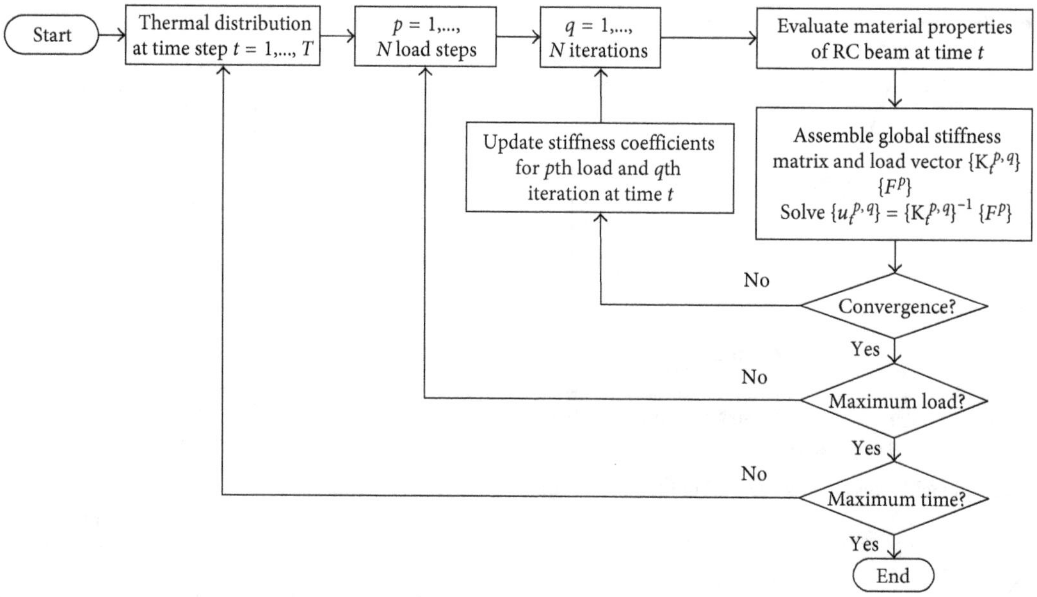

FIGURE 6: Framework of proposed numerical strategy.

presented in Section 2 will be implemented to simulate the mechanical responses of RC beams due to the simplicity and low time consumption of the proposed one-dimensional model. The temperature distribution of the cross section obtained at a given thermal load step will be utilized to update the constitutive equations of concrete and steel reinforcement based on Eurocode 2 [2]. The cross section has been divided into $m \times n$ layers along the depth and width of the RC beam, as shown in Figure 5, and the temperature in each block is assumed to remain constant to simplify the process of structural analysis. The temperature of each element T_{ij} can be approximately evaluated by averaging the temperatures of four nodes of the element according to the temperature distribution of the cross section provided at a transient time. Thus, the temperature-dependent material properties of each block can be obtained according to the average temperature T_{ij} at its location by applying the empirical equations in Eurocode 2. They will be applied to determine the stiffness matrix of the RC beam based on the proposed spectral model.

To describe the procedure of numerical analysis in detail, a flowchart of the proposed strategy is illustrated in Figure 6. At the beginning, the thermal distribution of a specimen under a fire scenario is evaluated by using a 2D FE model, and the external forces are applied to the RC beam with gradually increased loading steps as the temperature distribution remains constant. Thus, the problem of the fire performance of an RC beam has been transferred to a structural analysis with a new stress-strain relationship that has been updated by a given thermal distribution, and the proposed one-dimensional spectral model will be deployed to predict the mechanical behaviors of the RC beam.

In the structural analysis, when the applied loadings are at low level, the elastic moduli of concrete and reinforcement remain available for calculation. However, the nonlinear behaviors of materials would occur with the increase in load. This phenomenon is simulated by using equivalent secant elastic moduli for concrete and steel reinforcement, which would be simultaneously updated with the development of local strain. Hence, the parameters A_{11}, B_{11}, D_{11}, and A_{22} of DSM in (12) that related to tangent elastic moduli are no longer constant but in function of strain at every point. During the updating process, the temperature-dependent constitutive relations are defined by empirical formulas of Eurocode for each temperature scenario provided by the previous transient thermal analysis. As previously mentioned, the strain of steel bar equals the strain of concrete at the same location since perfect bond performance is assumed in this study. According to this, the DSM is implicitly updated to simulate the progressive cracking of concrete during iterations until force equilibrium is reached. Consequently, the displacement vector of structure is obtained under one fire scenario, and the same procedure can be performed for the next thermal step.

Following the numerical approach illustrated in Figure 6, the process of calculation is performed in an implicit manner to obtain the nodal displacement vector and the related load-carrying capacity by the iterative procedure until convergence is achieved. By doing this, the structural responses can be evaluated by gradually increasing the load on the RC beam until failure occurs after fire exposure. Alternatively, the mechanical behaviors can also be determined under a given load with the development of time-temperature exposure.

6. Validation of the Proposed Numerical Method

6.1. Experimental Tests. To validate the accuracy and capability of the proposed numerical method for the structural analysis of an RC beam under fire exposure, experimental tests performed by Kumar and Kumar [12] were taken as a reference. In Kumar's tests, a set of RC beams were cast to study their structural performance when exposed to fire with

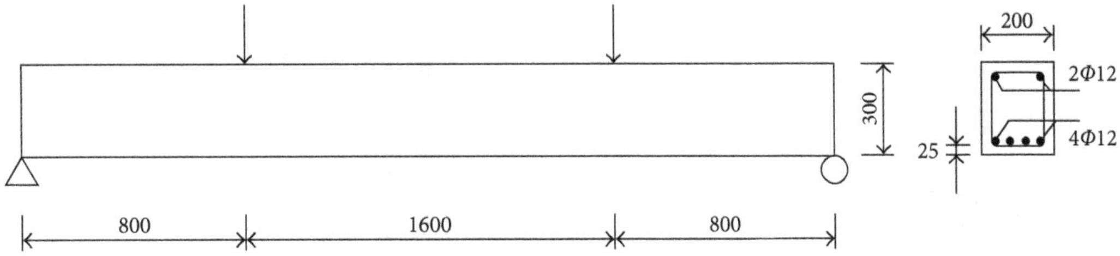

FIGURE 7: Experimental test setup.

variable duration. The width, depth, and length of the reinforced beams were 200 mm, 300 mm, and 3950 mm, respectively, while the span between supports was 3200 mm. Four-point loading was progressively applied to the beam with a shear span of 800 mm; the experimental setup is shown in Figure 7. The material and geometrical properties of the specimens are illustrated in Table 1.

Four specimens were selected to perform fire tests in a furnace with a constant ambient temperature of 20°C at the top surface of the beams, whereas the bottom and two lateral sides were exposed to the following temperature-time relationship according to ISO 834 standard:

$$T(t) = T(t_0) + 345\log(8t - 1), \tag{17}$$

where $T(t_0)$ denotes the initial ambient temperature (20°C). They are exposed to fire for 1, 1.5, 2, and 2.5 hours. The ambient temperature remains at 20°C as the specimens cool down, and the four-point loading tests were performed on these specimens including one unheated reference beam up to failure in order to check their residual resistances. Two loading platens were employed to control the deflection of RC beams during the tests. Only experimental data of the four specimens (reference beam and beam exposed to fire for 1, 1.5, and 2 hours) are available because the beam with 2.5 hours of fire exposure failed to be tested due to spalling of concrete after cooling down.

A 3D FE model was developed by Ozbolt in [23] to study the mechanical behavior of RC beams tested by Kumar and Kumar in [12]. In his work, a transient three-dimensional thermomechanical model was implemented into a three-dimensional FE code, and it was performed under a framework of continuum mechanics and irreversible thermodynamics. The constitutive law of concrete is the temperature-dependent microplane model whereas the mechanical properties would not influence the thermal distribution. Besides, damage and cracking is simulated by applying smeared crack model. The numerical prediction of the more complex FE model agrees quite well with experimental results. Thus, they will be used to validate the efficiency of the simplified spectral model.

TABLE 1: Material and geometrical properties of specimens.

	Specimens	
Concrete	Elastic modulus (MPa)	19,600
	Compressive strength (MPa)	17.10
	Tensile strength (MPa)	2.00
Steel bar	Diameter (mm)	12
	Yield stress (MPa)	480
	Ultimate strength	550
RC beam	Span (mm)	3200
	Width (mm)	200
	Height (mm)	300

6.2. Results and Discussion. As presented in the previous section, this study has two main steps, including transient thermal and structural analysis. The beams were exposed to elevated temperatures on three sides (bottom and two lateral faces) in the fire tests; thus, the thermal distribution of an RC beam can be evaluated by implementing an FEM analysis at each time step. In the FE model, the element size was chosen to be 5 mm resulting in 60 elements along the depth cross section, and thermal loads are applied on the bottom and two lateral surfaces. By doing this, the temperature distribution throughout the entire cross section can be evaluated. In order to validate the accuracy of proposed 2D FE model, the numerical thermal results should be compared with measured experimental data; however, the thermal data of experimental tests were not presented in [12]. To solve this problem, the experimental thermal results measured from [22] were selected to compare the numerical thermal results in this work, for the reason that the geometrical and material properties of RC beams used in [12] and [22] are very similar. Therefore, the numerical and experimental results in the middle layer of the RC beam at different depths of 10, 25, and 100 mm are compared in Figure 8, and the numerical results have to be found to be in close agreement with existing test results.

To further understand the thermal responses of an RC beam, temperature distributions along different levels of depth at 0.5, 1, 1.5, and 2 hours fire exposure are illustrated in Figure 9, and the numerical results are similar to the data obtained in the literature [22]. The temperature in each block of the cross section shown in Figure 5 can be calculated by taking the average temperature of the four nodes of the element. Nonlinear stress-strain relationships of concrete and steel reinforcement at any locations of the cross section are able to be identified according to the obtained thermal data. After that, a structural analysis is performed using the proposed spectral model, and the numerical simulation follows the framework described in Figure 6.

According to the formulation of proposed spectral model, one element is enough to represent a section as long as no discontinuity of geometrical and material properties exists throughout the entire segment. In this study, in order to capture the mechanical behaviors of the RC beam with the increase in load under different fire scenarios, more accurate results would be achieved by applying more spectral elements since the progressive degradation of the RC beam due

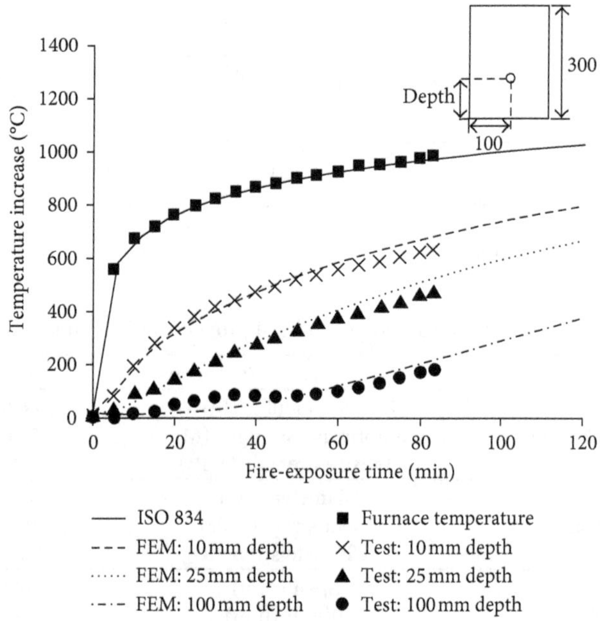

FIGURE 8: Temperatures evaluated by FEM at various locations with fire duration.

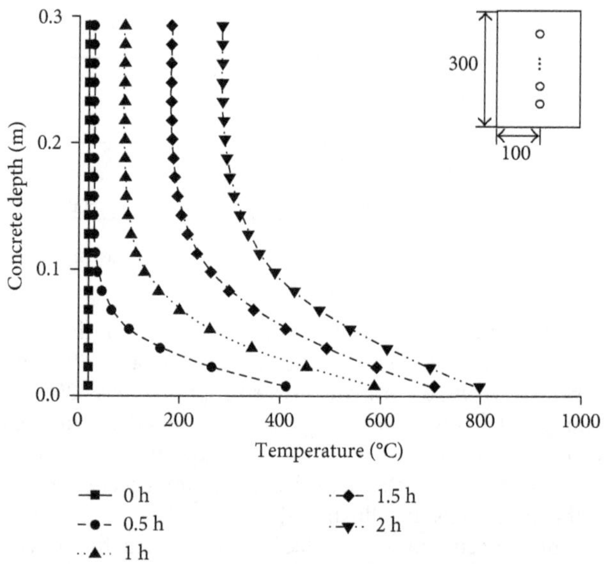

FIGURE 9: Temperature distributions over concrete depth at the central layer of midspan with fire duration.

to material nonlinearities can be simulated more appropriately with a finer mesh. If more elements were applied, the precision of numerical results would be higher, but the increment may not compensate the remarkable increment in the computational burden. Therefore, the number of elements can be defined by searching a balance between accuracy and computational costs according to the desired degree of precision.

Here, an element size of the proposed spectral model of 200 mm is selected for the structural analysis resulting in only 16 spectral elements in this study to simulate the mechanical behavior of RC beams with a length of 3.2 m. Thus, the computational costs are significantly reduced compared with the conventional FE model. To validate the accuracy of the proposed numerical approach, the numerical results of an RC beam for four fire scenarios (no heating and heating after 1, 1.5, and 2 hours) are illustrated in Figures 10(a)–10(d) to compare with the results of the experimental tests and FE analysis.

The load-deflection curve of the reference beam is shown in Figure 10(a), and the numerical results correspond quite well with the results of the experimental tests and FE model. Concrete cracking and the subsequent steel yielding occur at the loading points of 50 kN and 150 kN. The cracking loads of the RC beams after 1 h, 1.5 h, and 2 h decreased to 40, 35, and 30 kN according to the numerical results predicted by the proposed model, as shown in Figures 10(b)–10(d). This is attributed to the loss of concrete strength with the increase in temperature. The steel yields loading are 140, 110, and 70 kN for the three fire scenarios, which demonstrate the reduction of steel strength exposed to thermal loads. The numerical results indicate agreement with the experimental tests and FE model in Figures 10(a)–10(c). As shown in Figure 10(d), a comparatively large difference exists between the proposed model and the FE model after 2 h of fire exposure. This might be attributed to the fact that the interfacial bond behavior between concrete and reinforcement can be appropriately modeled by applying FEM, and it would significantly influence the global structural performances under elevated temperatures. In this study, the reinforcement is assumed to be perfectly bonded to the concrete. This assumption is appropriate to predict the degradation of load-carrying capacity of an RC beam exposed to fire taking into account that the deterioration of material strength is the critical factor, and the proposed one-dimensional numerical model is established to perform the structural analysis according to the simplification. Despite this, the loss of precision can be compensated by the efficiency of the proposed spectral model since one of the main objectives is to model the mechanical behaviors of the RC beam under fire exposure in an efficient way.

In the proposed spectral approach, the nonlinear behavior of the material is introduced by using equivalent secant elastic modulus which would be updated implicitly during iteration. To avoid convergence difficulties, the external load is divided into load levels, and an iterative secant procedure is performed to search the solution until force equilibrium is achieved for each load level as illustrated in Figure 6. According to this load-control strategy, the nodal displacement would increase with the increase in load, and consequently, convergence would be failed to achieve as long as applied load exceeds the bearing capacity of the RC beam. Therefore, monotonic increase in bearing capacity can be observed in Figure 10 because of the implementation of load-control strategy in this study. In spite of this, the accuracy of numerical results can be considered to be satisfactory since a limited number of elements were employed in the proposed model. Hence, the proposed spectral model is appropriate for determining the mechanical behaviors of RC beams exposed to fire.

To gain a better understanding of the structural deformation of RC beams after fire exposure, the midspan

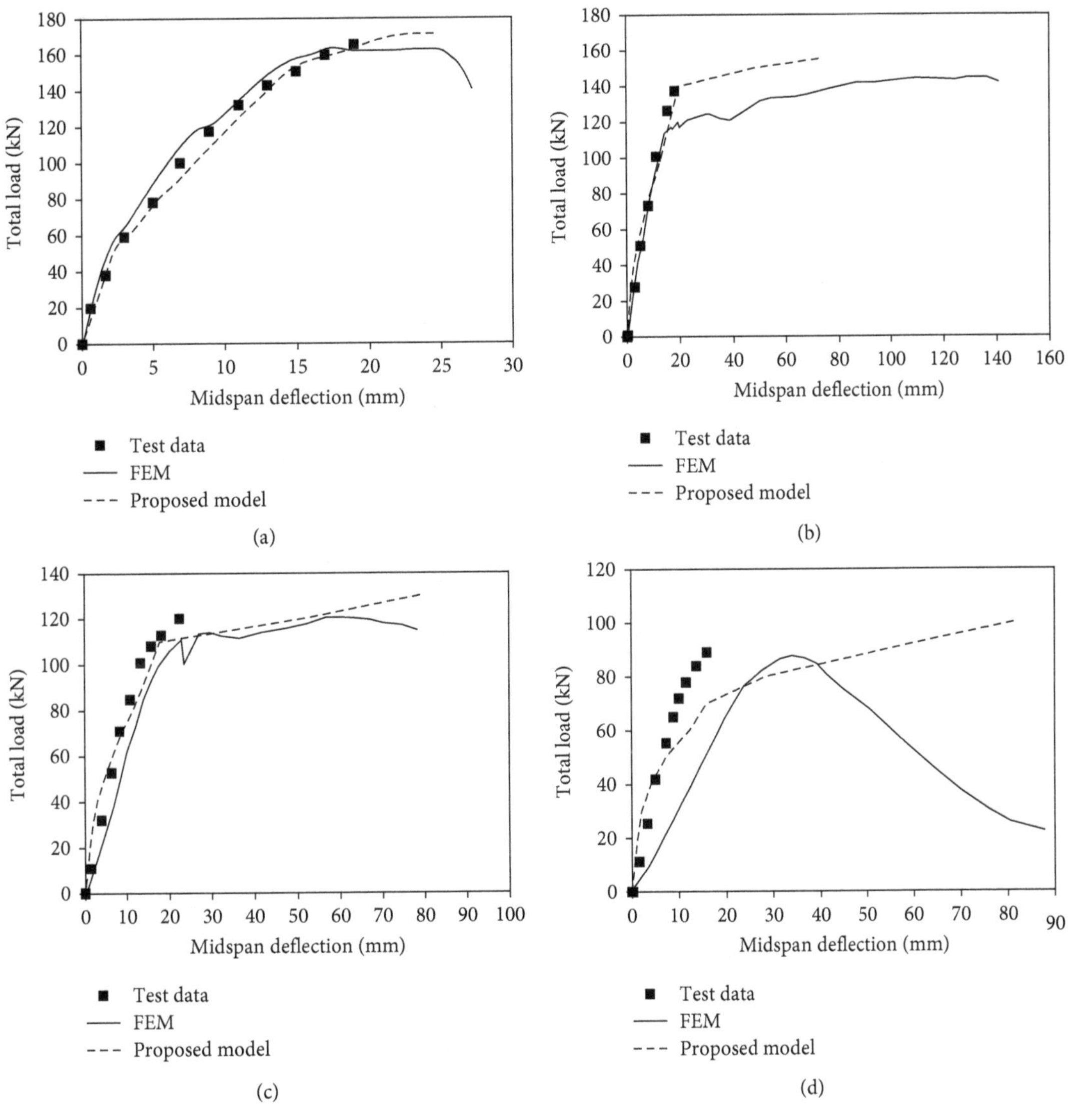

FIGURE 10: Load-midspan deflection (a–d): reference beam; fire exposure duration of 1 h, 1.5 h, and 2 h.

deflections under external loadings of 60, 70 and 80 kN are illustrated in Figures 11(a)–11(c). With the development of fire exposure, the experimental results exhibit large midspan displacements, which increase from 3.1, 4.2 and 5.1 mm after 0 h of fire exposure to 8.1, 10 and 12.5 mm, respectively, after 2 h of fire exposure for the three loading scenarios. As expected, the same phenomena can be observed from the results of the FEM analysis and proposed numerical approach, which clearly demonstrate the reduction of resistance capacity of RC beams under thermal loads. A sharp increase in midspan deflections is observed from 1.5 h to 2 h fire duration according to the proposed model and the FEM. This is mainly due to the rapid loss of steel strength as long as temperature reached 600°C, as shown in Figures 3 and 4. As illustrated in Figure 9, the temperature at the location of tensile reinforcement (25 mm from the bottom) has exceeded this value from 1.5 h to 2 h of fire exposure. Thus, the damages of concrete and steel bars caused by high temperatures would lead to large deformations, as plotted in Figure 11. As well known, the mechanical responses of the RC beam are highly dependent on the temperature distribution obtained from transient thermal analysis. A typical cross section of the RC beam is selected by neglecting longitudinal heat transfer, and thermal results are obtained by applying 2D thermal analysis. According to this simplification, differences between numerical prediction and experimental results might appear due to the errors cumulated from previous thermal analysis, especially under longtime fire exposure, for example, Figure 11 after 2 h exposure. Generally, the predicted data are consistent with the measured data.

Figure 12 shows the variation of ultimate load with the increase in fire duration. According to the proposed model, the residual load-carrying capacity progressively reduces to 50% of the unheated reference beam after approximately 2 h of exposure, which corresponds with the test data and FEM. To provide a quantitative overview of time-dependent stiffness, the secant stiffness, which is evaluated at 75% of

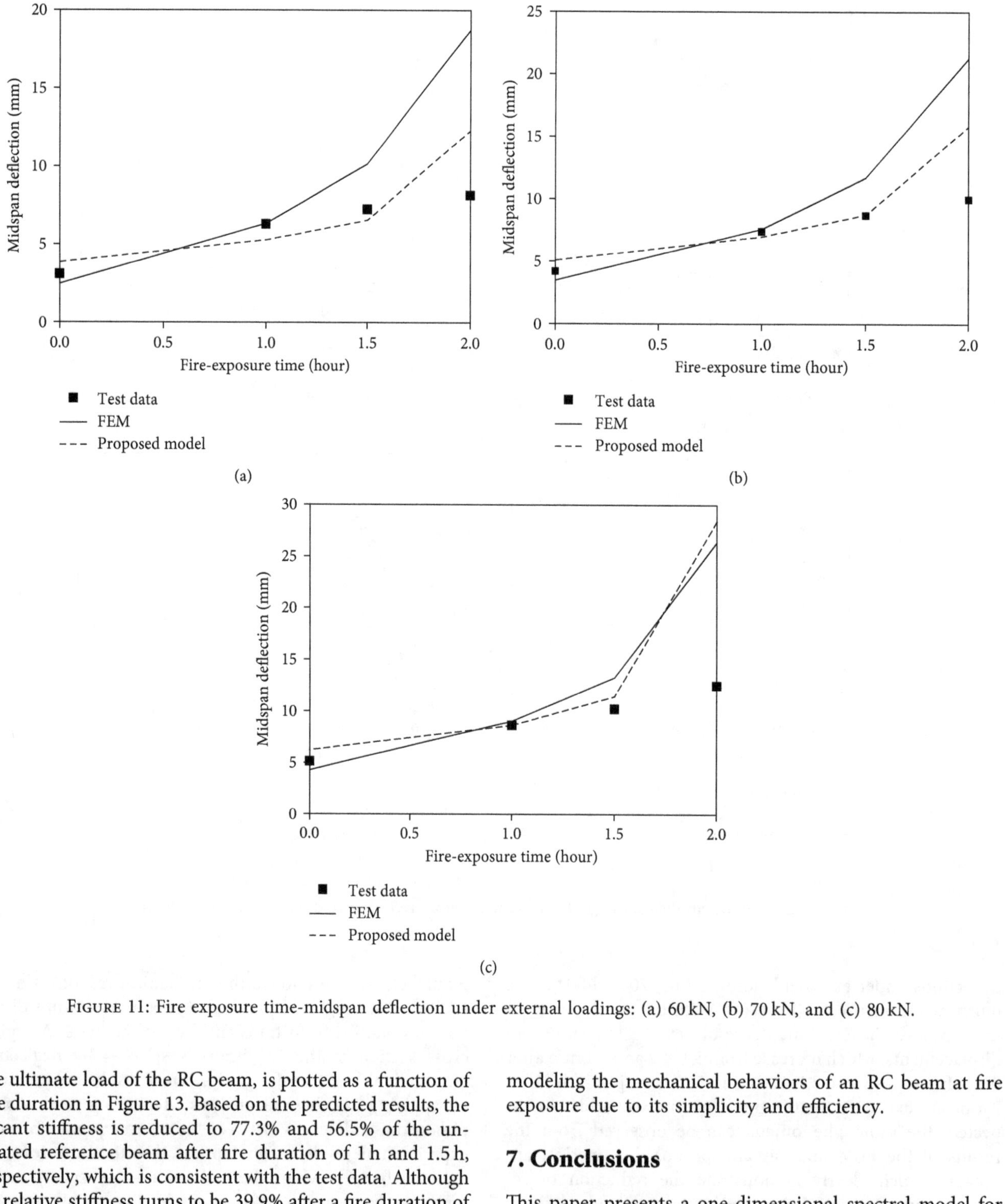

FIGURE 11: Fire exposure time-midspan deflection under external loadings: (a) 60 kN, (b) 70 kN, and (c) 80 kN.

the ultimate load of the RC beam, is plotted as a function of fire duration in Figure 13. Based on the predicted results, the secant stiffness is reduced to 77.3% and 56.5% of the unheated reference beam after fire duration of 1 h and 1.5 h, respectively, which is consistent with the test data. Although its relative stiffness turns to be 39.9% after a fire duration of 2 h, which is larger than the relative stiffness in the experimental tests, considering that the numerical results are highly dependent on the empirical model for material properties at elevated temperatures, it is very acceptable considering the fact that only 16 spectral elements are deployed in this study to determine the structural performances, and the results correspond quite well with the data obtained from the literature [23]. Therefore, the proposed numerical approach is a promising tool for appropriately modeling the mechanical behaviors of an RC beam at fire exposure due to its simplicity and efficiency.

7. Conclusions

This paper presents a one-dimensional spectral model for simulating the behavior of RC beams that are subjected to standard ISO 834 fire exposure. This model is cooperated with a transient thermal analysis by using a 2D FE model. The computational costs are significantly reduced since a limited number of elements were implemented under the framework of the proposed numerical approach in this study. The experimental data of four fire scenarios are utilized for comparison: no heating (reference beam), 1 h, 1.5 h, and 2 h heating, and four-point loading tests were

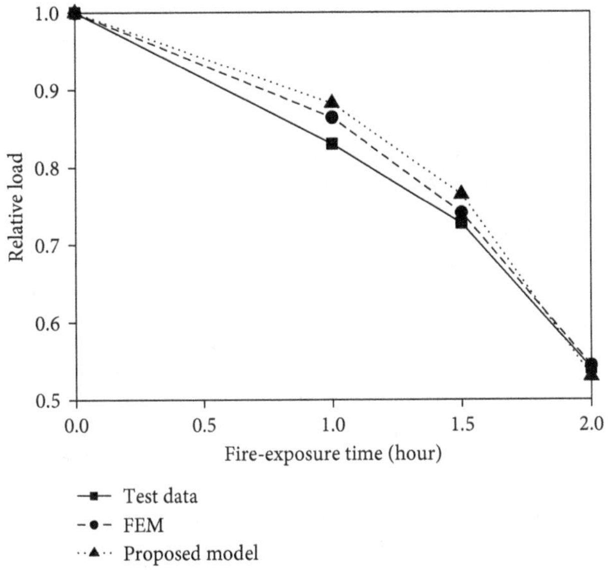

FIGURE 12: Ultimate load reduction.

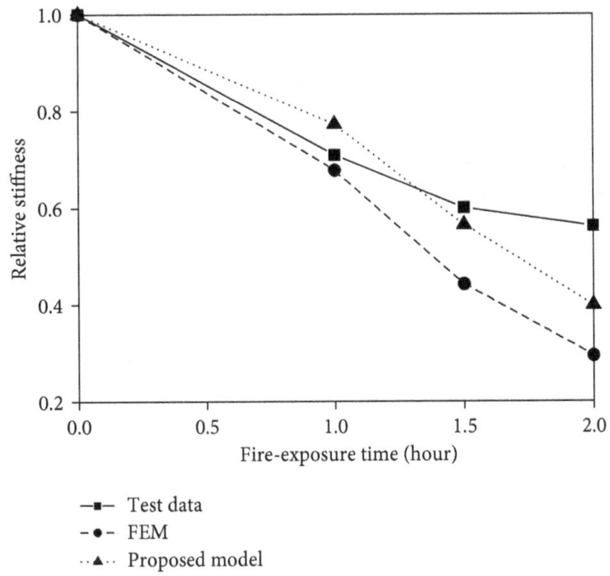

FIGURE 13: Secant stiffness reduction.

applied to the specimens after cooling down to obtain their mechanical responses after fire exposure. The results of a more complex FE model are also introduced to validate the accuracy and efficiency of the proposed model. Several conclusions can be drawn from this work in the following:

(1) One cross section of an RC beam was chosen to represent its typical thermal field using a 2D FE model, and the obtained temperature distribution agrees well with the results in the literature.

(2) A one-dimensional spectral model was developed to model the RC beam, which cooperates with the thermal distribution obtained from the FE analysis to calculate the structural behavior of an RC beam subjected to a four-point loading test for different levels of fire exposure.

(3) Few elements were deployed in the numerical simulation, since one element would be sufficient for modeling a section as long as its geometrical and material properties remain unchanged along the length of beam. The computational burdens are significantly reduced due to the simplicity of the proposed numerical approach compared with conventional FE model.

(4) The interfacial bond-slip behavior between concrete and steel reinforcement was disregarded in this study, and a perfect bond model was assumed to perform the numerical simulation in a simplified way. Despite this, the predicted results exhibit good agreement with the test data and FEM analysis, which validate the accuracy and capability of the proposed numerical approach.

(5) The predicted load-deflection curves are found to be in good agreement with the experimental and FEM data. The residual resistances and stiffness of the RC beam are quantitatively evaluated in this study, which demonstrate the damages of structures due to fire exposure.

Conflicts of Interest

The authors declare that they have no conflicts of interest.

Acknowledgments

The authors acknowledge the financial supports from the National Natural Science Foundation of China (Project no. 51708062), 111 Project of China (Grant no. B18062), Fundamental Research Funds for the Central Universities of China (Project nos. 106112016CDJXY200001 and 2018CDQYTM0044), and Spanish Ministry of Economy and Competitiveness (Project no. BIA2017-84975-C2-1-P).

References

[1] J. A. Purkiss, *Fire Safety Engineering, Design of Structures*, Butterworth Heinemann, Oxford, UK, 2nd edition, 2007.

[2] EN 1992-1-2, *Eurocode 2: Design of Concrete Structures. Part 1-2: General Rules-Structural Fire Design*, European Committee for Standardization, Brussels, Belgium, 2004.

[3] FIP/CEB, *FIP/CEB Report on Methods of Assessment of the Fire Resistance of Concrete Structural Members*, Cement and Concrete Association, Slough, UK, 2004.

[4] ACI 216.1, *Code Requirements for Determining Fire Resistance of Concrete and Masonry Construction Assemblies*, American Concrete Institute, Detroit, MI, USA, 2007.

[5] ISO 834-1, *Fire Resistance Tests-Elements of Building Construction. Part 1: General Requirement*, International Organization for Standardization, Geneva, Switzerland, 1999.

[6] Z. P. Bazant and M. F. Kaplan, *Concrete at High Temperatures: Material Properties and Mathematical Models*, Longman, Harlow, UK, 1996.

[7] B. Zhang and N. Bicanic, "Residual fracture toughness of normal- and high-strength gravel concrete after heating to 600°C," *ACI Materials Journal*, vol. 99, pp. 217–226, 2000.

[8] P. D. Morley and R. Royles, "Response of the bond in reinforced concrete to high temperature," *Magazine of Concrete Research*, vol. 35, no. 123, pp. 67–74, 1983.
[9] T. Pothisiri and P. Panedpojaman, "Modeling of bonding between steel rebar and concrete at elevated temperatures," *Construction and Building Materials*, vol. 27, no. 1, pp. 130–140, 2012.
[10] V. K. R. Kodur and A. Agrawal, "Effect of temperature induced bond degradation on fire response of reinforced concrete beams," *Engineering Structures*, vol. 142, pp. 98–109, 2017.
[11] H. J. Wu, T. T. Lie, and J. Y. Hu, "Fire resistance of beam-slab specimens—experimental studies," Internal Report No. 641, Institute for Research in Construction, National Research Council Canada, Ottawa, ON, Canada, 1993.
[12] A. Kumar and V. Kumar, "Behaviour of RCC beams after exposure to elevated temperatures," *Journal of the Institution of Engineers (India), Civil Engineering Division*, vol. 84, pp. 165–170, 2003.
[13] M. B. Dwaikat and V. K. R. Kodur, "Response of restrained concrete beams under design fire exposure," *Journal of Structural Engineering*, vol. 135, no. 11, pp. 1408–1417, 2009.
[14] J.-r. Ren, H.-g. Chen, T. Sun, H. Song, and M.-s. Wang, "Flexural behaviour of combined FA/GGBFS geopolymer concrete beams after exposure to elevated temperatures," *Advances in Materials Science and Engineering*, vol. 2017, Article ID 6854043, 9 pages, 2017.
[15] B. Ellingwood and T. D. Lin, "Flexural and shear behaviour of concrete beams during fires," *Journal of Structural Engineering*, vol. 117, no. 2, pp. 440–458, 1991.
[16] D. D. Capua and A. R. Mari, "Nonlinear analysis of reinforced concrete cross-sections exposed to fire," *Fire Safety Journal*, vol. 42, no. 2, pp. 139–149, 2007.
[17] J. H. Hsu and C. S. Lin, "Effect of fire on the residual mechanical properties and structural performance of reinforced concrete beams," *Journal of Fire Protection Engineering*, vol. 18, no. 4, pp. 245–274, 2008.
[18] V. R. Kodur and A. Agrawal, "An approach for evaluating residual capacity of reinforced concrete beams exposed to fire," *Engineering Structures*, vol. 110, pp. 293–306, 2016.
[19] S. Bratina, I. Planinc, M. Saje, and T. Goran, "Non-linear fire-resistance analysis of reinforced concrete beams," *Structural Engineering and Mechanics*, vol. 16, no. 6, pp. 695–712, 2003.
[20] B. M. Luccioni, M. I. Figueroa, and R. F. Danesi, "Thermo-mechanic model for concrete exposed to elevated temperatures," *Engineering Structures*, vol. 25, no. 6, pp. 729–742, 2003.
[21] Z. H. Huang, I. W. Burgess, and R. J. Plank, "Three-dimensional analysis of reinforced concrete beam-column structures in fire," *Journal of Structural Engineering*, vol. 135, no. 10, pp. 1201–1212, 2009.
[22] W. Y. Gao, J. G. Dai, J. G. Teng, and G. M. Chen, "Finite element modeling of reinforced concrete beams exposed to fire," *Engineering Structures*, vol. 52, pp. 488–501, 2013.
[23] J. Ozbolt, J. Bosnjak, G. Periskic, and A. Sharma, "3D numerical analysis of reinforced concrete beams exposed to elevated temperature," *Engineering Structures*, vol. 58, pp. 166–174, 2014.
[24] H. Dong, W. Cao, J. Bian, and J. Zhang, "The fire resistance performance of recycled aggregate concrete columns with different concrete compressive strengths," *Materials*, vol. 7, no. 12, pp. 7843–7860, 2014.
[25] V. K. R. Kodur, M. S. Dwaikat, and N. Raut, "Macroscopic FE model for tracing the fire response of reinforced concrete structures," *Engineering Structures*, vol. 31, no. 10, pp. 2368–2379, 2009.
[26] B. Wu and J. Z. Lu, "A numerical study of the behaviour of restrained RC beams at elevated temperatures," *Fire Safety Journal*, vol. 44, no. 4, pp. 522–531, 2009.
[27] R. Perera and D. Bueso-Inchausti, "A unified approach for the static and dynamic analyses of intermediate debonding in FRP-strengthened reinforced concrete beams," *Composite Structures*, vol. 92, no. 11, pp. 2728–2737, 2010.
[28] R. Sun, E. Sevillano, and R. Perera, "A discrete spectral model for intermediate crack debonding in FRP-strengthened RC beams," *Composites Part B: Engineering*, vol. 69, pp. 562–575, 2015.
[29] M. B. Dwaikat and V. K. R. Kodur, "A numerical approach for modeling the fire induced restraint effects in reinforced concrete beams," *Fire Safety Journal*, vol. 43, no. 4, pp. 291–307, 2008.
[30] P. Bamonte and F. L. Monte, "Reinforced concrete columns exposed to standard fire: comparison among different constitutive models for concrete at high temperature," *Fire Safety Journal*, vol. 71, pp. 310–323, 2015.
[31] S. Bratina, B. Cas, M. Saje, and I. Planinc, "Numerical modelling of behaviour of reinforced concrete columns in fire and comparison with Eurocode 2," *International Journal of Solids and Structures*, vol. 42, no. 21-22, pp. 5715–5733, 2005.
[32] C. Lopez, J. P. Firmo, J. R. Correia, and C. Tiago, "Fire protection systems for reinforced concrete slabs strengthened with CFRP laminates," *Construction and Building Materials*, vol. 47, pp. 324–333, 2013.
[33] J. P. Firmo, M. R. T. Arruda, and J. R. Correia, "Numerical simulation of the fire behaviour of thermally insulated reinforced concrete beams strengthened with EBR-CFRP strips," *Composite Structures*, vol. 126, pp. 360–370, 2015.
[34] M. B. Dwaikat and V. K. R. Kodur, "Hydrothermal model for predicting fire-induced spalling in concrete structural systems," *Fire Safety Journal*, vol. 44, no. 3, pp. 425–434, 2009.
[35] M. Terro, "Numerical modeling of the behavior of concrete structures in fire," *ACI Structural Journal*, vol. 95, pp. 183–192, 1998.

8

Study on Aseismic Characteristics of Tibetan Ancient Timber Structure

Juan Wang, Junxiao He, Na Yang, and Qingshan Yang

Beijing's Key Laboratory of Structural Wind Engineering and Urban Wind Environment, School of Civil Engineering, Beijing Jiaotong University, Beijing 100044, China

Correspondence should be addressed to Juan Wang; juanwang@bjtu.edu.cn

Academic Editor: Gianluca Cicala

Tibetan ancient timber structure has great historical, cultural, artistic, and scientific values. The structural configuration of members such as wall, roof, slab, and timber skeleton with semirigid joints is studied in this paper based on field survey results. The aseismic characteristics of Tibetan ancient timber structure have been summarized and introduced. One of the unique features of the Tibetan ancient timber structure is its special beam-column joint which has a semirigid behavior. Numerical simulations of a typical 3-storey structural unit in the Potala Palace are studied. The acceleration responses of the structure under the action of El-Centro seismic waves are studied. Results show that the arrangement of the joints is helpful for seismic resistance of the structure. Most Tibetan ancient timber structures are suffering from different types of damage to certain extent which is vulnerable to seismic actions. Typical damage problems of the structural components are summarized and the main causes of this damage are analyzed. Different rehabilitation methods that can be implemented are discussed, providing references for maintenance of the structures.

1. Introduction

Timber structure can be frequently found in historic buildings in Tibet. As a treasure of architectural heritage in China, Tibetan ancient timber structure has its own distinguished features due to the special natural environment, history, and culture in Tibet. Pictures of typical Tibetan ancient timber structure are shown in Figure 1.

Study of Chinese ancient architecture began in the 1930s. Liang and other architects published a lot of field investigation reports which have big influences to later studies on the Chinese ancient structures [1]. It is however, not until 1990s, that some scientists began to employ structural analysis in their research on ancient timber structures. Ma studied the building technologies of ancient Chinese timber structures [2]. Wang's research focused on the distribution and transmission of vertical loads in timber structural members [3].

Due to an urgent need to protect the historic buildings, many rehabilitation projects on existing ancient timber buildings have been carried out. Scholars have done exploratory tests and studies to provide references for the maintenance and protection of ancient timber structure. Systematic studies on the structural mechanics of ancient architectures have been carried out at Xi'an Jiaotong University since 1982. Structural mechanics, computational approaches, and modern experimental methods were firstly brought into the structural analysis of ancient timber structures [4]. Fang et al. conducted on-site measurements and model tests on the front tower over the North Gate of the Xi'an City Wall. The first several natural frequencies of the structure were found. Damping ratio and the hysteresis curves were obtained. A semirigid Finite Element Model (FEM) is developed to simulate the behavior of Dou Gong (corbel brackets) and Mortise-Tenon joints. A three-dimensional FEM was established to study the dynamic characteristics of the structure. The FEM program SAFATS was developed to conduct dynamic and earthquake response analysis for this kind of timber structure [5]. Research group of Xi'an University of Architecture and Technology have taken a series of studies and tests on ancient timber structure since 1999. Xue et al. analyzed the earthquake responses on historic palace timber building, and they pointed out that the structures possessed good aseismic resisting behaviors [6]. The research group conducted static and low cyclic loading tests of structural components, such as Dou Gong and Mortise-Tenon joint and shaking table test

(a) Building facade

(b) Inside the building

FIGURE 1: Pictures of typical Tibetan ancient timber structure.

of a hall to study the aseismic mechanism of ancient timber structure [7–10].

The aforementioned literatures are mostly on traditional ancient timber structure. For Tibetan ancient buildings, most studies were focused on the historical and artistic features of the building and only a few were on the structural features. Jiang and Gasu are involved in the first large-scale maintenance project of the Potala Palace, and they studied the failure modes of the structure and its components with mechanical characteristics of the structure [11]. Chen presented the material and construction details of Tibetan architectures [12]. Guo and Yang conducted field test on a typical Tibetan ancient structure instrumented with displacement and acceleration transducers to study the effect due to crowd walking and running [13]. The measured data is then used to analyze the dynamic property such as fundamental frequency and damping ratio of the test structure. The acceleration time histories were used to evaluate the vibration serviceability of the floor slab. Yan analyzed the static properties of the beam-column through tests and numerical simulations [14]. The moment-rotation relation of the joint is derived from the test results. Yang et al. studied the timber material of Tibetan ancient timber structure. The main mechanical properties of three kinds of timber including the old Tibetan Populus cathayana, new Tibetan Populus cathayana, and new Tibetan Pine were obtained by tests. The principles and assumptions made in the determination of their strengths and time-dependent laws on the design strength of timber in ancient Tibetan structures were discussed [15].

In this paper, the structural configuration of Tibetan ancient timber structure with connections, such as the column base, chock block, short stick and mortise-tenon connection, is studied based on a large number of field survey results. The differences of construction and force flow between Han-style and Tibetan buildings are introduced. The aseismic characteristics of Tibetan ancient timber structure have been introduced. Semirigid beam-column joint, which is one of the most important features of Tibetan ancient timber structure, influences significantly the global behavior of the structural system. Numerical simulations of a typical 3-storey structural unit in Potala Palace are studied. The acceleration responses of the structure under the action of El-Centro seismic waves are studied. Most Tibetan ancient timber structures are suffering from different degrees of structural damage which are weakness to the seismic resistance. Typical damage problems of the structural components are summarized and the main causes of this damage are analyzed. Different rehabilitation methods that can serve as references for maintenance of the Tibetan ancient timber structure are also discussed.

2. Structural Configuration

Figures 2 and 3 show the two kinds of typical Han-style timber structures with post and lintel skeletons. Timber frame with laminated beams and columns shown in Figure 2 is mainly in north of China. The timber skeleton of this kind of structure consists of posts and cross beams rising toward the ridge in diminishing lengths. The purlins-horizontal members that support the rafters are positioned along the stepped shoulders of the skeleton. The rafters are short, stretching down only from purlin to purlin [1]. Figure 3 is the timber frame with column and ties usually used in south of China [16]. The main feature of this kind of structure is that the purlins are supported directly by columns. The cross tie connecting the columns is called Chuan, only having the cross link function.

Typical Tibetan timber structure shown in Figure 4 is quite different from Han-style ancient timber structures. For either kind of Han-style timber structure, timber skeleton is the main load bearing system and the wall only has an enclosure function. In Tibetan architecture, wall and timber frame are both load bearing components as shown in Figure 4(a) [17] and Figure 4(b). In a space separated by walls, the vertical row frame series composed of beams and columns support rafters at close spacing and upper plat roof. Figure 4(c) shows the typical frame of Tibetan ancient timber structure. The components such as beam, cap block, chock block, and column are connected with each other by short sticks. Chock blocks are used at the top of column for supporting the beam.

3. Aseismic Characteristics of Tibetan Ancient Timber Structure

Typical Han-style timber structures with mortise and tenon joints connecting structural components can be defined as a kind of flexible structure. The rotation and slip of joints can

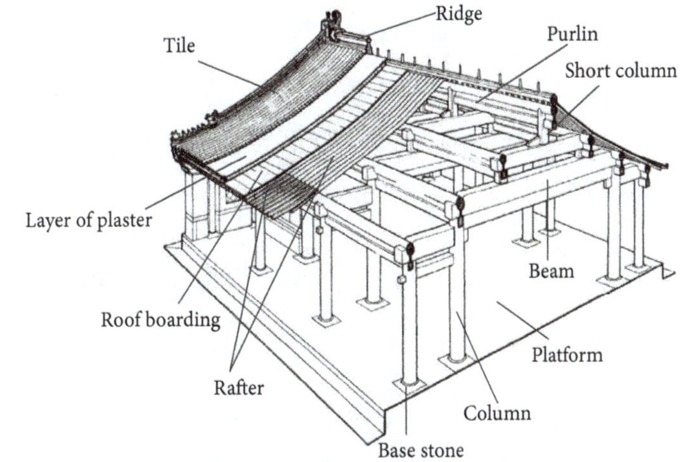

(a) 3D view of timber frame with laminated beams and columns

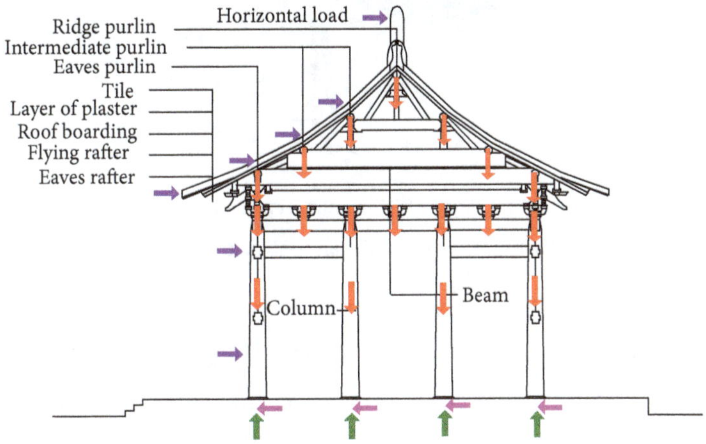

(b) Cross section of timber frame with laminated beams and columns

FIGURE 2: Timber frame with laminated beams and columns.

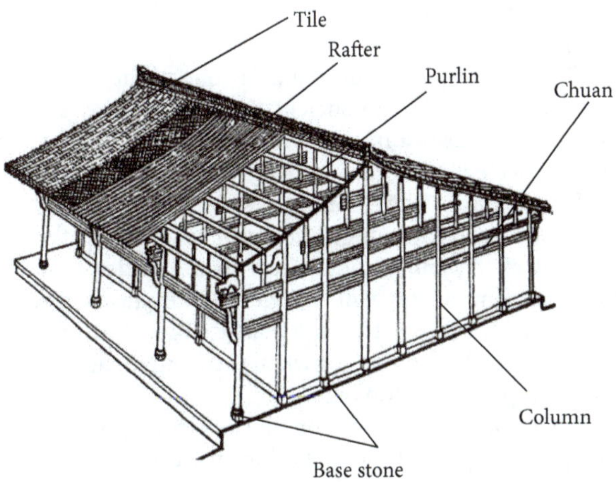

(a) 3D view of timber frame with columns and ties

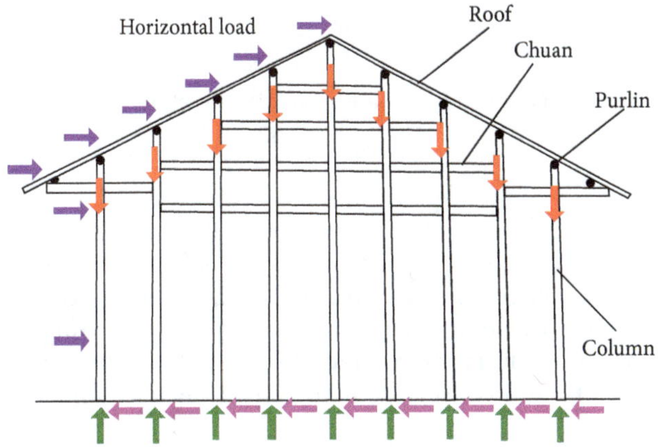

(b) Cross section of timber frame with columns and ties

FIGURE 3: Timber frame with columns and ties.

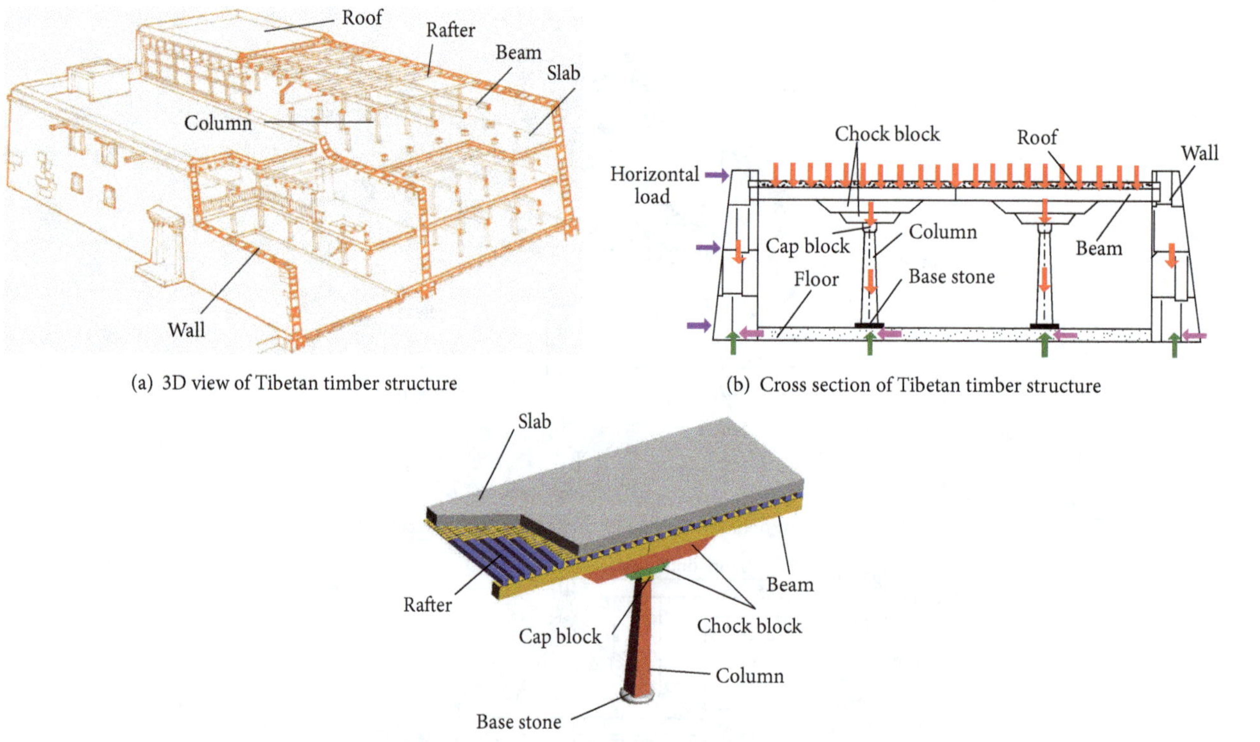

(a) 3D view of Tibetan timber structure

(b) Cross section of Tibetan timber structure

(c) Frame of Tibetan timber structure

FIGURE 4: Typical Tibetan timber structure.

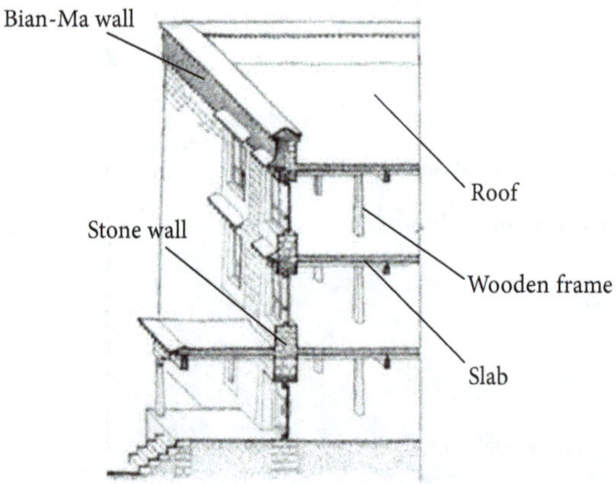

FIGURE 5: Rigid-flexible combination structure.

dissipate a large amount of energy when taking up seismic load. For Tibetan timber structure, it can be defined as a kind of rigid-flexible combination structure as shown in Figure 5 [12]. Timber frame together with masonry wall and strong roof not only ensure the bearing capacity of the structure, but also make the structure have certain ductility under earthquake.

Figure 6 shows the masonry outer wall with inward tilt [18]. It makes the gravity center of the wall lower to let the wall be more stable. Bian-Ma wall shown in Figure 7 is usually used at the top of the outer wall. It is made of willows fixed by pieces of wood, reducing the weight of the upper wall [12]. Bian-Ma wall is painted in red color, decorating the facade of the building.

Figures 8(a) and 8(b) show the construction layers of roof and slab, respectively. Slab can also be constructed by the same layers of roof, but with thinner Aga soil. The strong roof and slabs let the overall stiffness of the structure be strengthened.

It is found in Figure 9(a) that, similar to outer wall, columns are usually with inward tilt, making the timber skeleton more stable. The column base is not inserted into but is resting directly on top of a stone base as shown in Figure 9(b).

Semirigid beam-column joint is one of the most important features of Tibetan ancient timber structure. The components such as beam, cap block, chock block, and column are connected with each other by short sticks as shown in Figure 10(a). The short sticks not only have the role of geometric positioning, but also can improve the shear stiffness of the joint. Under shear load, the deformation of the joint components will experience four stages, which are slip, elastic, elastic-plastic, and plastic stages. Yielding failure of the short sticks will occur when the relative slippage of the joint components is too big.

There are strict limits on the size and proportions of the joint components. The width of the cap block is slightly bigger than the width of the top of column as shown in Figure 10(b). Two chock blocks are layered above the cap block and the length of the lower one is shorter than the upper one as shown in Figure 10(c). The width of the beam above them is slightly bigger than the upper chock block. The two

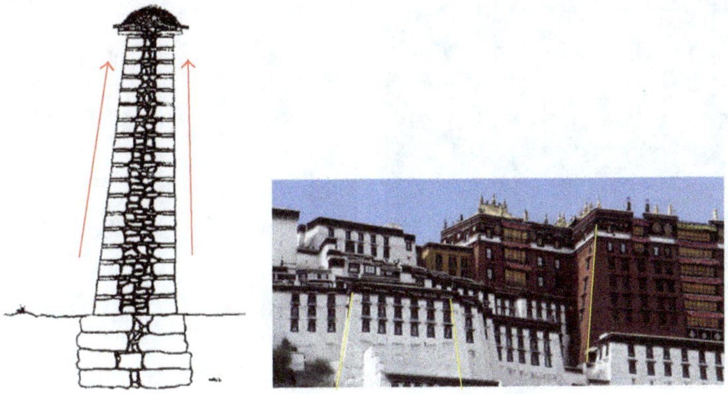

(a) Diagrammatic sketch of the wall

(b) Picture of the wall

FIGURE 6: Outer wall with inward tilt.

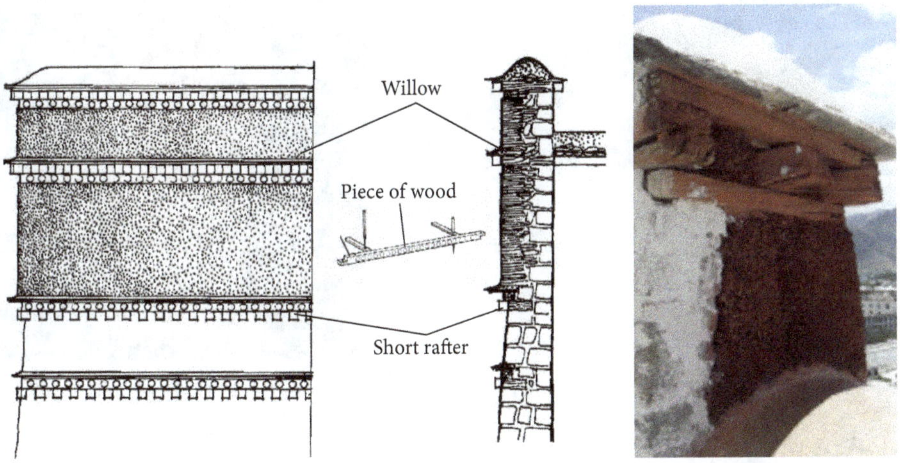

(a) Diagrammatic sketch of the wall

(b) Picture of the wall

FIGURE 7: Bian-Ma wall.

① Aga soil
② Sand and pebble
③ Board
④ Dry stick
⑤ Rafter
⑥ Beam

(a) Construction layers of roof

① Wood floor
② Wooden grid
③ Dry stick
④ Rafter
⑤ Beam

(b) Construction layers of slab

FIGURE 8: Construction layers of roof and slab.

(a) Column with inward tilt

(b) Base stone of the column

FIGURE 9: Column of the timber skeleton.

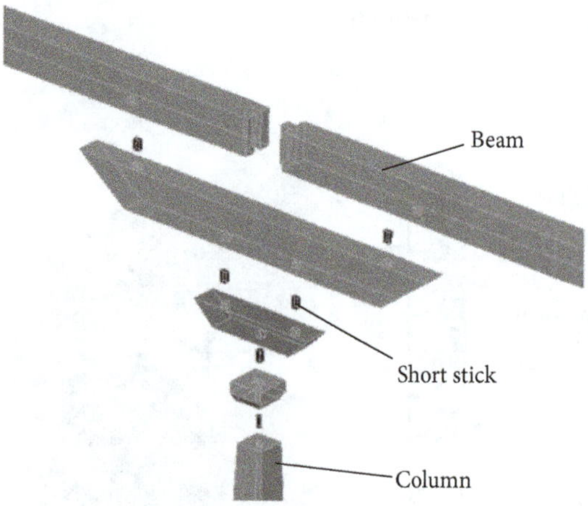

(a) Inner connections between joint components

(b) Joint components

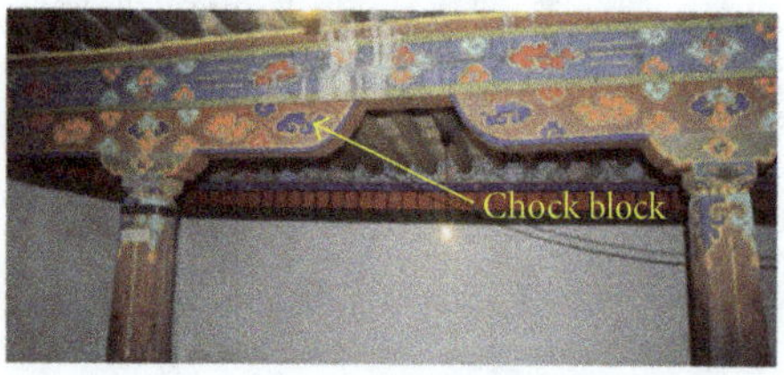

(c) Chock block

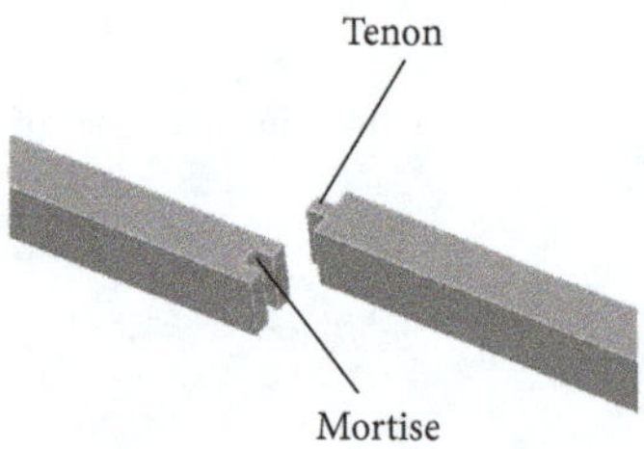

(d) Mortise and Tenon connection

FIGURE 10: Semirigid beam-column joint.

chock blocks improve the shear and flexural capacity of the joint.

The mortise and tenon connection is a semirigid connection, and it is used between beams as shown in Figure 10(d). Usually the mortise cannot completely fit the tenon due to manufacturing errors. At the beginning of loading, the joint can be considered as a hinge joint. Mortise and tenon grip each other tightly as the load increases. The joint is no longer hinged if slippage exists between mortise and tenon. If the load continues to increase, the joint will finally reach its limit state with the tenon pulled out of the mortise. Under the action of earthquake, energy dissipation of the joint occurs by friction and compression between mortise and tenon, such that the earthquake responses of the structure can be reduced.

4. Characteristics of Semirigid Beam-Column Joint

4.1. Numerical Simulations. The FEM of a typical 3-storey structural unit in the Potala Palace is developed using the ANSYS software package. The geometric size of the building and its members is measured on site. Figure 11(a) shows a picture of the structure. Figures 11(b) and 11(c) are the plan and elevation of the structure, respectively. The FEM

(a) A picture of the building

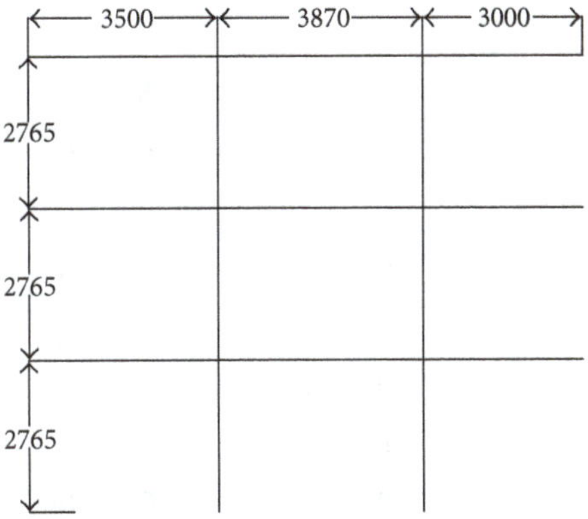

(b) Building elevation

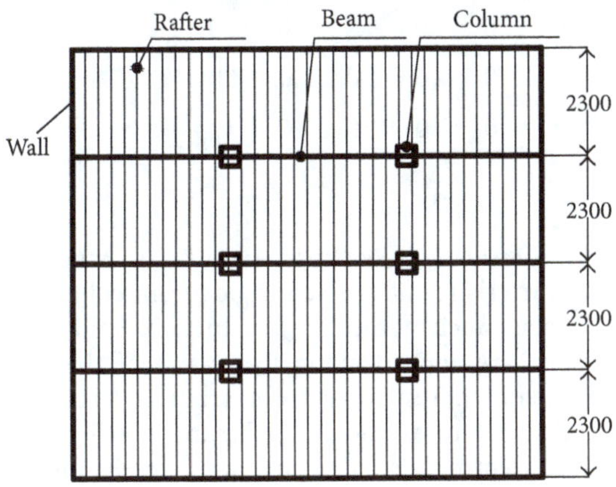

(c) Building plan

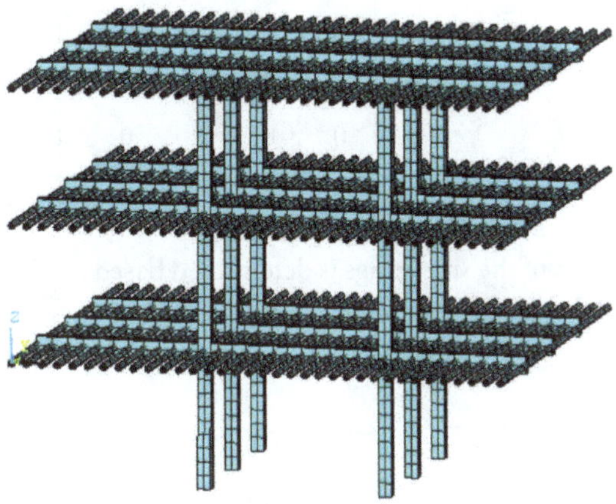

(d) FEM of the building

FIGURE 11: Information of the FEM.

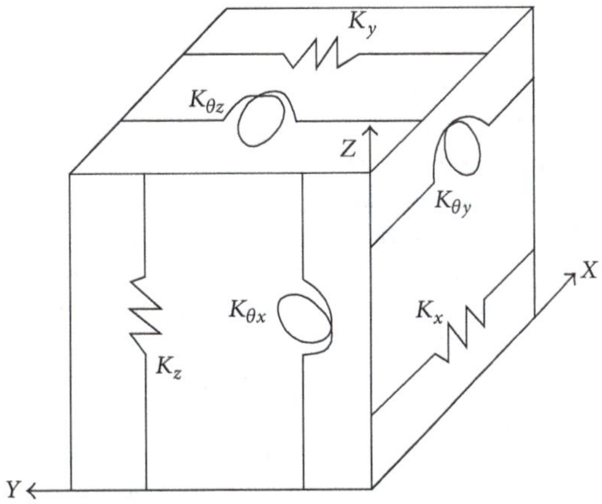

FIGURE 12: Spring element.

as shown in Figure 11(d) consists of three kinds of structural components which are beam, column, and rafter. The cross-sectional dimensions of beam are $b = 0.22$ m and $h = 0.35$ m. The cross section of column is square with 0.27 m edges. The rafter has circular cross section with 0.07 m radius.

The beams, columns, and rafters are all simulated as beam element with six degrees of freedom at each node. As shown in Figure 12, the beam-column joint is simulated by an element composed of 6 separate springs [19]. The stiffness matrix for the element is defined in (1), where k_x, k_y, and k_z represent the translational stiffness values in x, y, and z directions, respectively, while $k_{\theta x}$, $k_{\theta y}$, and $k_{\theta z}$ represent the rotational stiffness around x, y, and z-axis, respectively.

$$k_e = \begin{bmatrix} k_x & 0 & 0 & 0 & 0 & 0 & -k_x & 0 & 0 & 0 & 0 & 0 \\ 0 & k_{\theta x} & 0 & 0 & 0 & 0 & 0 & -k_{\theta x} & 0 & 0 & 0 & 0 \\ 0 & 0 & k_y & 0 & 0 & 0 & 0 & 0 & -k_y & 0 & 0 & 0 \\ 0 & 0 & 0 & k_{\theta y} & 0 & 0 & 0 & 0 & 0 & -k_{\theta y} & 0 & 0 \\ 0 & 0 & 0 & 0 & k_z & 0 & 0 & 0 & 0 & 0 & -k_z & 0 \\ 0 & 0 & 0 & 0 & 0 & k_{\theta z} & 0 & 0 & 0 & 0 & 0 & -k_{\theta z} \\ -k_x & 0 & 0 & 0 & 0 & 0 & k_x & 0 & 0 & 0 & 0 & 0 \\ 0 & -k_{\theta x} & 0 & 0 & 0 & 0 & 0 & k_{\theta x} & 0 & 0 & 0 & 0 \\ 0 & 0 & -k_y & 0 & 0 & 0 & 0 & 0 & k_y & 0 & 0 & 0 \\ 0 & 0 & 0 & -k_{\theta y} & 0 & 0 & 0 & 0 & 0 & k_{\theta y} & 0 & 0 \\ 0 & 0 & 0 & 0 & -k_z & 0 & 0 & 0 & 0 & 0 & k_z & 0 \\ 0 & 0 & 0 & 0 & 0 & -k_{\theta z} & 0 & 0 & 0 & 0 & 0 & k_{\theta y} \end{bmatrix} \quad (1)$$

The stiffness of the six springs is determined based on the numerical and experimental studies [20], where

$$\begin{aligned} k_x &= 10^{10}, \\ k_y &= k_z = 1.579 \times 10^6, \\ k_{\theta x} &= k_{\theta z} = 9.266 \times 10^6. \end{aligned} \quad (2)$$

The stiffness of $k_{\theta y}$ is defined as a bilinear model shown in Figure 13 [14].

The FEM includes 2586 beam elements and 216 springs. The total element number and node number are 2802 and 3051, respectively. The wall in Tibetan ancient timber structure is very thick and it is made of big stone and soil. Since there is no information on the physical properties of the composition material of the wall, in the FEM the wall is assumed as support to the timber frame. Connection between the beam and wall is considered rigid with moment fixity because the beam end is inserted into the wall. As the rafter is considered to have rotational degrees of freedom, only translation in x, y, and z directions have to be restrained. The slab and sticks are simulated as distributed masses on rafter.

The timber material used in the structure is Tibetan Cathay Poplar, and for their properties please refer to [21]. The mass density and elastic modulus of material are 418 kg/m^3 and 6.435 Gpa, respectively. The stress-strain behavior of timber is assumed to be bilinear. Poisson's ratio is 0.2 and the yield stress is 31.01 Mpa. Von Mises yield criterion is applied in the simulation.

4.2. Modal Analysis. The natural frequencies of the structure are obtained using the method of Block Lanczos. The fundamental frequency of the model with semirigid beam-column joints is 2.035 Hz which is quite close to the measured result 2.05 Hz [13]. It is proved that the FEM with semirigid joints is suitable for modeling the structure.

4.3. Earthquake Response Analysis. To study the aseismic characteristics of the semirigid beam-column joint of the Tibetan ancient timber structure, N-S El-Centro seismic waves are applied along the x-axis of the structure. The acceleration responses from arbitrarily selected points of the structure as shown in Figure 14 when under the N-S El-Centro seismic waves with acceleration peak value 70 gal (small earthquakes) and 400 gal (large earthquakes) are studied. The FEM with rigid joints is taken as a reference to evaluate the dynamic and aseismic performance of the model with semirigid joints. In the FEM with rigid joints, the translational stiffness values k_x, k_y, and k_z are set as 10^{15} N/m and the rotational stiffness values $k_{\theta x}$, $k_{\theta y}$, and $k_{\theta z}$ are set as 10^{15} Nm/rad.

Figures 15(a) and 15(b) are the acceleration responses of Node 2921 (column) and Node 2644 (beam) in the FEM with semirigid joints under El-Centro waves (Duration 10 s, time interval 0.02 s) with acceleration peak values 70 gal and 400 gal, respectively. Node 2921 and Node 2644 have the same coordinates and are on two sides of a semirigid joint with 6 springs. The average of the absolute value of the acceleration response history is used as a measure for comparison of the two sets of acceleration responses. It can be found in Table 1 that the average values of Node 2921 are bigger than that of Node 2644. It is revealed that the semirigid beam-column connections are just like dampers that can reduce the earthquake responses by friction and extrusion.

Figures 15(c) and 15(d) show the acceleration responses of Node 2798 in the two FEMs under El-Centro waves with acceleration peak values 70 gal and 400 gal, respectively. It is noted that the peak acceleration responses of Node 2798 in the FEM with semirigid joints are lower than those in the FEM with rigid joints. Figures 15(e) and 15(f) are acceleration responses of Node 2920 in the two FEMs and the results are similar to Figures 15(c) and 15(d). It can be found in Table 1 that the average values of Node 2920 are bigger than that of

TABLE 1: Average of the absolute value of the acceleration response history for the arbitrarily selected points of the model.

Node	FEM	Average of the absolute value of the acceleration response history (m/s^2)	
		Seismic wave (70 gal)	Seismic wave (400 gal)
2644	With semirigid joint	0.1265	0.7519
2921	With semirigid joint	0.1312	0.8654
2920	With semirigid joint	0.1980	0.9950
2920	With rigid joint	0.2120	0.9224
2798	With semirigid joint	0.1265	0.7206
2798	With rigid joint	0.2096	1.0017

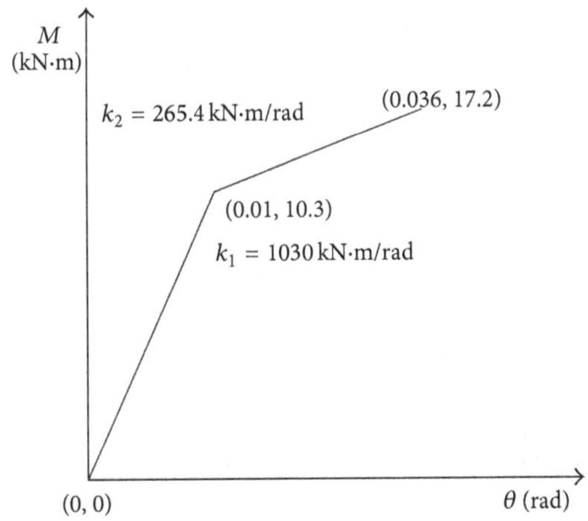

FIGURE 13: Moment-rotation relationship for $k_{\theta y}$.

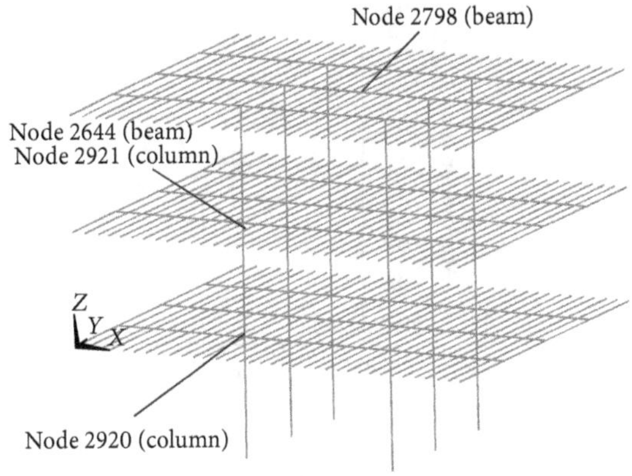

FIGURE 14: Arbitrarily selected points of the model.

Node 2644. Results reveal that the semirigid beam-column joints are helpful for seismic resistance of the Tibetan ancient timber structure.

5. Structural Damage and Rehabilitation Methods

Most Tibetan ancient timber structures are suffering from different degrees of structural damage which are adverse to the seismic resistance. Typical damage problems of the structural components are summarized below and the main causes of this damage are analyzed. Besides, different rehabilitation methods for maintenance of Tibetan ancient timber structure are discussed.

5.1. Typical Structural Damage. According to the provisions of technical specifications on maintenance and rehabilitation of Chinese ancient timber structure, the following requirements of the structural components should be satisfied [21]:

(1) Column: material should be in good condition; degrees of cracking and corrosion at the base should be within the allowance.

(2) Mortise and tenon connection: the mortise and tenon should be connected tightly; the extent of pulling out of the tenon from the mortise, cracking, and the crookedness of member should be within the allowance.

(3) Beam: material should be in good condition; excessive deformation or big cracks are not acceptable. The end of beam should be well-connected with other structural components.

(4) Rafter: material should be in good condition; rafters should be firmly connected to the floor slab.

(5) Overall tilt in horizontal direction of the structure should be within the allowance.

Field surveys on structural damage in typical Tibetan ancient timber structures, such as the Patala Palace, Jokhang Temple, and Norbulingka in the city of Lhasa, have been carried out [22]. Approximately one hundred rooms have been surveyed. The typical structural damage is summarized into the following four categories based on these survey results.

(1) Damage of Column. Corrosion of column at the base and crack are the main damage of column, as shown in Figures 16(a) and 16(b). The cross section of the corrupted base would be smaller. When there is an earthquake, the stress in column will substantially increase at the bottom end with possible insufficient axial capacity and large lateral displacement of column.

(2) Damage of Beam. Crack and large deflection are the main damage in beam, as shown in Figures 17(a) and 17(b). Cracks are associated with a smaller cross section of the beam, and thus the sectional stress and deflection of beam increase.

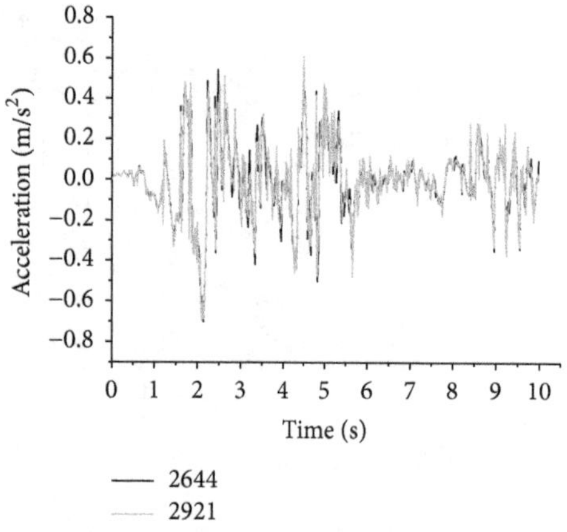

(a) Acceleration responses of Nodes 2644 and 2921 under El-Centro wave with acceleration peak value 70 gal

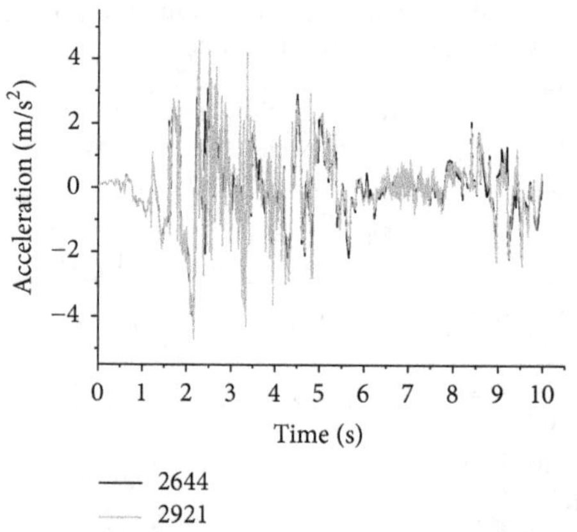

(b) Acceleration responses of Nodes 2644 and 2921 under El-Centro wave with acceleration peak value 400 gal

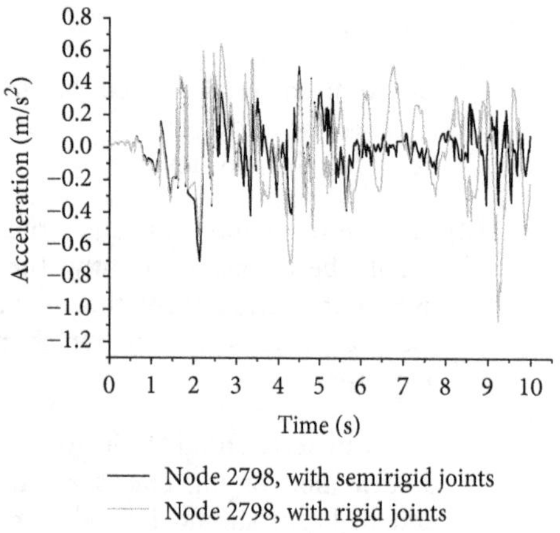

(c) Acceleration responses of Node 2798 in the two FEMs under El-Centro wave with acceleration peak value 70 gal

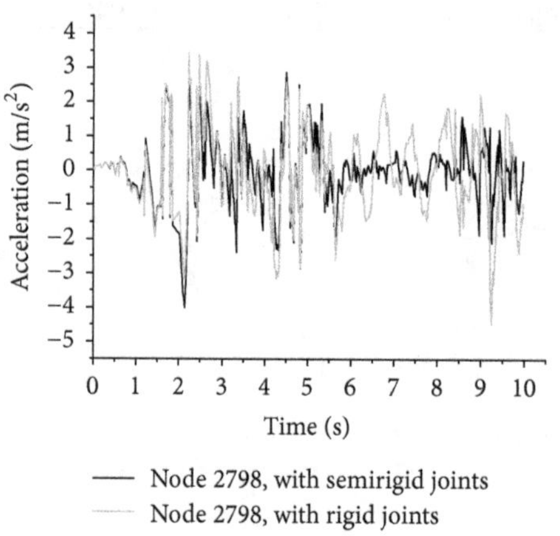

(d) Acceleration responses of Node 2798 in the two FEMs under El-Centro wave with acceleration peak value 400 gal

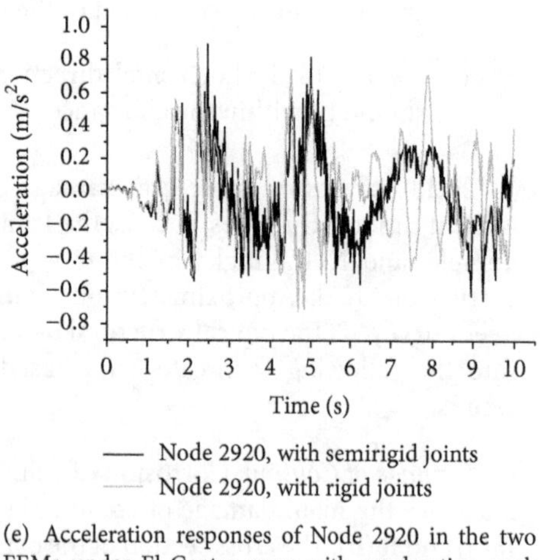

(e) Acceleration responses of Node 2920 in the two FEMs under El-Centro wave with acceleration peak value 70 gal

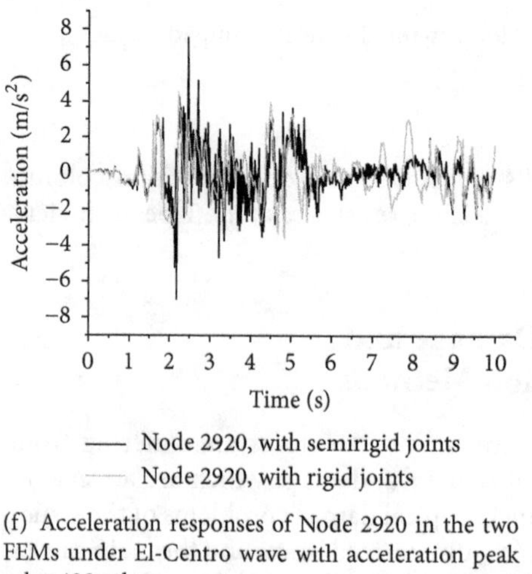

(f) Acceleration responses of Node 2920 in the two FEMs under El-Centro wave with acceleration peak value 400 gal

FIGURE 15: Acceleration responses at the selected points under El-Centro seismic waves.

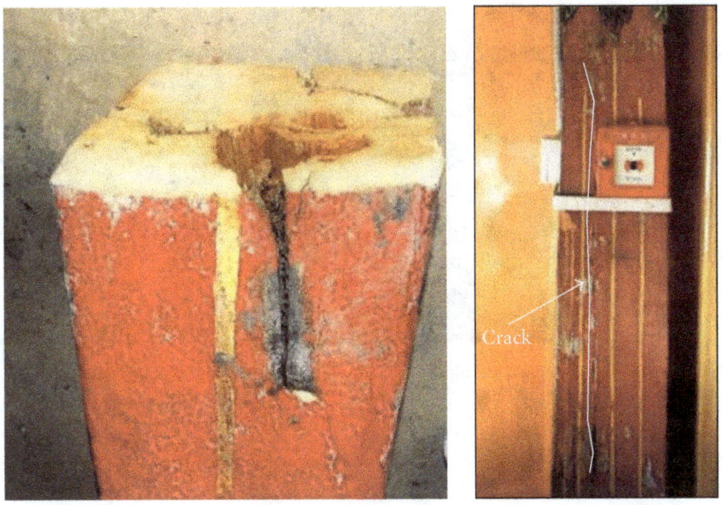

(a) Corrosion of column at the base (b) Crack in column

FIGURE 16: Damage of column.

(a) Cracks in beam (b) Large deflection

FIGURE 17: Damage of beam.

When earthquake happens, the damaged beams are easier to fail.

(3) Damage in the Cap Block and Chock Block. Radial cracks often exist in the cap blocks as shown in Figure 18(a). Out-of-plane distortion is a general type of damage in the chock blocks, as shown in Figure 18(b) that may cause eccentric compression on column and influence the overall stability of the structure.

(4) Damage of Rafter. Crack and corrosion are the main damage in rafter, as shown in Figures 19(a) and 19(b). Cracks accelerate the internal decay of the rafter. Damp and insect induced corruptions make the cross-sections of corrupted rafters become smaller. When earthquake happens, pebbles and stones in the upper floor slab are easier to get loose and fall out.

5.2. Analysis on the Causes of Damage. The damage of structural components is mostly caused by a combination of factors which are summarized into the following categories.

(1) Material Properties. The elastic modulus of timber is low compared to other building materials, and it decreases with time. Under long-term loading, large deformation of structural components may occur, for example, Figure 17(b). Shear strength of timber material is small and shear failure occurs in the structural component, for example, Figures 17(a) and 19(a). Timber is easily corrupted in humid environment, for example, Figure 16(a). Shrinkage of the timber during the drying process will produce cracks, for example, Figures 16(b) and 18(a). In addition, timber is vulnerable to insect effects that may cause damage in the component, for example, Figure 19(b).

(2) Member Arrangement Configuration. Timber frames are pieced together by different components without nails or rivets. Different structure components are connected to each other by short sticks or mortise and tenon connections. The connections need to have open holes which may generate stress concentration causing cracks and additional deformations of the structural components. The distance between adjacent beams and columns is about 1/3 of the height of column, because of the presence of the cap block and chock block. If the upper load distribution is not uniform or the structural components themselves have defects, out-of-plane distortion will happen in the chock blocks, for example, Figure 18(b).

(a) Crack in cap block

(b) Out-of-plane distortion in the chock blocks

FIGURE 18: Damage of cap block and chock block.

(a) Cracks in rafter

(b) Corrosion

FIGURE 19: Damage of rafter.

(3) Construction. The base of the column is directly resting on the stone base which is embedded in "Aga" soil forming the floor slab. The column bottom is easily corrupted in the humid environment, for example, Figure 16(a). Similarly, the end of rafter is inserted directly into the wall, and it is also easily corrupted with poor ventilation. For beam-column joint, if the size of the open hole for inserting the short sticks is too large, the joint will have initial deformation after installation and it may induce out-of-plane distortion in the chock blocks, for example, Figure 18(b).

(4) Environment Conditions. Tibet is an earthquake-prone area and yet many Tibetan ancient timber structures remained centuries after their construction. This indicates that they have good seismic performance. However, earthquakes cause damage and deformations of the structures. Wind, rain, and snow also have adverse effects. Humid environment will cause corrosion to the structural components. With the recent development of tourism in Tibet, many ancient buildings are opened to visitors, and the adverse effects of crowd load on the structure will also contribute to the corruption of the structural members.

5.3. Rehabilitation Methods. To ensure the seismic resistance of Tibetan ancient timber structure, some rehabilitation methods are discussed to provide reference for maintenance of the different structural components.

(1) Rehabilitation Methods for Cracks. Traditional metal hoop rehabilitation method can be used for big cracks as shown in Figures 20(a) and 20(b). If the crack is relatively small, wooden bar is used to fill into the crack as shown in Figure 20(c). CFRP which has strong tensile strength can be used to prevent the deformation of the cracked components and to improve the load bearing capacity of the cracked components [23].

(2) Joint Rehabilitation Methods. The rehabilitation of joints in ancient timber structure must satisfy the following requirements [24]: (a) The semirigidity of the joint that can absorb earthquake energy should be maintained; (b) the inner connection between the joint components should be strengthened in order to improve the structural rigidity. Methods using armor plate and CFRP are usually used to reinforce the joints of timber structure [25].

(3) Rehabilitation Method for Beam with Large Deflection. One extra column can be added as an additional support at mid-span of the beam as shown in Figure 21. Internal force redistribution of the beam will take place with the new support. The deflection of beam can be reduced while the bearing capacity of beam can be improved.

(4) Rehabilitation of the Corrupted Components. Conventional method can be used for corrupted column, for example, to replace the corrupted part with new material and

(a) Metal hoop rehabilitation method on column

(b) Metal hoop rehabilitation method on rafter

(c) Bar filling method on column

FIGURE 20: Rehabilitation methods for structural components with cracks.

FIGURE 21: Rehabilitation method for beam with large deflection.

use metal hoop to wrap the connection. Glue applied on the corrupted part can be used for reinforcing the corrupted beam. The replacement of corrupted rafters is a little bit complicated as shown in Figures 22(a) and 22(b). The upper floors will have to be removed first. The new rafters replace the old ones, and then the upper floors are put back in position.

6. Conclusions

This paper studied the structural configurations of Tibetan ancient timber structure, such as wall, roof and slab, and timber skeleton with semirigid joints. It is found that these construction characteristics enhance the seismic resistance of Tibetan ancient timber structure. One of the unique features of Tibetan ancient timber structure is its special beam-column joint which has a semirigid behavior. Numerical simulations of a typical 3-storey structural unit in the Potala Palace are studied. The acceleration responses at some arbitrarily selected points of the structure under the action of N-S El-Centro seismic waves are studied. Peak acceleration responses of the structure model with semirigid joints are much lower than those from the model with rigid joints. The joints work like dampers which can reduce the earthquake responses by friction and extension of members. Results reveal that the semirigid joint is helpful for seismic resistance of Tibetan ancient timber structure. This paper also summarizes the structural damage which is adverse for seismic resistance. They are mainly categorized into problems with the material properties, member arrangement configuration, construction, and environment conditions. Different rehabilitation methods of structural components are discussed providing references for maintenance of the Tibetan ancient timber structure.

Competing Interests

The authors declare that there is no conflict of interests regarding the publication of this paper.

Acknowledgments

This work is supported by the National Natural Science Foundation of China (nos. 51408038 and 51338001) and the

(a) New rafters

(b) Floor removing

FIGURE 22: Rehabilitation of corrupted rafters.

Talent Fund of Beijing Jiaotong University (no. 2014RC011). The authors would also like to thank the research project funding of the 111 Project of China (no. B13002), the National Key Technology R&D Program (no. 2015BAK01B02), and the Beijing Natural Science Foundation (no. 8151003).

References

[1] S. C. Liang, *A Pictorical History of Chinese Architecture*, MIT Press, Cambridge, Mass, USA, 1984.

[2] B. J. Ma, *Wood Construction Techniques of Chinese Ancient Architecture*, Science Press, Beijing, China, 1991.

[3] T. Wang, *Static Mechanics of Ancient Timber Structures*, Historical Relic Press, Beijing, China, 1992.

[4] M. H. Yu, Y. Oda, and D. P. Fang, "Advances in structural mechanics of Chinese ancient architectures," *Advances in Mechanics*, vol. 36, no. 1, pp. 43–64, 2006.

[5] D. P. Fang, S. Iwasaki, M. H. Yu, Q. P. Shen, Y. Miyamoto, and H. Hikosaka, "Ancient Chinese timber architecture. II: dynamic characteristics," *Journal of Structural Engineering*, vol. 127, no. 11, pp. 1358–1363, 2001.

[6] J. Y. Xue, P. C. Zhang, and H. T. Zhao, "Study on the aseismic mechanism of historic timber structural building," *Journal of Xi'an University of Architecture & Technology*, vol. 32, no. 1, pp. 8–11, 2000.

[7] P. C. Zhang, *Study on structure and its seismic behavior development of Chinese ancient timber structure buildings [Ph.D. thesis]*, Xi'an University of Architecture & Technology, Xi'an, China, 2003.

[8] L. Sui, *Study on aseismic mechanics and analysis on Tou-Kung of Chinese ancient buildings [M.S. thesis]*, Xi'an University of Architecture & Technology, Xi'an, China, 2006.

[9] K. Yao, H.-T. Zhao, and H.-P. Ge, "Experimental studies on the characteristic of mortise-tenon joint in historic timber buildings," *Engineering Mechanics*, vol. 23, no. 10, pp. 168–173, 2006.

[10] S. D. Xue, J. Zhao, and X. Y. Gao, *Seismic Design of Building*, Science Press, Beijing, China, 2007.

[11] H. Y. Jiang and P. Gasu, *Potala(Lhasa, China)-Conservation and Restoration*, Cultural Relics Publishing House, Beijing, China, 1994.

[12] Y. D. Chen, *Tibetan Architecture in China*, China Architecture & Building Press, Beijing, China, 2007.

[13] T. Guo and N. Yang, "Dynamic crowd loading test on Tibetan ancient structures," in *Proceedings of the 4th International Symposium on Environment Vibrations-Prediction, Monitoring, Mitigation and Evaluation*, pp. 591–598, Beijing Jiaotong University, Beijing, China, October 2009.

[14] H. C. Yan, *Experimental research and finite element analysis of timber beam-column connection of Tibetan ancient architecture [M.S. thesis]*, Beijing Jiaotong University, Beijing, China, 2011.

[15] N. Yang, P. Li, S. S. Law, and Q. Yang, "Experimental research on mechanical properties of timber in ancient tibetan building," *Journal of Materials in Civil Engineering*, vol. 24, no. 6, pp. 635–643, 2012.

[16] G. X. Pan, *A History of Chinese Architecture*, China Architecture & Building Press, Beijing, China, 2015.

[17] The Institute for the History of Natural Sciences and Chinese Academy of Sciences, *History of Chinese Ancient Architecture Technology*, Science Press, Beijing, China, 2000.

[18] A. W. Luo Dan, C. Duo, and P. Ci, *Overview of Tibetan Architecture in Tibet*, Sichuan Publishing Group, Chengdu, China, 2007.

[19] D. P. Fang, S. Iwasaki, M. H. Yu, Q. P. Shen, Y. Miyamoto, and H. Hikosaka, "Ancient Chinese timber architecture. I: Experimental Study," *Journal of Structural Engineering*, vol. 127, no. 11, pp. 1348–1357, 2001.

[20] S. Cang, R. X. Zhu, C. X. Ren, and M. H. Yu, "Dynamic analysis of ancient wood structures with mortise and tenon connection," *Journal of Ningbo University (NSEE)*, vol. 17, no. 3, pp. 332–335, 2004.

[21] GB50165-92, "Technical code for maintenance and strengthening of ancient timber buildings," Tech. Rep., 1992.

[22] P. Li, *Mechanical mechanism of beam-column joints in wooden frame of tibetan ancient architecture [M.S. thesis]*, Beijing Jiaotong University, Beijing, China, 2009.

[23] Q. Xu and L. Zhu, "An experimental study on partially-damaged wood columns repaired and strengthened with CFRP," *Journal of China Civil Engineering*, vol. 40, no. 8, pp. 41–46, 2007.

[24] Q. Zhou, W. M. Yan, and J. B. Ji, "Typical aseismic construction problems in ancient Chinese wooden buildings of the Ming and Qing Dynasties," *Science of Conservation and Archaeology*, vol. 23, no. 2, pp. 36–48, 2011.

9

Performance of Retrofitted Self-Compacting Concrete-Filled Steel Tube Beams Using External Steel Plates

Ahmed A. M. AL-Shaar[1,2] **and Mehmet Tolga Göğüş**[1]

[1]*Civil Engineering Department, University of Gaziantep, 27310 Gaziantep, Turkey*
[2]*Civil Engineering Department, AL-Nahrain University, 10072 Baghdad, Iraq*

Correspondence should be addressed to Ahmed A. M. AL-Shaar; ahmedcst77@gmail.com

Academic Editor: Hongchao Kou

Self-compacting concrete-filled steel tube (SCCFST) beams, similar to other structural members, necessitate retrofitting for many causes. However, research on SCCFST beams externally retrofitted by bolted steel plates has seldom been explored in the literature. This paper aims at experimentally investigating the retrofitting performance of square self-compacting concrete-filled steel tube (SCCFST) beams using bolted steel plates with three different retrofitting schemes including varied configurations and two different steel plate lengths under flexure. A total of 18 specimens which consist of 12 retrofitted SCCFST beams, three unretrofitted (control) SCCFST beams, and three hollow steel tubes were used. The flexural behaviour of the retrofitted SCCFST beams was examined regarding flexural strength, failure modes, and moment versus deflection curves, energy absorption, and ductility. Experimental results revealed that the implemented retrofitting schemes efficiently improve the moment carrying capacity and stiffness of the retrofitted SCCFST beams compared to the control beams. The increment in flexural strength ranged from 1% to 46%. Furthermore, the adopted retrofitting schemes were able to restore the energy absorption and ductility of the damaged beams in the range of 35% to 75% of the original beam ductility. Furthermore, a theoretical model was suggested to predict the moment capacity of the retrofitted SCCFST beams. The theoretical model results were in good agreement with the test results.

1. Introduction

Concrete-filled steel tube (CFST) members have been used over the world for high-rise buildings, bridges, and other structures involving very large applied moments, particularly in zones of high seismic risk [1–5]. CFST members provide several merits over the conventional tubular or reinforced concrete members, including high strength and stiffness, fire resistances, large energy absorption capacities, and favourable ductility [6, 7].

In the last years, self-compacting concrete (SCC) arrived as a revolution in the field of concrete technology. SCC was originated in Japan in the eighties and then spread to other countries [8–10]. SCC is a high-performance concrete that can discharge under its weight and fill in the formwork without any compaction efforts and without showing defects due to segregation and bleeding. Hence, reduces the noise level and construction time and cost, whereas still achieving good compaction [11]. Since the steel tubes have confined sections, SCC has a high potential to be the most appropriate concrete type in CFST members to secure that the best compaction is provided without any further precautions [12]. The adoption of SCC in CFST members can enhance the constructional quality of concrete, hence ensuring the performance of these composite members. The past experimental research revealed that the behaviour of CFST members filled with SCC is similar to those filled with normal concrete [13, 14].

Similar to other structural members, CFST members may necessitate retrofitting or upgrading for various causes. For instance, CFST members may require being repaired because of degradation due to environmental factors, fire, ageing, and earthquake accidents. Also, they may request strengthening in order to increment the load-carrying capacity or to recover the desired structural behaviour due to design faults, constructional errors, updating of design standard, and modification in the practical usage [15–21].

Hence, CFST members may need to be retrofitted or substituted to compensate for the inadequate capacity.

Fundamentally, there are two different techniques available for external retrofitting of the structural member. The first method is to attach steel plates to the external surfaces of the structural members [16–18, 22–24]. However, the second method is to bond advanced composite materials, such as carbon fiber-reinforced polymers (CFRPs) [15, 19–21, 25–29].

Recently, the applications of FRP composites in civil structures have been more acceptable due to their higher strength, higher stiffness, and low weight compared to steel plates [20, 21]. Nevertheless, there are some disadvantages regarding the implementation of FRP materials. For example, the cost of these materials is still relatively high besides the need for skilled labour [17, 22, 30]. Furthermore, there are uncertainties regarding the long-term performance of these materials and the bonding between the composite materials and steel [20, 26, 27]. On the other hand, the implementation of external steel plating is a prevalent successful retrofitting method that was widely experienced in practice [22, 27]. The advantages of this method include the convenience and relative simplicity of application, the wide availability and low prices of the steel material, and the ductile behaviour and high deformation capacity of the steel plates compared to the composite materials which characterized by brittle nature [17, 22]. In this research, the external steel plating method was used which is expected to increment or recover the flexural strength and stiffness of the SCCFST beams, reducing their service-load deflection and behaving in a ductile manner.

Search in the available literature revealed that many research works had been done on the strengthening and repair of CFST beams. However, most of these investigations have been performed on normal CFST beams, which externally retrofitted by CFRP laminates (e.g., [19–21, 27]). On the contrary, studies that examined the behaviour of SSCFST beams retrofitted with bolted steel plate are too scarce. However, there are many studies that have been conducted on reinforced concrete beams strengthened by steel plates (e.g., [16–18, 22, 30–33]).

Al Zand et al. [21] presented an experimental study on the strengthening performance of circular and rectangular CFST beams using unidirectional CFRP sheets. The results demonstrated that the moment carrying capacity, energy absorption capacity, and flexural stiffness of the strengthened beams considerably enhanced with the increase of CFRP layers. Sundarraja and Prabhu [27] experimentally examined the suitability of CFRP fabrics in the external strengthening of square CFST beams. Results indicated that the flexural strength and stiffness of the strengthened beams incremented when the number of CFRP layers was increased except in the case of beams strengthened by partial wrapping. On the other hand, the ductility index decreased as the number of CFRP layers increased.

Peng et al. [17] studied the behaviour of damaged rectangular reinforced concrete (RC) beams strengthened by bolted steel plates considering the effect of corrosion. The test results showed that adopting bolted steel plates in strengthening enhances the strength and ductility of the retrofitted beams. Also, it was found that the maximum load of strengthened damaged beams increased as the steel plate thickness increased. Aykac et al. [22] experimentally investigated the flexural strengthening and repair behaviour of rectangular RC beams retrofitted using external bolted steel plates. The results revealed that the ductility of the tested beams increased as the plate thickness decreased. The experiments proved that the anchorage of steel plates to the beams using steel bolts was a useful technique for obtaining sufficient strength and ductility and preclude the premature peeling failure of the plates in beams strengthened by thick solid plates.

2. Research Significance and Objectives

In view of the literature above, it can be concluded that research in the field of SCCFST beam retrofitted with external bolted steel plates is still rarely found in the literature. Furthermore, up to the best knowledge of the authors, no any previous study was carried out on the retrofitting of SCCFST beams with external steel plates. Therefore, it is evident that the research in this field is still in the primary stage and far from sufficient, especially if the differences in mechanical properties between the steel plates and the FRP composites were taken into consideration. More efforts are required in this field to fill this gap in the literature, to provide a better understanding of the flexural behaviour of such retrofitted composite beams, and to provide novel experimental information for the engineering practice. Therefore, the primary aims of this study are threefold: first, to generate novel new test data on the retrofitting of SCCFST beams with bolted steel plates; second, to investigate experimentally the flexural performance of retrofitted SCCFST beams regarding moment resistance, failure modes, ductility, and moment versus deflection curves; and finally, to examine the effect of the different parameter on the flexural behaviour of retrofitted SCCFST beams, such as different steel plate arrangements and lengths including full and partial retrofitting schemes.

3. Experimental Investigation

3.1. Material Properties

3.1.1. Self-Compacting Concrete (SCC). In this research, self-compacting concrete (SCC) was adopted as infill material due to its superior workability. The mix design is prepared to attain a targeted compressive strength of 50 MPa with a water to binder (cement and fly ash) ratio of 0.33. The adopted SCC mix was prepared with a binder material consisting of the Portland cement PC (CEM I 42.5R) and fly ash (class F). The river gravel was used as a coarse aggregate with 12 mm maximum particle size. While, the fine aggregate was a mixture of river sand and crushed sand with a maximum size of 4 mm and 2 mm, respectively. Finally, superplasticizer (Glenium 51) was adopted to obtain the required workability of the SCC mix. The mix proportions of the SCC are given in Table 1.

In order to assess and check the workability properties of the adopted SCC, the fresh characteristics of SCC were examined by slump flow and V-funnel tests, according to the EFNARC committee recommendations [34]. The slump flow test represents a substantial check for the flowability of

Table 1: Mix proportions and properties of self-compacting concrete (SCC).

Cement (kg/m^3)	Fly ash (kg/m^3)	Coarse aggregate (kg/m^3)	River sand (kg/m^3)	Crushed sand (kg/m^3)	Water (kg/m^3)	SP (kg/m^3)	f'_c (MPa)	Density (kg/m^3)
430	140	730	670	223	188.1	3.00	59.48	2378.57

Table 2: Workability characteristics of self-compacting concrete (SCC).

Tests	Result	Workability classes
Slump flow diameter (DF) (mm)	650	SF1 (550–650)
Slump flow time (T_{50}) (seconds)	2.5	VS2/VF2 (>2)
V-funnel flow time (V_f) (seconds)	8.5	VS2/VF2 (>8)

Figure 1: (a) Concrete cylinders before testing. (b) Concrete cylinders after testing.

SCC in unconfined zones, segregation resistance, uniformity of the fresh SCC mix, and the consistency to satisfy the specification. On the other hand, the V-funnel test describes the viscosity and deformability of SCC.

The workability characteristics of the fresh SCC mixture used in this research are listed in Table 2, together with the limitations of the EFNARC committee [34].

Three self-compacting concrete cylinders (100 × 200 mm) were cast and cured in water for 28 days and tested at the same time of the corresponding beam specimens as shown in Figure 1. The concrete cylinder samples were tested according to the ASTM C39/C39M-14a standard [35]. At the test day, the average cylinder compressive strength of concrete was determined as 59.48 MPa.

3.1.2. Steel Tubes. This study includes 18 beam specimens divided into three groups. All the steel tubes were made from three cold-formed square hollow sections with 6 m length. The nominal dimensions of the square steel tube sections were 100 × 100 × 3 mm. Three coupons from each tube were cut from the three faces of the hollow steel sections to determine the actual material properties as shown in Figure 2.

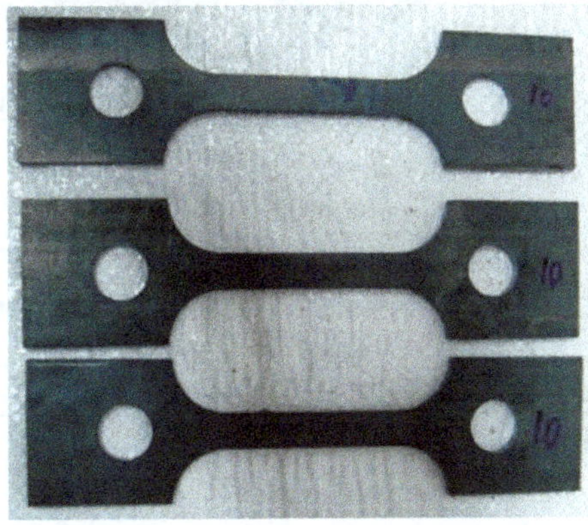

Figure 2: Steel coupon specimen.

Table 3: Properties of steel tubes.

Group name	Tube cross section $B \times H \times t$ (mm)	Yield strength (f_y) (MPa)	Ultimate strength (f_u) (MPa)
A	100 × 100 × 3	287.52	382.17
B	100 × 100 × 3	288.97	381.72
C	100 × 100 × 3	285.53	377.64

Table 4: Properties of steel plates.

Steel plate thickness (mm)	Yield strength (f_y) (MPa)	Ultimate strength (f_u) (Mpa)
3.0	322.32	436.95

Tensile coupon tests were performed and tested according to the recommendations of ASTM E8/E8M-15a [36]. The average yield and ultimate tensile stress for the steel tubes are listed in Table 3.

3.1.3. Steel Plates and Bolts. Rectangular steel plates with nominal thicknesses of 3 mm were used for the retrofitting of the SCCFST beams. The steel plates had a width of 94 mm and fabricated with two different lengths, 925 mm and 300 mm, respectively. Similar to the steel tubes, three coupons were taken from the adopted steel plates to conduct the tensile coupon according to the recommendations of ASTM E8/E8M-15a [36]. The average yield and ultimate tensile stress for the steel plates are listed in Table 4.

All of the steel plates were drilled with an electrical rotary drill to produce two rows of circular holes distributed in

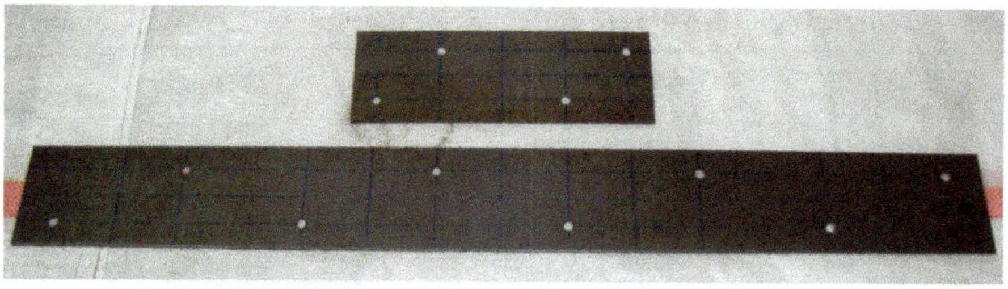

FIGURE 3: Typical perforated steel plates.

TABLE 5: Properties of bolts.

Bolt type	Diameter (mm)	Length (mm)	Yield strength (MPa)	Tensile strength (MPa)
Grade 8.8	6.3	40	640	800
Grade 12.9	7.9	45	1100	1220

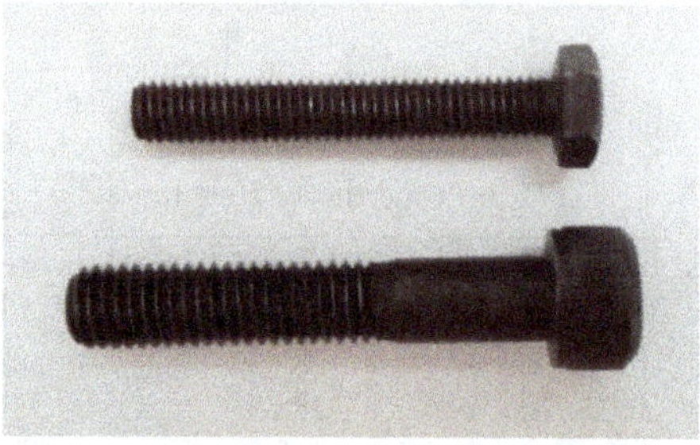

FIGURE 4: Typical bolts.

a staggered manner. All holes were performed with a suitable diameter and the center to center distance between holes was 125 mm and 50 mm, in the longitudinal and transverse directions, respectively. Figure 3 depicts the perforated steel plates.

Two types of bolts were utilized in this work. Grade 8.8 and 12.9 threaded bolts were used to fasten the steel plates to the bottom and top faces of the SCCFST beams. The details and mechanical properties of these bolts are presented in Table 5. Figure 4 shows photos of typical anchor bolts adopted for retrofitting.

3.2. Specimen Preparation. The three cold-formed steel tubes of six meters long were cut and machined to the desired length. The total span of each beam specimen is 950 mm with an effective span of 750 mm between the end supports. The inside faces of the beam specimens were brushed and cleaned out. The bottom ends of the hollow steel tubes were capped with suitable steel base plates, the SCC was poured from the top without applying any compaction, and the steel tubes were kept in a vertical position for a curing period of 28 days. After that, the SCC surface at the top end of the beam was smoothed with epoxy coating and welded with a steel base plate to form complete composite beams. Holes with a suitable diameter were predrilled through the steel flanges and the concrete core of the SCCFST beams. These predrilled holes were executed at the locations where the retrofitting steel plates were required to be anchored. The inside of the predrilled holes was cleaned carefully using air to get rid of any dust or loose parts. Epoxy resin with high-strength properties was used to fill the constructed holes partially. Then, the predrilled steel plates were attached to be coinciding with the beam flanges. Finally, the steel bolts were installed in the holes and fastened tightly to hold and anchor the steel plates firmly to the beam flanges. Figure 5 depicts the different stages of drilling and anchoring of steel plates to the SCCFST beams.

3.3. Beams and Retrofitting Scheme Details. A total of 18 beam specimens of square sections were prepared for testing in this study. The 18 beams consist of 12 SCCFST beams retrofitted with steel plates with two different lengths and three distinct retrofitting schemes, three SCCFST beams without any strengthening (control beams) for comparison purposes, and three hollow steel tube beams. The 12 retrofitted SCCFST beams were also divided into two branches. 6 SCCFST beams prepared as intact (not preloaded) beams. These beams were strengthened with steel plates and bolts (grade 8.8 bolts) and then tested without any preloading. These beams aim at exploring the influence of drilling holes in the body of the SCCFST beams on both the steel tube and concrete. On the other hand, the remaining 6 SCCFST beams were considered as retrofitted (preloaded) beams. That is to say, these beams were initially preloaded up to a certain level, to simulate the damage condition. After that, the retrofitting techniques including the steel plates and bolts (grade 12.9 bolts) were applied to these beams before their testing.

The beam specimens were classified into three different groups A, B, and C. These groups represent three different retrofitting schemes. Each group consists of five beams, namely, one hollow beam, one control beam, two intact beams, and two retrofitted beams. The dimensions and details of each beam specimen are listed in Table 6. Regarding group A beams, the retrofitting technique was achieved using external single steel plates applied to the bottom face of the SCCFST beams. On the other hand, beams of group B were retrofitted using external steel plates anchored to the top and bottom surfaces of the SCCFST beams. Finally, double steel plates were fastened to the bottom face of the SCCFST beams in group C.

Moreover, in each group, the retrofitting technique was utilized using two different steel plate lengths: full retrofitting length in which the steel plate was attached to the full length of the SCCFST beams and partial length in which the steel plates

FIGURE 5: Preparation stages for the retrofitted beams. (a) SCCFST beams. (b) Drilled SCCFST beams. (c) Typically drilled hole. (d) Application of epoxy and installation of bolts through holes.

TABLE 6: Dimensions and details of the beam specimens.

Group name	Beam designation	Beam dimensions $H \times B \times t$ (mm)	Steel plate dimensions $B_p \times t_p \times L_p$ (mm)	Details
Group A (single bottom plate)	HB-A	$100 \times 100 \times 3$	—	Hollow
	CB-A		—	Control
	IB-PL-SB		$94 \times 3 \times 925$	Intact
	IB-FL-SB		$94 \times 3 \times 300$	Intact
	RB-PL-SB		$94 \times 3 \times 925$	Retrofit
	RB-FL-SB		$94 \times 3 \times 300$	Retrofit
Group B (top and bottom plates)	HB-B	$100 \times 100 \times 3$	—	Hollow
	CB-B		—	Control
	IB-PL-TB		$94 \times 3 \times 925$	Intact
	IB-FL-TB		$94 \times 3 \times 300$	Intact
	RB-PL-TB		$94 \times 3 \times 925$	Retrofit
	RB-FL-TB		$94 \times 3 \times 300$	Retrofit
Group C (double bottom plates)	HB-C	$100 \times 100 \times 3$	—	Hollow
	CB-C		—	Control
	IB-PL-DB		$94 \times 3 \times 925$	Intact
	IB-FL-DB		$94 \times 3 \times 300$	Intact
	RB-PL-DB		$94 \times 3 \times 925$	Retrofit
	RB-FL-DB		$94 \times 3 \times 300$	Retrofit

Note: H, B, t: beam depth, width, and thickness; $B_p \times t_p \times L_p$: steel plate width, thickness, and length.

Figure 6: Retrofitting schemes of SCCFST beams. (a) Bottom flange view. (b) Side view. (c) Groups A and C retrofitting schemes. (d) Group B retrofitting scheme. (e) Double bottom steel plates. (f) Top and bottom steel plates.

connected to only one third (middle third) of the beam length. Figure 6 illustrates the details of the various retrofitting schemes adopted to the SCCFST beam specimens.

The following labelling system was used to designate each beam specimen in Table 6. For instance, HB and CB indicate hollow steel tube and control (unstrengthened) SCCFST beams, respectively. RB-PL-SB and RB-FL-SB represent damaged beams (RB) retrofitted by single bottom (SB) steel plates with partial length (PL) or full length (FL), respectively. Whereas, IB-PL-SB and RB-FL-SB refer to the strengthened intact beams (IB). Similarly, SCCFST beams in groups B and C have the same designations of the above beams, except the symbols (TB) in group B and (DB) in group C which refer to retrofitting with top and bottom (TB) steel plates and double bottom steel (DB) plates, respectively.

3.4. Test Procedure. All the beam specimens were tested under pure bending moment up to failure. All beam specimens had a span length of 950 mm (an effective span of 750 mm) and were placed in a simply supported condition as shown in Figure 7. The 500 KN capacity testing machine was used to conduct the test. The beams were tested under the four-point loading method providing a constant bending zone of 250 mm. The force was applied using a displacement control method with a loading rate of 1 mm/min. The implementation of displacement control method allowed the experiments to be continued into the post-peak stage and enabled better control when recording the postultimate behaviour [37]. Deflections along the beam span were measured by three linear variable displacement transducers (LVDTs). One was placed at the midspan of the specimen;

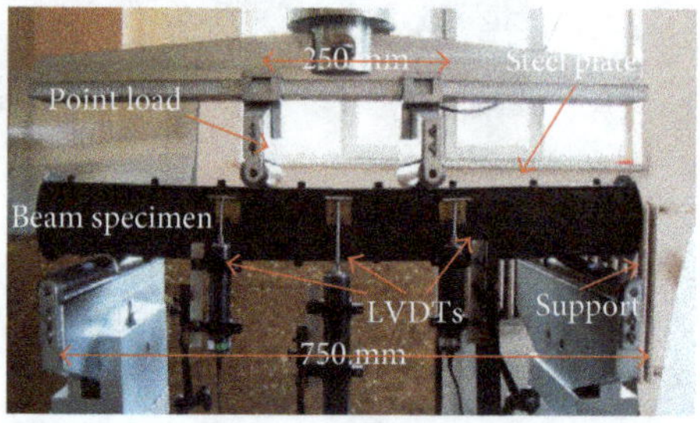

Figure 7: Beam test setup.

the other two were placed under concentrated loads with the shifting of 50 mm from the left and right of the concentrated load positions as shown in Figure 7. The above testing procedure was applied directly to the hollow, control, and intact SCCFST beams. However, for the retrofitted SCCFST beams, a preloading stage was implemented to simulate the damage condition. To achieve this goal, SCCFST beams were preloaded before retrofitting. The preloading stage was done using the same loading setup as explained above. In the beginning, the beams were loaded until reaching a moment capacity of 90% of the ultimate moment capacity recorded for the corresponding control beam.

4. Test Results and Discussion

4.1. Failure Modes. All the tested composite beam specimens reached their ultimate moment capacity with no lateral movement signs of the cross section or any other instability form. The typical failure modes of all the SCCFST beams are shown in Figure 8.

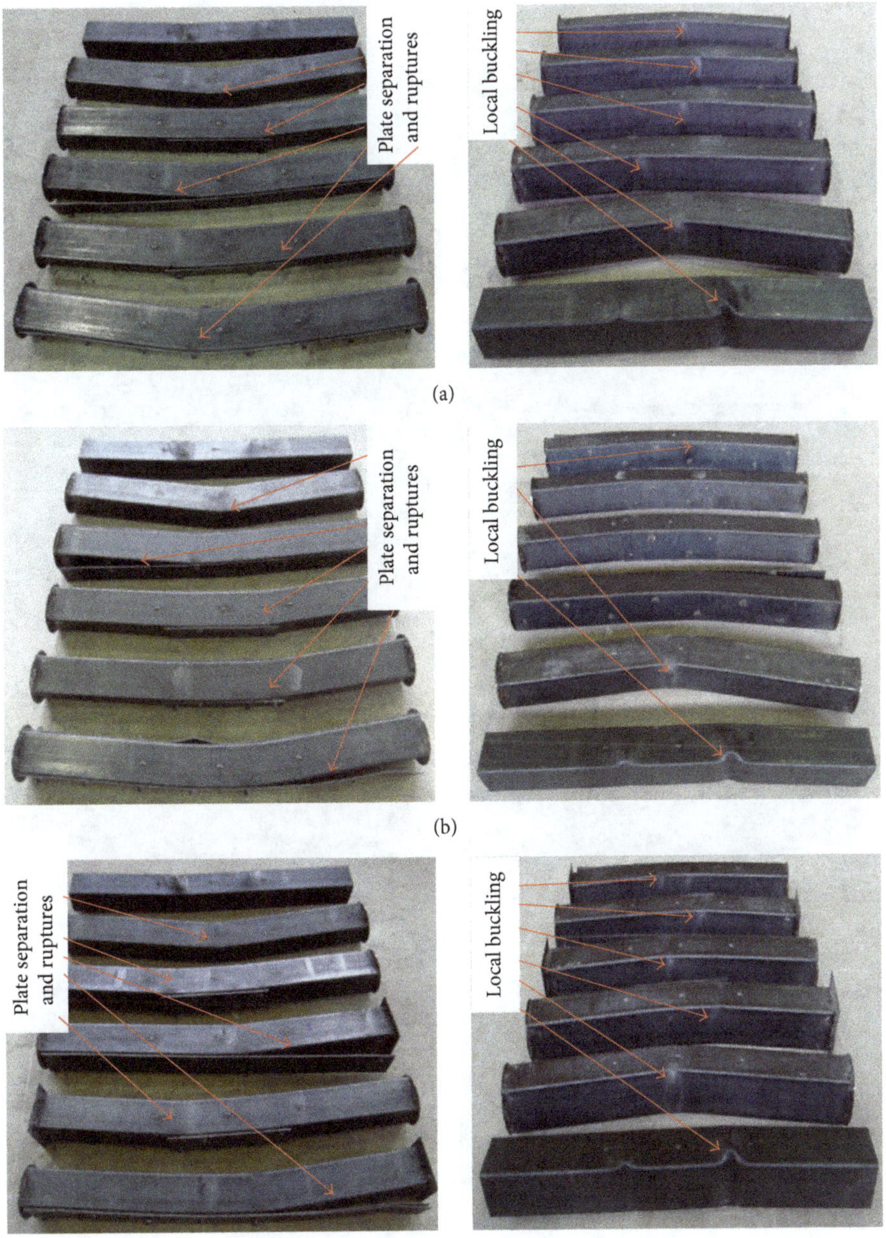

FIGURE 8: Failure modes of tested beams (general). (a) Group A retrofitting scheme (single bottom plate: SB). (b) Group B retrofitting scheme (top and bottom plates: TB). (c) Group C retrofitting scheme (double bottom plates: DB).

The hollow steel tube beams in all groups failed by the inward local buckling mode under the two loading points at the top flange. On the other hand, SCCFST control beams failed by the outward local buckling mode of the steel tube at the top compression flange with a single outward local buckling at the midspan section within the constant bending moment zone. It can be expected that the concrete at the local buckling locations was crushed. Finally, the control beams failed by initiation of rupture at the bottom tensile steel flange at the same location of the observed local buckling as shown in Figure 9(a). It was obvious that the concrete core at the rupture zone was cracked. The difference in failure modes between SSCFST beams and the hollow beams indicated that the SCCFST core was able to change the failure mode of the hollow steel beam from inward to outward local buckling at a greater flexural strength and deformation capacity.

Regarding beams of group A (single bottom steel plate retrofitting scheme), the typical failure mode of all the intact and retrofitted beams with different retrofitting lengths (beams retrofitted partially or entirely along their length) was outward local buckling of the top compression steel flange as shown in Figure 8(a). Similar to the control beams, outward local buckling was observed either close to the midspan section or near the locations of the applied loads within the constant bending zone. In addition to the local

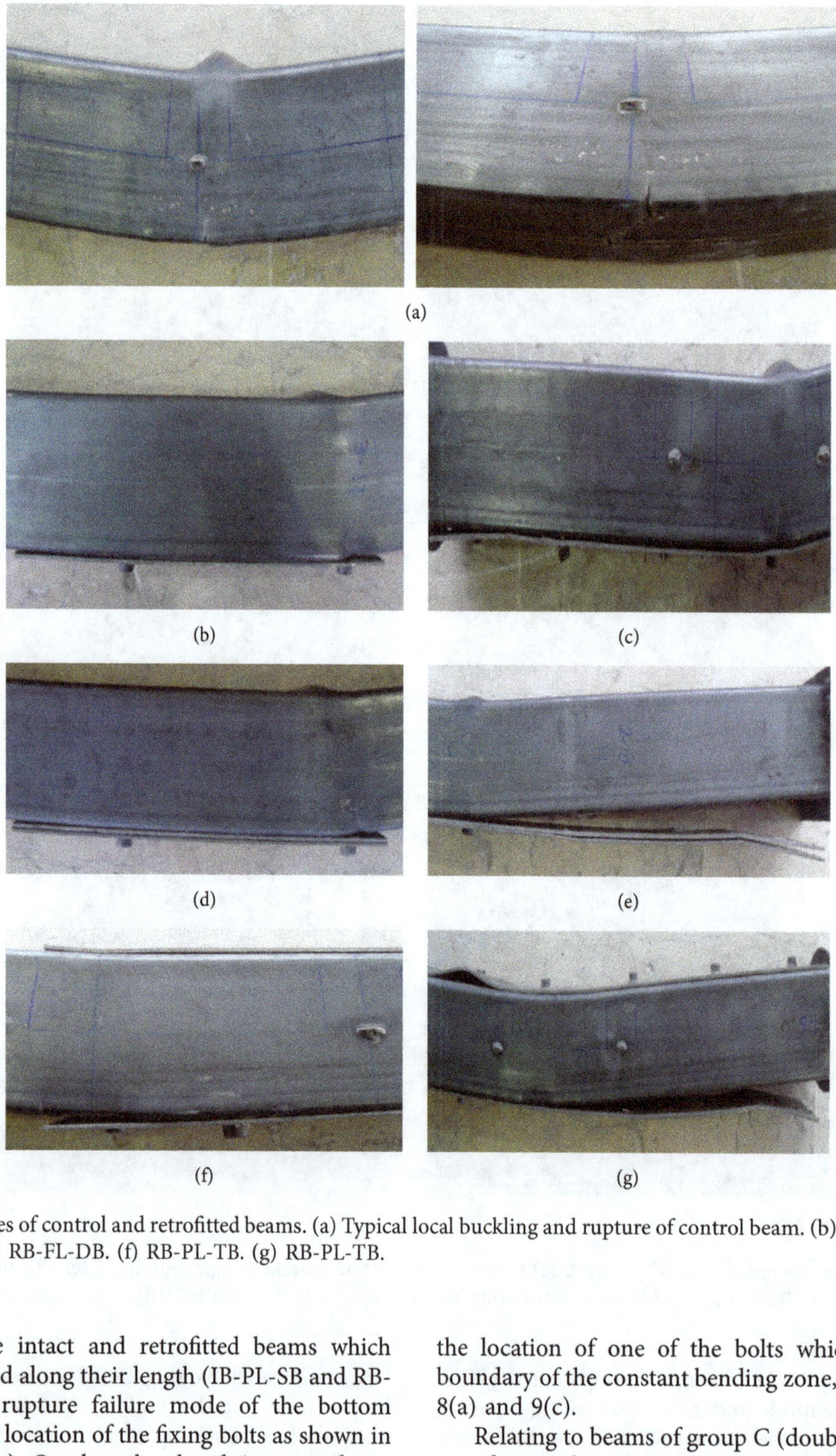

FIGURE 9: Failure modes of control and retrofitted beams. (a) Typical local buckling and rupture of control beam. (b) RB-PL-SB. (c) RB-FL-SB. (d) RB-PL-DB. (e) RB-FL-DB. (f) RB-PL-TB. (g) RB-PL-TB.

buckling mode, the intact and retrofitted beams which partially strengthened along their length (IB-PL-SB and RB-PL-SB) revealed a rupture failure mode of the bottom tension flange at the location of the fixing bolts as shown in Figures 8(a) and 9(b). On the other hand, intact and retrofitted beams which fully strengthened along their length (IB-FL-SB and RB-FL-SB) revealed a different failure trend. Beams IB-FL-SB and RB-FL-SB underwent ruptures of some of their anchoring bolts within the beam shear span. This failure mode resulted in delamination and separation of the attached retrofitting single plate from the beam bottom flange on one side due to the successive bolt rupture. Finally, the local buckling, bolt rupture, and steel plate separation developed the rupture failure of the tension bottom flange at the location of one of the bolts which are close to the boundary of the constant bending zone, as shown in Figures 8(a) and 9(c).

Relating to beams of group C (double bottom steel plate retrofitting scheme), Figure 8(c) showed that all beams failed with the typical local buckling mode of the top compression flange in a comparable mode to those recorded in the control beams and beams of group A, as shown in Figure 8(c). Furthermore, concerning the failure modes of the bolted double bottom steel plates and the bottom tension flange of the tested beams, all beams in this group (C) showed similar failure mechanism as that stated previously for their counterparts in group A, as shown in Figures 8(c), 9(d), and 9(e). Hence, it can be indicated that beams of groups A and C

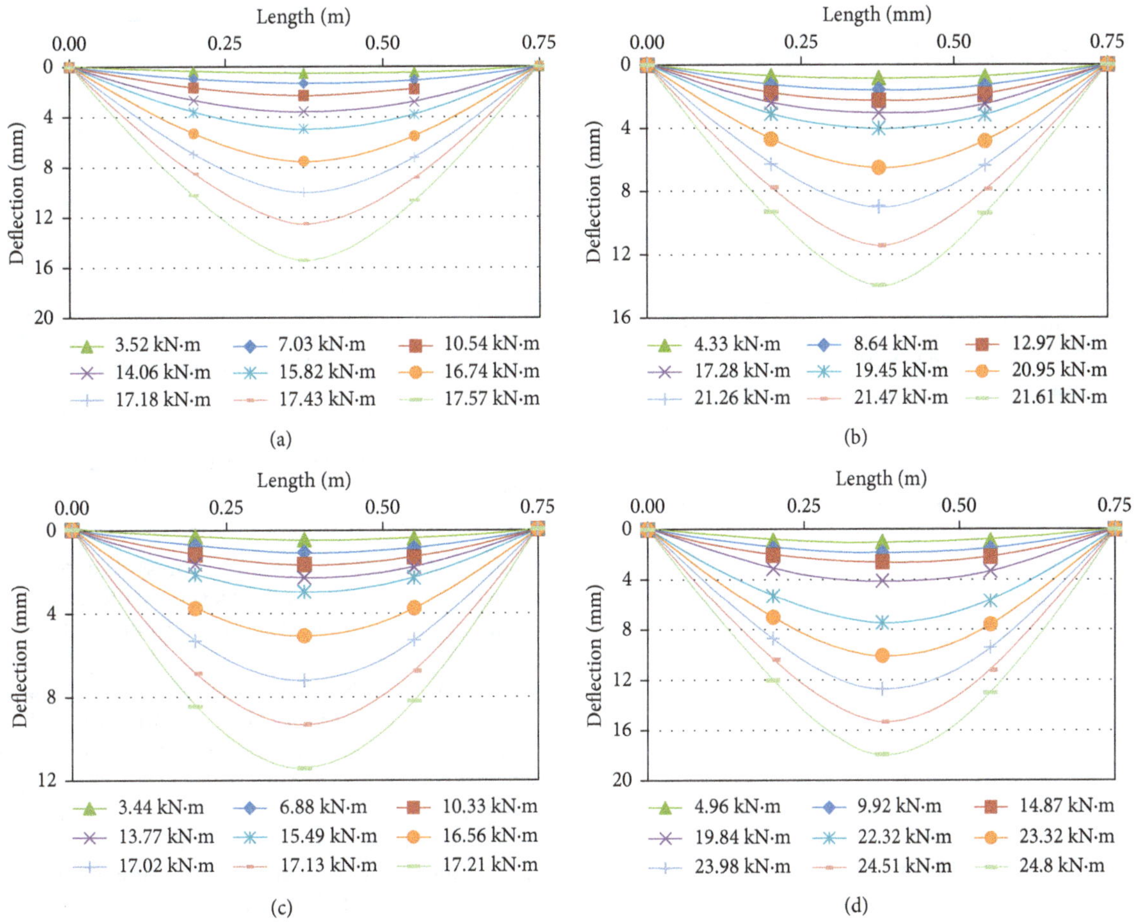

FIGURE 10: Deflection curves along different loading stages. (a) CB (control beam). (b) RB-FL-SB. (c) RB-PL-DB. (d) RB-FL-TB.

had almost identical failure modes regardless the difference of retrofitting scheme (single or double bottom steel plates).

Finally, considering beams of group B (top and bottom steel plate retrofitting scheme), beams of this group revealed a characterized different failure mode compared to their counterparts in other groups. The major difference was that no local buckling was recognized at the top steel flange of all the beams, as shown in Figure 8(b). This may be attributed to the existence of the top retrofitting steel plate. Furthermore, another interesting observation was recorded. An upward local buckling appeared at the top retrofitting steel plate of the beam (RB-FL-TB) which was retrofitted along its full length, as shown in Figure 9(f). This phenomenon was unique in this beam specimen. Nevertheless, regarding the tension failure modes, all beams in this group (B) were exhibited similar failure mechanism as that recorded for their counterparts in the other groups, as depicted in Figures 8(b), 9(f), and 9(g). Finally, it can be concluded that the failure modes of beams in group B were to some extent featured from those noticed in other groups. The adoption of anchored top steel plates, fastened to the top flange of the SCCFST beam, was an effective retrofitting technique. This technique was able to change the predominated local buckling mode of CFST beams and hence lead to relatively increase the flexural strength of retrofitted beams compared to the other retrofitting schemes (group A and C).

Figure 10 shows the deflection curves along the length of typical control beam and typical retrofitted beams for each group versus different bending moment levels. From these figures, it was detected that the curves are symmetrical, and the deflection curves take the shape of half sine wave curves.

4.2. Moment (M) versus Midspan Deflection (D_m) Curves. The moment (M) versus midspan deflection (D_m) curves of all the tested beams are depicted in Figures 11(a)–11(c) for the three retrofitting scheme groups A, B, and C, respectively. Also, Figure 12 represents the typical moment versus midspan deflection curves of the preloaded beams compared to the control beams.

All the tested control SCCFST beams (CB-A, CB-B, and CB-C) exhibited an elastic behaviour during the initial loading stage, followed by inelastic behaviour with a gradual reduction in flexural stiffness until reaching the ultimate moment capacity. Then, the control beams showed a plastic behaviour with large deformation capacity and associated with almost constant flexural strength up to the initiation of bottom steel rupture. The reduction in moment capacity at the failure point was less than 4% of the ultimate moment capacity. This behaviour reflects the superior ductile behaviour of the SCCFST beams.

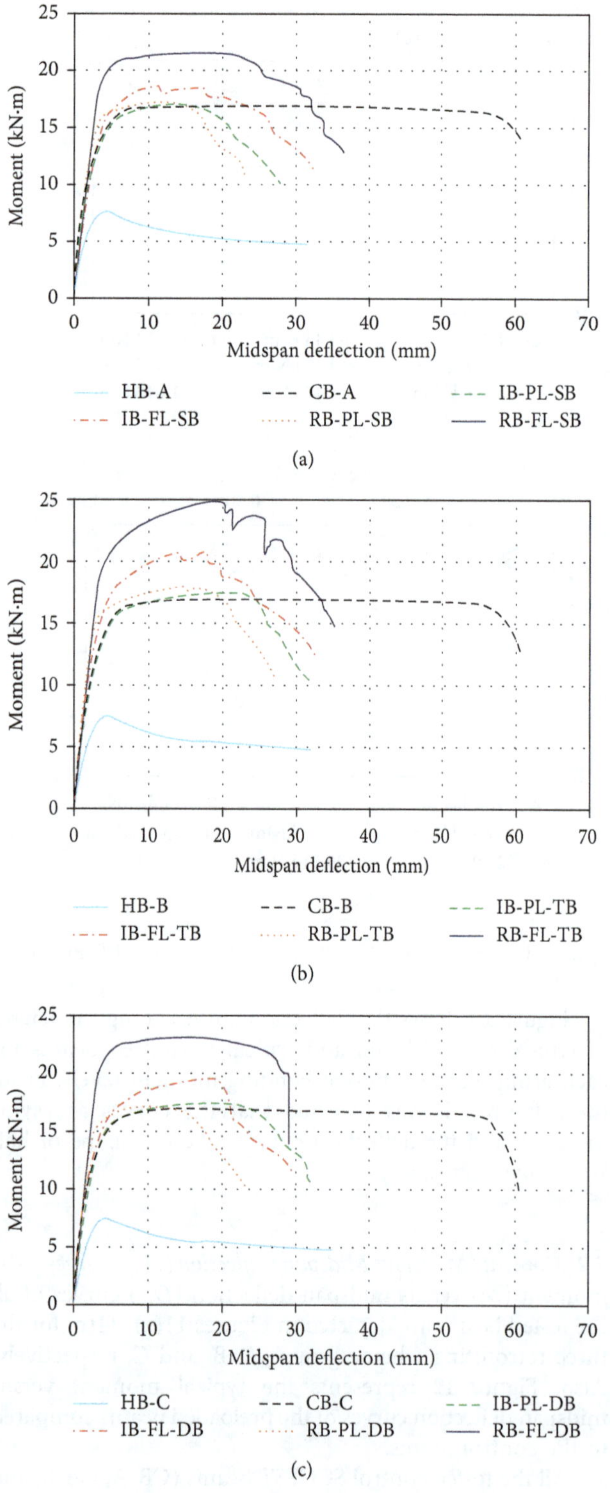

FIGURE 11: Moment versus midspan deflection curves. (a) Beams of group A. (b) Beams of group B. (c) Beams of group C.

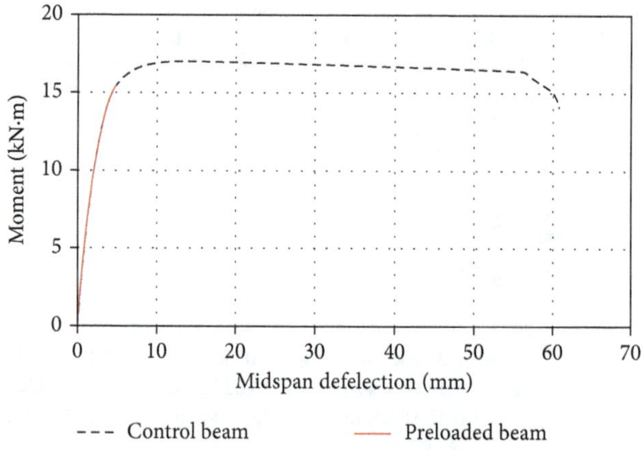

FIGURE 12: Typical moment versus midspan deflection curve of preloaded beams.

Figure 12 indicates that the preloaded beams followed exactly the same path of the control beams up to their predetermined loading stage. This response reflected the ability to conduct a healthy comparison between the retrofitted beams and their counterpart control beams. The preloaded beams were loaded up to an average bending moment of 15.322 kN·m which represents about 90% of the average ultimate moment achieved by the control beam.

In general, all the intact and retrofitted SCCFST beams behaved in an elastic behaviour followed by an inelastic response with a gradual reduction in flexural stiffness up to the ultimate moment capacity. This performance is close to that showed by the control beams. However, some differences in behaviour were detected from the moment versus midspan deflection curves, as will be described in the next sections.

All the intact SCCFST beams followed the similar elastic and inelastic behaviour of the control SCCFST beams with almost similar flexural stiffness. However, the major difference was in the plastic behaviour. The intact specimens showed limited plastic zone compared to the control beams. This may attribute to the reduction in the cross-sectional area of the bottom tensile flanges of the beams and strengthen steel plates and some deficiency that may occur in the concrete core due to the predrilled holes. Intact beams (IB-PL-SB, IB-PL-TB, and IB-PL-DB) which are partially strengthened showed similar flexural strength and stiffness compared to that obtained for the control beams. However, beams (IB-FL-SB, IB-FL-TB, and IB-FL-DB) which are fully strengthened (steel plates applied over the full beam length) showed some greater flexural strength compared to both control beams and partially strengthened beams.

Generally, all the retrofitted SCCFST beams exhibited similar elastic and inelastic response to those of the control SCCFST beams. However, the retrofitted beams revealed greater flexural stiffness. The partial retrofitted beams (RB-PL-SB, RB-PL-TB, and RB-PL-DB) showed close ultimate flexural strength to those obtained by the control beams and their counterpart intact beams (IB-PL-SB, IB-PL-TB, and IB-PL-DB). However, these beams also had a limited plastic zone and hence lower ductility compared to the control beams, similar to their intact counterpart beams. The fully retrofitted beams (RB-PL-SB, RB-PL-TB, and RB-PL-DB) exhibited larger flexural strength and stiffness compared to all other tested beams in this experimental work. This potential enhancement in flexural strength and stiffness was more booming in retrofitted beams that belong to

Table 7: Ultimate moments and percentage of increase in moment capacity.

Group name	Designation of beams	Moment capacity (kN·m)	% increase in moment capacity compared to CB (kN·m)
Group A (single bottom plate)	HB-A	7.62	—
	CB-A	17.04	—
	IB-PL-SB	16.99	0.00
	IB-FL-SB	18.69	9.70
	RB-PL-SB	17.20	0.92
	RB-FL-SB	21.61	26.80
Group B (top and bottom plates)	HB-B	7.56	—
	CB-B	17.04	—
	IB-PL-TB	17.57	3.14
	IB-FL-TB	20.84	22.29
	RB-PL-TB	17.92	5.21
	RB-FL-TB	24.80	45.55
Group C (double bottom plates)	HB-C	7.51	—
	CB-C	17.04	—
	IB-PL-DB	17.47	2.54
	IB-FL-DB	19.60	15.03
	RB-PL-DB	17.21	1.00
	RB-FL-DB	23.13	35.72

the strengthening scheme of group B (RB-FL-TB beam). Furthermore, the fully retrofitted beams (RB-PL-SB, RB-PL-TB, and RB-PL-DB) behaved in a more ductile mode compared to all other retrofitted and intact beams. However, the ductility of these beams was still lower than those obtained for the control beams. Finally, it can be concluded that all the retrofitted beams were able to restore and even to develop enhanced behaviour in terms of flexural strength and stiffness compared to their original undamaged condition.

4.3. Moment Capacities. The ultimate experimental bending moments (M_u) for all beam specimens in all groups and percentage of increase in moment capacity of the retrofitted beams compared to the control beams are tabulated in Table 7 and depicted in Figure 13.

For group A beams (Figure 13(a)), the percentage increase in the ultimate moment of intact SCCFST beams ranged from about 0% for beam IB-PL-SB to 9.70% for beam IB-FL-SB. On the other hand, retrofitted beams achieved a percentage increment in ultimate flexural strength ranged from 0.92% for beam RB-PL-SB to 26.80% for beam RB-FL-SB compared to the control beam. It is clear that the maximum advantage of retrofitting the intact SCCFST beams was achieved by using steel plates along the full beam length. However, a slight increase in the ultimate moment was obtained from beams retrofitted with steel plates along the partial beam length. This may attribute to the fact that steel plates with full length covered the full span of intact beams and hence increase both the moment capacity and stiffness. While, for beams with a partially strengthened span, the steel plates were attached only to the pure bending zone and it was not extended to the shear span.

For group B beams (Figure 13(b)), the percentage enhancement in the ultimate moment due to the strengthening of the SCCFST intact beams was varied from 3.14% for beam IB-PL-TB to 22.29% for beam IB-FL-TB. Regarding the retrofitted SCCFST beams, the percentage improvement in the ultimate moment capacity due to the retrofitting of the SCCFST damaged (preloaded) beams was varied from 5.21% for beam RB-PL-TB to 45.55% for beam RB-FL-TB. Similar to group A, it showed that the maximum benefit of upgrading was obtained by retrofitting the beams along their full length compared to those retrofitted along their partial length. It is worth to note that, that intact and retrofitted beams in this group (group B) attained the highest ultimate moment capacity compared to the beams in other groups (groups A and C) as shown in Figure 13(d). Therefore, it can be stated that the best enhancement in ultimate flexural strength was gained due to the adopting of group B retrofitting technique (top and bottom steel plates).

For group C beams (Figure 13(c)), the percentage increase in the ultimate moment due to the strengthening of SCCFST intact beams ranged from 2.54% for beam IB-PL-DB to 21.59% for beam IB-FL-DB. However, the percentage increment in the ultimate moment due to the retrofitting of SCCFST damaged beams ranged from 1.00% for beam RB-PL-DB to 35.72% for beam IB-FL-DB. Similar to groups A and B, it is evident that the maximum avail of retrofitting of SCCFST beam was achieved by adopting full retrofitting scheme rather than partial retrofitting schemes. Furthermore, the retrofitting scheme used in this group (double bottom plates) provides a moderate improvement in ultimate flexural strength compared to the other retrofitting schemes adopted in groups A (single bottom plate) and B (top and bottom plates).

Finally, it can be indicated that all the adopted retrofitting schemes were effective in restoring the flexural strength of the damaged beams and even enhance the ultimate flexural strength, especially for beams retrofitted along their full span length. Moreover, among the

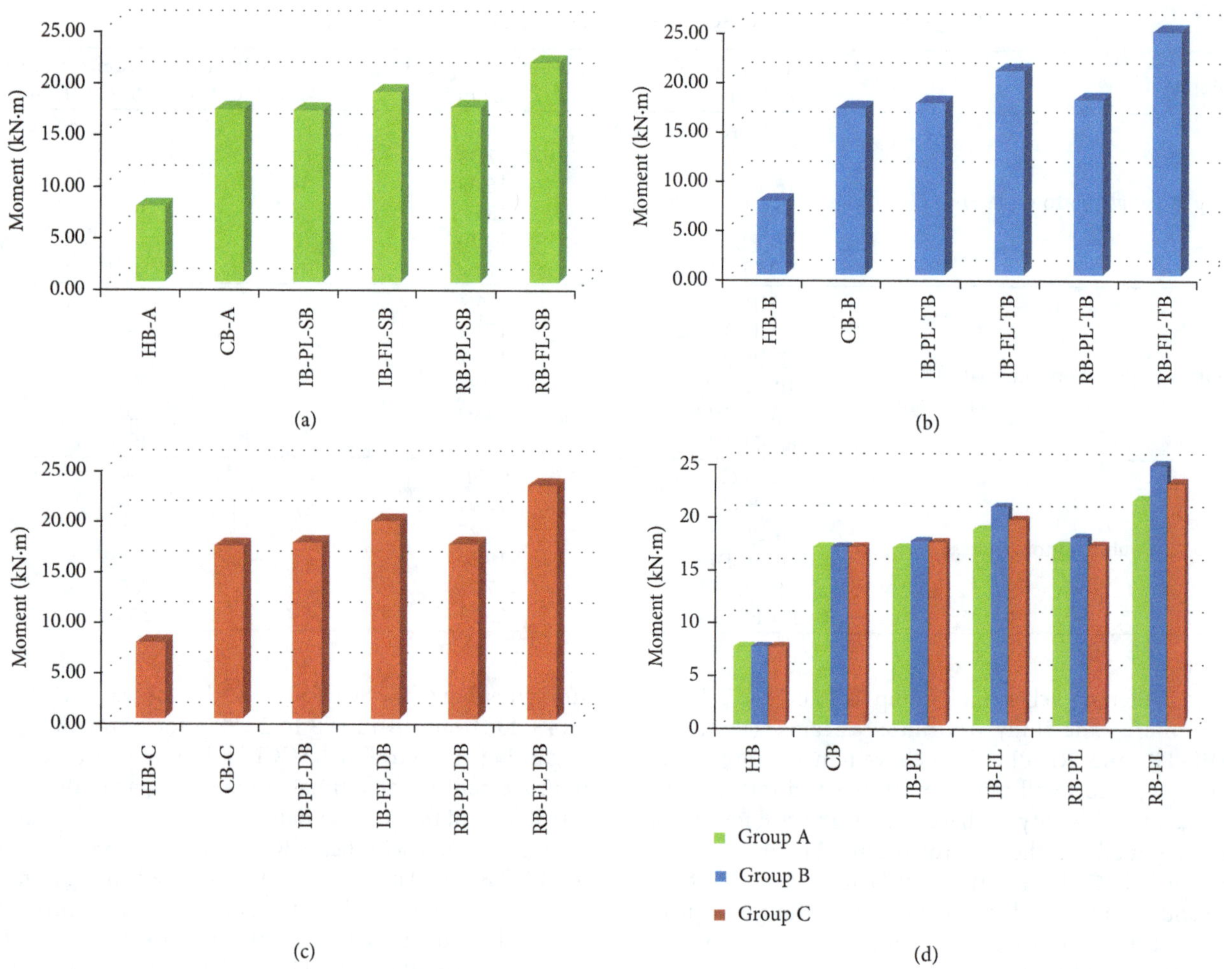

Figure 13: Ultimate moment capacity. (a) Group A beams. (b) Group B beams. (c) Group C beams. (d) Comparison.

three different upgrading schemes, it was found that the retrofitting method of damaged beams using top and bottom steel plates was the most effective method.

4.4. Ductility Capacities. Ductile performance of beams implies the ability to absorb as much energy as possible before failure. The ductility of the tested beams was evaluated by the modulus of toughness (MOT), which represents the entire area under the load-deflection curves. Furthermore, the relative ductility index (RDI) of the tested beams was calculated which represents the ratio of ductility achieved by the retrofitted and intact beams to the ductility of the control beams. The results of the relative ductility index and modulus of toughness are listed in Table 8 and illustrated in Figure 14.

For group A beams, the RDI of intact beams ranged from 0.41 for beam IB-PL-SB to 0.53 for beam IB-FL-SB which referred to about 53% (average value) reduction in ductility compared to the control beam. However, for the retrofitted beams, the RDI ranged from 0.35 for beam RB-PL-SB to 0.70 for beam IB-FL-SB which indicated that the retrofitted beams were able to restore about 35% to 70% of the original ductility of the control beam.

For group B beams, the RDI of intact beams varied from 0.49 for beam IB-PL-SB to 0.57 for beam IB-FL-SB which referred to about 47% (average value) reduction in ductility compared to the control beam. On the other hand, for the retrofitted beams, the RDI ranged from 0.43 for beam RB-PL-TB to 0.75 for beam IB-FL-TB which implied that the retrofitted beams were able to recover about 43% to 75% of the original ductility of the control beam. It is worth noting that the retrofitted beams in this group developed the highest ductility compared to their counterparts in other groups. This pointed that, the retrofitting scheme adopting in this group (top and bottom steel plates) demonstrated the most efficient retrofitting technique.

For group C beams, the RDI of intact beams was about 0.50 (average) which refereed to about 50% lower ductility compared to the control beam. However, regarding the retrofitted beams, the RDI ranged from 0.35 for beam RB-PL-DB to 0.63 for beam RB-FL-DB which indicated that the retrofitted beams were capable of getting back about 35% to 63% of the original ductility of the control beam. However, retrofitted beams of this group attained the lowest ductility relative to beams in other groups.

Finally, similar to the moment capacities, it is obvious that the full retrofitting scheme is much efficient than the partial

TABLE 8: Modulus of toughness and relative ductility.

Group name	Designation of beams	Modulus of toughness (kN·mm)	Relative ductility index (RDI)
Group A (single bottom plate)	CB-A	7841.9	1.00
	IB-PL-SB	3201.1	0.41
	IB-FL-SB	4120.9	0.53
	RB-PL-SB	2776.5	0.35
	RB-FL-SB	5511.5	0.70
Group B (top and bottom plates)	CB-B	7832.3	1.00
	IB-PL-TB	3824.8	0.49
	IB-FL-TB	4470.1	0.57
	RB-PL-TB	3338.2	0.43
	RB-FL-TB	5864.4	0.75
Group C (double bottom plates)	CB-C	7754.4	1.00
	IB-PL-DB	3980.2	0.51
	IB-FL-DB	3801.3	0.49
	RB-PL-DB	2742.1	0.35
	RB-FL-DB	4884.5	0.63

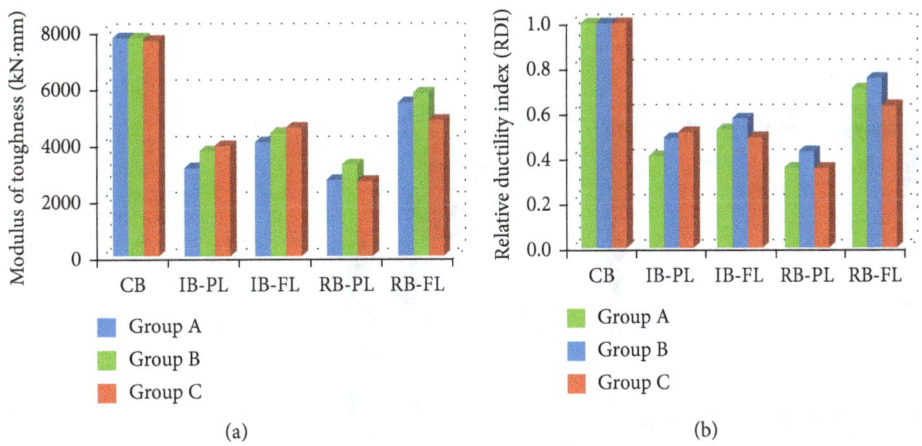

FIGURE 14: Ductility capacities. (a) Modulus of toughness. (b) Relative ductility index.

retrofitting scheme regarding the improvement in ductility and energy absorption capacity of the retrofitted beams.

5. Theoretical Model for the Ultimate Moment Capacity

The American specification AISC 360-10 [38] provides a theoretical model for predicting the ultimate moment capacity of normal concrete-filled steel tube (NCFST) members based on the plastic stress distribution method. Furthermore, Lai et al. [3] reported more details about the calculations of the moment capacities for the NCFST members based on the original model described by AISC 360-10. However, the main limitation of this model is that it was developed to predict the moment capacity of NCFST beams without external retrofitting steel plates. That is to say, the AISC 360-10 model did not take into consideration the effect of the external retrofitting steel plates. It is expected that the existence of external steel plates will increase the ultimate moment capacity of the CFST beams. However, this increase cannot be accounted for by using the AISC 360-10 model. So that, the adoption of the AISC 360-10 model will result in a considerable underestimation (conservative) in predicting the ultimate flexural strength of the retrofitted SCCFST beams using external steel plates.

In this study, a new theoretical model based on the AISC 360-10 [38] model and Lai et al. [3] calculation equations were developed. The new model is developed in such a way to take into consideration the effect of external retrofitting steel plates on the prediction of the ultimate moment capacity of SCCFST beams. The developed model utilizes the plastic stress distribution method as described in the AISC 360-10 specification to compute the plastic moment (M_p) capacity of the beam cross section as shown in Figure 15. This plastic moment strength represents the nominal flexural (M_n) capacity of retrofitted rectangular SCCFST beams with compact sections using external steel plates. The plastic stress method assumes that the steel and concrete materials have a rigid plastic behaviour. This approach considers that the steel stress in tension and compression is equal to the yield stress (f_y), while the concrete stress is equal to 0.85 f'_c in compression and equal to zero in tension. In this model, the effect of retrofitted steel plates was taken into consideration by considering that the external steel plates at the compression and tension sides can reach their yielding stress (f_y) in a similar way to that assumed for the steel section.

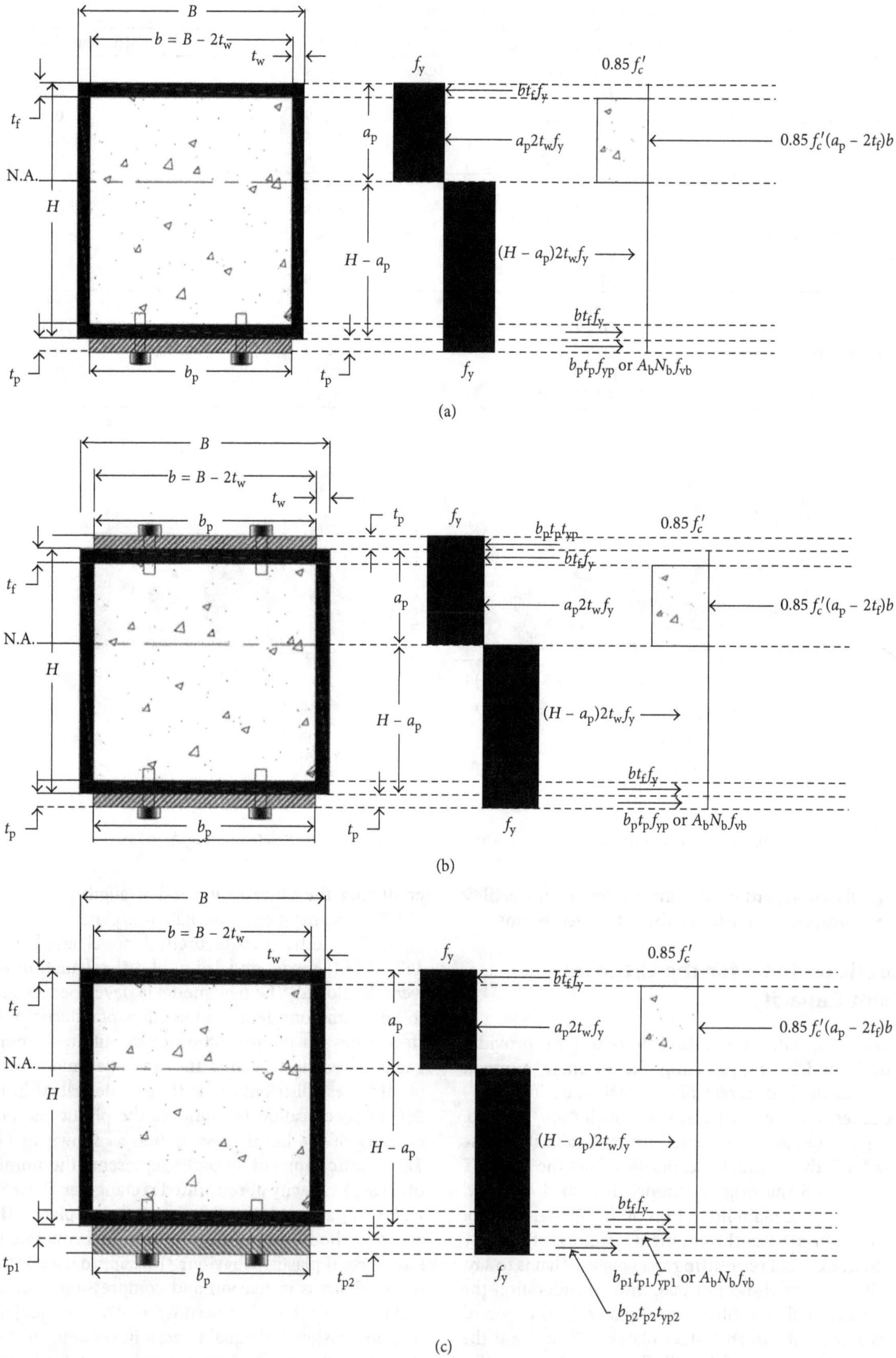

FIGURE 15: Theoretical model for calculating the ultimate moment capacity. (a) Stress blocks and forces for group A beams. (b) Stress blocks and forces for group B beams. (c) Stress blocks and forces for group C beams.

Hence, it ensures that the SSCFST beam and the retrofitted external steel plates can achieve a ductile behaviour. Therefore, the tension or compression forces obtained from the bottom or top external steel plates can be predicted based on the yielding tensile stress of the steel plates as follows:

$$F_{\mathrm{p}} = f_{\mathrm{yp}}\, A_{\mathrm{gp}}, \tag{1}$$

where F_{p} is the force in the external steel plates and f_{yp} and A_{gp} are the yielding stress and cross-sectional area of the external steel plates, respectively.

The above equation was found to be convenient for computing the ultimate moment capacity of SCCFST beams retrofitted with external steel plates using grade 12.9 bolts (bolts with the higher strength). However, for SCCFST beams retrofitted with external steel plates using grade 8.8 bolts (bolts with the lower strength), another condition was considered for the steel plates on the tension side (bottom steel plates). For steel plates in tension with grade 8.8 bolts, the most control criteria are the shear strength of the fixing bolts rather than the yielding strength of the steel plates. So that, the expected tension forces obtained from the bottom external steel plates can be predicted based on the shear strength of the fixing bolts as follows:

$$F_{\mathrm{b}} = f_{\mathrm{nv}}\, A_{\mathrm{b}}\, N_{\mathrm{b}}, \tag{2}$$

where F_{b} is the force in the external bottom steel plates and f_{nv}, A_{b}, and N_{b} are the shearing stress, cross-sectional area, and number of the bolts, respectively.

Figures 15(a)–15(c) illustrate the stress blocks and forces of the concrete and steel at the ultimate loading stage for the three different retrofitting schemes, single bottom plate, top and bottom plates, and double bottom plates, respectively. The depth of the neutral axis (a_{p}) from the compression face is computed by achieving axial force equilibrium over the cross section. The plastic moment capacity (M_{p}) is computed as the summation of bending moments caused by the concrete and steel forces around the neutral axis location (a_{p}). The obtaining equations for calculating a_{p} and M_{p} are given below for the three different retrofitting schemes:

(a) For retrofitted SCCFST beams using external single bottom steel plate (group A) with grade 12.9 bolts, a_{p} and M_{p} can be calculated as follows:

$$a_{\mathrm{p}} = \frac{2f_{\mathrm{y}}Ht_{\mathrm{w}} + f_{\mathrm{yp}}b_{\mathrm{p}}t_{\mathrm{p}} + 0.85f'_{\mathrm{c}}bt_{\mathrm{f}}}{4f_{\mathrm{y}}t_{\mathrm{w}} + 0.85f'_{\mathrm{c}}b},$$

$$M_{\mathrm{p}} = f_{\mathrm{y}}bt_{\mathrm{f}}(H-t_{\mathrm{f}}) + f_{\mathrm{y}}t_{w}(H-a_{\mathrm{p}})^2 + f_{\mathrm{y}}t_{\mathrm{w}}a_{\mathrm{p}}^2 + \frac{0.85}{2}f'_{\mathrm{c}}b(a_{\mathrm{p}}-t_{\mathrm{f}})^2 + f_{\mathrm{yp}}b_{\mathrm{p}}t_{\mathrm{p}}\left(H-a_{\mathrm{p}}+\frac{t_{\mathrm{p}}}{2}\right). \tag{3}$$

(b) For retrofitted SCCFST beams using external single bottom steel plate (group A) with grade 8.8 bolts, a_{p} and M_{p} can be calculated as follows:

$$a_{\mathrm{p}} = \frac{2f_{\mathrm{y}}Ht_{\mathrm{w}} + f_{\mathrm{nv}}A_{\mathrm{b}}N_{\mathrm{b}} + 0.85f'_{\mathrm{c}}bt_{\mathrm{f}}}{4f_{\mathrm{y}}t_{\mathrm{w}} + 0.85f'_{\mathrm{c}}b},$$

$$M_{\mathrm{p}} = f_{\mathrm{y}}bt_{\mathrm{f}}(H-t_{\mathrm{f}}) + f_{\mathrm{y}}t_{\mathrm{w}}(H-a_{\mathrm{p}})^2 + f_{\mathrm{y}}t_{\mathrm{w}}a_{\mathrm{p}}^2 + \frac{0.85}{2}f'_{\mathrm{c}}b(a_{\mathrm{p}}-t_{\mathrm{f}})^2 + f_{\mathrm{nv}}A_{\mathrm{b}}N_{\mathrm{b}}\left(H-a_{\mathrm{p}}+\frac{t_{\mathrm{p}}}{2}\right). \tag{4}$$

(c) For retrofitted SCCFST beams using external top and bottom steel plates (group B) with grade 12.9 bolts, a_{p} and M_{p} can be calculated as follows:

$$a_{p} = \frac{2f_{\mathrm{y}}Ht_{\mathrm{w}} + 0.85f'_{\mathrm{c}}bt_{\mathrm{f}}}{4f_{\mathrm{y}}t_{\mathrm{w}} + 0.85f'_{\mathrm{c}}b},$$

$$M_{\mathrm{p}} = f_{\mathrm{y}}bt_{\mathrm{f}}(H-t_{\mathrm{f}}) + f_{\mathrm{y}}t_{\mathrm{w}}(H-a_{\mathrm{p}})^2 + f_{\mathrm{y}}t_{\mathrm{w}}a_{\mathrm{p}}^2 + \frac{0.85}{2}f'_{\mathrm{c}}b(a_{\mathrm{p}}-t_{\mathrm{f}})^2 + f_{\mathrm{yp}}b_{\mathrm{p}}t_{\mathrm{p}}\left(H-a_{\mathrm{p}}+\frac{t_{\mathrm{p}}}{2}\right) + f_{\mathrm{yp}}b_{\mathrm{p}}t_{\mathrm{p}}\left(a_{\mathrm{p}}+\frac{t_{\mathrm{p}}}{2}\right). \tag{5}$$

(d) For retrofitted SCCFST beams using external top and bottom steel plates (group B) with grade 8.8 bolts, a_{p} and M_{p} can be calculated as follows:

$$a_{\mathrm{p}} = \frac{2f_{\mathrm{y}}Ht_{\mathrm{w}} + f_{\mathrm{nv}}A_{\mathrm{b}}N_{\mathrm{b}} - f_{\mathrm{yp}}b_{\mathrm{p}}t_{\mathrm{p}} + 0.85f'_{\mathrm{c}}bt_{\mathrm{f}}}{4f_{\mathrm{y}}t_{\mathrm{w}} + 0.85f'_{\mathrm{c}}b},$$

$$M_{\mathrm{p}} = f_{\mathrm{y}}bt_{\mathrm{f}}(H-t_{\mathrm{f}}) + f_{\mathrm{y}}t_{\mathrm{w}}(H-a_{\mathrm{p}})^2 + f_{\mathrm{y}}t_{\mathrm{w}}a_{\mathrm{p}}^2 + \frac{0.85}{2}f'_{\mathrm{c}}b(a_{\mathrm{p}}-t_{\mathrm{f}})^2 + f_{\mathrm{nv}}A_{\mathrm{b}}N_{\mathrm{b}}(H-a_{\mathrm{p}}) + f_{\mathrm{yp}}b_{\mathrm{p}}t_{\mathrm{p}}\left(a_{\mathrm{p}}+\frac{t_{\mathrm{p}}}{2}\right). \tag{6}$$

(e) For retrofitted SCCFST beams using external double bottom steel plates (group C) with grade 12.9 bolts, a_{p} and M_{p} can be calculated as follows:

$$a_{\mathrm{p}} = \frac{2f_{\mathrm{y}}Ht_{\mathrm{w}} + f_{\mathrm{yp1}}b_{\mathrm{p1}}t_{\mathrm{p1}} + f_{\mathrm{yp2}}b_{\mathrm{p2}}t_{\mathrm{p2}} + 0.85f'_{\mathrm{c}}bt_{\mathrm{f}}}{4f_{\mathrm{y}}t_{\mathrm{w}} + 0.85f'_{\mathrm{c}}b},$$

$$M_{\mathrm{p}} = f_{\mathrm{y}}bt_{\mathrm{f}}(H-t_{\mathrm{f}}) + f_{\mathrm{y}}t_{\mathrm{w}}(H-a_{\mathrm{p}})^2 + f_{\mathrm{y}}t_{\mathrm{w}}a_{\mathrm{p}}^2 + \frac{0.85}{2}f'_{\mathrm{c}}b(a_{\mathrm{p}}-t_{\mathrm{f}})^2 + f_{\mathrm{yp1}}b_{\mathrm{p1}}t_{\mathrm{p1}}\left(H-a_{\mathrm{p}}+\frac{t_{\mathrm{p1}}}{2}\right) + f_{\mathrm{yp2}}b_{\mathrm{p2}}t_{\mathrm{p2}}\left(H-a_{\mathrm{p}}+t_{\mathrm{p1}}+\frac{t_{\mathrm{p2}}}{2}\right). \tag{7}$$

TABLE 9: Comparisons of experimental moment capacities against the theoretical model and design code values.

Group name	Designation of beams	M_{ue} (kN·m)	M_{ua} (kN·m)	M_{ua}/M_{ue}	M_{uc} (kN·m)	M_{uc}/M_{ue}
Group A (single bottom plate)	CB-A	17.04	14.39	0.844	14.39	0.844
	IB-PL-SB	16.99	17.41	1.025	14.39	0.847
	IB-FL-SB	18.69	19.13	1.024	14.39	0.770
	RB-PL-SB	17.20	16.43	0.955	14.39	0.837
	RB-FL-SB	21.61	20.54	0.950	14.39	0.666
	Mean			0.960		0.793
	SD			0.074		0.078
Group B (top and bottom plates)	CB-B	17.04	14.45	0.848	14.45	0.848
	IB-PL-TB	17.57	18.46	1.051	14.45	0.822
	IB-FL-TB	20.84	21.89	1.050	14.45	0.693
	RB-PL-TB	17.92	17.15	0.957	14.45	0.806
	RB-FL-TB	24.80	23.82	0.960	14.45	0.583
	Mean			0.973		0.751
	SD			0.084		0.111
Group C (double bottom plates)	CB-C	17.04	14.29	0.839	14.29	0.839
	IB-PL-DB	17.47	16.86	0.965	14.29	0.818
	IB-FL-DB	19.60	18.94	0.966	14.29	0.729
	RB-PL-DB	17.21	18.39	1.069	14.29	0.830
	RB-FL-DB	23.13	24.85	1.074	14.29	0.618
	Mean			0.983		0.767
	SD			0.096		0.094
Total	Mean			0.972		0.770
	SD			0.079		0.090

(f) For retrofitted SCCFST beams using external top and bottom steel plates (group C) with grade 8.8 bolts, a_p and M_p can be calculated as follows:

$$a_p = \frac{2f_y H t_w + f_{nv}A_b N_b + 0.85 f'_c b t_f}{4 f_y t_w + 0.85 f'_c b},$$

$$M_p = f_y b t_f (H - t_f) + f_y t_w (H - a_p)^2 + f_y t_w a_p^2 + \frac{0.85}{2} f'_c b (a_p - t_f)^2 + f_{nv} A_b N_b (H - a_p). \quad (8)$$

Finally, regarding the SCCFST beams retrofitted using partial length steel plates for the three different retrofitting schemes (groups A, B, and C). The experimental test results revealed that partial length retrofitting scheme (beams retrofitted along their pure bending zone only) has an inconsiderable effect on the ultimate moment capacity. Therefore, the plastic moment capacity (M_p) of these SCCFST beams is calculated based on M_p calculated according to the previously mentioned equations multiplied by a reduction factor of 0.91 and 0.8, for beams in group A with bolt grades 8.8 and 12.9, respectively. For beams in group B, the reduction factors are equal to 0.84 and 0.72. However, for the group C beams, the reduction factors are 0.89 and 0.74, for beams retrofitted by steel plates with bolt grades 8.8 and 12.9, respectively. These reduction factors were estimated based on the experimental test results obtained from this study.

5.1. Test Results against the Theoretical Model and Design Code. The predicted flexural capacities (M_{ua}) of the retrofitted SCCFST beams utilizing the aforementioned developed theoretical model and the American specification AISC 360-10 method (M_{uc}) are compared with the current experimental results (M_{ue}) as shown in Table 9. Furthermore, the mean value and the standard deviation (SD) of the ratios M_{ua}/M_{ue} and M_{uc}/M_{ue} are listed in Table 9.

The results in Table 9 revealed that the developed model in this study could predict the ultimate moment capacity of retrofitted SCCFST beams accurately. On the other hand, the results indicated that the AISC 360-10 model underestimates (conservative) the ultimate moment capacity of these beams. This underestimation is expected because the AISC 360-10 model did not take into consideration the effect of external steel plates on the flexural strength of retrofitted SCCFST beams. However, the new theoretical model which is suggested in this study includes the influence of these bolted steel plates.

For instance, the theoretical model was able to predict the moment capacity of retrofitted SCCFST beams of group A with an average value of 0.960 and with a standard deviation (SD) of 0.074. On the other hand, the AISC 360-10 code underestimated the moment capacity by an average of about 21% and with SD equal to 0.078. Similarly, regarding the retrofitted beams of group B, the theoretical model predicted the moment capacity with an average value of 0.973 and with a SD of 0.084. However, the AISC 360-10 code underestimated the moment capacity by an average of about 25% and with SD equal to 0.078. Finally, the theoretical model gave an average estimation of about 0.983 and with a SD of 0.096 for retrofitted beams within group C. While, the AISC 360-10 specification again underestimated the moment resistance of retrofitted beams by about 23%.

It is worth to note that the AISC 360-10 method can predict the moment capacity of SCCFST beams retrofitted with partial length steel plate and with lower bolt strength, more accurately than beams retrofitted with full-length plates and with higher strength bolts. For example, the AISC 360-10 method estimated the flexural strength of beams IB-PL-SB, IB-PL-TB, and IB-PL-DB, with an average underestimation of 17%. However, regarding beams RB-FL-SB, RB-FL-TB, and RB-FL-DB, the level of conservation was about 38%.

Finally, comparison results indicated that the moment capacity of beams retrofitted using partial length steel plates could be computed by neglecting the effect of the bolted steel plates. It is clear that when the SCCFST beams retrofitted with steel plates along their partial length (along with the constant bending zone span) only, the enhancement in ultimate moment capacity is indeed insignificant. So that, both the theoretical model of this study and the AISC 360-10 model can be used to estimate the moment resistance of such retrofitted beams.

6. Conclusions

This paper highlights the results of an experimental study investigating the flexural performance of SCCFST beams retrofitted using externally bolted steel plates with three different schemes including varied configurations. The moment capacities, ductility, failure modes, and moment-deflection curves of beam specimens were reported and discussed. Based on the experimental results and observations, the following conclusions can be stated:

(i) The typical failure modes of the fully retrofitted SCCFST beams were characterized by outward local buckling of the beam top steel flange, ruptures of some of their anchoring bolts within the beam shear span, delamination and separation of the attached retrofitting plates, and finally rupture failure of the tension bottom flanges.

(ii) Different failure modes were observed for SCCFST beams which retrofitted by top and bottom steel plates (group B), compared to their counterparts in other groups. The basic difference was that no local buckling was recognized at the top steel flange of all the beams. Furthermore, an upward local buckling appeared at the top retrofitting steel plate of the beam RB-FL-TB.

(iii) The moment versus midspan deflection curves of the control and retrofitted SCCFST beams in all groups showed elastic behaviour at the first stage of loading. Then, the curves follow an inelastic path with a gradual reduction in stiffness until reaching the maximum bending capacity.

(iv) Test results revealed that the percentage increase in the ultimate moment capacity due to the retrofitting of SCCFST beams using the different three schemes ranged from 1.00% to 46.00%. It is worth to note that retrofitting scheme of group B beams (top and bottom bolted steel plates) was the most efficient technique regarding the flexural strength enhancement.

(v) The relative ductility index (RDI) results indicated that all the retrofitted beams were able to restore about 35% to 75% of the original ductility of the control beams. However, the most effective retrofitting method regarding ductility recovering was the retrofitting scheme of group B (top and bottom bolted steel plates).

(vi) It was found that the full retrofitting scheme (steel plates applied along the entire beam length) is much efficient than the partial retrofitting scheme (steel plates applied along constant bending zone only) regarding improving the ultimate moment resistance, stiffness, and ductility capacity of the retrofitted beams.

(vii) A new theoretical model based on the AISC 360-10 method was developed. The new model is developed in such a way to take into consideration the effect of external retrofitting steel plates on the prediction of the ultimate moment capacity of SCCFST beams. The comparison results revealed that this model could predict the ultimate moment capacity of retrofitted SCCFST beams accurately with a total average of 97%.

(viii) Finally, it can be concluded that all the adopted retrofitting schemes were effective in restoring the flexural strength and ductility of the damaged beams and even enhance the ultimate flexural strength, especially for beams retrofitted along their full span length with top and bottom plates.

Conflicts of Interest

The authors declare that they have no conflicts of interest.

Acknowledgments

The authors acknowledge the financial support of the Scientific Research Unit of the University of Gaziantep on the Project MF 14.05. The authors also would like to thank the valuable support of the BASF/Turkey (AKAS) in supplying the concrete admixtures.

References

[1] Z.-B. Wang, Z. Tao, L.-H. Han, B. Uy, D. Lam, and W.-H. Kang, "Strength, stiffness and ductility of concrete-filled steel columns under axial compression," *Engineering Structures*, vol. 135, pp. 209–221, 2017.

[2] M. Elchalakani, A. Karrech, M. F. Hassanein, and B. Yang, "Plastic and yield slenderness limits for circular concrete filled tubes subjected to static pure bending," *Thin-Walled Structures*, vol. 109, pp. 50–64, 2016.

[3] Z. Lai, A. H. Varma, and K. Zhang, "Noncompact and slender rectangular CFT members: experimental database, analysis, and design," *Journal of Constructional Steel Research*, vol. 101, pp. 455–468, 2014.

[4] G. Li, D. Liu, Z. Yang, and C. Zhang, "Flexural behavior of high strength concrete filled high strength square steel tube," *Journal of Constructional Steel Research*, vol. 128, pp. 732–744, 2017.

[5] M. V Chitawadagi and M. C. Narasimhan, "Strength deformation behaviour of circular concrete filled steel tubes subjected to pure bending," *Journal of Constructional Steel Research*, vol. 65, no. 8-9, pp. 1836–1845, 2009.

[6] L.-H. Han, W. Li, and R. Bjorhovde, "Developments and advanced applications of concrete-filled steel tubular (CFST) structures: members," *Journal of Constructional Steel Research*, vol. 100, pp. 211–228, 2014.
[7] H. Abbas, Y. Al-Salloum, S. Alsayed, M. Alhaddad, and R. Iqbal, "Post-heating response of concrete-filled circular steel columns," *KSCE Journal of Civil Engineering*, vol. 21, no. 4, pp. 1367–1378, 2017.
[8] L.-H. Han, H. Lu, G.-H. Yao, and F.-Y. Liao, "Further study on the flexural behaviour of concrete-filled steel tubes," *Journal of Constructional Steel Research*, vol. 62, no. 6, pp. 554–565, 2006.
[9] M. Mahgub, A. Ashour, D. Lam, and X. Dai, "Tests of self-compacting concrete filled elliptical steel tube columns," *Thin-Walled Structures*, vol. 110, pp. 27–34, 2017.
[10] P. Kumar, K. Sudalaimani, and M. Shanmugasundaram, "An investigation on self-compacting concrete using ultrafine natural steatite powder as replacement to cement," *Advances in Materials Science and Engineering*, vol. 2017, Article ID 8949041, 8 pages, 2017.
[11] M. Lachemi, K. M. A. Hossain, and V. B. Lambros, "Axial load behavior of self-consolidating concrete-filled steel tube columns in construction and service stages," *ACI Structural Journal*, vol. 103, no. 1, p. 38, 2006.
[12] B. J. M. AL-Eliwi, T. Ekmekyapar, R. H. Faraj, M. T. Göğüş, and A. A. M. AL-Shaar, "Performance of lightweight aggregate and self-compacted concrete-filled steel tube columns," *Steel and Composite Structures*, vol. 25, no. 3, pp. 299–314, 2017.
[13] L.-H. Han and G.-H. Yao, "Experimental behaviour of thin-walled hollow structural steel (HSS) columns filled with self-consolidating concrete (SCC)," *Thin-Walled Structures*, vol. 42, no. 9, pp. 1357–1377, 2004.
[14] L.-H. Han, G.-H. Yao, and X.-L. Zhao, "Tests and calculations for hollow structural steel (HSS) stub columns filled with self-consolidating concrete (SCC)," *Journal of Constructional Steel Research*, vol. 61, no. 9, pp. 1241–1269, 2005.
[15] Y. T. Obaidat, S. Heyden, O. Dahlblom, G. Abu-Farsakh, and Y. Abdel-Jawad, "Retrofitting of reinforced concrete beams using composite laminates," *Construction and Building Materials*, vol. 25, no. 2, pp. 591–597, 2011.
[16] C.-J. Jiang, Z.-D. Lu, and L.-Z. Li, "Shear performance of fire-damaged reinforced concrete beams repaired by a bolted side-plating technique," *Journal of Structural Engineering*, vol. 143, no. 5, p. 4017007, 2017.
[17] J. Peng, H. Tang, and J. Zhang, "Structural behavior of corroded reinforced concrete beams strengthened with steel plate," *Journal of Performance of Constructed Facilities*, vol. 31, no. 4, p. 4017013, 2017.
[18] F. Sevuk and G. Arslan, "Retrofit of damaged reinforced concrete beams by using steel plate," in *Proceedings of the Structures Congress 2005: Metropolis and Beyond*, pp. 1–8, New York, NY, USA, April 2005.
[19] Z.-B. Wang, Q. Yu, and Z. Tao, "Behaviour of CFRP externally-reinforced circular CFST members under combined tension and bending," *Journal of Constructional Steel Research*, vol. 106, pp. 122–137, 2015.
[20] M. C. Sundarraja and G. G. Prabhu, "Finite element modelling of CFRP jacketed CFST members under flexural loading," *Thin-Walled Structures*, vol. 49, no. 12, pp. 1483–1491, 2011.
[21] A. W. Al Zand, W. H. W. Badaruzzaman, A. A. Mutalib, and S. J. Hilo, "The enhanced performance of CFST beams using different strengthening schemes involving unidirectional CFRP sheets: an experimental study," *Engineering Structures*, vol. 128, pp. 184–198, 2016.
[22] S. Aykac, I. Kalkan, B. Aykac, S. Karahan, and S. Kayar, "Strengthening and repair of reinforced concrete beams using external steel plates," *Journal of Structural Engineering*, vol. 139, no. 6, pp. 929–939, 2012.
[23] R. A. Barnes, P. S. Baglin, G. C. Mays, and N. K. Subedi, "External steel plate systems for the shear strengthening of reinforced concrete beams," *Engineering Structures*, vol. 23, no. 9, pp. 1162–1176, 2001.
[24] X. Xu, Z. Lu, L. Li, and C. Jiang, "Numerical study on the local buckling behaviour of bolted steel plates in steel jacketing," *Advances in Materials Science and Engineering*, vol. 2017, Article ID 1352084, 15 pages, 2017.
[25] T. A. El-Maaddawy, "Mechanically fastened composites for retrofitting corrosion-damaged reinforced-concrete beams: experimental investigation," *Journal of Composites for Construction*, vol. 18, no. 2, p. 4013041, 2013.
[26] Y. Guan, B. Jiang, and Y. Jiang, "Experimental study on RC beams strengthened in shear with the FRP-bolt strengthening technology," in *Proceedings of the Design, Construction, Rehabilitation, and Maintenance of Bridges*, pp. 57–64, Hunan, China, June 2011.
[27] M. Sundarraja and G. Prabhu, "Experimental investigation on strengthening of CFST members under flexure using CFRP fabric," *Arabian Journal for Science and Engineering*, vol. 39, no. 2, pp. 659–668, 2014.
[28] N. Yazdani, F. Haque, and I. Hasan, "Flexural and shear behavior of FRP strengthened AASHTO type concrete bridge girders," *Journal of Engineering*, vol. 2016, Article ID 5201910, 10 pages, 2016.
[29] J. Vercher, E. Gil, Á. Mas, C. Lerma, and M. E. Torner, "Flexural strengthening of damaged T-joists with severe corrosion using CFRP sheets," *Advances in Materials Science and Engineering*, vol. 2017, Article ID 6030357, 9 pages, 2017.
[30] A. Alfeehan, "Strengthening of RC beams by external steel plate using mechanical connection technique," *Journal of Engineering and Development*, vol. 18, no. 1, pp. 202–215, 2014.
[31] R. K. L. Su and Y. Zhu, "Experimental and numerical studies of external steel plate strengthened reinforced concrete coupling beams," *Engineering Structures*, vol. 27, no. 10, pp. 1537–1550, 2005.
[32] R. K. L. Su, B. Cheng, L. Wang, W. H. Siu, and Y. Zhu, "Use of bolted steel plates for strengthening of reinforced concrete beams and columns," *The IES Journal Part A: Civil & Structural Engineering*, vol. 4, no. 2, pp. 55–68, 2011.
[33] Y. Zhu and R. K. L. Su, "Behavior of strengthened reinforced concrete coupling beams by bolted steel plates, Part 2: evaluation of theoretical strength," *Structural Engineering and Mechanics*, vol. 34, no. 5, pp. 563–580, 2010.
[34] EFNARC, *The European Guidelines for Self-Compacting Concrete, Specification, Production and Use*, EFNARC, Europe, 2005.
[35] ASTM C39/C39M-14a, *Standard Test Method for Compressive Strength of Cylindrical Concrete Specimens*, ASTM International, West Conshohocken, PA, USA, 2014.
[36] ASTM E8/E8M-15a, *Standard Test Methods for Tension Testing of Metallic Materials*, ASTM International, West Conshohocken, PA, USA, 2015.
[37] J. Chen, J. Wang, F. Xie, and W. Jin, "Behavior of thin-walled dodecagonal section double skin concrete-filled steel tubes under bending," *Thin-Walled Structures*, vol. 98, pp. 293–300, 2016.
[38] AISC 360-10, *Specification for Structural Steel Buildings*, American Institute of Steel Construction (AISC), Chicago, IL, USA, 2010.

10

Properties of Ready Mixed Acrylic Paste for Exterior Insulation Using Pozzolanic Materials and Atomizing Slag

Sung-Yong Song,[1] Hwa-Sung Ryu,[2] Sang-Heon Shin,[2] Deuck-Mo Kim,[3] and Won-Jun Park[4]

[1]*Yoosong Engineering, Dongan-gu 14054, Anyang, Republic of Korea*
[2]*Hanyang Experiment and Consulting, Hanyang University, Ansan 15588, Republic of Korea*
[3]*Hwarok Construction Industry, Hanyang University, Ansan 15588, Republic of Korea*
[4]*Department of Architectural Engineering, Kangwon National University, Samcheok 25913, Republic of Korea*

Correspondence should be addressed to Hwa-Sung Ryu; rhsung73@hanyang.ac.kr

Academic Editor: Antonio Boccaccio

Ready mixed acrylic paste (RMAP) is a material used in exterior insulation systems of buildings; it consists of liquids and aggregates. However, it is considerably difficult to use an admixture to control the physical properties of the RMAP owing to problems associated with strength and workability. Thus, to improve the properties of RMAP, this study evaluates the properties of a mortar produced with RMAP as well as pozzolanic materials and atomizing slag, which is a by-product from the steelmaking industry. It was observed that if the density difference is large, a liquid RMAP may experience segregation. Through our experiments, we confirm that segregation resistance could be achieved even at 20% atomizing slag, which has a high density, by increasing the viscosity by approximately 1000 cP through the use of 10% fly ash and 3% metakaolin in the RMAP. Despite the increase in viscosity with the addition of the atomizing slag, the flow of the RMAP increased from 160 to 175 mm due to the spherical particle shape of the slag; in addition, the strength and water absorption coefficient also improved. In particular, the adhesion tension increased from 1.8 N/mm^2 to 2.4 N/mm^2 or higher.

1. Introduction

Among the various methods to improve energy efficiency of buildings, window and wall insulation systems with high airtightness and insulation are typically applied. In particular, wall insulation systems are classified into interior insulation, exterior insulation, and hollow insulation systems; among these, the exterior insulation approach has several economic and workability advantages over the others. Furthermore, the exterior insulation method causes less heat bridging in structures, thus allowing for low-cost insulation renovation. The exterior insulation method involves attaching insulating materials using cement mortar to the exterior wall of masonry and reinforced concrete structures and leveling the surface, thus integrating the insulating materials within the structures themselves [1–5]. The cement mortar that is used as an adhesive in exterior insulation systems can be largely divided into dry mortar, which is then used by adding water at the construction site, and ready mixed acrylic paste (RMAP), which is cement mortar already mixed with aggregates, water, polymers, and additives [6–8]. The drawback of dry mortar is that the workability and strength of the obtained cement mortar vary considerably based on the amount of added water; in contrast, because RMAP already has the aggregates, water, and polymers mixed in, the workability can be predetermined, as the polymers are already dispersed in the cement mortar. However, a disadvantage of the RMAP is that it is difficult to use cement-based admixtures with it for performance improvement because RMAP with semiliquid is mixed with cement in the ratio of 1 : 1 in the construction field; thus, only aggregates, water, and polymers can be used in RMAP, and a separate admixture is difficult to use.

Admixtures are effective materials for dry shrinkage control, reduction of hydration heat, and improvement of packing density, among others; in particular, admixtures are divided into cement-based materials, which harden by directly reacting with water, and pozzolanic materials, which react with cement over time. Pozzolanic materials, including fly ash, metakaolin, and silica fumes, react with Na^{2+}, K^{+}, and Ca^{2+} in the cement, thus generating hydrates [9, 10]. Hence, if there is no or a small number of reactive cations in the liquid when pozzolanic materials are applied to the RMAP, the pozzolanic material might not react, thus staying unreacted till later; therefore, applying pozzolanic materials to the RMAP might be effective in improving the physical characteristics of structures.

Furthermore, slags generated as industrial by-products in the steelmaking industry can lead to the problem of alkali leachate and fugitive dust if they are stored in an open storage yard. Therefore, to address this issue, the slags that are generated at high temperature are transformed into spherical particles of composite oxide, having a diameter of 4 mm or less, through an atomizing treatment. The slag produced through this process is called the "atomizing slag," which is a functional aggregate that suppresses the abnormal expansion inside cement by fixing free CaO, effectively improving the fluidity of cement compounds. However, the application of atomizing slags is limited because they have high densities and pose a risk of segregation during long-term storage when used in the RMAP [11–13].

Therefore, in order to improve the physical and mechanical performance of RMAP, its properties were evaluated with the addition of pozzolanic materials and atomizing slags. First, pozzolanic materials, fly ash and metakaolin, are used to improve physical performance by improving the cement structure through the pozzolanic reaction. Then, atomizing slags are used to improve cement fluidity by replacing the aggregates in RMAP. To that end, aggregates of RMAP were produced by using pozzolanic materials and then stored in a sealed curing state at the temperature of 30°C for six months, after which the viscosity change in the cement was measured [14]. After mixing the pozzolanic materials and atomizing slags with the cement, the compressive strength analysis, flexural strength analysis, adhesion tension analysis, water absorption coefficient analysis, X-ray diffraction (XRD), thermal analysis, and mercury instrusion porosimetry (MIP) were conducted. Furthermore, after replacing the aggregates in the RMAP with atomizing slags, segregation during storage, fluidity, compressive strength, flexural strength, and adhesion tension were analyzed.

2. Experiment

2.1. Materials

2.1.1. Pozzolanic Materials. Pozzolanic materials, fly ash and metakaolin, were used to improve the performance of the RMAP; their chemical compositions are listed in Table 1. In our study, the fly ash that was used had a density and fineness of 2.2 g/cm^3 and 2,800 cm^2/g, respectively, with the percentage of SiO_2 + Al_2O_3 + Fe_2O_3 being approximately 78.346%. Furthermore, the metakaolin used in our study had a density and Blaine value of 2.5 g/cm^3 and 120,000 cm^3/g, respectively.

2.1.2. Cement. The cement used for our mortar experiment was Class 1 ordinary Portland cement (OPC) whose chemical composition and physical properties are listed in Tables 2 and 3, respectively.

2.1.3. Fine Aggregates. For fine aggregates, quartz and atomizing slag were used in our study; their properties are listed in Table 4.

2.1.4. Polymers. The polymer used in our study was acrylic emulsion; its properties are listed in Table 5.

2.1.5. Methylcellulose. On contact with water, methylcellulose has the property of swelling up by absorbing the water. If a large quantity of methylcellulose is mixed with water, it rapidly absorbs water, transforming into a gel form, which causes the problem of low dispersibility. Thus, in this experiment, methylcellulose was dispersed in water slowly. The properties of methylcellulose are listed in Table 6.

2.1.6. pH Controller. To thicken the dispersed methylcellulose, 2-amino-2-methyl-1-propanol (hereinafter, AMP) was used as a pH controller; the properties of AMP are listed in Table 7. As mentioned above, when methylcellulose comes in contact with water, it swells and becomes highly viscous. If methylcellulose is not evenly dispersed, the swollen methylcellulose binds to undispersed methylcellulose and becomes vitrified. The delayed methylcellulose has been designed in such a manner that it does not swell immediately upon contact with water; however, its viscosity increases as the water temperature increases. Therefore, if a pH controller is used after methylcellulose is evenly dispersed, the pH increases to 9 or higher and methylcellulose swells rapidly, thus increasing the viscosity of the liquid.

2.2. Experimental Procedure

2.2.1. Overview. Table 8 lists the measurement method used for each part of our experiment. In our experiment, three types of characteristics were primarily evaluated; these characteristics can be summarized as follows:

(1) Stability of pozzolanic materials was evaluated using an analysis test of the cement paste with RMAP, but without fine aggregates

(2) Viscosity based on the mixing of pozzolanic materials and aggregates after production of RMAP

(3) Characteristics of the cement mortar with RMAP produced using aggregates and pozzolanic materials

Table 1: Chemical composition of the pozzolanic materials.

	SiO_2	Al_2O_3	CaO	MgO	SO_3	Fe_2O_3	P_2O_5	K_2O	Na_2O	LOI
Fly ash	50.88	22.33	6.702	2.18	0.795	5.136	0.237	0.98	1.69	9.07
Metakaolin	49.3	47.1	0.05	2.18	0.01	0.51	0.05	0.05	0.15	0.6

Table 2: Chemical composition of the OPC.

SiO_2	Al_2O_3	CaO	MgO	SO_3	Fe_2O_3	P_2O_5	K_2O	Na_2O	LOI
21.09	5.24	63.57	2.18	2.03	3.52	0.14	0.85	0.11	0.27

Table 3: Physical properties of the OPC.

Specific gravity (°C)	Blaine specific surface area (cm^2/g)	Setting time (min)		Compressive strength of mortar (MPa)		
		Initial setting	Final setting	3 days	7 days	28 days
3.15	3400	400	520	20	23	38

Table 4: Properties of the fine aggregates.

Type	Maximum size (mm)	Unit weight (kg/l)	Specific gravity (20°C)	Water absorption (%)
Quartz	0.14–0.3	1.5	2.62	1.19
Atomizing slag	0.14–0.3	2.2	3.56	0.42

Table 5: Properties of the acrylic emulsion.

Color	Solid content (%)	Viscosity (mPa·s)	pH (20°C)
White	48	102	9.0

Table 6: Properties of methylcellulose.

Appearance	Particle size	Viscosity at 20°C 1% solution	Water content (%)
Whitish powder	<600 μm (98%)	2,500–3,500 mPa·s	Max 5%

Table 7: Properties of the pH controller (AMP) used in the experiment.

pH of 0.1 M aqueous AMP solution	Molecular weight	Specific gravity at 25°C	Freezing point	pKa at 25°C
11.3	89.1	0.942	−2°C	9.75

Table 8: Measurement methods for different parts of the experiment.

Part of the experiment	Sample	Measurement items	Standards
RMAP with pozzolanic materials	Cement paste	pH	ASTM D1293-99 [15]
		XRD	ASTM D 1365-18 [16]
		TG-DTA	ASTM D 3850-12 [17]
	RMAP	Viscosity (cP)	ASTM D 2196 [18]
		Segregation	ASTM C1610 [19]
	Cement mortar	Viscosity (cP)	ASTM D 2196
		Flow	ASTM C 230/230M-14 [20]
		Compressive strength	ASTM C 349 [21]
		Flexural strength	ASTM C 348 [22]
		Adhesion tension	ASTM D 7234-12 [23]
		Water absorption coefficient	ASTM C 1794-15 [24]
		MIP	ASTM D 3850-12 [25]

2.2.2. Semi Premix Mortar Using Pozzolanic Materials. The RMAP mixing method employed in our study is outlined as follows. First, water, polymers, and methylcellulose are evenly mixed together for 2 min. Then, the pH controller is added to the mixture, and the viscosity of methylcellulose is increased by stirring the mixture. Then, the pozzolanic materials, quartz, and atomizing slags are added to the thickened mixture, which is further mixed for 2 min to produce RMAP. Figure 1 shows the flowchart of the RMAP mixing process.

2.2.3. Mixing Method Used for RMAP. Table 9 lists the composition of the paste and mortar used in our experiment. First, to evaluate their six-month stability, the paste and RMAP specimens were cured at a temperature of 30°C and relative density of 60% and then stored at 20°C. Then, the polymer-cement mortar specimen was produced at a temperature of 20°C.

In particular, the paste specimen was produced by mixing 100 g of OPC in the paste mixture for 3 min; then, pH analysis, XRD, and TG-DTA (thermogravimetry-differential thermal analysis) were performed. For the measurement of the pH of the cement paste, 10 ml of the mixture was collected and immersed in 100 g of distilled water, after which the pH was measured thrice: 2, 4, and 6 months after immersion. For XRD analysis, 30 g of the cured paste specimen was collected after 3 and 28 days and crushed to particles of 300 μm or smaller sizes, or crushed and immersed in acetone and dried. Then, the specimen was analyzed using Cu radiation at 45 kV and 40 mA using an X-ray diffractometer (Shimadzu XD-R1). The scanning rate was 5°/min with a range of 5–60°.

The RMAP specimen was produced by mixing quartz and atomizing slags for 3 min; then, the viscosity of the produced specimen was measured.

For the cement mortar, after cement was added to the RMAP, they were mixed for 3 min. Then, the mixture was left standing for 2 min and mixed again for 2 min after then. The flow of the produced specimen was measured in accordance with ASTM C 230. The specimen of each mixture was poured in a 4 cm × 4 cm × 16 cm mold and stripped after curing for 18 hours. The compressive and flexural strengths were measured three times at 3, 7, and 28 days in accordance with ASTM C 348. In addition, the adhesion tension and water absorption coefficient were measured in accordance with ASTM D 7234-12 and ASTM C 1794-15, respectively.

3. Results and Analysis

3.1. Storage Stability after Mixing Pozzolanic Admixture and Atomizing Slag

3.1.1. Characteristics of the RMAP. Table 10 lists the pH and viscosity of RMAP. After mixing plain RMAP with fly ash and metakaolin, the pH of the RMAP increases sharply from 7 to 11 because of the addition of the pH controller. Though the pH of the specimen decreases slightly after fly ash is mixed into it, this pH change was not considerable. Furthermore, during the curing process, the pH hardly changed. Thus, it can be deduced that there was almost no change in pH by the addition of fly ash and metakaolin. In the cement paste, fly ash and metakaolin react directly with $Ca(OH)_2$ present in the cement to form hydrates, thus lowering the pH; however, it is believed that, in the RMAP, the expected pH decrease because of this pozzolanic reaction was not observed because the chemical structure of AMP, which is used as a pH controller, raises the pH through the alcohol group at its terminal.

The viscosity of the semi premix paste after mixing and curing the RMAP for 6 months was approximately 743–860 cP. In particular, the viscosity of the plain specimen immediately after mixing was approximately 800 cP but slightly increased to 830 cP after curing for 6 months. In a similar manner, the viscosity of the specimen produced by mixing fly ash and metakaolin also slightly increased; it is considered that, after the dispersion process of methylcellulose, the undispersed methylcellulose absorbed water over time and diffused in the liquid, thus slightly increasing the viscosity of the mix.

When the changes in viscosity based on the mixing ratios of fly ash and metakaolin were examined, it was observed that the viscosity slightly decreased after the addition of fly ash and increased when metakaolin was added; it is considered that metakaolin increased the viscosity because it has a large specific surface area, thus requiring a large amount of water for fluidity.

3.1.2. XRD and TG-DTA of the RMAP. Figures 2 and 3 show the XRD graph for the RMAP according to the ages. It is observed that the XRD curves are similar with no significant differences between the semi premix paste immediately after mixing and the RMAP after curing for six months.

Figure 4 shows the TG-DTA graph of the RMAP mixed with cement. In this graph, the RMAP immediately after mixing and the semi premix paste after curing for six months show similar changes in the amount of calcium hydroxide. Table 11 lists the quantities of calcium hydroxide in the specimen by curing age; in particular, the amount of calcium hydroxide does not show a significant difference between the values immediately after mixing and curing the specimen for six months. This observation indicates that there was no pozzolanic reaction in the liquid during storage. Furthermore, the amount of calcium hydroxide in the specimen with metakaolin was lower than that in the one with fly ash; after 28 days, the amount of calcium hydroxide was approximately 65% that in the plain specimen.

3.1.3. Viscosity and Fluidity of Cement Mortar Based on Atomizing Slag Replacement Ratio. Table 12 lists the viscosity graph values for the RMAP based on the replacement ratio of atomizing slags. As the replacement ratio of the atomizing slag increased, the viscosity decreased and the fluidity increased. However, after six months, the specimen mixed with a high-density atomizing slag showed segregation. It is considered that the atomizing slag with coarse particles does not have a significant effect on the viscosity of the cement paste; however, the fluidity of the cement mortar

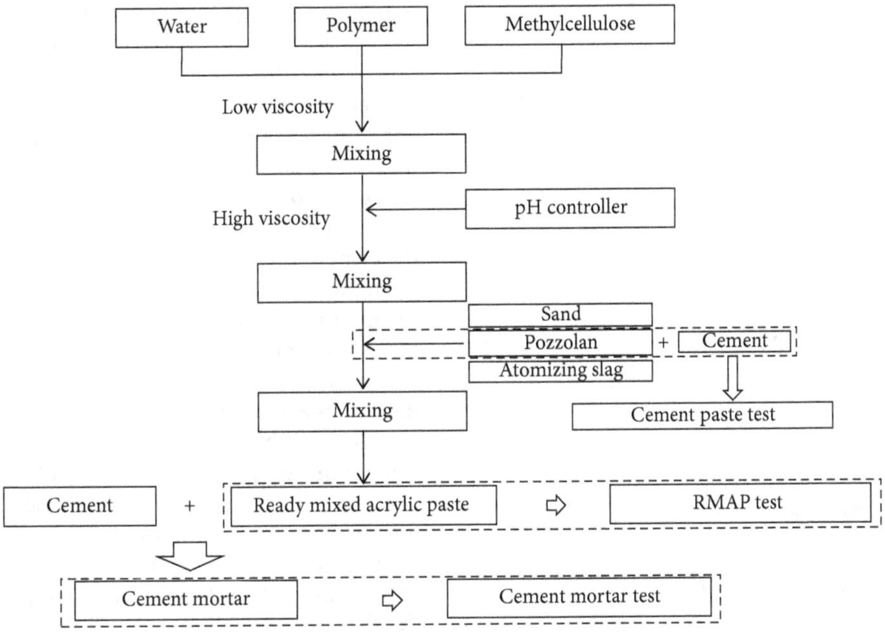

FIGURE 1: Flowchart of the RMAP mixing process.

TABLE 9: Compositions of the paste and mortar.

Mixture number	Polymer (g)	Water (g)	Methyl cellulose (g)	pH controller (g)	Fly ash (g)	Metakaolin (g)	Quartz (g)	Atomizing slag (g)	OPC (g)
Plain	6.5	38.5	0.5	0.1	—	—	54	—	100
F5M0	6.5	38.5	0.5	0.1	5	—	49	—	100
F10M0	6.5	38.5	0.5	0.1	10	—	44	—	100
F15M0	6.5	38.5	0.5	0.1	15	—	39	—	100
F10M1	6.5	38.5	0.5	0.1	10	1	43	—	100
F10M3	6.5	38.5	0.5	0.1	10	3	41	—	100
F10M5	6.5	38.5	0.5	0.1	10	5	39	—	100
P-AS10	6.5	38.5	0.5	0.1			31	10	100
P-AS20	6.5	38.5	0.5	0.1			21	20	100
FM-AS10	6.5	38.5	0.5	0.1	10	3	31	10	100
FM-AS20	6.5	38.5	0.5	0.1	10	3	21	20	100
Specimen			Cement past test		RMAP Cement mortar				

TABLE 10: pH and viscosity of RMAP.

		Plain	F5M0	F10M0	F15M0	F10M1	F10M3	F10M5
pH	Initial	11	11.2	11	11.2	11	11	11
	6 months	11	11.0	10.9	11.0	10.8	11	10.7
Viscosity (cP)	Initial	800	780	765	743	770	790	810
	6 months	830	810	784	761	810	835	860

increases owing to the ball bearing phenomenon caused by spherical particles.

3.2. Mechanical Properties after Mixing the Pozzolanic Admixture and Atomizing Slag. Figure 5 shows the adhesion tension graph for the cement mortar produced using RMAP. The adhesion tension at 3 days of age increased until the fly ash replacement ratio of 10% but decreased after 15%. When metakaolin was used instead, the adhesion tension increased up to a 3% replacement ratio.

At 7 days of age, the adhesion tension showed a similar trend to that at 3 days of age until the fly ash replacement ratio of 10%; however, in the case of metakaolin, the adhesion tension increased until the replacement ratio of 5%. Furthermore, at 14 days of age, strength values showed an increasing trend similar to that in the case of the mortar of 7 days of age. In the case of compressive strength, the strength

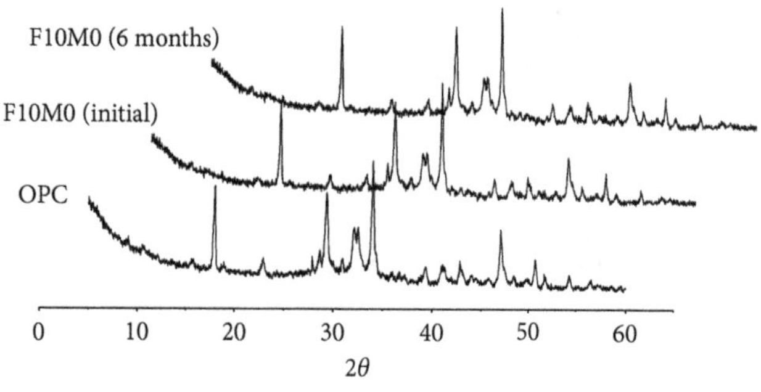

Figure 2: XRD graph for the semi premix paste after 3 days.

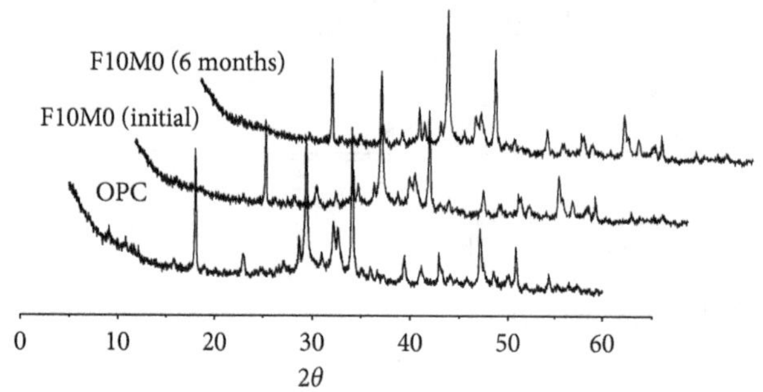

Figure 3: XRD graph for the semi premix paste after 28 days.

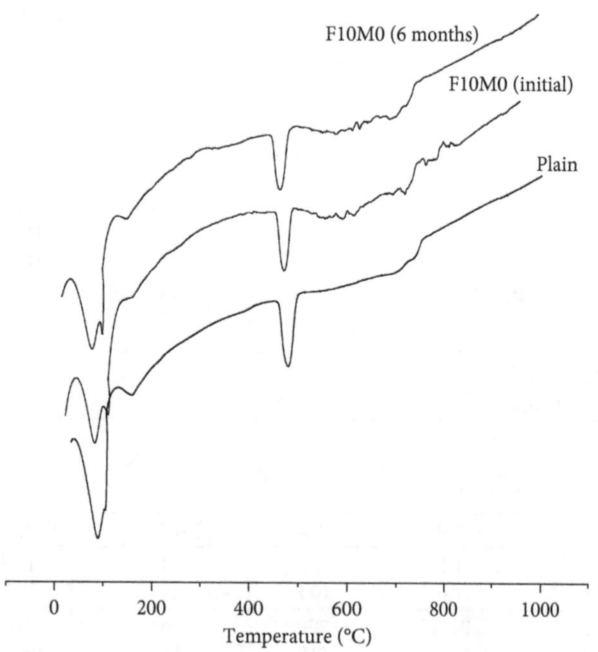

Figure 4: TG-DTA graph for RMAP with cement after 3 days.

increased as the replacement ratios of fly ash and metakaolin increased; in contrast, in the case of adhesion tension, the largest value was observed at 10% fly ash and 3% metakaolin.

Furthermore, the compressive strength increased with increasing age as the internal voids are filled. However, the adhesion tension is realized owing to anchoring from the hardening of the cement paste that penetrates into the voids of the adhered surface. Anchoring is essential for the manifestation of adhesion tension; adhesion may deteriorate if the specimen is fractured owing to shrinking deformation due to dry shrinkage after the cement specimen hardens. Therefore, it is considered that a specimen with a high initial adhesion tension may show a decrease in anchoring failure due to dry shrinkage, thus showing a strength trend that is different from that of the compressive strength.

Figure 6 shows the graph for the compressive strength of the cement mortar produced using RMAP. The compressive strength of the specimen mixed with fly ash increased until the mixing ratio of 5% and decreased after that. In contrast, in the case of metakaolin, the compressive strength increased even at 1% mixing ratio and showed the highest value at 28 days of age when the mixing ratio was 5%.

Figure 7 shows the graph for the flexural strength of the cement mortar produced using RMAP. At 3 days of age, the compressive strength of the specimen mixed with fly ash increased until a 5% mixing ratio and decreased after that. At 28 days of age, however, the flexural strength increased until 10% mixing ratio. In contrast, when metakaolin was mixed, the flexural strength improved at 3 days of age; however, it was lower than that of the plain specimen. Nevertheless, at 28 days of age, the flexural strength was high in all mixtures of metakaolin.

3.3. Mechanical Properties of the RMAP Based on the Replacement Ratio of Atomizing Slag. Figure 8 shows the adhesion tension graph for cement mortar based on the replacement ratio of the atomizing slag. As the mixing ratio of the atomizing slag increased, the adhesion tension at 3 days of age was slightly higher than that of the specimen using quartz, and a similar trend was observed until 28 days of age. For specimens modified with fly ash and metakaolin, the adhesion tension at 3 days of age was higher by 0.1 N/mm^2 than the unmodified specimen.

Figure 9 shows the compressive strength graph for the cement mortar based on the replacement ratio of the atomizing slag. As the mixing ratio of the atomizing slag increased, the specimen showed higher compressive strengths compared with the plain specimen at 3 and 28 days of age. However, the specimens modified with fly ash and metakaolin showed lower compressive strengths at 3 days of age but higher compressive strengths at 28 days of age. Because the atomizing slag is a spherical particle, it is considered that the strength is increased due to the improvement of the filling performance due to the increase of fluidity [26, 27].

Figure 10 shows the flexural strength graph for the cement mortar based on the replacement ratio of the atomizing slag. As the mixing ratio of the atomizing slag increased, the specimen showed higher flexural strengths compared with the plain specimen at 3 and 28 days of age. However, the specimens modified with fly ash and metakaolin showed lower flexural strengths at 3 days of age but higher compressive strengths at 28 days of age.

3.4. Coefficient of Water Absorption for RMAP Modified with the Admixture. Figures 11 and 12 show the graph for the

TABLE 11: Amount of $Ca(OH)_2$ according to curing age.

Specimen	Amount of $Ca(OH)_2$ 3 days Absolute value (%)	3 days Equivalent to plain specimen (%)	28 days Absolute value (%)	28 days Equivalent to plain specimen (%)
Plain	−2.426	100	−1.956	100
F10M0 (initial)	−1.768	72.87	−1.508	77.10
F10M0 (6 months)	−1.81	74.60	−1.592	81.39
F10M3 (initial)	−1.419	58.49	−1.276	65.23
F10M3 (6 months)	−1.451	59.81	−1.329	67.94

TABLE 12: pH and viscosity of the RMAP.

		Plain	P-AS10	P-AS20	FM-AS10	FM-AS20
Unit weight (kg)		2105	2160	2210	2140	2190
Segregation	Initial	Not detected	Not detected	Not detected	Not detected	Not detected
	6 months	Not detected	Detected	Detected	Not detected	Not detected
Viscosity (cP)	Initial	10,000	9,200	8,700	11,200	11,600
	6 months	10,500	9,300	8,750	11,500	11,300
Flow (mm)		160	165	170	170	175

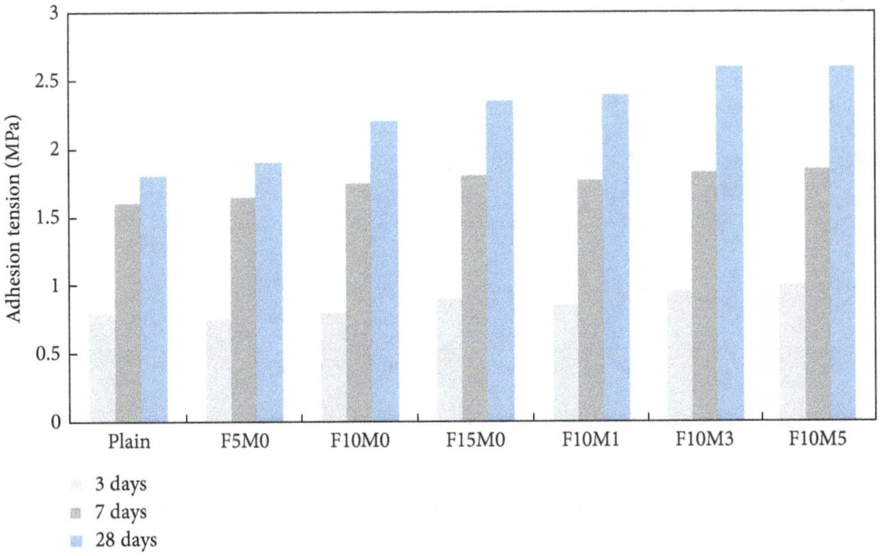

FIGURE 5: Adhesion tension for the cement mortar with RMAP.

coefficient of water absorption for the cement mortar modified with the admixture. In particular, the coefficient of water absorption showed a continuously decreasing trend until 15% fly ash. In contrast, when metakaolin was mixed, the coefficient of water absorption decreased as the mixing ratio increased. Furthermore, when the atomizing slag was mixed, the change in the coefficient of water absorption was not large.

3.5. MIP Analysis. Figure 13 shows the void characteristics of the specimen composed of RMAP and cement. The internal voids decreased after the pozzolanic admixture was added. Thus, the pozzolanic admixture is considered to be the cause of the decreasing coefficient of water absorption.

4. Conclusions

In our study, RMAP specimens were produced using industrial by-products; the characteristics of the obtained polymer-cement mortar produced using them were examined. Consequently, the following conclusions are obtained based on our study.

When the specimens were produced using pozzolanic admixtures in the RMAP, the properties immediately after mixing the RMAP were identical to those of the specimen after curing for six months; in addition, the properties of the RMAP with pozzolanic admixtures in the liquid state showed no deterioration.

In regard to the different pozzolanic admixtures used with the RMAP, using fly ash decreased the viscosity of the

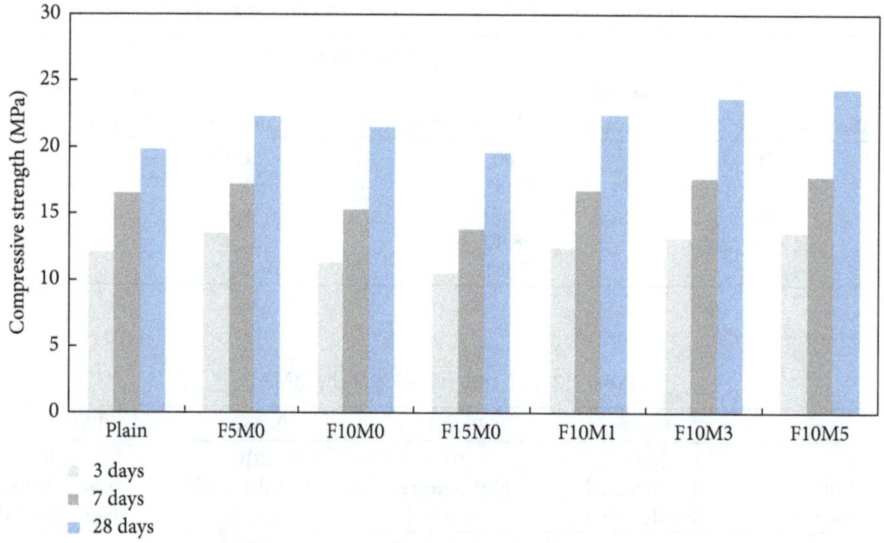

FIGURE 6: Compressive strength of the cement mortar with RMAP.

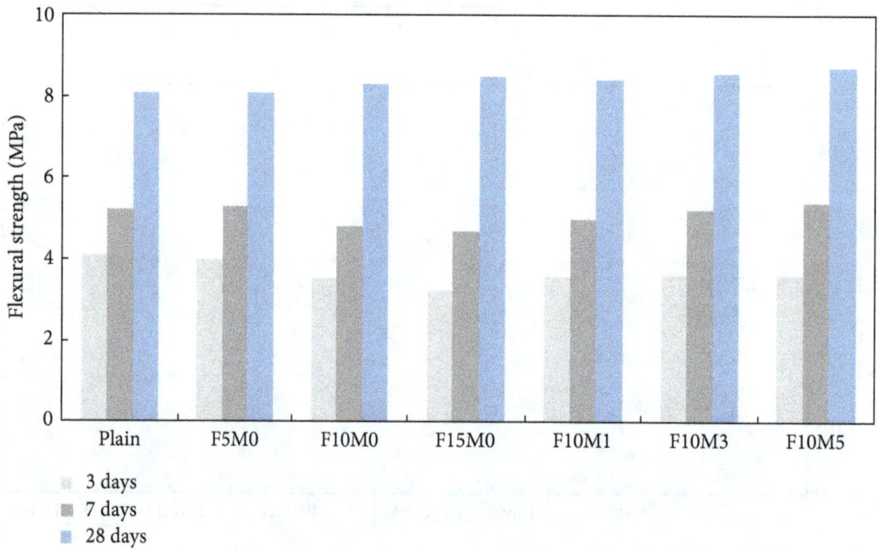

FIGURE 7: Flexural strength of the cement mortar with RMAP.

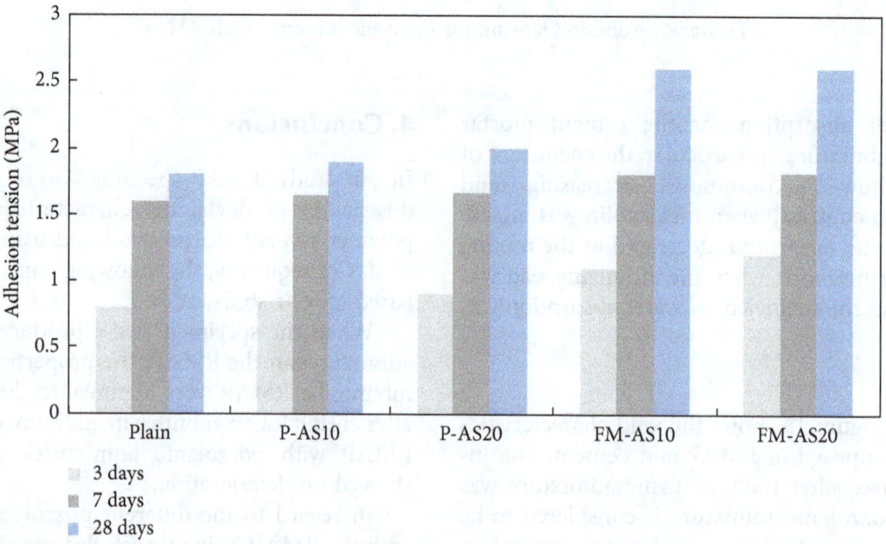

FIGURE 8: Adhesion tension for the cement mortar using atomizing slag-modified RMAP.

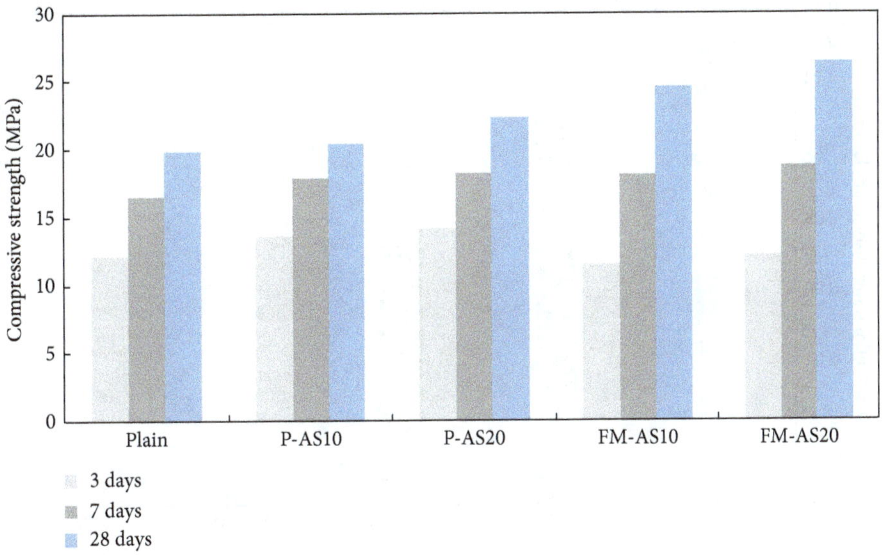

FIGURE 9: Compressive strength of cement mortar using atomizing slag-modified RMAP.

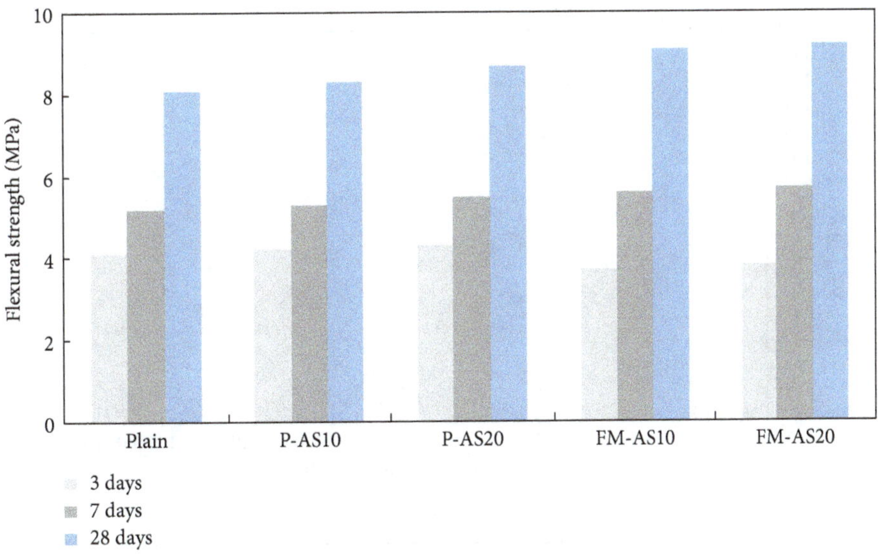

FIGURE 10: Flexural strength of cement mortar using atomizing slag-modified RMAP.

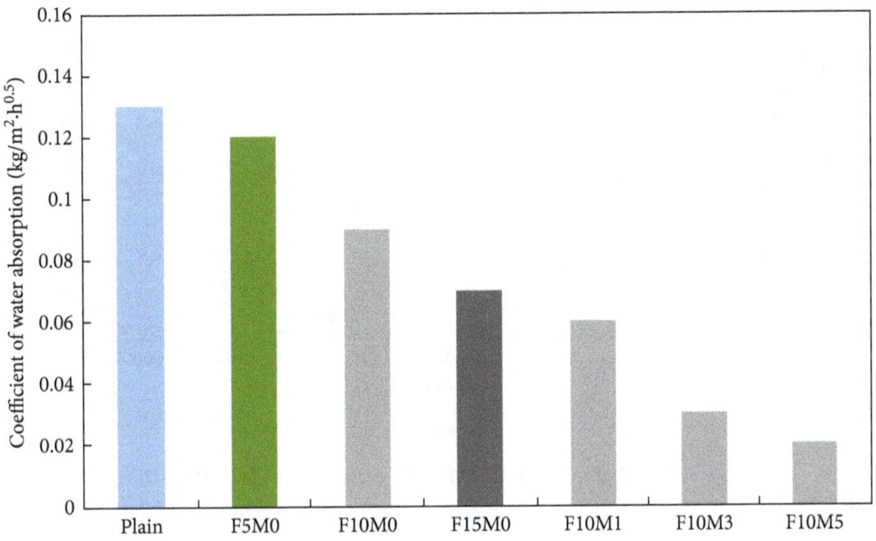

FIGURE 11: Coefficient of water absorption of the cement mortar with RMAP.

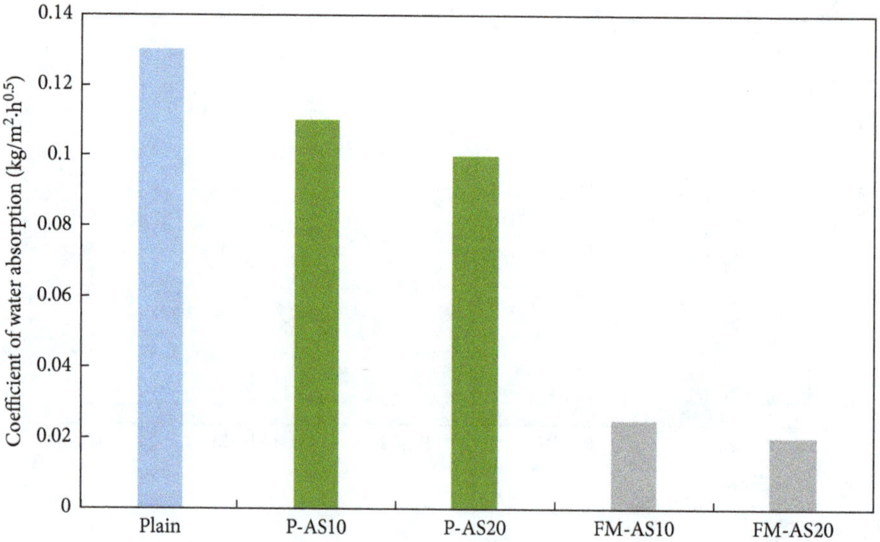

FIGURE 12: Coefficient of water absorption of the atomizing slag-modified RMAP.

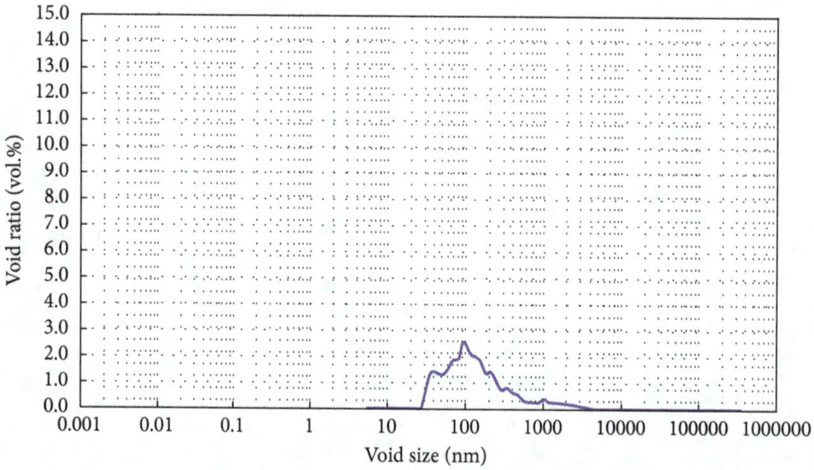

(a)

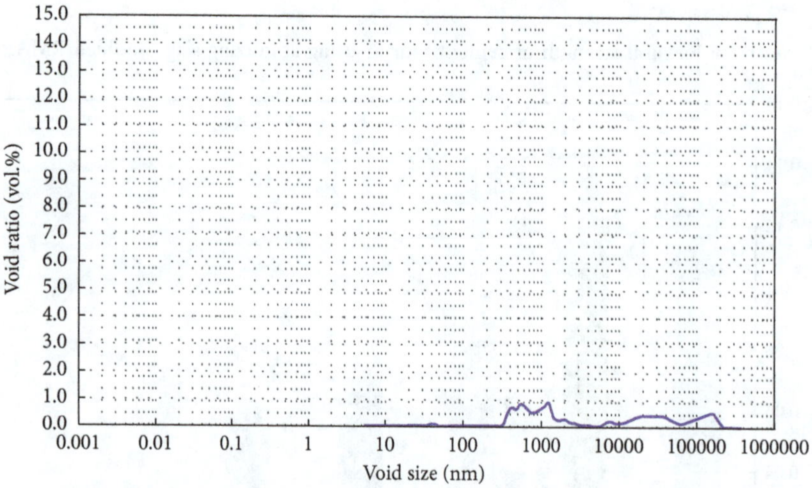

(b)

FIGURE 13: MIP graphs for RMAP with cement. (a) Plain. (b) F10M3.

specimen, whereas metakaolin increased it. Thus, a product that has similar viscosity characteristics as the plain specimen could be produced by their combination.

The RMAP with the atomizing slag experienced segregation because of high specimen density six months after mixing. However, segregation could be suppressed by increasing the viscosity by the addition of fly ash and metakaolin. The RMAP with pozzolanic admixtures showed higher strengths than the plain specimen at every age considered in the study as well as in the case when a mixture of both fly ash and metakaolin was used. In the case of compressive and flexural strengths, the specimen with fly ash showed higher strengths than the plain specimen at early ages; however, the specimen using metakaolin showed higher strengths than the plain specimen even in the long term. It was also observed that the use of pozzolanic admixtures in the RMAP can ensure the stability of adhesion tension.

When the atomizing slag was used, the viscosity decreased and fluidity increased. The mixture with the atomizing slag showed strength-increasing effects for adhesion tension, flexural strength, and compressive strength. Thus, using pozzolanic admixtures could provide higher performance compared to plain specimens. The RMAP with pozzolanic admixtures showed a decrease in the coefficient of water absorption due to the effect of the filling of internal voids, which was confirmed using MIP analysis as well.

Conflicts of Interest

The authors declare that there are no conflicts of interest regarding the publication of this paper.

Acknowledgments

This research was supported by grant number 18CTAP-C130219-02 through the Korea Agency for Infrastructure Technology Advancement funded by the Ministry of Land, Infrastructure and Transport of the Korean government.

References

[1] J. L. Dentz, *Evaluating Exterior Insulation and Finish Systems for Deep Energy Retrofits*, New York State Research and Development Authority, Albany, NY, USA, 2017.

[2] European Organization for Technical Approvals (EOTA), *ETAG 004: Guideline for European Technical Approval of External Thermal Insulation Composite Systems with Rendering*, 2000.

[3] W. C. Brown, M. Lawton, and A. Lang, "Stucco-Clad wall-drying experiment," BERC Report, Canada Mortgage and Housing Corporation, Ottawa, ON, Canada, Section 4, 1999.

[4] M. Bomberg, K. Kumaran, and K. Day, "Moisture management of EIFS walls- Part 1: the basis for evaluation," *Journal of Thermal Envelope and Building Science*, vol. 23, pp. 78–94, 2016.

[5] European Organisation for Technical Approvals (EOTA), *ETAG 004: Guideline for European Technical Approval of External Thermal Insulation Composite Systems with Rendering, Section 5.1.3*, 2000.

[6] http://www.terraco-eifs.com/basecoats.

[7] http://www.dryvit.com/products/adhesivesbase-coats/primus/.

[8] http://www.parex.com/products.pl.

[9] C. Shi and R. L. Day, "Pozzolanic reaction in the presence of chemical activators: Part II—reaction products and mechanism," *Cement and Concrete Research*, vol. 30, pp. 607–613, 2000.

[10] P. T. Bui, Y. Ogawa, K. Nakarai, and K. Kawai, "A study on pozzolanic reaction of fly ash cement paste activated by an injection of alkali solution," *Construction and Building Materials*, vol. 94, pp. 28–34, 2015.

[11] N. Wang, H. Peng, X. Ling, J. Kang, and M. Xu, "Experimental investigation of slag particles of ligament mode disintegration in spinning disk atomizing," *Energy Procedia*, vol. 105, pp. 622–627, 2017.

[12] E. H. Hwang, C. H. Lee, and J. M. Kim, "Physical properties of polymer concrete composite using rapid-cooled steel slag (I)," *Applied Chemistry for Engineering*, vol. 23, pp. 210–216, 2012.

[13] D. G. Montgomery and G. Wang, "Instant-chilled steel slag aggregate in concrete (strength related properties)," *Cement and Concrete Research*, vol. 21, pp. 1083–1091, 1991.

[14] L. Wang, K. Uji, and A. Ueno, "Evaluation on reaction efficient of fly ash based on threshold value of replacement ratio," *Cement Science and Concrete Technology*, vol. 71, pp. 645–652, 2017.

[15] ASTM, *ASTM D 1293-99, Standard Test Methods for pH of Water*, 1999.

[16] ASTM, *ASTM D 1365-18, Standard Test Method for Determination of the Proportion of Phases in Portland Cement and Portland-Cement Clinker Using X-Ray Powder Diffraction Analysis*, 2018.

[17] ASTM, *ASTM D 3850-12, Standard Test Method for Rapid Thermal Degradation of Solid Electrical Insulating Materials By Thermogravimetric Method*, 2012.

[18] ASTM, *ASTM D 2196, Standard Test Methods for Rheological Properties of Non-Newtonian Materials by Rotational Viscometer*, 2015.

[19] ASTM, *ASTM C 1610, Standard Test Method for Static Segregation of Self-Consolidating Concrete Using Column Technique*, 2017.

[20] ASTM, *ASTM C 230/230M-14, Standard Specification for Flow Table for Use in Tests of Hydraulic Cement*, 2014.

[21] ASTM, *ASTM C 349 (Test Method for Compressive Strength of Hydraulic-Cement Mortars (Using Portions of Prisms Broken in Flexure))*, 2014.

[22] ASTM, *ASTM C 348-14, Standard Test Method for Flexural Strength of Hydraulic-Cement Mortars*, 2014.

[23] ASTM, *ASTM D 7234-12, Standard Test Method for Pull-Off Adhesion Strength of Coatings on Concrete Using Portable Pull-Off Adhesion Testers*, 2012.

[24] ASTM, *ASTM C 1794-15, Standard Test Methods for Determination of the Water Absorption Coefficient by Partial Immersion*, 2015.

[25] ASTM, *ASTM D 3850-12, Standard Test Method for Rapid Thermal Degradation of Solid Electrical Insulating Materials by Thermogravimetric Method (TGA)*, 2012.

11

Effect of Dosage of Alkaline Activator on the Properties of Alkali-Activated Slag Pastes

Zhenzhen Jiao,[1,2,3] **Ying Wang**,[1,2,3] **Wenzhong Zheng**,[1,2,3] **and Wenxuan Huang**[1,2,3]

[1]*Key Lab of Structures Dynamic Behavior and Control of the Ministry of Education, Harbin Institute of Technology, Harbin 150090, China*
[2]*Key Lab of Smart Prevention and Mitigation of Civil Engineering Disasters of the Ministry of Industry and Information Technology, Harbin Institute of Technology, Harbin 150090, China*
[3]*School of Civil Engineering, Harbin Institute of Technology, Harbin 150090, China*

Correspondence should be addressed to Wenzhong Zheng; zhengwenzhonghit@163.com

Academic Editor: Francesco Ruffino

This study focused on the engineering properties of alkali-activated slag (AAS) pastes prepared by mixing an activator consisting of sodium silicate and sodium hydroxide at room temperature. The water-to-slag ratio of AAS paste was kept constant at 0.35 by mass. AAS pastes were prepared using the activator with five different silicate moduli of 1, 1.2, 1.4, 1.6, and 1.8 and three different Na_2O contents of 6%, 8%, and 10%. The results showed that both the silicate moduli and Na_2O contents had significant effects on the engineering properties of AAS pastes. All the AAS pastes exhibited properties such as fast setting, good workability, and high early compressive strength. The final setting time varied from 9 to 36 min, and the fluidity was in the range of 147–226 mm. The 1 d compressive strength of all the AAS pastes, which could be easily achieved, had values above 55 MPa, whereas the highest strength obtained was 102 MPa with the silicate modulus of 1 and Na_2O content of 8% at room temperature. The drying shrinkage increased as the silicate modulus increased. Furthermore, the hydration products and microstructures of AAS pastes were explained according to the microanalysis methods.

1. Introduction

Due to the fast development of the construction industry all over the world, the demand for building materials has grown strong, especially in the developing countries. It is estimated that 85% of the total CO_2 emissions originate from the cement industry [1]. Therefore, as the demand and subsequent production of cement increase, the CO_2 emission increases, which can cause the greenhouse effect [2]. Blast furnace slag is an industrial by-product, produced during the process of making pig iron, and consists of calcium-rich and silicon-rich glass having latent hydraulic activity [3, 4]. In addition, large amounts of slag are generated worldwide. For example, China produces a total of more than 2.4 million tonnes of slag annually, out of which, only a small portion (20–30%) can be used efficiently, whereas a large portion (70–80%) is deemed as the solid industrial waste [5]. Thus, in order to protect the environment and save resources, new environment friendly building materials must be explored. AAS was prepared by an alkaline activator and aluminosilicate materials. AAS cements are environment friendly cementitious materials, through which the industrial waste can be recycled in an effective way. In recent years, AAS cements have received significantly more attention as potential sustainable alternatives to Portland cement [6], which can convert industrial waste into green building materials.

Alkali-activated cementitious materials are greatly affected by the chemical composition of the alkaline activator [7–9]. In this regard, factors such as type and contents of the activator [6, 8, 10–12], curing conditions [13–16], and mineral admixtures [17–19] in the alkali activation process were considered important. Živica [20] reported that AAS used by the NaOH solution, Na_2CO_3 solution, and water glass of three different types of activators showed an important acceleration for setting as compared to the ordinary Portland cement, and the setting time of slag activated by sodium silicate was the shortest. Jiao et al. [21] reported that the addition of Na_2CO_3 in the NaOH solution can prolong the setting time, increase the later compressive strength, and reduce the drying shrinkage of AAS paste compared with the alkaline activator of only NaOH. Chang [22] studied the effects of concentrations of SiO_2 and Na_2O on the setting time and concluded that phosphoric acid can be used as an effective retarder. Wu et al. [23] reported that the activators with higher pH values can efficiently enhance the hydration process of AAS paste with more silicon and aluminum ions dissolving due to the weaker strengths of Si-O and Al-O bonds in the alkaline environment. Ravikumar and Neithalath [24] reported the effect of alkalinity of an alkali activator on the mechanical performance and microstructures of AAS materials. The results suggested that a higher alkalinity was helpful in improving the compressive strength, whereas the major hydration product was calcium silicate hydrate. Ahmari et al. [25] studied the development of compressive strength of alkali-activated binder based on copper mine tailings at 60–120°C of curing temperature. They showed that the strength first increased with the increase in initial temperature and then decreased when the temperature crossed a certain value. Lee and Lee [26] studied the reaction mechanism of a material produced by blending together sodium silicate and sodium hydroxide, which showed that a higher amount of slag resulted in a higher quantity of C-S-H gel and less aluminosilicate gel, which were produced as the hydration products. Many researchers [27–29] reported that the Ca/Si ratio of calcium silicate hydrate in AAS was lower than that in ordinary Portland cement.

It is well known that AAS has excellent mechanical performance, good durability, and low porosity as compared to Portland cement. Gebregziabiher et al. [30] reported that the compressive strength of AAS with sodium silicate was 25 and 70 MPa at 1 and 28 days, respectively, and that there was almost no increase thereafter until 56 days. Chi [31] showed that the resistance of AAS to sulfate attack was better than that of Portland cement, as shown by a lower weight loss and reduction in compressive strength of AAS concrete. Shi [32] found that the porosity of AAS was lower than that of the Portland cement, whereas the pore structure was also finer. However, there are also some disadvantages of rapid setting and high drying shrinkage of AAS pastes. Gu et al. [33] reported that the setting time of AAS was about 50 min at room temperature and tried to prolong the setting time by lowering the temperature. They also concluded that the drying shrinkage of AAS was 2-3 times that of the ordinary Portland cement.

Although many researchers focused on the properties of AAS, there is still a significant need for further investigations. The fast setting time and high early compressive strength of AAS were used for rapid repair construction and strengthening of concrete structures. The drying shrinkage is an important aspect for the durability of AAS pastes. In order to be better used for the construction, the more appropriate setting time, higher early compressive strength, and less drying shrinkage of AAS were needed to be studied. The composition of slag had a key influence on the properties and reaction mechanism of AAS [10]. The main alkaline activator was the mixture of sodium silicate and sodium hydroxide, which was controlled by the silicate modulus and Na_2O content. Therefore, it is essential to investigate the effects of an optimum activator on the engineering properties of AAS pastes. The objective of this paper is to investigate the effects of silicate moduli and Na_2O contents on fresh properties (setting time and fluidity) and hardened performance (compressive strength, drying shrinkage, and mass loss) of AAS pastes. The reaction mechanism of AAS was revealed using techniques of X-ray diffraction (XRD), Fourier transform infrared (FTIR) spectroscopy, scanning electron microscopy with energy dispersive X-ray spectroscopy (SEM/EDS), and mercury intrusion porosimetry (MIP).

2. Materials and Methods

2.1. Materials. Blast furnace slag was procured from the Tangshan steel plant, China, and its chemical composition is presented in Table 1. The specific surface area of slag was found to be $424\,m^2/kg$, whereas its specific gravity was determined to be 2.91 [21]. The specific surface area was tested by an electric Brinell permeability surface area analyzer (DBT-127; Wuxi Jianyi Instrument and Machinery Co. Ltd., China), according to the Chinese standard GB/T 8074-2008 [34]. The same sodium hydroxide was used in the previous study [21]. Sodium silicate solution consisting of 10.3% Na_2O, 32.1% SiO_2, and 57.6% H_2O was supplied by Julide Chemical Co. Ltd., Langfang, China. The alkaline activator of AAS pastes was prepared using sodium silicate and NaOH solutions.

2.2. Mixing Proportions. The AAS pastes were prepared from the slag in alkaline activators, whereas the mixing proportions are listed in Table 2. The water-to-slag ratio was kept constant at a value of 0.35 (by weight). To prepare AAS specimens, the activator was introduced into the slag powder in a bowl of cement paste mixer, in which the mixture was mixed at both low and high speeds for around 1.5 min. There are five silicate moduli and three different Na_2O contents for activators. The silicate moduli (Ms) consisted of 1, 1.2, 1.4, 1.6, and 1.8, whereas the Na_2O contents were 6%, 8%, and 10% (by weight of the slag). The codes "M1N6," "M1N8," and "M1N10" represented the silicate moduli of 1 and Na_2O contents of 6%, 8%, and 10%, respectively, and so on.

2.3. Methods. The effects of silicate modulus and Na_2O content on the setting times, fluidity, compressive strength, drying shrinkage, and microstructures of AAS pastes were investigated in this paper. The following methods have also been used by the previous study [21].

2.3.1. Test for the Setting Times. The setting times of alkali-activated slag/fly ash were measured by Vicat in the previous

TABLE 1: Chemical composition of slag.

Oxide	SiO_2	Al_2O_3	CaO	Fe_2O_3	K_2O	MgO	Na_2O	SO_3	Others
Percentage (%)	32.83	17.19	36.69	0.38	0.37	8.20	0.65	1.94	1.75

TABLE 2: The mixing proportions used for this study.

Code	Slag (g)	Sodium silicate (g)	NaOH (g)	H_2O (g)
M1N6	1000.0	181.2	53.2	233.7
M1N8	1000.0	241.6	71.0	194.9
M1N10	1000.0	301.9	88.7	156.1
M1.2N6	1000.0	217.4	48.4	213.9
M1.2N8	1000.0	289.9	64.5	168.5
M1.2N10	1000.0	362.3	80.6	123.2
M1.4N6	1000.0	253.6	43.5	194.1
M1.4N8	1000.0	338.2	58.1	142.1
M1.4N10	1000.0	422.7	72.6	90.2
M1.6N6	1000.0	289.9	38.7	174.3
M1.6N8	1000.0	386.5	51.6	115.8
M1.6N10	1000.0	483.1	64.5	57.2
M1.8N6	1000.0	326.1	33.9	154.5
M1.8N8	1000.0	434.8	45.2	89.4
M1.8N10	1000.0	543.5	56.5	24.2

studies [22, 35–37]. The fresh pastes were immediately poured into a truncated cone mold (h = 40 mm, d_{top} = 65 mm, d_{bottom} = 70 mm) to measure the setting time at room temperature. In accordance with the GB/T 1346-2011 standard [38], the setting times of AAS pastes were tested using a Vicat apparatus. The initial setting time was defined as the time between the initial contact of slag with the activating solution and that when the penetration value reached 25 mm (based upon the ASTM C191 standard). The final setting time corresponded to the point when the cyclic annular needle could not display a complete circular print on the AAS paste surface.

2.3.2. Fluidity Test. Based on the GB/T 8077-2012 standard [39], the fluidity of designed mixtures was measured using a truncated cone (h = 60 mm, d_{top} = 36 mm, d_{bottom} = 60 mm) at room temperature. Firstly, the fresh pastes were filled into a mold. After about 15 s, the mold was lifted vertically. After another 30 s, the diameters of the paste spread were measured along two perpendicular directions. The average diameter was interpreted as the fluidity of the pastes.

2.3.3. Compressive Strength Test. The paste samples were poured into prismatic molds of 40 mm × 40 mm × 160 mm dimensions, followed by shaking them manually. Afterwards, plastic foils were used to seal the surfaces of specimens to avoid any moisture losses. Then, the specimens were put in a standard room (20 ± 2°C and RH > 95%) for 1 day. After demolding them, the specimens were kept in the standard room until tests for the age of strength were conducted. Compressive strength values were tested at 1, 3, 7, 14, and 28 days by means of a YAW-300 compressive strength analysis apparatus at a loading rate of 1.0 kN/s. The mean values of six samples were interpreted as the compressive strength of AAS pastes for each age.

2.3.4. Drying Shrinkage and Mass Loss Tests. According to the Chinese standard JCT 603-2004 [40], the fresh pastes were poured into prismatic molds of 25 mm × 25 mm × 280 mm dimensions. The further procedure was the same as described in Section 2.3.3 until the demolding of specimens. The lengths of specimens were first determined, and then they were put in isothermal chambers (20 ± 2°C and RH = 50 ± 5%). The drying shrinkage and mass of specimens were tested every day until 28 days. The drying shrinkages were monitored by using a comparator with a measuring accuracy of 0.001 mm. Meanwhile, the mass values were measured by means of an electronic balance having an accuracy of 0.1 g.

2.3.5. Microstructure Analytical Methods. After compressive strength tests at the corresponding curing time, the samples from the crushed prisms were kept in a mixture of alcohol and acetone (1 : 1 v/v, respectively) for one week to stop the hydration process. Afterwards, the samples were taken out of the solution and dried in an oven at 60°C under vacuum. Some of the samples grounded into powder were passed through an 80 μm sieve and, then, were used for XRD and FTIR spectroscopy analyses. Rest of the samples were used for SEM/EDS and MIP tests. In order to determine the hydration products and reaction mechanism of AAS pastes, the phase and characteristics of the bonds of raw material and products were determined using XRD (D8 ADVANCE manufactured by AXS, Germany) and FTIR spectroscopy (FTIR-650 spectrometer from China) analyses, respectively. The morphology and elemental analysis of the hydration products were determined using the SEM/EDS (ZEISS MERLIN Compact FE-SEM, Germany) technique. The pore structure of AAS was analyzed using MIP (AutoPore IV 9500, Micromeritics Instrument Ltd., USA).

3. Results and Discussion

3.1. Setting Times. Figure 1 shows the effects of silicate moduli and Na_2O contents on the initial and final setting times of AAS pastes at room temperature. When the Na_2O contents increased from 6% to 10%, the initial and final setting times increased at the silicate modulus from 1.2 to 1.8, respectively. However, the setting times decreased for the silicate modulus of 1. Moreover, the optimum silicate modulus was 1.4 for longer setting times. For a higher Na_2O content and silicate moduli in the alkaline activator, Ca from the decomposition of slag tends to form $Ca(OH)_2$ rather than C-S-H due to the excess of OH^- concentration in the earlier stage, resulting in the prolongation of time of Ca^{2+} precipitation [6, 41, 42]. The setting time of AAS pastes changed slightly, showing that all of the pastes can be used as

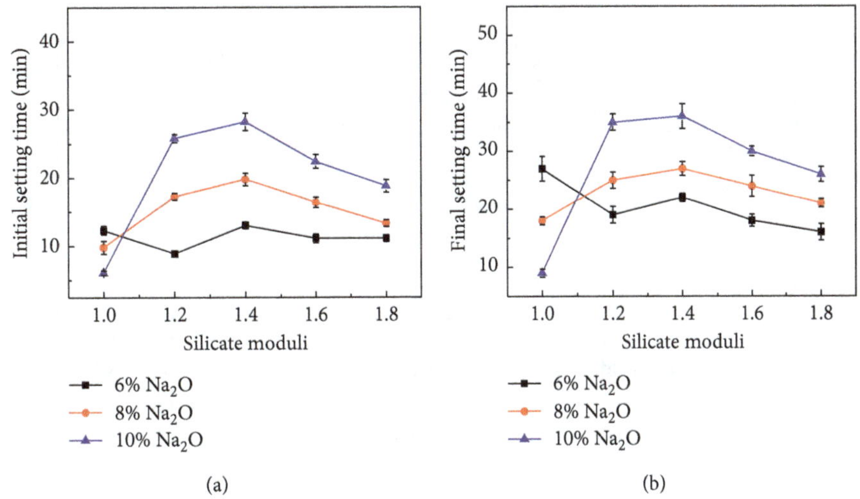

FIGURE 1: The setting times of AAS pastes: (a) initial setting time; (b) final setting time.

the building material of urgent construction or repair due to their fast setting property. These faster setting properties of AAS were ascribed to the rapid formation of initial C-S-H in the sodium silicate activator [10]. Another study [43] reported that the initial and final setting times of AAS were 76 and 106 min, respectively, when the activator of water glass was used. In short, silicate moduli and Na_2O contents of the alkaline activator have a combined effect on the setting properties of AAS pastes.

3.2. Fluidity. Based upon the silicate moduli and Na_2O contents, the fluidity values of fresh pastes are shown in Figure 2. The fluidity of AAS pastes increased with the increase in silicate moduli, suggesting that increasing the silicate modulus can effectively improve the functioning of AAS pastes, which increased slightly once the silicate modulus went beyond the value of 1.4. The slag activated by 8% Na_2O presented higher fluidity than those activated by 6% and 10%. The M1N10 specimen has no fluidity as a lot of reaction products are produced in a very short time, which is related to the shortest setting time of M1N10 (Figure 1). A previous research [44] has reported that the nature and concentration of the activator have vital effects on the rheological performance of AAS pastes. They proved that their rheology depends on both the silicate modulus and Na_2O concentration for AAS pastes activated with water glass. Wang et al. [1] reported that the slump flow increased as the alkali agent increased from 4% to 8% in liquid-to-solid ratios of 0.35, 0.40, and 0.45 in the alkali-activated ladle furnace slag geopolymer. These results were mainly attributed to the stickiness of alkali metal silicate solution.

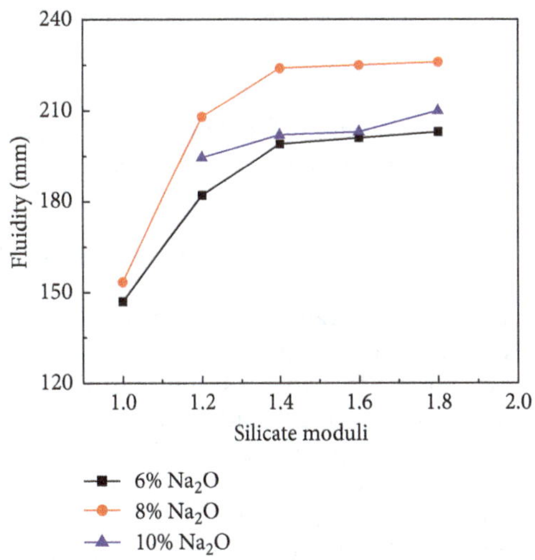

FIGURE 2: The effects of the silicate moduli and Na_2O contents on the fluidity of AAS pastes.

3.3. Compressive Strength. The effects of silicate moduli and Na_2O contents on the compressive strength of AAS pastes within the studied periods are presented in Figure 3. For the Na_2O content of 8%, highest compressive strength was observed, whereas beyond this content of Na_2O (e.g., 10%), the strength decreased. Some previous studies [45, 46] have shown that the excess content of Na_2O in the alkali activator is used for charge balancing the Al^{3+} in the process of polymerization reaction. On the contrary, when the Na_2O contents were 6% and 8%, the specimens with the silicate moduli of 1.4 and 1.6 exhibited relatively good compressive strengths at the early and later stages of the process. Moreover, all the specimens showed higher early strengths (more than 55 MPa after 1 d), whereas the highest 1-day compressive strength obtained was 102 MPa at room temperature. The compressive strength values after 1 day lied within the range of 55%–91%, which were close to those obtained after 28 days. This indicated that the sodium silicate solution-activated slag pastes can enhance the early compressive strength development. The high early compressive strength can be suitable for emergency construction and repair works. Aydin and Baradan [10] showed that the C-S-H gel was rapidly formed for improving the early strength. Puertas and Torres-Carrasco [47] concluded that the gel

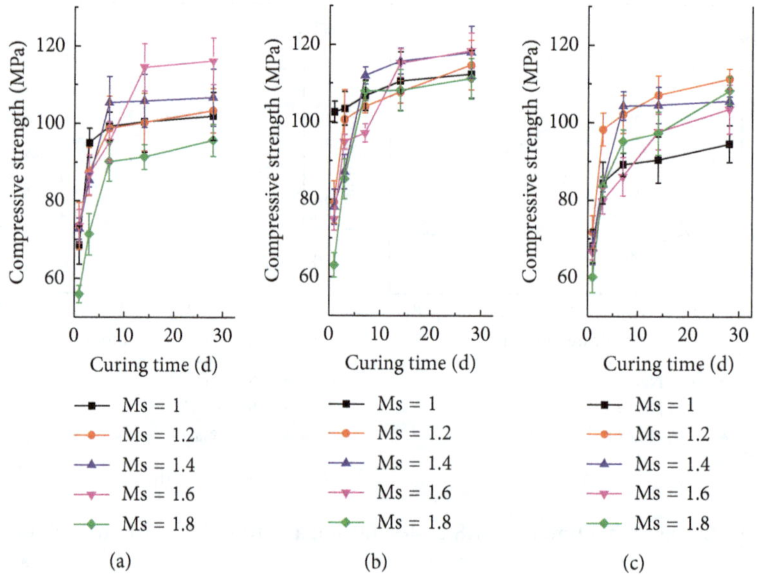

FIGURE 3: The effects of silicate moduli and Na_2O contents on compressive strength of AAS pastes. (a) 6% Na_2O. (b) 8% Na_2O. (c) 10% Na_2O.

formed by the activation of sodium silicate had a more condensed structure during the early stages.

3.4. Drying Shrinkage. The effects of silicate moduli and Na_2O contents on the drying shrinkage before 28 days are shown in Figure 4. As can be observed from Figure 4, the drying shrinkage of AAS pastes increased as the silicate modulus increased. The drying shrinkage values of specimens with Ms = 1.2, 1.4, 1.6, and 1.8 were 1.12, 1.43, 1.59, and 2.01 times that of specimens with Ms = 1 with 8% Na_2O content after 28 days, respectively. In addition, when the silicate modulus was 1 and 1.2, the drying shrinkage increased as the Na_2O contents increased from 6% to 8%, after which the value decreased for 10% Na_2O content. However, the drying shrinkage increased with the increase in Na_2O content for the silicate moduli of 1.4 and 1.8. A previous study [1] has shown that the drying shrinkage increased with the increase in alkali agent from 4% to 8% at the liquid-to-slag ratios of 0.35, 0.40, and 0.45. In addition, Komnitsas et al. [45] reported that the excess KOH did not react with the raw material. Neto et al. [48] concluded that the higher content of sodium silicate resulted in a higher drying shrinkage of AAS pastes. Based upon these results, it can be concluded that the silicate moduli played a more important role in the drying shrinkage of AAS pastes than the Na_2O contents.

3.5. Mass Loss. Figure 5 shows the mass loss (expressed as the mass percent of the total specimen) of AAS pastes for different Na_2O contents (6%, 8%, and 10%) and silicate moduli (Ms = 1, 1.2, 1.4, 1.6, and 1.8) during the drying shrinkage testing within 28 days. Regardless of the Na_2O content of the specimens, the mass loss increased as the silicate modulus increased. In addition, the mass loss decreased with the increase in the content of Na_2O. A previous study [13] has shown that the water loss decreased as the content of Na_2O increased from 4% to 6% in AAS cement. This may indicate that more free water existed in specimens having a lower content of Na_2O and higher silicate modulus. However, Lee et al. [49] have reported contrary results and concluded that the higher dosage of sodium silicate resulted in the loss of less evaporable water, as more water took part in the polymerization reaction. According to a previous study [50], there was a strong relationship between the drying shrinkage and mass loss for Portland cement. However, for AAS pastes, the relationship was not clear, although the drying shrinkage and mass loss of AAS pastes followed the same trend due to the effect of silicate moduli, whereas the mass loss was found not to be directly related to the drying shrinkage due to the effect of Na_2O content.

3.6. XRD Analysis. Figure 6 shows the XRD diffractograms of AAS pastes for various Na_2O contents and silicate moduli at an early age. All the hydration products of the specimens showed some weak peaks, which can be identified as omphacite ((NaCa) $(AlMg)SiO_6$), larnite (Ca_2SiO_4), poorly crystalline C-S-H, and calcite ($CaCO_3$). In addition, calcium aluminum silicate ($Ca_2Al_2SiO_7$) was only found in the M1N10 specimen, whereas the peaks of C-S-H and calcite overlapped with each other. There was almost no significant difference between the crystal phases of the hydration products in the specimens for different contents of Na_2O or silicate moduli. These results were in accordance with those obtained in a previous study [51], in which the researchers also concluded that there were no obvious peaks, indicating that significant quantities of crystalline hydration products were not formed at room temperature. However, Rashad et al. [52] concluded that the intensity of amorphous hump increased as the Na_2O content increased from 3.5% to 6.5%, which resulted in a higher compressive strength.

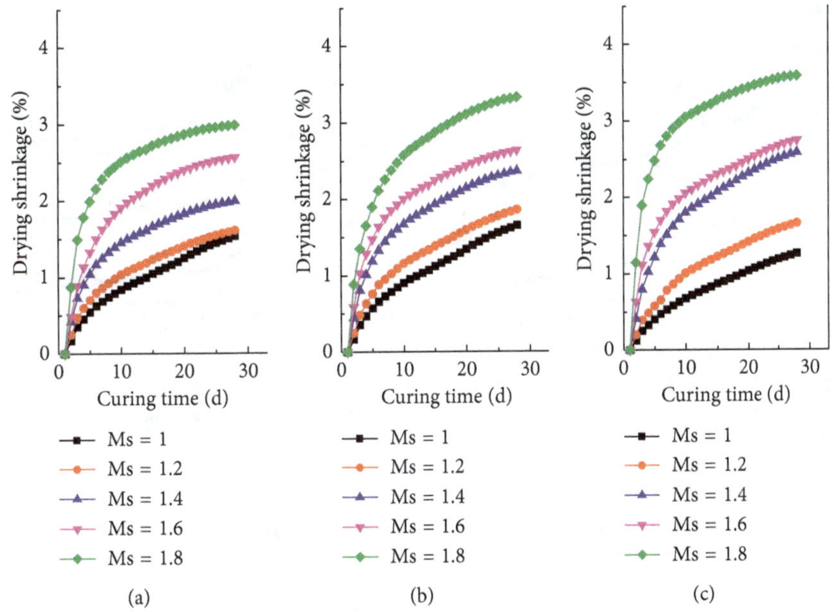

FIGURE 4: The effects of the silicate moduli and Na_2O contents on drying shrinkage of AAS pastes. (a) 6% Na_2O. (b) 8% Na_2O. (c) 10% Na_2O.

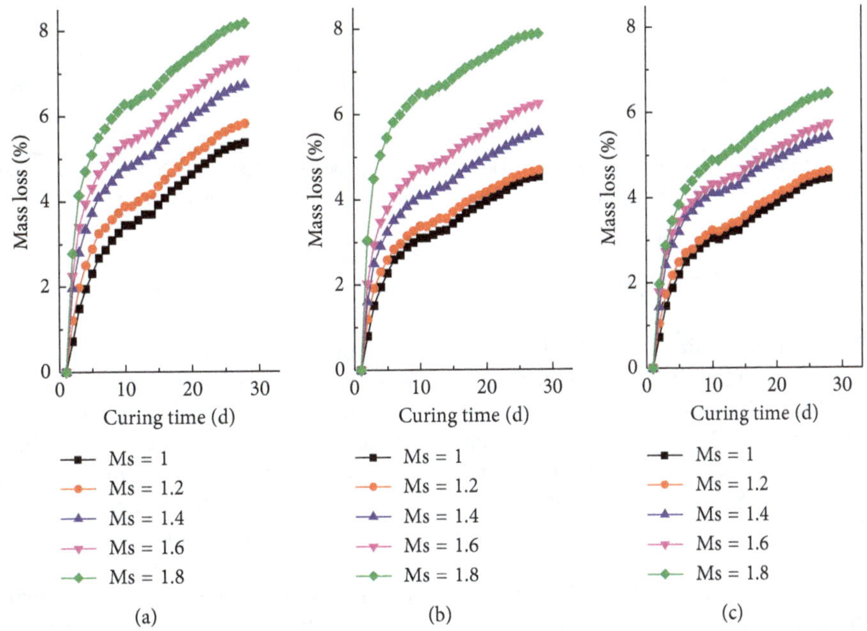

FIGURE 5: The effects of the silicate moduli and Na_2O contents on the mass loss of AAS pastes. (a) 6% Na_2O. (b) 8% Na_2O. (c) 10% Na_2O.

Additionally, there was no indication to prove the existence of $Ca(OH)_2$ compared to the ordinary Portland cement paste [33].

3.7. FTIR Spectroscopy Analysis. The FTIR spectra of AAS pastes are shown in Figure 7. In pastes, the reaction products had the stretching and bending vibrations of H-O-H groups at 3458 cm^{-1} and 1651 cm^{-1}, respectively. Both the wave numbers shifted towards the larger wavelength as compared to the raw slag (at 3435 cm^{-1}), indicating that more hydration products were formed with the progress of the polymerization process after one day. The band located at 1454 cm^{-1} was related to the existence of CO_3^{2-} and corresponded to the production of calcite [53], which was in agreement with the XRD results discussed in Section 3.6. The asymmetric stretching vibration of Si-O bonds was located at 964 cm^{-1}, which represented C-S-H in the hydration product. The bands at 675 cm^{-1} were assigned to Al-O bonds, which implied that Al replaced Si in the silicon-oxygen tetrahedron [54]. The wavenumber of the Al-O bond was lower after the activation than that of the unreacted slag (at 694 cm^{-1}), which indicated that the reaction of AAS included the dissolution of Al^{3+} from the slag and the

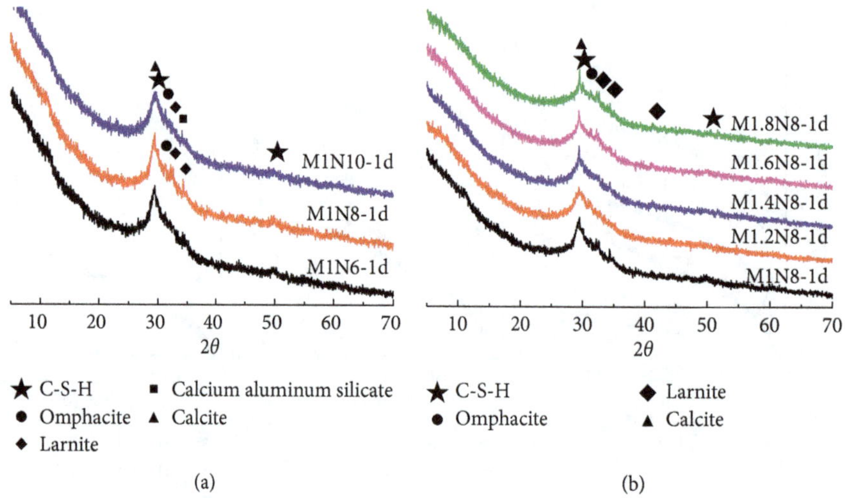

Figure 6: The XRD diffractograms of AAS pastes at different Na_2O contents (a) and silicate moduli (b).

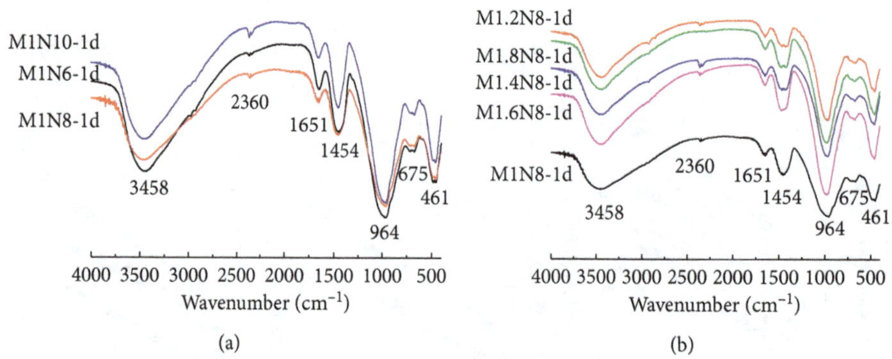

Figure 7: FTIR spectra of AAS pastes: (a) Na_2O contents; (b) silicate moduli.

recombination of Al^{3+} with silicate anions [33]. The bands at 461 cm^{-1} were attributed to Si-O bending. The results were in accordance with the information of C-(A)-S-H, which was identified using the XRD analysis. It can be concluded that the sodium silicate activator had a significant effect on the degree of polymerization of AAS pastes at an early age.

3.8. SEM/EDS Analysis. The microstructures of AAS pastes synthesized using different contents of Na_2O and silicate moduli are shown in Figures 8 and 9. As can be observed from Figures 8(a) and 8(c), the hardened AAS pastes with 6% and 8% Na_2O consisted of rod-like particles on the surface of the gel, which were the main reaction products (such as C-(A)-S-H and N-A-S-H) [28, 55]. Moreover, the AAS paste of M1.2N8 was denser than that of M1.2N6, which was in accordance with the result obtained regarding the development of compressive strength and discussed in Section 3.3. This was due to that the strength of M1.2N8 was higher than that of M1.2N6 after 28 days. The products of AAS pastes mainly consisted of Si, Ca, Na, and Al, while the amount of Mg produced was relatively low. The chemical elements of hydration products were tested using energy dispersive X-ray spectroscopy (EDS), and the results of Ca/Si were found to be 0.961 and 1.171 for M1.2N6 and M1.2N8 specimens, respectively. It may be concluded that the activator has higher pH for larger Na_2O contents, and vice versa. The increase in pH can accelerate the solubility of Ca in the dissolved slag. The silicate modulus has a more pronounced effect on the degree of polymerization of AAS. In addition, silicate modulus also determined the structure and morphology of reaction products [56]. SEM images of AAS pastes with different silicate moduli (Ms = 1, 1.2, 1.4, 1.6, and 1.8) and Na_2O content of 8% are shown in Figures 8(b)–8(f), respectively. The network of fine particles and layered silicate minerals were noticeably observed for the M1N8 specimen. The M1.2N8 sample showed rod-like morphology of the reaction products, and the matrices looked rather dense. The alkali-activated products presented a homogeneous and close-grained structure (calcium silicate hydrate gel), whereas the Ca/Si ratio of the gel was found to be 1.4 for the M1.4N8 specimen. In addition, a network of fine particles was also observed in the M1.6N8 specimen. Meanwhile, the paste was close-grained hydration products and made the structure more compact compared with the M1N8 specimen. The surface of the M1.8N8 specimen was covered with a significant number of rod-like particles and porous structure, resulting in weak bonding strength. These results were in agreement with the development of mechanical properties, as the compressive strength of the M1.8N8

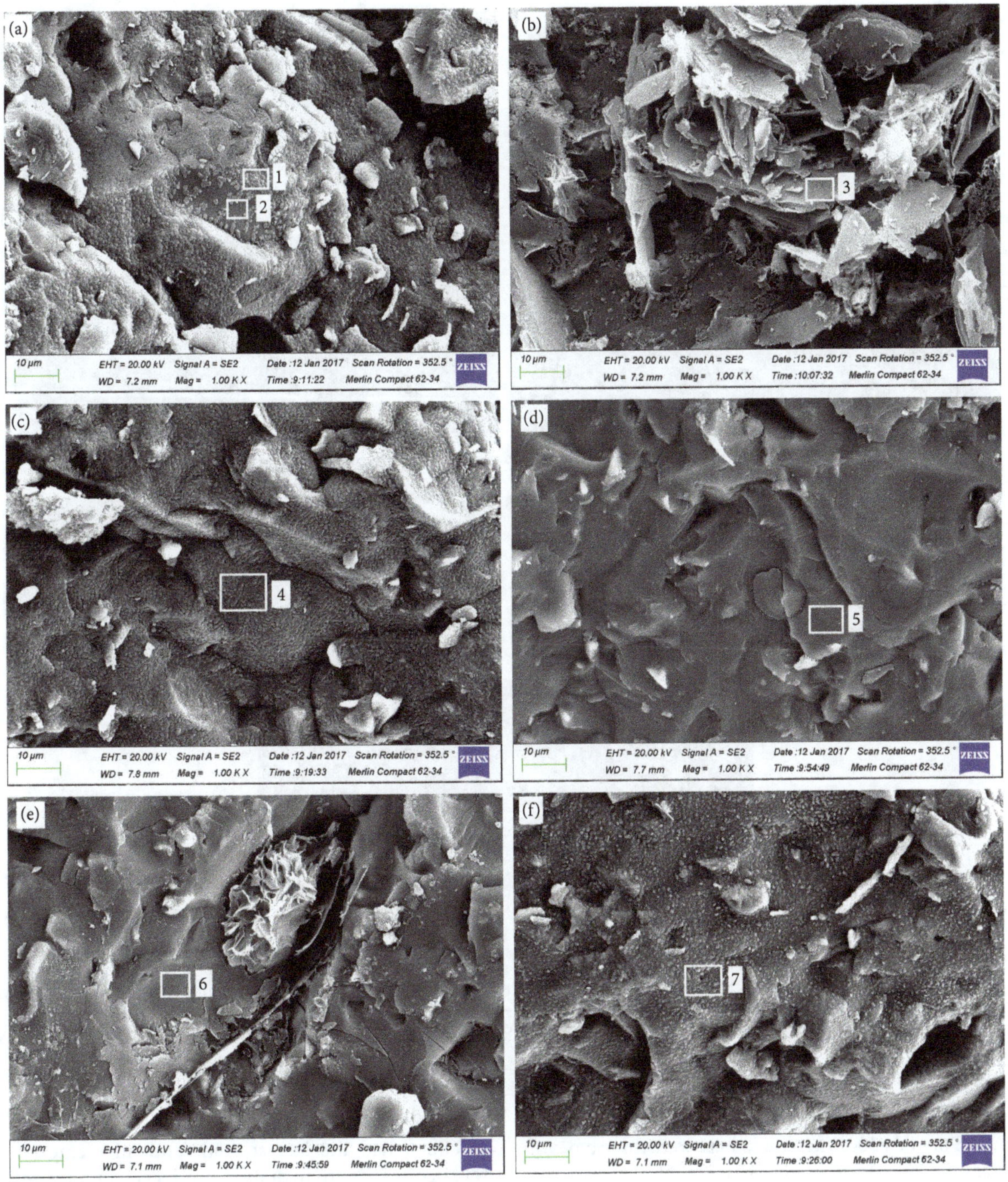

FIGURE 8: SEM images of AAS pastes for 28 days: (a) M1.2N6. (b) M1N8. (c) M1.2N8. (d) M1.4N8. (e) M1.6N8. (f) M1.8N8.

specimen was lower than those of M1.4N8 and M1.6N8 specimens. Rattanasak et al. [57] reported that the difference in gel formation affected the binding capacity and strength of geopolymer binder. Moreover, there were little crystalline products in all the samples.

The EDS analysis revealed that the hydration products have lots of O, Na, Al, Si, and Ca elements, whereas the Ca/Si ratio was in the range of 0.8–1.4. The main product of AAS pastes was C-S-H (or N-A-S-H) gel, which was amorphous at room temperature. In addition, Na and Al elements were found to be present in the structure of calcium silicate hydrate gel, as observed by the EDS. Meanwhile, the atomic percentages of Na and Al were found to be 1.45–27.77% and 2.99–11.06%, respectively. A previous study [58] concluded that a higher silicate modulus of the activator can accelerate the rate of polymerization and form lower crystalline products.

3.9. *MIP Analysis.* Table 3 presents the pore sizes of AAS pastes. The pore size was tested between 5 and 360000 nm

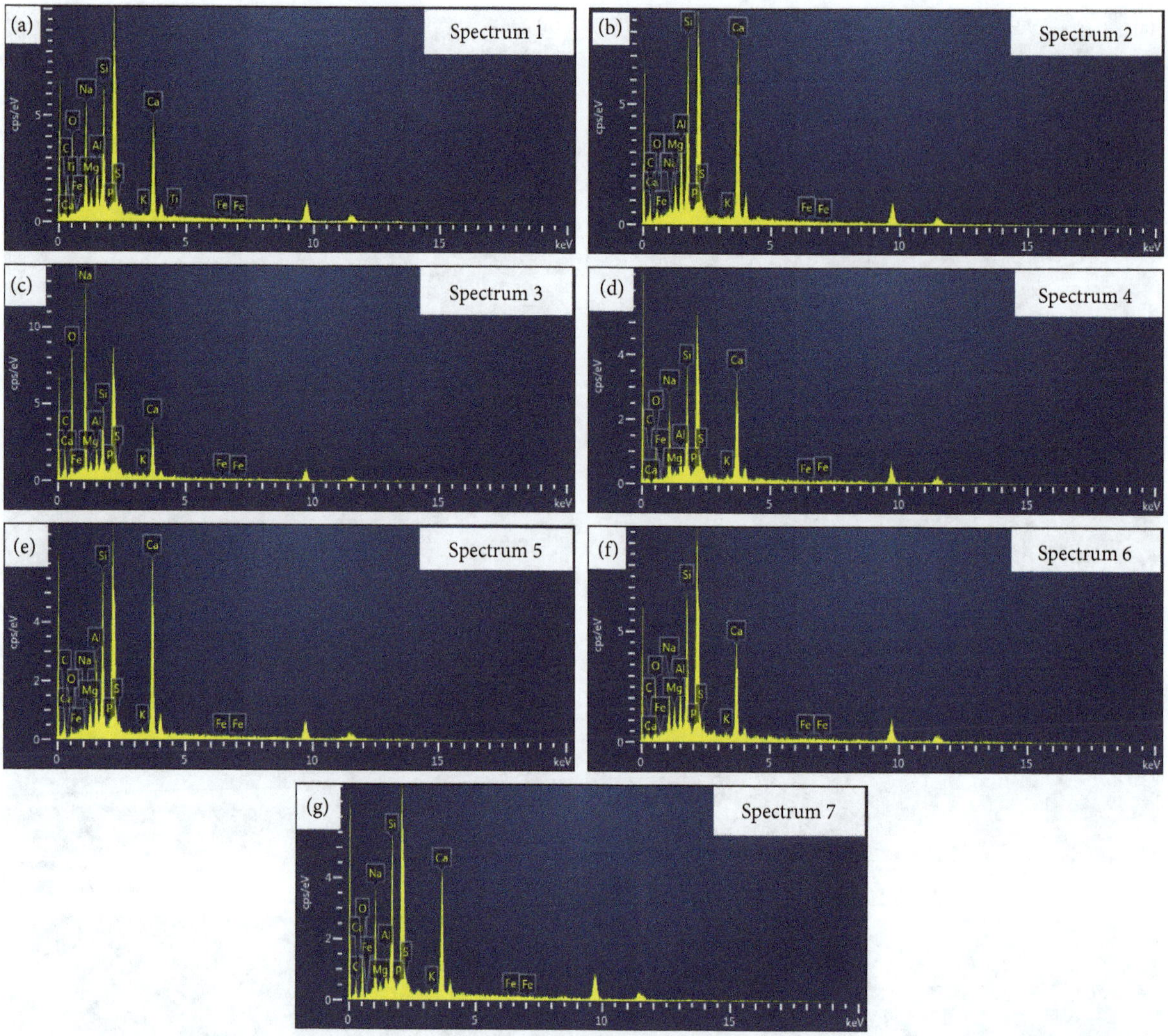

Figure 9: EDS images of AAS pastes corresponding to Figure 8: (a, b) M1.2N6. (c) M1N8. (d) M1.2N8. (e) M1.4N8. (f) M1.6N8. (g) M1.8N8.

Table 3: The classifications of pores in the AAS pastes (according to the International Union of Pure and Applied Chemistry system).

	Micropores	Mesopores	Macropores	Voids and microcracks
Pore size (mm)	<2.5	2.5–50	50–10000	>10000

using AutoPore IV 9500. Therefore, there were three types of pore sizes in the AAS pastes, namely, the mesopores, macropores, and voids and microcracks. The effect of silicate modulus on the pore size distribution of AAS pastes is shown in Figure 10. Figure 10(a) shows that the pore volume decreased with the increase in silicate modulus, except for the M1.8N8 specimen, indicating that a lower silicate modulus resulted in a larger pore volume. Figure 10(b) indicated that the volume of mesopores increased, whereas the volume of macropores decreased with the increase in silicate modulus. The larger volume of mesopores resulted in higher drying shrinkage values [33], which was consistent with the development of drying shrinkage, as discussed in Section 3.4. It can also be seen that the increase in silicate modulus led to a lower volume of macropores along with smaller voids and microcracks (M1.8N8 specimen had the largest voids and microcracks), which revealed that more hydration products had filled the larger pores, thus making it denser with the increase in silicate modulus at the same curing ages [59]. The porosities and pore size distribution of M1.6N8 and M1.4N8 were almost the same, which could explain the fact that both of them had relatively higher compressive strengths as mentioned above in Section 3.3.

4. Conclusions

Based on the experimental results, the following main conclusions can be drawn from the study:

(1) The initial and final setting times were in the range of 6.1–28.2 min and 9–36 min, respectively. The fluidity of AAS pastes was between 147 and 226 mm.

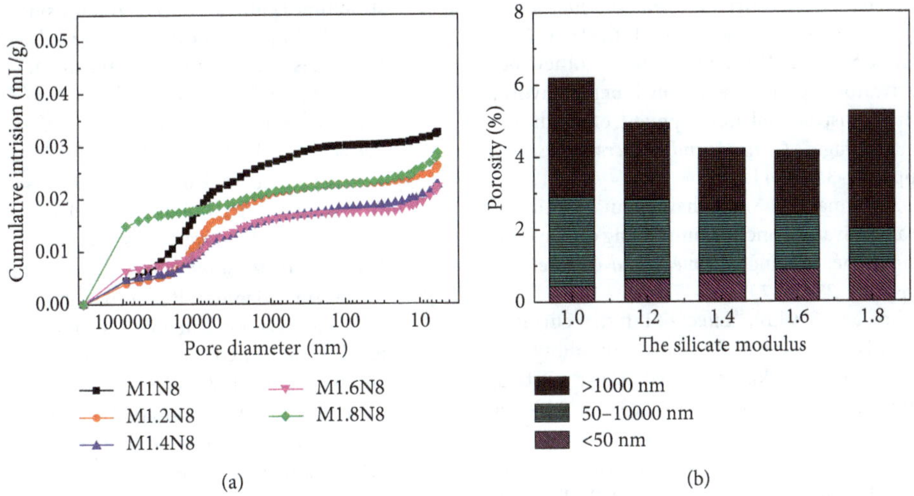

FIGURE 10: Pore size distribution of AAS pastes with different silicate moduli: (a) cumulative intrusion; (b) porosity.

(2) AAS pastes also exhibited higher early compressive strengths, whereas the highest 1-day compressive strength was found to be 102 MPa at room temperature.

(3) Both the drying shrinkage and mass loss increased as the silicate modulus increased during the curing process. However, with the increase of Na_2O content, the drying shrinkage increased, while the mass loss decreased when the silicate modulus went beyond the value of 1.2. The mechanism of drying shrinkage was not entirely due to the mass loss and had a strong relationship between the volume of micropores and hydration products.

(4) The dominant reaction products of AAS pastes were amorphous C-S-H and omphacite. The hydration products have high amounts of O, Na, Al, Si, and Ca elements, whereas the products of the M1.4N8 specimen exhibited a homogeneous and close-grained microstructure. Furthermore, the Ca/Si ratio was found to be ca. 1.4.

(5) Lower values of porosity and volume of voids and microcracks were observed for the specimen with silicate moduli of 1.4 and 1.6 having 8% Na_2O, which explained the excellent mechanical performance of the specimen.

Conflicts of Interest

The authors declare that there are no conflicts of interest regarding the publication of this paper.

Acknowledgments

This study was supported by the National Natural Science Foundation of China (51478142).

References

[1] W. C. Wang, H. Y. Wang, and H. C. Tsai, "Study on engineering properties of alkali-activated ladle furnace slag geopolymer," *Construction and Building Materials*, vol. 123, pp. 800–805, 2016.

[2] A. M. Rashad, "A comprehensive overview about the influence of different additives on the properties of alkali-activated slag—a guide for civil engineer," *Construction and Building Materials*, vol. 47, pp. 29–55, 2013.

[3] Y. J. Zhang, L. C. Liu, L. L. Ni, and B. L. Wang, "A facile and low-cost synthesis of granulated blast furnace slag-based cementitious material coupled with Fe_2O_3 catalyst for treatment of dye wastewater," *Applied Catalysis B: Environmental*, vol. 138-139, pp. 9–16, 2013.

[4] N. Mobasher, S. A. Bernal, and J. L. Provis, "Structural evolution of an alkali sulfate activated slag cement," *Journal of Nuclear Materials*, vol. 468, pp. 97–104, 2016.

[5] F. Puertas, S. Martínez-Ramírez, S. Alonso, and T. Vázquez, "Alkali-activated fly ash/slag cements," *Cement and Concrete Research*, vol. 30, no. 10, pp. 1625–1632, 2000.

[6] S. A. Bernal, "Effect of the activator dose on the compressive strength and accelerated carbonation resistance of alkali silicate-activated slag/metakaolin blended materials," *Construction and Building Materials*, vol. 98, pp. 217–226, 2015.

[7] Y. F. Gong, Y. H. Fang, Y. R. Yan, and L. Q. Chen, "Investigation on alkali activated recycled cement mortar powder cementitious material," *Materials Research Innovations*, vol. 18, no. 2, pp. 784–787, 2014.

[8] L. Xin, X. Jin-yu, L. Weimin, and B. Erlei, "Effect of alkali-activator types on the dynamic compressive deformation behavior of geopolymer concrete," *Materials Letters*, vol. 124, pp. 310–312, 2014.

[9] T. Phoo-ngernkham, A. Maegawa, N. Mishima, S. Hatanaka, and P. Chindaprasirt, "Effects of sodium hydroxide and sodium silicate solutions on compressive and shear bond strengths of FA-GBFS geopolymer," *Construction and Building Materials*, vol. 91, pp. 1-8, 2015.

[10] S. Aydin and B. Baradan, "Effect of activator type and content on properties of alkali-activated slag mortars," *Composites Part B: Engineering*, vol. 57, pp. 166–172, 2014.

[11] C. Elibol and O. Sengul, "Effects of activator properties and ferrochrome slag aggregates on the properties of alkali-

activated blast furnace slag mortars," *Arabian Journal for Science and Engineering*, vol. 41, no. 4, pp. 1561–1571, 2015.

[12] M. Ben Haha, G. Le Saout, F. Winnefeld, and B. Lothenbach, "Influence of activator type on hydration kinetics, hydrate assemblage and microstructural development of alkali activated blast-furnace slags," *Cement and Concrete Research*, vol. 41, no. 3, pp. 301–310, 2011.

[13] R. J. Thomas, D. Lezama, and S. Peethamparan, "On drying shrinkage in alkali-activated concrete: improving dimensional stability by aging or heat-curing," *Cement and Concrete Research*, vol. 91, pp. 13–23, 2017.

[14] C. Song, Y. C. Choi, and S. Choi, "Effect of internal curing by superabsorbent polymers—internal relative humidity and autogenous shrinkage of alkali-activated slag mortars," *Construction and Building Materials*, vol. 123, pp. 198–206, 2016.

[15] A. M. Rashad, D. M. Sadek, and H. A. Hassan, "An investigation on blast-furnace stag as fine aggregate in alkali-activated slag mortars subjected to elevated temperatures," *Journal of Cleaner Production*, vol. 112, pp. 1086–1096, 2016.

[16] C. Bilim, O. Karahan, C. D. Atiş, and S. İlkentapar, "Influence of admixtures on the properties of alkali-activated slag mortars subjected to different curing conditions," *Materials and Design*, vol. 44, pp. 540–547, 2013.

[17] F. Messina, C. Ferone, F. Colangelo, and R. Cioffi, "Low temperature alkaline activation of weathered fly ash: influence of mineral admixtures on early age performance," *Construction and Building Materials*, vol. 86, pp. 169–177, 2015.

[18] M. Murmu and S. P. Singh, "Influence of mineral admixtures on strength and hydration products of lime-activated slag cement," *Advances in Cement Research*, vol. 26, no. 6, pp. 334–343, 2014.

[19] Z. Li, Z. Pan, Y. Liu et al., "Effects of mineral admixtures and lime on disintegration of alkali-activated slag exposed to 50°C," *Construction and Building Materials*, vol. 70, pp. 254–261, 2014.

[20] V. Živica, "Effects of type and dosage of alkaline activator and temperature on the properties of alkali-activated slag mixtures," *Construction and Building Materials*, vol. 21, no. 7, pp. 1463–1469, 2007.

[21] Z. Jiao, Y. Wang, W. Zheng, and W. Huang, "Effect of dosage of sodium carbonate on the strength and drying shrinkage of sodium hydroxide based alkali-activated slag paste," *Construction and Building Materials*, vol. 179, pp. 11–24, 2018.

[22] J. J. Chang, "A study on the setting characteristics of sodium silicate-activated slag pastes," *Cement and Concrete Research*, vol. 33, no. 7, pp. 1005–1011, 2003.

[23] X. Wu, W. Jiang, and D. M. Roy, "Early activation and properties of slag cement," *Cement and Concrete Research*, vol. 20, no. 6, pp. 961–974, 1990.

[24] D. Ravikumar and N. Neithalath, "Effects of activator characteristics on the reaction product formation in slag binders activated using alkali silicate powder and NaOH," *Cement and Concrete Composites*, vol. 34, no. 7, pp. 809–818, 2012.

[25] S. Ahmari, L. Zhang, and J. Zhang, "Effects of activator type/concentration and curing temperature on alkali-activated binder based on copper mine tailings," *Journal of Materials Science*, vol. 47, no. 16, pp. 5933–5945, 2012.

[26] N. K. Lee and H. K. Lee, "Reactivity and reaction products of alkali-activated, fly ash/slag paste," *Construction and Building Materials*, vol. 81, pp. 303–312, 2015.

[27] R. San Nicolas, S. A. Bernal, R. Mejía de Gutiérrez, J. S. J. van Deventer, and J. L. Provis, "Distinctive microstructural features of aged sodium silicate-activated slag concretes," *Cement and Concrete Research*, vol. 65, pp. 41–51, 2014.

[28] F. Puertas, M. Palacios, H. Manzano, J. S. Dolado, A. Rico, and J. Rodríguez, "A model for the C-A-S-H gel formed in alkali-activated slag cements," *Journal of the European Ceramic Society*, vol. 31, no. 12, pp. 2043–2056, 2011.

[29] N. Marjanović, M. Komljenović, Z. Baščarević, V. Nikolić, and R. Petrović, "Physical–mechanical and microstructural properties of alkali-activated fly ash–blast furnace slag blends," *Ceramics International*, vol. 41, no. 1, pp. 1421–1435, 2015.

[30] B. S. Gebregziabiher, R. Thomas, and S. Peethamparan, "Very early-age reaction kinetics and microstructural development in alkali-activated slag," *Cement and Concrete Composites*, vol. 55, pp. 91–102, 2015.

[31] M. Chi, "Effects of dosage of alkali-activated solution and curing conditions on the properties and durability of alkali-activated slag concrete," *Construction and Building Materials*, vol. 35, pp. 240–245, 2012.

[32] C. Shi, "Strength, pore structure and permeability of alkali-activated slag mortars," *Cement and Concrete Research*, vol. 26, no. 12, pp. 1789–1799, 1996.

[33] Y. M. Gu, Y. H. Fang, D. You, Y. F. Gong, and C. H. Zhu, "Properties and microstructure of alkali-activated slag cement cured at below- and about-normal temperature," *Construction and Building Materials*, vol. 79, pp. 1–8, 2015.

[34] China Standardization Administration, *Gb/T 8074-2008, Testing Method for Specific Surface of Cement-Blaine Method*, China Standards Press, Beijing, China, 2008.

[35] C. Wang, P. Wen, M. Wang, Q. Fan, and X. Wang, "Preparation and characterization of road alkali-activated blast furnace slag paste," *Construction and Building Materials*, vol. 181, pp. 175–184, 2018.

[36] W. Shen, Y. Wang, T. Zhang, M. Zhou, J. Li, and X. Cui, "Magnesia modification of alkali-activated slag fly ash cement," *Journal of Wuhan University of Technology-Materials Science Edition*, vol. 26, no. 1, pp. 121–125, 2011.

[37] M. Hojati and A. Radlińska, "Shrinkage and strength development of alkali-activated fly ash-slag binary cements," *Construction and Building Materials*, vol. 150, pp. 808–816, 2017.

[38] Standardization Administration of the People's Republic of China, *GB/T 1346-2011: Test Methods for Water Requirement of Normal Consistency, Setting Time and Soundness of the Portland Cement*, China Standards Press, Beijing, China, 2011.

[39] China Standardization Administration, *GB/T 8077-2012, Methods for Testing Uniformity of Concrete Admixture*, China Standards Press, Beijing, China, 2012.

[40] National Development and Reform Commission of People's Republic of China, *JC/T 603-2004: Standard Test Method for Drying Shrinkage of Mortar*, China Building Materials Industry Press, Beijing, China, 2004.

[41] C. K. Yip, G. C. Lukey, and J. S. J. van Deventer, "The coexistence of geopolymeric gel and calcium silicate hydrate at the early stage of alkaline activation," *Cement and Concrete Research*, vol. 35, no. 9, pp. 1688–1697, 2005.

[42] O. Burciaga-Díaz and J. I. Escalante-García, "Structure, mechanisms of reaction, and strength of an alkali-activated blast-furnace slag," *Journal of the American Ceramic Society*, vol. 96, no. 12, pp. 3939–3948, 2013.

[43] A. Fernandez-Jimenez and F. Puertas, "Setting of alkali-activated slag cement. Influence of activator nature," *Advances in Cement Research*, vol. 13, no. 3, pp. 115–121, 2001.

[44] F. Puertas, C. Varga, and M. M. Alonso, "Rheology of alkali-activated slag pastes. Effect of the nature and concentration of the activating solution," *Cement and Concrete Composites*, vol. 53, pp. 279–288, 2014.

[45] K. Komnitsas, D. Zaharaki, and V. Perdikatsis, "Effect of synthesis parameters on the compressive strength of low-calcium ferronickel slag inorganic polymers," *Journal of Hazardous Materials*, vol. 161, no. 2-3, pp. 760–768, 2009.

[46] A. Allahverdi, E. N. Kani, and M. Yazdanipour, "Effects of blast-furnace slag on natural pozzolan-based geopolymer cement," *Ceram-Sil*, vol. 55, no. 1, pp. 68–78, 2011.

[47] F. Puertas and M. Torres-Carrasco, "Use of glass waste as an activator in the preparation of alkali-activated slag. Mechanical strength and paste characterisation," *Cement and Concrete Research*, vol. 57, pp. 95–104, 2014.

[48] A. A. M. Neto, M. A. Cincotto, and W. Repette, "Drying and autogenous shrinkage of pastes and mortars with activated slag cement," *Cement and Concrete Research*, vol. 38, no. 4, pp. 565–574, 2008.

[49] N. K. Lee, J. G. Jang, and H. K. Lee, "Shrinkage characteristics of alkali-activated fly ash/slag paste and mortar at early ages," *Cement and Concrete Composites*, vol. 53, pp. 239–248, 2014.

[50] W. Hansen, "Drying shrinkage mechanisms in Portland cement paste," *Journal of the American Ceramic Society*, vol. 70, no. 5, pp. 323–328, 1987.

[51] A. R. Brough and A. Atkinson, "Sodium silicate-based, alkali-activated slag mortars: Part I. Strength, hydration and microstructure," *Cement and Concrete Research*, vol. 32, no. 6, pp. 865–879, 2002.

[52] A. M. Rashad, S. R. Zeedan, and A. A. Hassan, "Influence of the activator concentration of sodium silicate on the thermal properties of alkali-activated slag pastes," *Construction and Building Materials*, vol. 102, pp. 811–820, 2016.

[53] A. F. Abdalqader, F. Jin, and A. Al-Tabbaa, "Characterisation of reactive magnesia and sodium carbonate-activated fly ash/slag paste blends," *Construction and Building Materials*, vol. 93, pp. 506–513, 2015.

[54] H. El-Didamony, A. A. Amer, and H. Abd Ela-ziz, "Properties and durability of alkali-activated slag pastes immersed in sea water," *Ceramics International*, vol. 38, no. 5, pp. 3773–3780, 2012.

[55] J. I. Escalante-Garcia, A. F. Fuentes, A. Gorokhovsky, P. E. Fraire-Luna, and G. Mendoza-Suarez, "Hydration products and reactivity of blast-furnace slag activated by various alkalis," *Journal of the American Ceramic Society*, vol. 86, no. 12, pp. 2148–2153, 2003.

[56] S. A. Bernal, R. M. de Gutierrez, J. L. Provis, and V. Rose, "Effect of silicate modulus and metakaolin incorporation on the carbonation of alkali silicate-activated slags," *Cement and Concrete Research*, vol. 40, no. 6, pp. 898–907, 2010.

[57] U. Rattanasak, K. Pankhet, and P. Chindaprasirt, "Effect of chemical admixtures on properties of high-calcium fly ash geopolymer," *International Journal of Minerals, Metallurgy, and Materials*, vol. 18, no. 3, pp. 364–369, 2011.

[58] M. Komljenović, Z. Baščarević, and V. Bradić, "Mechanical and microstructural properties of alkali-activated fly ash geopolymers," *Journal of Hazardous Materials*, vol. 181, no. 1–3, pp. 35–42, 2010.

[59] Y. Ma, J. Hu, and G. Ye, "The pore structure and permeability of alkali activated fly ash," *Fuel*, vol. 104, pp. 771–780, 2013.

12

Effects of Mixing and Curing Temperature on the Strength Development and Pore Structure of Fly Ash Blended Mass Concrete

Ki-Bong Park[1] and Takafumi Noguchi[2]

[1]*Department of Architectural Engineering, College of Engineering, Kangwon National University, Chuncheon-si, Republic of Korea*
[2]*Department of Architecture, Graduate School of Engineering, The University of Tokyo, Tokyo, Japan*

Correspondence should be addressed to Ki-Bong Park; kbpark@kangwon.ac.kr

Academic Editor: Xiao-Jian Gao

The aim of this work is to know clearly the effects of temperature in response to curing condition, hydration heat, and outside weather conditions on the strength development of high-performance concrete. The concrete walls were designed using three different sizes and three different types of concrete. The experiments were conducted under typical summer and winter weather conditions. Temperature histories at different locations in the walls were recorded and the strength developments of concrete at those locations were measured. The main factors investigated that influence the strength developments of the obtained samples were the bound water contents, the hydration products, and the pore structure. Testing results indicated that the elevated summer temperatures did not affect the early-age strength gain of concrete made using ordinary Portland cement. Strength development was significantly increased at early ages in concrete made using belite-rich Portland cement or with the addition of fly ash. The elevated temperatures resulted in a long-term strength loss in both belite-rich and fly ash containing concrete. The long-term strength loss was caused by a reduction in the degree of hydration and an increase in the total porosity and amount of smaller pores in the material.

1. Introduction

High-performance concrete is being used increasingly in the construction of high-rise buildings, bridges and marine structures. If high strength concrete is used for columns or other large section members of massive concrete structures, the center temperature of the members will rise rapidly at early ages due to hydration heat and the high temperatures will remain in the member for a relatively long period of time due to the low thermal conductivity of concrete. High temperatures generally accelerate the early-age strength development of concrete, while reducing the long-term strength development. These elevated temperatures will result in physical and chemical transformations in the concrete [1–7]. Various researchers have studied the microstructure and hydration to explain these effects. It has been reported that the loss of long-term strength is induced by both physical effects and chemical effects. The physical effects are an increase in porosity and an increased incidence of microcracking in cement paste, the latter caused by large differences in the thermal expansion coefficients of the concrete constitutes [8–18]. The chemical effects include changes in the structure of the hydration products and the loss of water in concrete [19–24]. Most of the research on cement paste, mortar, and concrete was conducted with the specimens hydrated at room temperature, on a level surface and during the early stages of hydration. There is not enough information available on high-performance concrete curd at elevated temperatures, in mass structures. Studies on the effects of outside weather conditions on the strength development of high-performance concrete in mass structures are very scarce in the technical literature.

This research provided the data needed to set a limitation on the maximum permissible internal temperature of a mass structure, such as a nuclear facility structure, a bridge foundation, or a marine structure, to ensure the structure is strong

Table 1: Bogue composition of cement used in concretes.

Cement	Blaine(cm^2/g)	Density(g/cm^3)	Mineralogical properties (%)			
			C_3S	C_2S	C_3A	C_4AF
OPC	3280	3.16	52	24	9	9
BPC	4080	3.20	29	54	3	8

Table 2: Properties of fly ash.

Ig. loss (%)	Specific gravity	Fineness: 45 μm sieve residual quantity (%)	Fineness: Blaine (cm^2/g)	Percentage of flow (%)	SiO_2 (%)	CaO (%)	Al_2O_3 (%)
0.9	2.27	2.3	3890	110	33.2	42.3	14.1

Table 3: Mix proportions and properties in fresh concrete.

Concrete	W/C	S/A	Unit content (kg/m^3): Water	Cement	Fly ash	Sand	Gravel
OPC-S	0.33	0.49	170	516	—	787	840
BPC-S	0.33	0.53	170	516	—	854	776
FPC-S	0.33	0.50	170	310	206	774	776
OPC-W	0.33	0.49	170	516	—	787	840
BPC-W	0.33	0.53	170	516	—	854	776
FPC-W	0.33	0.50	170	310	206	774	776
0.49	0.49	0.49	0.49	0.49	0.49	0.49	0.49
Concrete	S/A	Chemical agent[*1]	Chemical agent[*2]	Slump (cm)	Flow (cm)	Air (%)	Temp. (°C)
OPC-S	0.49	C × 1.2%	C × 0.001%	22.0	38.5	4.9	31
BPC-S	0.53	C × 1.3%	—		63.5	4.8	30
FPC-S	0.50	C × 1.0%	—		63.0	4.4	30.5
OPC-W	0.49	C × 1.2%	C × 0.001%	23.5	42.0	4.1	8.5
BPC-W	0.53	C × 1.3%	—		61.5	4.2	10
FPC-W	0.50	C × 1.0%	—		61.2	4.2	8.5

[*1]High range water reducing agent, SP8 SX_2. [*2]Air reducing agent, MA404.

and durable. In this paper we will report on some experimental work investigating the compressive strength gain at varying temperatures during the mixing, placing, and curing of concrete in mass structures. Concrete walls were designed having three different depths, being 1.5 m, 0.8 m, and 0.3 m, composed of three different types of concrete, and processed under two different climate conditions. Temperature histories at different locations in the walls were recorded and the compressive strength, the degree of hydration products, and the microstructure of the concrete at those locations are presented.

2. Experimental Design

2.1. Mixing and Materials. Three types of concrete were used in this study: concrete made with ordinary Portland cement (OPC), with belite-rich Portland cement (BPC), and with ordinary Portland cement supplemented with 40% fly ash (FPC). This level of fly ash has been increasingly used for high-performance concrete in mass structures. The properties of the cement are shown in Table 1. Fly ash conforming to JIS A 6201 had the properties as given in Table 2. The aggregate consisted of crushed sandstone (maximum size: 20 mm, density: 2.65 g/cm^3, absorption: 0.72%, and fineness modulus: 6.0) and construction sand (density: 2.58 g/cm3, absorption: 2.07%, and fineness modulus: 2.69). An air reducing agent and a high range water reducing agent were used as admixtures. Their amounts are given in Table 3. The concrete was mixed in a twin shaft-type mixer (200 L). After first mixing the mortar for 50 sec., the coarse aggregate was added and the resulting concrete was mixed for another 90 seconds. The mixing was conducted in the summer and in the winter to provide the two types of weather conditions. In Table 3, the composition and the properties of the fresh concrete are presented.

2.2. Design of Model Walls and Curing. To investigate the variation of long-term strength in actual concrete walls subjected to different curing temperatures, three model walls were designed. The model walls are shown in Figure 1. The depths of the walls were similar to those in an actual nuclear power plant, namely, 1.5 m, 0.8 m, and 0.3 m. In order to simulate an actual long wall, the circumference of the wall's

TABLE 4: Compressive strength data of concretes.

Concrete (summer)	Curingstates	Compressive strength (MPa)				Concrete (Winter)	Curing states	Compressive strength (MPa)			
		3 days	7 days	28 days	91 days			3 days	7 days	28 days	91 days
OPC-S	Standard	47.9	59.2	69.7	78.9	OPC-W	Standard	51.4	65.5	78.8	82.4
	Sealed	43.8	54.6	60.5	69.7		Sealed	46.0	57.7	68.4	74.5
	1.5 m-C	51.7	55.8	60.3	60.9		1.5 m-C	64.0	71.7	73.7	74.0
	1.5 m-S	47.9	49.2	59.2	60.3		1.5 m-S	56.7	58.3	72.1	72.9
	0.8 m-C			54.9	58.7		0.8 m-C			72.1	74.6
	0.8 m-S			56.3	59.5		0.8 m-S			70.3	74.0
	0.3 m-C			57.8	62.6		0.3 m-C			67.8	69.8
BPC-S	Standard	33.9	45.8	74.4	87.5	BPC-W	Standard	34.6	49.4	81.2	90.1
	Sealed	33.8	41.0	63.9	81.0		Sealed	33.3	43.0	66.6	77.2
	1.5 m-C	60.7	63.8	64.6	69.3		1.5 m-C	56.2	64.4	75.1	78.6
	1.5 m-S	48.5	58.0	62.6	65.7		1.5 m-S	37.0	44.4	61.5	73.0
	0.8 m-C			67.5	70.1		0.8 m-C			66.8	70.1
	0.8 m-S			64.2	69.4		0.8 m-S			60.6	73.8
	0.3 m-C			67.7	70.4		0.3 m-C			58.7	72.2
FPC-S	Standard	26.6	38.7	55.0	70.1	FPC-W	Standard	28.4	42.5	59.1	75.7
	Sealed	22.8	34.5	48.9	65.4		Sealed	28.3	35.2	47.7	59.6
	1.5 m-C	46.9	52.7	55.5	54.0		1.5 m-C	38.3	55.9	57.6	62.7
	1.5 m-S	39.9	46.8	52.3	55.5		1.5 m-S	26.8	39.7	47.7	57.8
	0.8 m-C			50.8	56.1		0.8 m-C			51.2	64.3
	0.8 m-S			53.5	58.7		0.8 m-S			48.3	56.1
	0.3 m-C			55.0	59.2		0.3 m-C			46.0	64.6

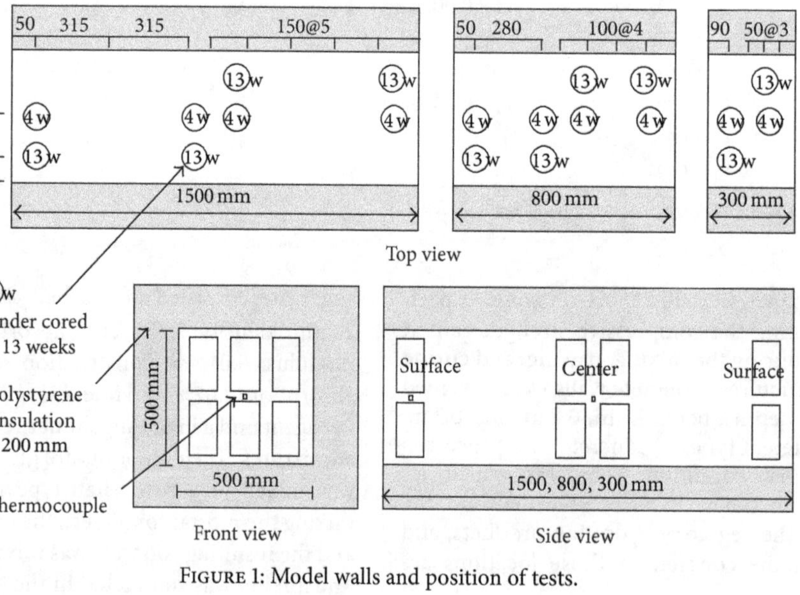

FIGURE 1: Model walls and position of tests.

depths direction was insulated with a layer of 200 mm thick polystyrene and the square surfaces of the wall were exposed to the open air. The walls were cast in 20 mm plywood formwork. The formwork was removed at 72 hours after casting. Core cylinders (100 × 200 mm) were taken from the walls at 3, 7, 28, and 91 days and used for compressive strength, pore structure, and hydration products measurements. For comparison, concrete cylinders were also cast in steel molds. All cylinders were removed from the steel molds 24 hours after casting. Sealed cylinders were sealed in polyethylene sheeting and then stored in a curing room at 20°C. The standard cured cylinders were stored in water at 20°C.

2.3. Testing Procedure. Temperature histories at different locations in the walls were recorded using thermocouples, shown in Figure 1. Cores, standard, and sealed cylinders were

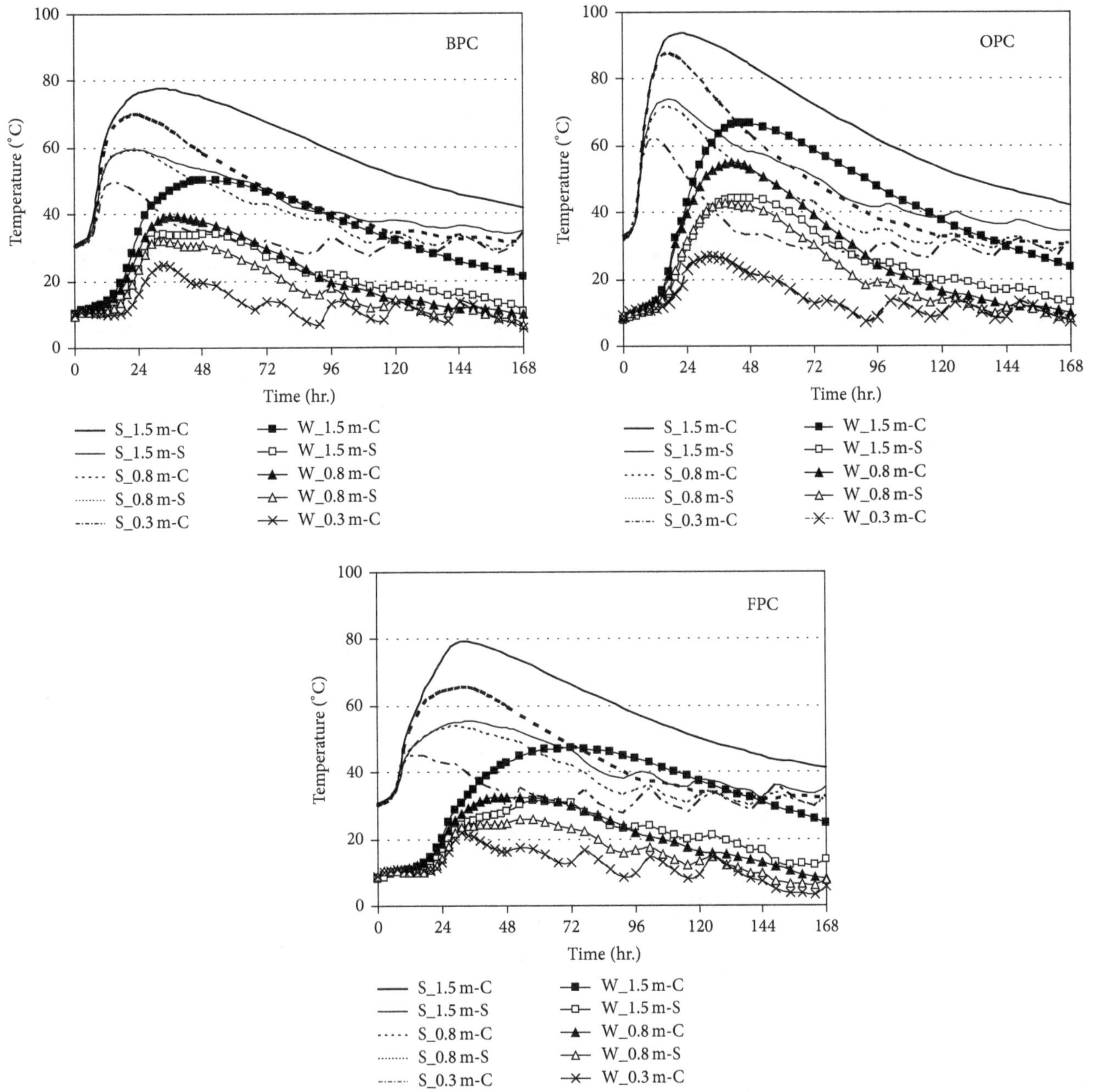

FIGURE 2: Temperature history in model walls. (S_1.5 m-C = S: summer, 1.5 m: size, C: center, W_1.5 m-S = W: winter, 1.5: size, and S: surface).

tested for compressive strength at 3, 7, 28, and 91 days. A portion of the concrete cylinders were cut into approximately 5 mm cubes using a diamond saw. These fragments were immediately immersed in acetone to prevent further hydration. Thereafter, all fragments were D-dried for about 2 weeks. Then the fragments were carefully ground by hand until a sample powder was obtained, which could pass through a 45 μm sieve and was suitable for the determination of calcium hydroxide (CH) contents and bound water contents. The amount of calcium hydroxide was determined by differential scanning calorimeter (DSC). The amount of calcium hydroxide was determined by ignition loss experiments. The porosity and pore size distribution were determined using mercury intrusion porosimetry (MIP). The pressure applied ranged from zero to 240 MPa. The 5 mm cubes for MIP measurements were also D-dried for about 2 weeks before testing.

3. Test Results and Discussion

3.1. Temperature Development. The temperature histories in the concrete walls are shown in Figure 2. In summer (temp. 32°C), the maximum temperature at the center of the 1.5 m deep walls made with OPC, BPC, and FPC was 94°C, 78°C, and 80°C respectively. The maximum temperature was achieved at 22 hours after casting in OPC and, at 31 hours

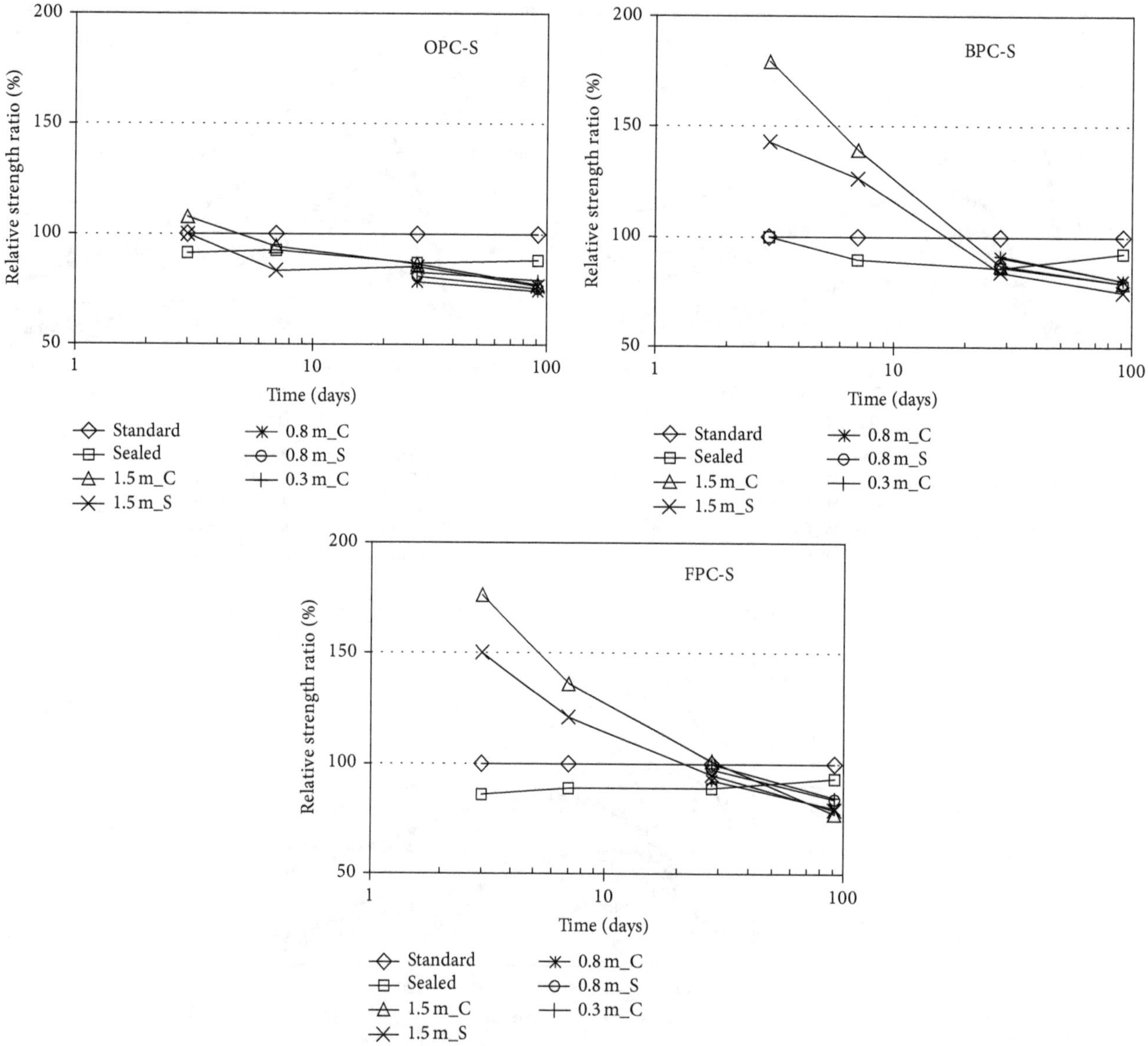

FIGURE 3: Effect of temperature on strength development of concrete at summer.

after casting in BPC, and FPC was achieved at 34 hours. The maximum temperature at the surface of the same members was 74°C, 60°C, and 55°C respectively. The temperature differential between the center and the surface of the concrete was greatest in the FPC mix. The temperature rise in FPC walls was smaller than that of BPC walls, except at the center of 1.5 m deep wall. The 1.5 m FPC walls showed a significant increase in temperature due to an increase in the rate of the pozzolanic reaction of fly ash. This can be attributed to the latent hydraulic properties inherent in fly ash. According to Fraay et al. [25], the glass material in fly ash is broken down when the pH value of the pore water is at least about 13.2. The increase in alkalinity required for the fly ash reaction is achieved through the reaction of the Portland cement. Accordingly, the slower hydration results in a more gradual development in temperature. Additionally, the high elevated temperature in the FPC wall was sustained for a relatively long period of time. Under winter condition (temp. 9°C), the temperature histories obtained in each wall's location reflected the difference in outside temperature between summer and winter. The initial mixing temperature significantly affected the maximum peak temperature and the retardation period.

3.2. Compressive Strength Development. The compressive strengths of core, standard, and sealed specimens of concrete made under summer and winter conditions are listed in Table 4. The compressive strength of the standard specimen cured in water at 20°C was higher than that of the sealed specimen. It is generally known that a low water to ratio in concrete mix leads to the possibility of self-desiccation and limitations on continued cement hydration, which explains the differing strength between specimens. Thus, the availability of external water is required to enable

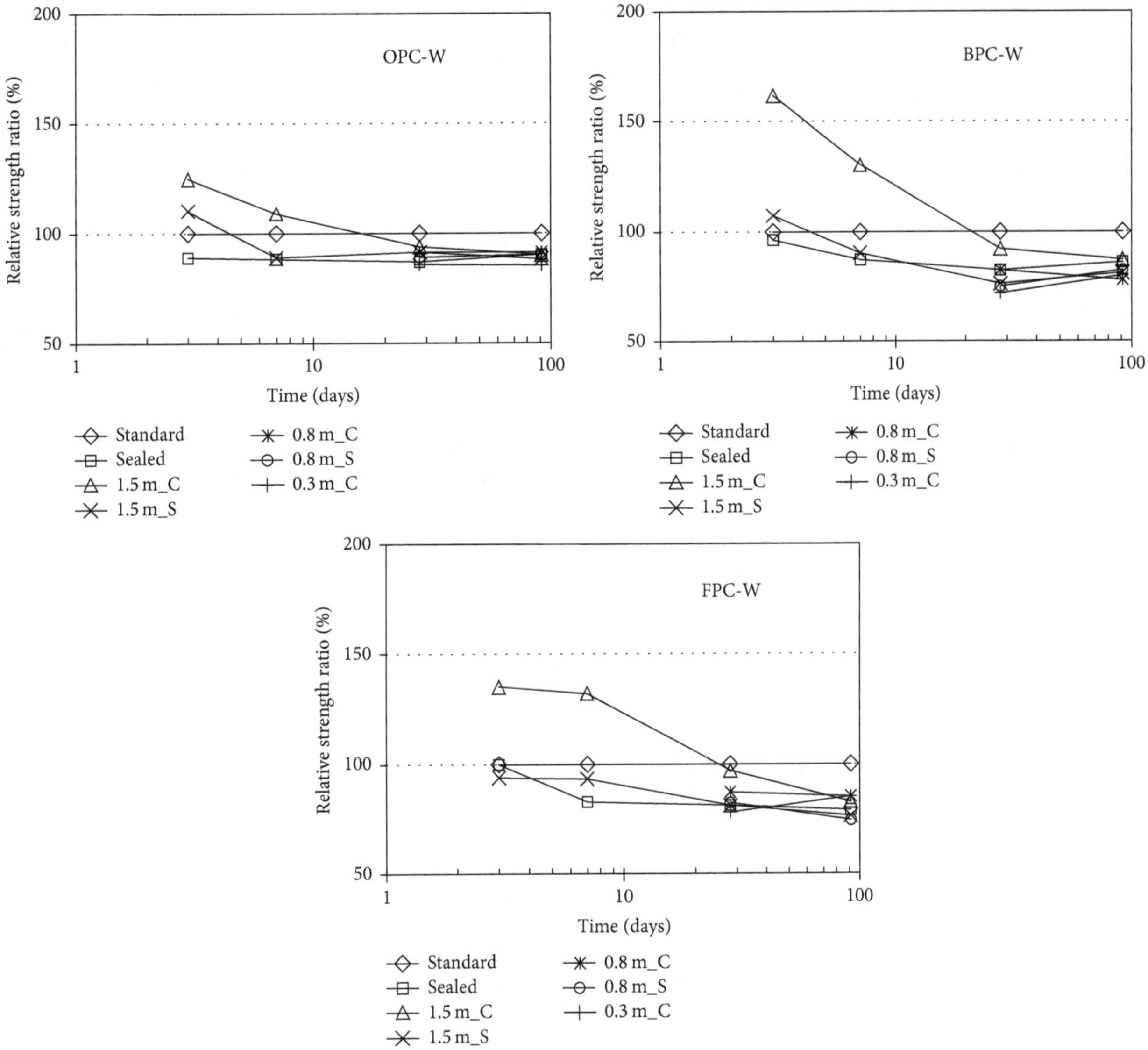

Figure 4: Effect of temperature on strength development of concrete at winter.

hydration to continue uninhibited. The strength development of concrete made under winter weather conditions was faster than that of concrete made under summer conditions, both for the standard and sealed specimens. These results are very interesting because we usually thought the high mixing temperature increases the compressive strength at early ages. The compressive strength of the core specimen was higher than that of the standard specimen at early ages due to the higher temperatures attained in core. However, the 1.5 m depth wall made under summer conditions using OPC behaved differently. In this sample, the elevated temperature did not affect the strength development at any age. In OPC, the strength of the core specimen made under winter conditions was significantly higher than the comparable specimen made under summer conditions. Such was not the case for the concrete made with BPC and FPC. In these samples, the strength of the core specimens made under winter conditions was lower than those made under summer conditions. In the summer, core strength of BPC samples was significantly increased at all ages regardless of the size of the walls. However, in the winter, the strength development of BPC was lower than that of OPC, except for the core specimen at the center of the 1.5 m deep wall at the longest ages.

The influence of temperature on the compressive strength can be clearly described in terms of a relative strength ratio defined against the strength of a standard specimen cured in water at 20°C. Using this measure, Figure 3 shows the effect of temperature on all of the compressive strength data for the three types of concrete. As shown in Figure 3, the center of the 1.5 m deep wall made under summer conditions with OPC has a 3-day compressive strength which is 108% that cured under standard conditions. By contrast, specimens from the center of 1.5 m wall made with BPC and FPC have

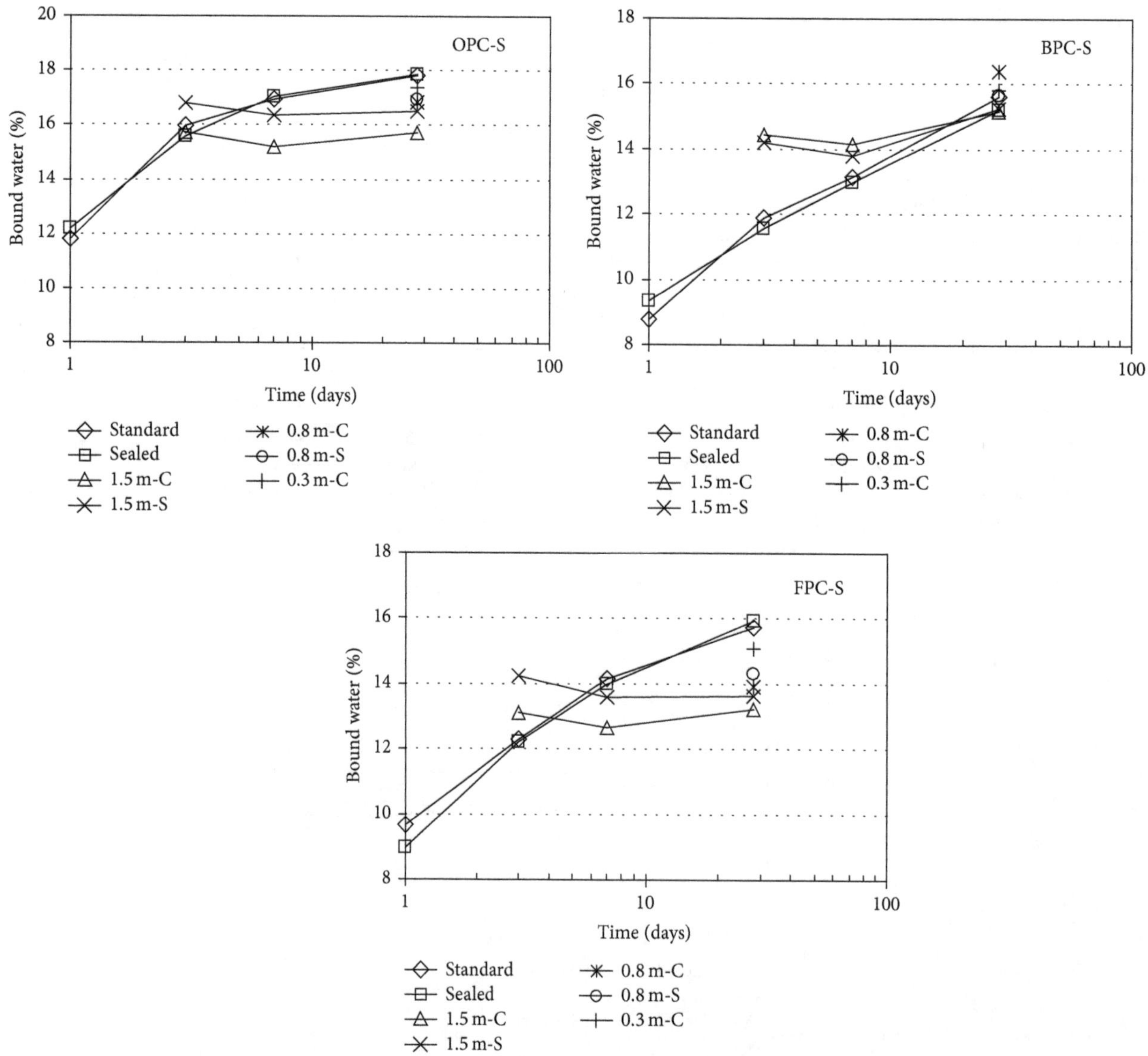

FIGURE 5: Effect of temperature on the bound water of concrete in summer experiments.

a 3-day compressive strength which was 180% and 176% of the standard. At 91 days, however the strength ratios dropped to 77%, 79%, and 77% of the standard, respectively, for the OPC, BPC, and FPC samples. The strength development of the core specimen made with OPC was not related to the maximum internal temperature at early ages. However, the use of belite-rich Portland cement and fly ash significantly increased the strength development at high temperatures. Figure 4 shows the results of compressive strength tests of concrete cured under winter weather conditions. The 3-day compressive strength at the center of 1.5 m deep wall made using OPC is 125% that concrete made under standard curing conditions. The center of 1.5 m wall made with BPC and FPC had 3-day compressive strengths which were 162% and 135% that of the standard sample. At 91 days. However, the strength ratios were 90%, 87%, and 83% for OPC, BPC, and FPC, respectively. The strength of winter cured concrete first fell down below that of the standard samples at 7 days.

3.3. Hydration Products. Figure 5 shows the change in the amount of bound water in the hydration products for the material produced under summer weather conditions. The amount of bound water in the core specimen cured at high temperature was larger than that of the standard specimen at early ages. However, after 7 days, in the OPC sample, and after 28 days, in the BPC and FPC samples, bound water content of these core specimens was lower than that of the standard specimen. Rodger and Groves [21] suggested that hydration at normal temperatures provides sufficient time for the hydration product to diffuse and precipitate relatively uniformly throughout the interstitial space between the cement grains. But accelerated hydration

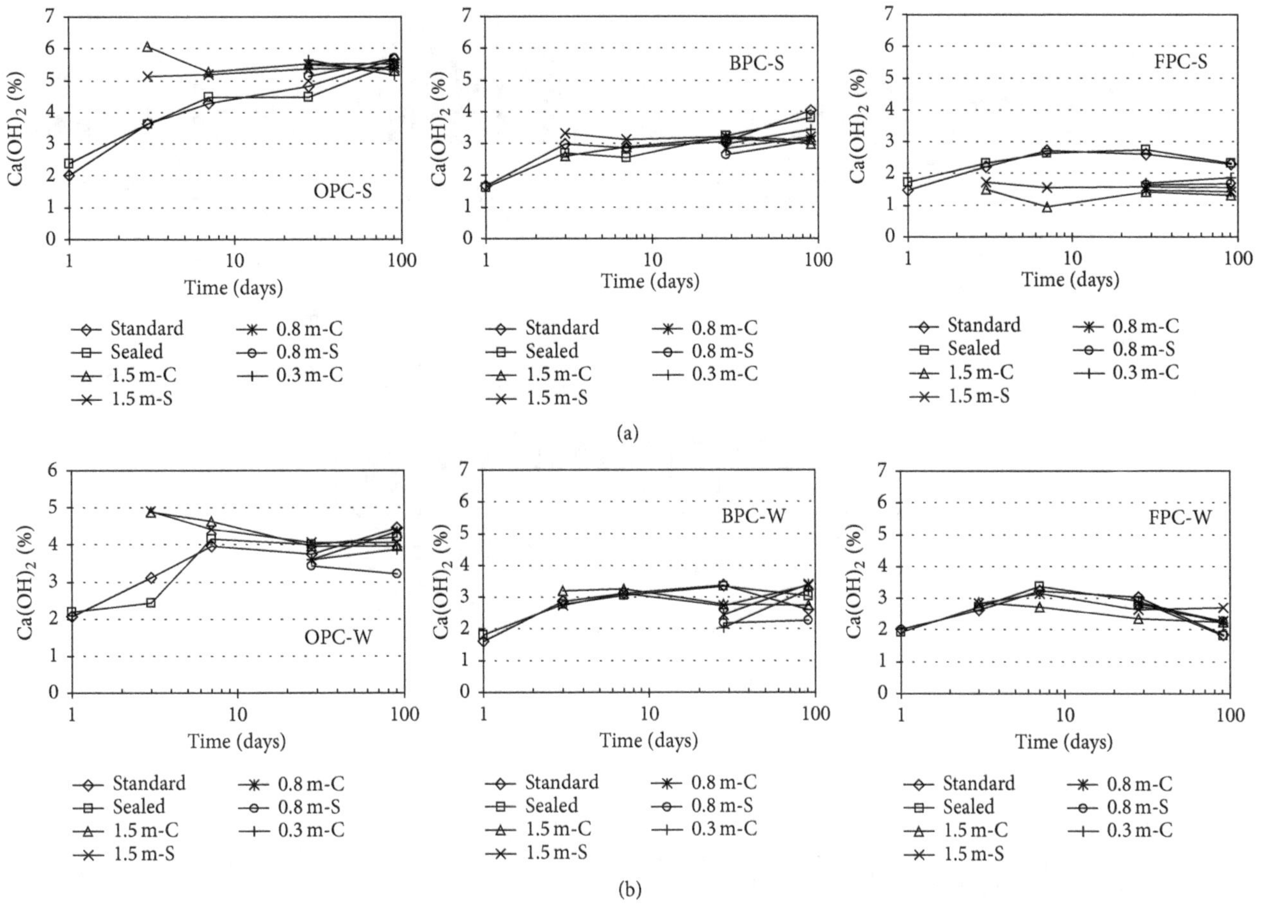

FIGURE 6: (a) Amount of calcium hydroxide of concrete in summer experiments. (b) Amount of calcium hydroxide of concrete in winter experiments.

by an increased curing temperature does not allow time for the diffusion of the hydrates. Consequently, it will produce a high nonuniform distribution of the hydrated products in the paste. The nonuniformity causes a reduction of strength at long curing times. The addition of fly ash increases the bound water content at early ages for the core specimen cured at high temperature perhaps because a rise in curing temperature speeds up the rates of the cement hydration and pozzolanic reactions. The pozzolanic reaction contributes to a decrease in the calcium ion concentration in the liquid phase, attributed to an acceleration of calcium dissolution from the cement grains [26].

The change in the amount of calcium hydroxide is shown in Figures 6(a) and 6(b). The amount of calcium hydroxide in the core specimen made with the OPC is higher than that of the standard specimen because it is directly related to the degree of hydration. In a mass structure, the higher temperature leads to an increase in the degree of hydration. However, no significant difference in calcium hydroxide content was found in any of the BPC specimens. The use of fly ash reduces the amount of the calcium hydroxide due to the pozzolanic reaction and so samples made with FPC showed less calcium hydroxide than the standard specimen. The pozzolanic reaction in FPC accelerates with an increase in the initial internal temperature and so the thicker wall exhibited less calcium hydroxide than the thinner wall.

3.4. Pore Structure. Figures 7(a) and 7(b) show the porosity results as determined by MIP. The total porosity of the OPC-based samples at center of the 1.5 m deep wall, cured under summer conditions, was 2.3% less than that of the standard specimen at 3 days. Comparable values for BPC and FPC samples showed 5.0% and 2.0% reductions, respectively. After 7 days, the total porosity of the core specimens began to increase with increased concrete temperature. The total porosity of the BPC and FPC materials exhibited a slight increase at 28 days and a significant increase at 91 days. The addition of pozzolanic materials increases the total porosity. Palardy et al. [18] explained that the dissolution of the calcium leaching originates mostly from the calcium hydroxide and the decomposition of ettringite at high temperature which will contribute to the increase in porosity.

The total porosity of concrete made under winter conditions was smaller than that of concrete made under summer conditions. In particular for OPC and BPC samples, the amount of larger pores, greater than 100 nm, was remarkably reduced when compared with summer cured samples. As

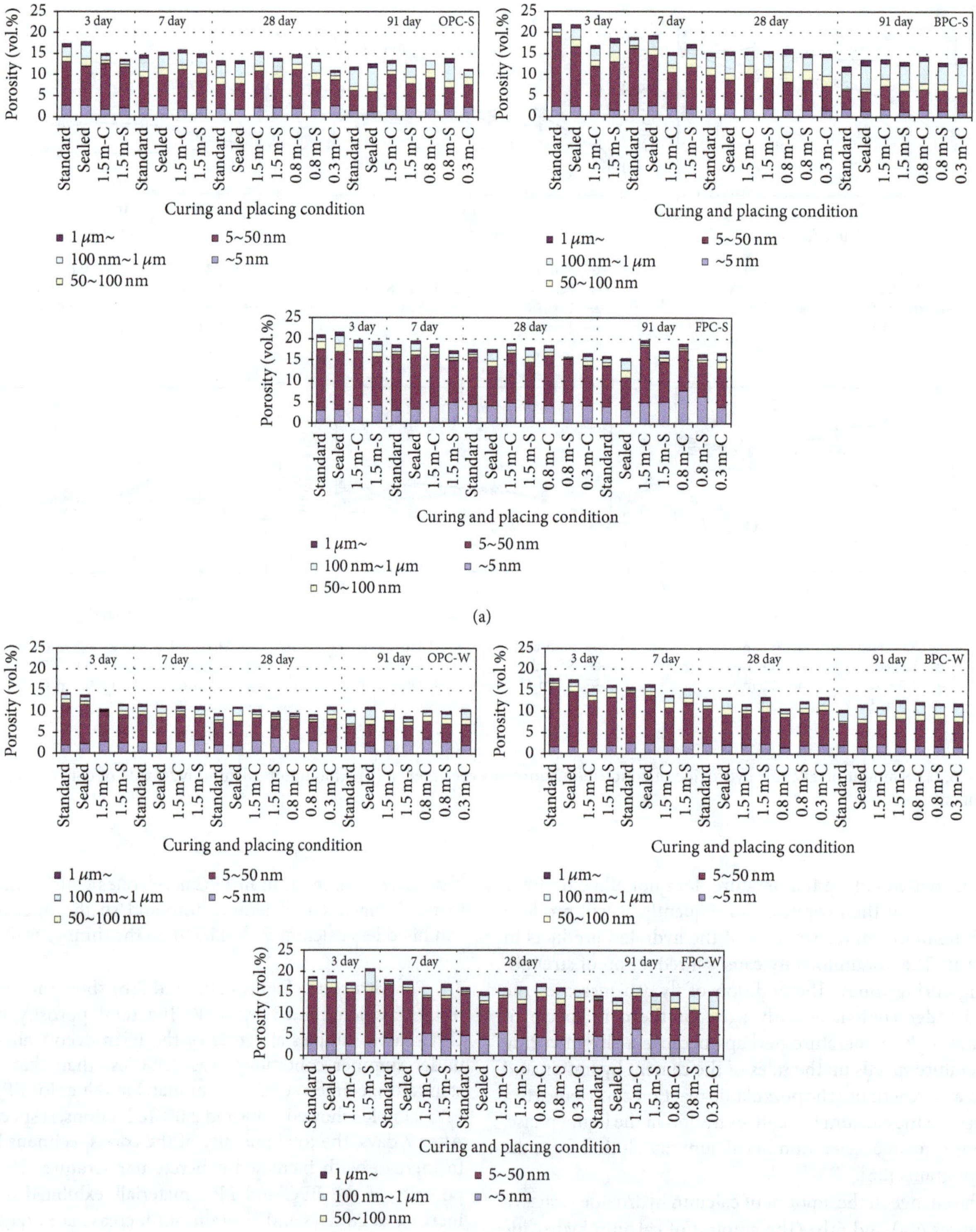

FIGURE 7: (a) The total porosity and pore size distribution of concrete in summer experiments. (b) The total porosity and pore size distribution of concrete in winter experiments.

shown in Figure 7, the pore size distribution also changed as the concrete temperature increased. With an increase in temperature, the number of larger pores decreased and the number of smaller pores increased, particularly for pores in a size range between 5 and 50 nm as measured by MIP. This suggests that as the hydration process proceeds, the pore sizes are reduced and the peak of the distribution curve shifts toward small pores.

4. Conclusions

The main conclusions of this paper can be summarized as follows.

(1) The type of cementitious material, the structural member size, and the curing climate were shown to significantly affect the initial internal temperature of mass concrete structures. Elevated temperatures did not affect the early-age strength gain of concrete made with ordinary Portland cement cured under summer weather conditions, but the concrete made with belite-rich Portland cement or fly ash showed a significant increase in early-age strength development. Elevated temperatures resulted in long-term strength loss for all materials.

(2) The bound water content of core specimens cured at elevated temperatures was larger than that of the standard specimen at early ages. However, the bound water content was lower than that of the standard specimen after 7 days for the OPC sample and after 28 days for the BPC and FPC samples. This trend could be correlated to a similar trend in compressive strength.

(3) The use of fly ash in mass concrete structures reduces the amount of calcium hydroxide due to the pozzolanic reaction. The pozzolanic reaction accelerates with an increase in the initial internal temperature and hence less calcium hydroxide was present under summer curing conditions.

(4) The total porosity of the core specimens cured under elevated temperatures was higher than that of the standard specimen. The porosity appeared to rise after 7 days in the OPC material and after 28 days in the BPC and FPC materials. The total porosity of the concrete made under winter weather conditions was smaller than that made under summer weather conditions. Due to the elevated internal temperatures in the structures, the peak of the pore size distribution curve was shifted toward the direction of smaller pores and the number of smaller pores was increased.

Conflicts of Interest

The authors declare no conflicts of interest.

Acknowledgments

This work was undertaken as a part of Japan Architecture Standard Specification 5 N project. This research was partly supported by Kangwon National University (Grant no. 120131429) in South Korea.

References

[1] K. Maekawa, T. Ishida, and T. Kishi, *Multi-Scale Modeling of Structural Concrete*, Taylor & Francis, London, UK, 2009.

[2] I. De La Varga, J. Castro, D. Bentz, and J. Weiss, "Application of internal curing for mixtures containing high volumes of fly ash," *Cement and Concrete Composites*, vol. 34, no. 9, pp. 1001–1008, 2012.

[3] C. Ozyildirim and W. J. Halstead, "Improved concrete quality with combinations of fly ash and silica fume," *ACI Materials Journal*, vol. 91, no. 6, pp. 587–594, 1994.

[4] A. Durán-Herrera, C. A. Juárez, P. Valdez, and D. P. Bentz, "Evaluation of sustainable high-volume fly ash concretes," *Cement and Concrete Composites*, vol. 33, no. 1, pp. 39–45, 2011.

[5] R. Siddique, "Performance characteristics of high-volume Class F fly ash concrete," *Cement and Concrete Research*, vol. 34, no. 3, pp. 487–493, 2004.

[6] W. F. Price and J. P. Hynes, "In-situ strength testing of high strength concrete," *Magazine of Concrete Research*, vol. 48, no. 3, pp. 189–197, 1996.

[7] K. Kovler, I. Schamban, S.-I. Igarashi, and A. Bentur, "Influence of mixture proportions and curing conditions on compressive strength of high-performance concrete," *ACI Structural Journal*, vol. 97, no. 1, pp. 21–26, 2000.

[8] X.-Y. Wang and K.-B. Park, "Analysis of compressive strength development of concrete containing high volume fly ash," *Construction and Building Materials*, vol. 98, pp. 810–819, 2015.

[9] B. Lothenbach, F. Winnefeld, C. Alder, E. Wieland, and P. Lunk, "Effect of temperature on the pore solution, microstructure and hydration products of Portland cement pastes," *Cement and Concrete Research*, vol. 37, no. 4, pp. 483–491, 2007.

[10] F. Deschner, B. Lothenbach, F. Winnefeld, and J. Neubauer, "Effect of temperature on the hydration of Portland cement blended with siliceous fly ash," *Cement and Concrete Research*, vol. 52, pp. 169–181, 2013.

[11] K. De Weerdt, M. B. Haha, G. Le Saout, K. O. Kjellsen, H. Justnes, and B. Lothenbach, "The effect of temperature on the hydration of composite cements containing limestone powder and fly ash," *Materials and Structures*, vol. 45, no. 7, pp. 1101–1114, 2012.

[12] A. Bentur, R. L. Berger, J. H. Kung, N. B. Milestone, and J. F. Young, "Structural properties of calcium silicate pastes: II, effect of curing temperature," *Journal of the American Ceramic Society*, vol. 62, no. 7-8, pp. 362–366, 1979.

[13] E. Berodier and K. Scrivener, "Evolution of pore structure in blended systems," *Cement and Concrete Research*, vol. 73, pp. 25–35, 2015.

[14] S. Jalali and M. Y. Abyaneh, "Prediction of final concrete strength in hot climates," *Magazine of Concrete Research*, vol. 47, no. 173, pp. 291–297, 1995.

[15] B. Lothenbach, K. Scrivener, and R. D. Hooton, "Supplementary cementitious materials," *Cement and Concrete Research*, vol. 41, no. 12, pp. 1244–1256, 2011.

[16] K. O. Kjellsen, R. J. Detwiler, and O. E. Gjørv, "Pore structure of plain cement pastes hydrated at different temperatures," *Cement and Concrete Research*, vol. 20, no. 6, pp. 927–933, 1990.

[17] I. Odler and Y. Chen, "Effect of cement composition on the expansion of heat-cured cement pastes," *Cement and Concrete Research*, vol. 25, no. 4, pp. 853–862, 1995.

[18] D. Palardy, M. Onofrei, and G. Ballivy, "Microstructural changes due to elevated temperature in cement based grouts," *Advanced Cement Based Materials*, vol. 8, no. 3-4, pp. 132–138, 1998.

[19] E. Gallucci, X. Zhang, and K. L. Scrivener, "Effect of temperature on the microstructure of calcium silicate hydrate (C-S-H)," *Cement and Concrete Research*, vol. 53, pp. 185–195, 2013.

[20] X.-Y. Wang and H.-S. Lee, "Evaluation of the mechanical properties of concrete considering the effects of temperature and aging," *Construction and Building Materials*, vol. 29, pp. 581–590, 2012.

[21] S. A. Rodger and G. W. Groves, "The microstructure of tricalcium silicate/pulverized-fuel ash blended cement pastes," *Advances in Cement Research*, vol. 1, no. 2, pp. 84–91, 1988.

[22] K. L. Scrivener and H. F. W. Taylor, "Delayed ettringite formation: a microstructural and microanalytical study," *Advances in Cement Research*, vol. 5, no. 20, pp. 139-146, 1993.

[23] C. D. Lawrence, "Mortar expansions due to delayed ettringite formation. Effects of curing period and temperature," *Cement and Concrete Research*, vol. 25, no. 4, pp. 903–914, 1995.

[24] J. Khunthongkeaw and S. Tangtermsirikul, "Model for simulating carbonation of fly ash concrete," *Journal of Materials in Civil Engineering*, vol. 17, no. 5, pp. 570–578, 2005.

[25] A. L. A. Fraay, J. M. Bijen, and Y. M. de Haan, "The reaction of fly ash in concrete a critical examination," *Cement and Concrete Research*, vol. 19, no. 2, pp. 235–246, 1989.

[26] R. J. Detwiler, C. A. Fapohunda, and J. Natale, "Use of supplementary cementing materials to increase the resistance to chloride ion penetration of concretes cured at elevated temperatures," *ACI Materials Journal*, vol. 91, no. 1, pp. 63–66, 1994.

13

Construction Time of Three Wall Types Made of Locally Sourced Materials

Wojciech Drozd, Agnieszka Leśniak, and Sebastian Zaworski

Institute of Construction Management, Tadeusz Kościuszko Cracow University of Technology, Warszawska 24 St., 31-155 Kraków, Poland

Correspondence should be addressed to Wojciech Drozd; wdrozd@ztob.pk.edu.pl

Academic Editor: Estokova Adriana

Similarly to any other industry, the construction sector puts emphasis on innovativeness, unconventional thinking, and alternative ideas. At present, when sustainable development, ecology, and awareness of people's impact on the environment grow in importance, low impact buildings can become an innovative alternative construction technology for the highly industrialized construction sector. The paper presents a comparative study of three walls made of available materials used locally, which can be classified as biosourced materials, in terms of construction time. The comparison of times necessary to make 1 m^2 of the wall allows us to decide which building technology is more advantageous in terms of the construction duration. A shorter construction time means lower labour costs and lower expenses for construction machines. In order to obtain answers to the questions posed, the authors made extensive searches of source data on the time-consuming building works which used locally sourced materials. Reference is made to "Temporary principles of erecting clay buildings" issued by the Institute of Housing Construction in Warsaw (Poland). Three types of walls made of locally sourced materials were studied: a wall made of clay blocks insulated with mineral wool boards, a wall made of clay compacted in formwork, and one insulated with mineral wool boards and wooden frame structure filled with straw bales and cladded with fibreboards. The layers have been chosen in such a manner that heat transfer coefficient values for the studied variants are as equal as possible (0.2 W/m^2 K), thus allowing a reliable comparative study.

1. Introduction

Modern material and technological solutions—concrete, steel, glass, and intelligent systems—have become the synonyms of modernity and luxury [1]. Buildings made of locally sourced materials are on the other extreme of modern construction. These solutions are based primarily on tradition and local, low-cost raw materials that do not require special treatment, which are readily available, such as soil, clay, straw, and sand. Human hands are the main building force, while the use of complicated techniques or expensive expertise is limited. Low impact buildings avoid complex technical solutions and support generally accessible raw materials. The construction market, dominated by giant construction companies, does not support the development of such solutions since they are not profitable. This makes it difficult to popularize such a building method on a large scale. Despite this, buildings made of locally sourced materials are becoming more common and perfectly fit into the idea of sustainable development.

The most important features of a building characterized by sustainable development can be described using four Rs: reduce, reuse, recycle, and recover [2]. Less material and energy are used to build such a building compared to conventional construction The materials used should be recycled and should allow for reuse after the end of the life of the building. Low impact buildings derive the materials from the surrounding environment. This supports local development and cultural independence of the region. Production of materials does not require high energy and high temperature processed and does not produce CO_2, so it consumes less energy needed to construct the building than conventional building construction. Transport of materials is only done locally [3]. And the structure itself is completely biodegradable, and after the end of its life, it does not leave harmful waste, hard to break down by the environment. Building made of locally sourced materials allows for carrying out a lot of the work on one's own. It is available to all and creates local jobs [4].

Research in the field of natural building technologies is not numerous. Among them in paper [5] a comparison of the mechanical performance of structural elements built in three basic techniques, Earth block (adobe) masonry, rammed Earth, and cob, is presented. Up to present, few studies are available concerning the mechanical behaviour of straw bales in buildings. Such study is presented in [6] and aims at investigating the behaviour of straw bales and leads to recommendations for required bale densities. In [7], the viability of straw-bale construction has recently been investigated, in particular, its resistance to moisture. Similarly in [8], two options for the use of straw to fill envelop walls were investigated in the Andean Patagonian region: the direct use of straw bales, whether in whole or in halves, and the manufacturing of straw-clay blocks. All straw options analysed result in significantly better thermal performance than current choices of fired bricks or concrete blocks, which are commonly used in the region. In turn, in [9] was evaluated a straw-bale house located in Bavaria, Germany. The experimental work included compression tests, moisture content, thermal stability of bales, and pH. Reference [10] examines the use and accuracy of a moisture probe used in the walls of a straw-bale building. The measurements from a number of moisture probes placed in the walls of a case study straw-bale building over a 2-year period are presented. Similarly [11] concludes results from a study on moisture monitoring in straw-bale construction and includes the development of an empirical equation which relates the straw moisture content to surrounding microclimate relative humidity and temperature. Reference [12] presents results from a study on the thermal conductivity of some plaster materials that could be used for straw-bale buildings. Walls constructed in the straw-bale technology can boast excellent health qualities, which are difficult to obtain in traditional technologies [13].

Up until today, few studies are available concerning the time of construction of walls made of locally sourced materials. In this paper, the construction time of 1 m^2 of three types of walls built in these technologies was studied. The layers of the walls were chosen in such a manner that the heat transfer coefficient values for the studied variants are as equal as possible (0.2 W/m^2 K), thus allowing a reliable comparative study:

(i) A wall made of clay blocks insulated with mineral wool boards

(ii) A wall made of clay compacted in formwork, insulated with mineral wool boards

(iii) A wooden frame structure filled with straw bales and cladded with fibreboards

Construction time allows us to tell which technology is more advantageous. In order to obtain answers to the questions posed, the authors made extensive searches of source data concerning the timing of building works using locally sourced materials. Literature does not actually present any time standards. The exception is "Temporary principles of erecting clay buildings" issued by the Institute of Housing Construction in Warsaw (Poland), which were used in the comparison.

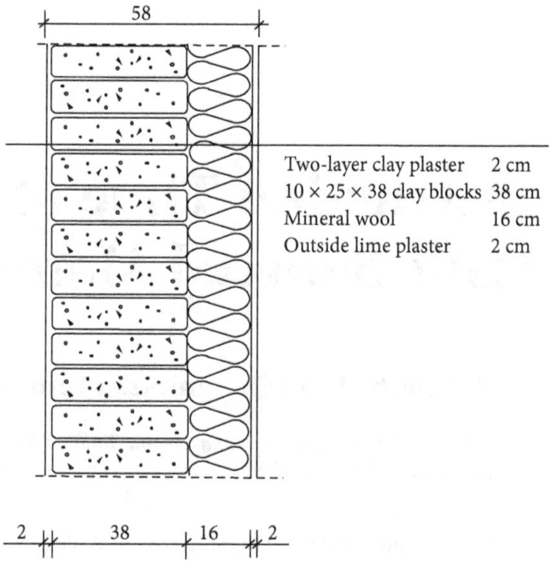

Figure 1: Cross section of the clay block wall.

2. Characteristics and Technical Parameters of Studied Walls

2.1. Wall Made of Clay Blocks Insulated with Mineral Wool Boards. The first studied structure was a wall made of 10 × 25 × 38 cm clay blocks insulated with mineral wool boards. The structural layer of the wall is 38 cm thick (Figure 1). On the outside, the wall will be insulated with 16 cm thick mineral wool boards and covered with lime plaster. On the inside, the wall will be covered with a two-layer clay plaster. Wood fibreboards have also been considered as insulation as more environmentally friendly, but this solution is presently too expensive to be compared with affordable EPS (expanded polystyrene).

2.2. Wall Made of Clay Compacted in Formwork, Insulated with Mineral Wool Boards. The other studied solution was the wall made of clay compacted in the formwork whose structural thickness was 30 cm (Figure 2). The formwork was demountable panels. The remaining wall layers were the same as in the clay block wall.

2.3. Wooden Frame Structure Filled with Straw Bales and Cladded with Fibreboards. The third variant was the wall made of small 31 × 41 × 70 cm straw bales placed in wooden frame structure (Figure 3). The frame structure will be erected in the timber-frame house technology where posts are made as frames, so-called ladders. The wooden frame skeleton will be cladded on both sides with 12-millimetre fibreboards for

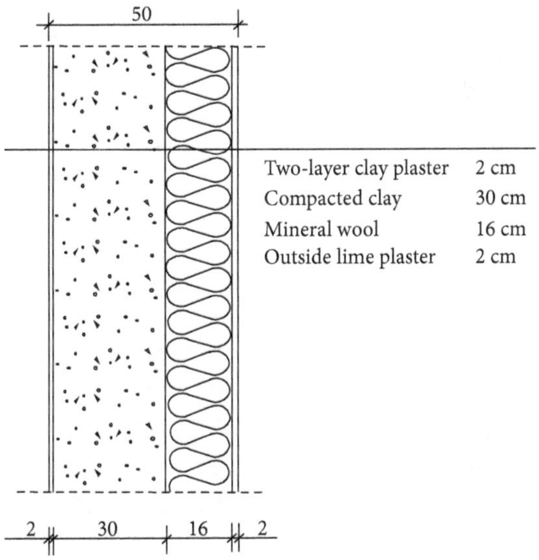

FIGURE 2: Cross section of the wall made of clay compacted in formwork.

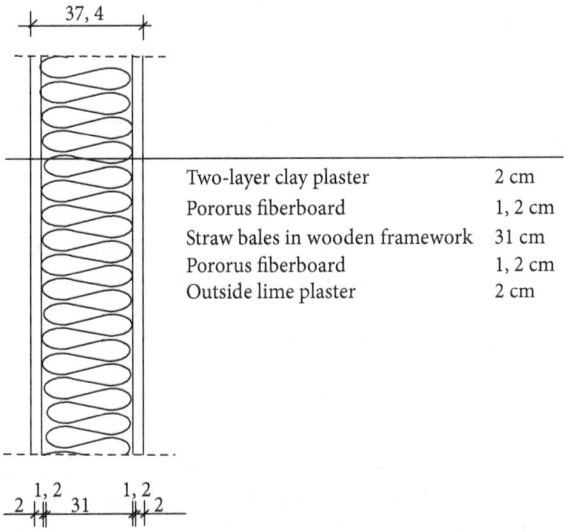

FIGURE 3: Cross section of the wall made in the straw-bale technology.

TABLE 1: Density and thermal conductivity of used materials.

	Material type	Density (kg/m^3)	Thermal conductivity λ (W/m K)
Insulation materials	Straw bales	100	0.073
	Porous fibreboard	300	0.060
	Mineral wool boards	160	0.037
Structural materials	Compacted clay and clay blocks	1800	0.900
	Pine wood	450	0.200
Materials	Lime plaster	1700	0.700
	Clay plaster	900	0.300

Source: own study, based on [14].

TABLE 2: Thermal properties of a clay block wall.

Layers	d (m)	λ (W/m K)	R (m^2 K/W)	U (W/m^2 K)
R_{si}			0.13	
Two-layer clay plaster	0.020	0.300	0.06666667	
Clay blocks	0.380	0.900	0.42222222	
Insulation: mineral wool	0.160	0.037	4.32432432	
Lime plaster	0.020	0.700	0.02857143	
R_{se}			0.04	
Total d	0.580	RT	5.012	0.200

Source: own study.

TABLE 3: Thermal properties of a compacted clay wall.

Layers	d (m)	λ (W/m K)	R (m^2 K/W)	U (W/m^2 K)
R_{si}			0.13	
Two-layer clay plaster	0.020	0.300	0.06666667	
Compacted clay	0.300	0.900	0.33333333	
Insulation: mineral wool	0.160	0.037	4.32432432	
Lime plaster	0.020	0.700	0.02857143	
R_{se}			0.04	
Total d	0.500	RT	4.923	0.203

Source: own study.

TABLE 4: Thermal properties of a straw-bale wall.

Layers	d (m)	λ (W/m K)	R (m^2 K/W)	U (W/m^2 K)
R_{si}			0.130	
Two-layer clay plaster	0.020	0.300	0.067	
Fibreboard	0.012	0.060	0.200	
Straw bales in frame structure	0.310	0.073	4.247	
Fibreboard	0.012	0.060	0.200	
Lime plaster	0.020	0.700	0.029	
R_{se}			0.040	
Total d	0.374	RT	4.912	0.204

Source: own study.

TABLE 5: Dependence of studied technologies on seasons and weather conditions.

Wall type	Works can be performed all year long	Works strongly dependent on weather conditions
Clay blocks	Yes	Yes
Compacted clay	No	Yes
Straw bale	No	Yes

Source: own study based on [14, 16, 17].

TABLE 6: Time necessary to make 10 × 25 × 38 cm clay blocks.

Clay preparation difficulty	Compound preparation	Making of elements	Squaring of elements	Putting in trestles	Total time	Total time
			m-h/100 pcs			m-h/m^2
Medium-cohesion 801–1400 g/5 cm^2	2.52	5.82	0.42	0.50	9.26	8.99 × 0.38 = 3.42
			m-h/m^3			
	2.45	5.65	0.41	0.49	8.99	

Source: own study.

TABLE 7: Time necessary to build the clay block wall (wall thickness 38 cm).

	Masonry works	Transport: Blocks	Transport: Mortar	Making of mortar	Auxiliary works	Total time
				m-h/m^3		
Ground floor	3.37	1.24	0.15	0.467	0.2	5.427
1st floor	3.37	1.88	0.20	0.467	0.21	6.127

Source: own study.

TABLE 8: Various works (wall thickness 38 cm).

Work	Unit	Time
Extra to the full wall for making corners	m-h/mb	0.23
Extra for making window reveals	m-h/mb	0.23

Source: own study.

TABLE 9: Time to make clay mortar for masonry works and plastering.

Making mortar with a machine	Manual sieving: Sand	Manual sieving: Clay	Total time	Total time (per m^3 of the wall)	Total time (plaster)
m-h/m^3				m-h/m^3	m-h/m^2
1.5	0.9	1.0	3.4	0.317	0.034

Source: own study.

good adhesion and improved thermal insulation. The wall will have a lime plaster on the outside and two-layer clay plaster on the inside.

3. Thermal Conductivity λ and Heat Transfer Coefficient U

Thermal conductivity λ (W/m K) is a measure of how well a material insulates the flow of heat. Porous, low-density materials have low thermal conductivity and hence are better insulators in comparison with denser materials.

While it is quite easy to obtain information on physical properties of typical construction materials, obtaining such information on locally sourced materials may prove a challenge.

The thermal conductivity of a straw bale and any other material depends on its compaction and moisture content. The values for straw bales in Table 1 are based on the studies conducted in November 2015 by the Building Research Institute in Warsaw (Poland) [15]. The study comprised evaluation of a 600 × 600 × 200 mm straw bale, with density of about 100 kg/m^3.

The thermal conductivity of compacted clay, clay blocks, and clay plaster depends on the clay composition and used admixtures. Table 1 gives values for highly compacted (compacted clay and clay blocks) and medium-compacted clay (clay plaster).

The heat transfer coefficient "U" is a measure of how well a building enclosure transfers heat. The "U value" is expressed in W/(m^{2}K) and denotes how many watts of thermal energy pass through 1 m^2 of enclosure when the temperature difference between the inside and the outside is equal to 1 K. The lower the heat transfer coefficient is, the better the insulation properties a particular enclosure has.

The heat transfer coefficient is expressed by [3]

$$U = \frac{1}{R_T}\ \left(\mathrm{W/m^2K}\right), \quad (1)$$

where R_T is the total thermal resistance of a building partition, and total thermal resistance R_T for homogeneous partitions is described by the formula:

$$R_T = R_{si} + R_1 + R_2 + \cdots + R_n + R_{se}\ \left(\mathrm{W/m^2K}\right), \quad (2)$$

where R_{si} is the heat transfer resistance on the inner surface, $R_1 + R_2 + \cdots + R_n$ is the design thermal resistance of each layer, and R_{se} is the heat transfer resistance on the outer surface.

Thermal resistance of a homogeneous layer with thickness d is obtained from the following equation:

$$R = \frac{d}{\lambda}\ \left(\mathrm{m^2K/W}\right), \quad (3)$$

where d is the thickness of the material and λ is the design thermal conductivity of the material.

Tables 2–4 present the calculations of a heat transfer coefficient for the studied walls. The layers have been chosen in such a manner that the "U" values for the studied variants are as equal as possible, thus allowing a reliable comparative study.

TABLE 10: Time necessary to make inside clay plasters (1 layer).

Mortar making	Plastering	Transport	Additional works	Scaffolding	Extras	Total time
			m-h/m^2			
0.034	0.284	0.022	0.013	0.061	0.038	0.452

Source: own study.

TABLE 11: Time necessary to make 1 m^2 of the clay block wall.

Operation	Standard time	
	m-h/m^2	%
Making clay blocks	3.416	40
Making inside clay plasters	0.904	11
Erecting clay block wall	2.339	28
Thermal insulation: mineral wool boards	1.154	14
Making outside lime plasters	0.640	8
Total	8.453	100

Source: own study.

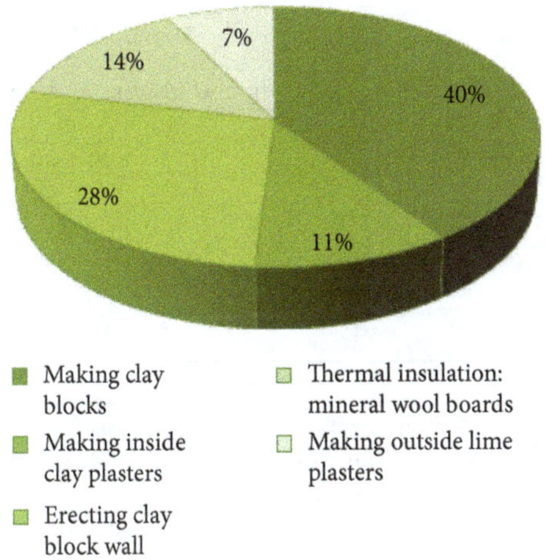

FIGURE 4: The percentage of the share of individual operations in making of 1 m^2 of the clay block wall.

4. Seasonality of Construction Works and Dependence on Weather Conditions

In temperate climate, the construction works and their effectiveness depend on the season and weather. Clay and straw are particularly sensitive to weather conditions and must be protected from water and subfreezing temperatures. After being put into the structure, the straw will be protected by the plaster, and clay must be protected from water at all times.

The construction works with locally sourced materials must be performed during a warm and relatively dry period, generally from late spring to early autumn. The works with conventional materials can be performed all year long if suitable precautions are taken, with exception of really low subfreezing temperatures. The exceptions among the locally sourced materials are dried clay blocks which also can be installed all year long if suitable precautions are taken.

In that area, conventional materials have a clear advantage over the alternative materials, although conventional materials must be protected from a long-term exposure to water and subfreezing temperatures. Such protection is however much simpler that in case of locally sourced materials.

Straw must not be wet when installed as it takes a long time to dry. Styrofoam, on the other hand, does not absorb water and is easy and quick to dry. Clay must be protected from precipitation at all times during the construction works by special roofing and by eaves when the building is in use. Dependence of studied technologies on seasons and weather conditions is given in Table 5.

5. Time of Making 1 m^2 of the Wall

5.1. Clay Block Wall. The study of time required to perform works in case of clay block walls refers to the "Temporary Principles of Erecting Clay Buildings" by the Institute of Housing Construction [16]. The standard times are expressed in m-h/m^3—man-hours per cubic metre of the clay block wall.

Standard times for making the clay blocks (Table 6) include maintenance of the storage yard for finished blocks, maintenance of machinery, tools, and fixtures, and squaring of the blocks and putting them in trestles. They do not, however, include storage yard levelling, construction of protective roofing, and excavation of drainage ditches.

Clay block walls are erected in the same manner as brick walls (Tables 7 and 8). The standard times include erecting walls using clay mortar, transport of blocks and mortar for a distance up to 20 m, and additional mortar mixing. The clay block walls are measured in m^3, and the openings in excess of 1 m^2 are subtracted from the wall volume.

Clay mortar should have the same composition as clay blocks (Table 9). The standard time for making clay mortar includes handling auxiliary materials at the distance of 10 m, mixing mortar with water, and operation and maintenance of machinery. The standard time does not include making a box for the mortar preparation and placement and relocation of machines used to make the mortar. The penultimate column of the Table 5 includes the time for making the mortar needed to erect 1 m^3 of the final clay block wall. The last column includes the time for preparing the mortar needed to make 1 m^2 of 1 cm thick plaster.

The clay plastering standard times include additional works: substrate wetting using a hose, surface scratching with a rake, application of plaster strips and skim coat and trowelling, and others (preparation and substrate cleaning) (Table 10). The clay plaster is made in layers, each 1 cm thick.

The time of making 1 m^2 thermal insulation using mineral wool boards has been taken from the Contractors Estimator KNR 33/2/4. The standard time for application of outside paster has been taken from the Contractors Estimator KNR 202/906/2.

Table 12: Time necessary to prepare the clay for compaction.

Clay preparation difficulty	Clay placement for mixing	Transport of admixtures	Processing with milling machine	Pouring into moulds	Total time
			m-h/m^3		
Medium-cohesion 801–1400 g/5 cm^2	0.51	0.52	0.08	0.35	1.46

Source: own study.

Table 13: Time necessary to make compacted clay walls.

	Formwork	Loading, transport, and unloading for formwork	Charge levelling	Compaction	Total time	Total time
		m-h/m^3				m-h/m^2
Ground floor	3.10	2.88	0.72	0.95	7.65	2.295
1st floor	3.38	2.92	0.75	0.95	8.00	2.40

Source: own study.

Table 14: Various works.

Various works	Unit of measure	Quantity
Placement of levelling layer under ceiling beams for load-bearing outside walls plus extra for more difficult formwork and compaction	m-h/m	1.00
Placement of levelling layer under ceiling beams for load-bearing inside walls plus extra for more difficult formwork and compaction	m-h/m	2.00
Placement of window templates, removal completion of works, plus extra for more difficult formwork and compaction		
Window templates above 0.5 m^2	m-h/pc	0.55
Window templates below 0.5 m^2	m-h/pc	0.40
Door templates above 0.5 m^2	m-h/pc	0.55

Source: own study.

Table 15: Time needed to build 1 m^2 of the compacted clay wall.

Operation	Standard time	
	m-h/m^2	%
Making inside clay plasters	0.904	15
Making compacted wall in formwork	3.248	55
Thermal insulation: mineral wool boards	1.154	19
Making outside lime plasters	0.604	11
Total	5.946	100

Source: own study.

The time to make 1 m^2 of the clay block wall (Table 11 and Figure 4) has been averaged due to various units of measure for intermediate operations (e.g., prefab lintels are expressed in m-h/m).

5.2. Compacted Clay Walls. The study of time required to make compacted clay walls also refers to the "Temporary Principles of Erecting Clay Buildings" by the Institute of Housing Construction [16]. The standard times are expressed in m-h/m^3—man-hours per cubic metre of the compacted clay wall.

Time to prepare the clay (Table 12) for compaction includes its mixing with cracking prevention admixtures and pouring into moulds after mixing. It does not include cutting of fibrous materials and protecting the materials from precipitation. The amount of prepared clay compound is measured in cubic metres. 1 m^3 of the compacted wall is an equivalent of about 1.55 m^3 of loose mixture.

Standard times for making compacted clay walls (Table 13) include transport of the mixture to the hoist at the distance of maximum 20 m, transport on the scaffolding at the distance of maximum 20 m, pouring into formwork, levelling of the poured layer, compaction, placement of window and door templates, and making of lime battens or placement of ceramic inserts. The standard times, however, do not include protecting the walls from weather conditions, making the insulation, handling of window and door templates, and making of the levelling layer as a cover for lintels (Table 14).

The unit of measure for walls in excess of 24 cm in thickness is m^3, and up to 24 cm in thickness is m^2. Openings larger than 1 m^2 should be subtracted from the wall volume.

The times of making the clay (Table 15 and Figure 5) mortar for inside plasters, mineral wall insulation, and outside plasters are identical as in case of the clay block wall. The time to make 1 m^2 of the compacted clay wall has been averaged due to various units of measure for intermediate operations (e.g., prefab lintels are expressed in m-h/m).

5.3. Straw-Bale Walls. The standard times for erecting walls in the straw-bale technology (Table 16 and Figure 6) have

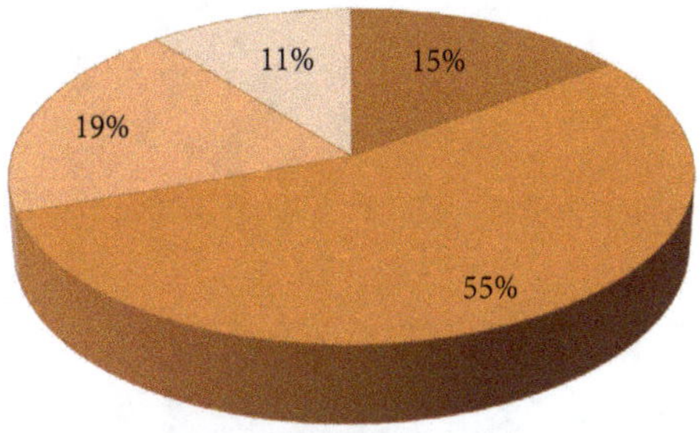

FIGURE 5: The percentage of the share of individual operations in making of 1 m^2 of the compacted clay wall.

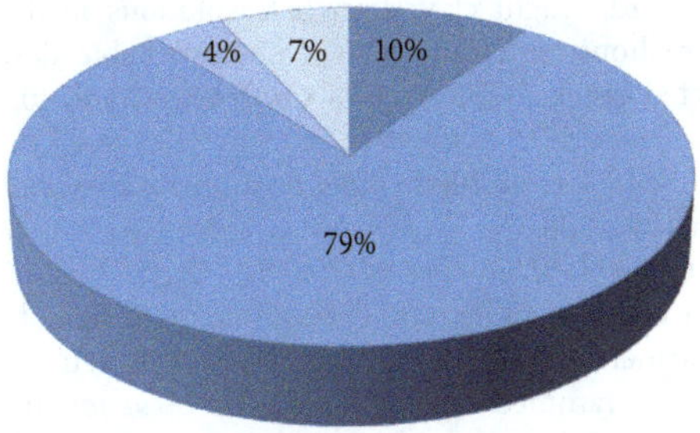

FIGURE 6: The percentage of the share of individual operations in making of 1 m^2 of the straw-bale wall.

TABLE 16: Time needed to build 1 m^2 of the straw-bale clay wall.

Operation	Standard time m-h/m^2	%
Making inside clay plasters	0.904	10
Erecting wooden framework structure	7.437	79
Insulation of straw bales and fibreboard cladding	0.366	4
Making outside lime plasters	0.640	7
Total	9.347	100

Source: own study.

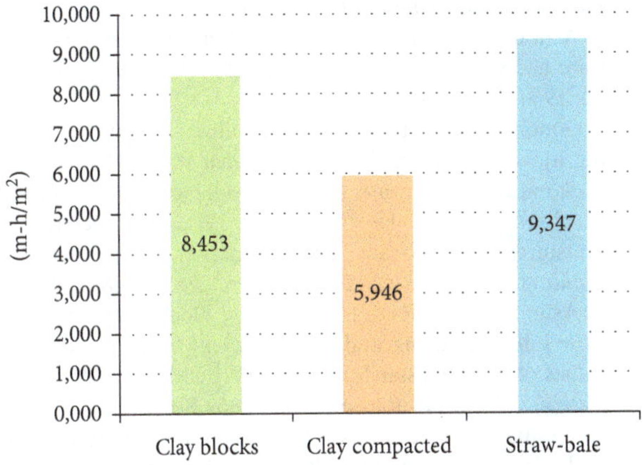

FIGURE 7: Time to make 1 m^2 of the wall in individual technologies.

been taken from the Contractors Estimator. The amount of labour needed to fill the wooden structure with straw bales has been determined analogously to the Contractors Estimator's item 202/613/6. The standard times for making inside clay plasters are identical as for the clay walls.

The time to make 1 m^2 of the compacted clay wall has been averaged due to various units of measure for intermediate operations.

6. Conclusions

The comparison of times necessary to make 1 m^2 of the wall allows us to decide which construction technology is more advantageous in terms of construction duration. The small surfaces in the wall or surfaces with numerous windows and doors, as well as the drying time between Earth wall making and rendering operation, are calculated by additional time; therefore, they were not taken into account in this paper. The shorter construction time means lower labour costs and lesser expenses for construction machines.

In the case of investors, the construction time is an important measure of project success, as the shorter payback periods mean a quicker opportunity to reinvest or reuse financial resources.

The construction time is equally important for people building a home for themselves. It is a huge advantage for them to be able to perform the works in a favourable weather and to move in and use the house earlier.

The studies and Figure 7 allow us to draw the following conclusions:

(i) Compacted clay walls have the shortest wall erection time from among the studied natural building technologies.

(ii) Time to make 1 m^2 of the clay block wall is significantly longer as a result of a very time-consuming clay block preparation process. In the paper, it was assumed that blocks are made in-house, using mainly materials available free at the construction site.

(iii) The straw-bale technology had the worst results in the comparison. The factor which determines the long time needed to make 1 m^2 of the straw-bale wall is the necessity to erect the wooden framework structure.

(iv) Two-layer clay plaster can be an alternative to other contemporary inside wall finishing solutions.

Conflicts of Interest

The authors declare that they have no conflicts of interest.

References

[1] Law on Construction Products of 16 April 2004, Dz.U. 2004 No. 92 Item 881.

[2] W. Drozd, "Light clay straw bale solutions in the contemporary housing as an element of sustainable development. Selected issues," *E3S Web of Conferences*, vol. 10, p. 00016, 2016.
[3] PN-EN ISO 6946:2008, *Building Components and Building Elements*, 2008.
[4] BN-62/6738-01, *Masses of Cement—Clay Fillers.*
[5] L. Miccoli, U. Müller, and P. Fontana, "Mechanical behaviour of earthen materials: a comparison between earth block masonry, rammed earth and cob," *Construction and Building Materials*, vol. 61, pp. 327-339, 2014.
[6] T. Lecompte and A. Le Duigou, "Mechanics of straw bales for building applications," *Journal of Building Engineering*, vol. 9, pp. 84-90, 2017.
[7] J. Robinson, H. K. Aoun, and M. Davison, "Determining moisture levels in straw bale construction," *Procedia Engineering*, vol. 171, pp. 1526-1534, 2017.
[8] A. D. González, "Energy and carbon embodied in straw and clay wall blocks produced locally in the Andean Patagonia," *Energy and Buildings*, vol. 70, pp. 15-22, 2014.
[9] T. Ashour, H. Georg, and W. Wu, "Performance of straw bale wall: a case of study," *Energy and Buildings*, vol. 43, no. 8, pp. 1960-1967, 2011.
[10] S. Goodhew, R. Grffiths, and T. Woolley, "An investigation of the moisture content in the walls of a straw-bale building," *Building and Environment*, vol. 39, no. 12, pp. 1443-1451, 2004.
[11] M. Lawrence, A. Heath, and P. Walker, "Determining moisture levels in straw bale construction," *Construction and Building Materials*, vol. 23, no. 8, pp. 2763-2768, 2009.
[12] T. Ashour, H. Wieland, H. Georg, F. J. Bockisch, and W. Wu, "The influence of natural reinforcement fibres on insulation values of earth plaster for straw bale buildings," *Materials & Design*, vol. 31, no. 10, pp. 4676-4685, 2010.
[13] A. Leśniak and K. Zima, "Comparison of traditional and ecological wall systems using the AHP method," in *Proceedings of the 15th International Multidisciplinary Scientific GeoConference Surveying Geology and Mining Ecology Management (SGEM 2015)*, vol. 3, pp. 157-164, Albena, Bulgaria, June 2015.
[14] G. Minke and B. Krick, *Straw Cube Handbook*, Cohabitat Foundation, Łódź, Poland, 2015.
[15] Research ITB (Institute of Building Technology), *Study of the Heat Conduction Coefficient of Straw Cubes*, Report no. LFS00-02236/15/Z00NF, Institute of Building Technology, Warsaw, Poland, 2015.
[16] Institute of Housing Construction, *Temporary Principles of Erecting Clay Buildings*, Institute of Housing Construction, Warsaw, Poland, 1955.
[17] M. Zatylny, *Formal and Legal Conditions for Natural Construction in Europe and Poland with Particular Emphasis on the Use of Straw Cubes*, United Nations Development Program (UNDP), New York, NY, USA, 2013.

14

Influences on Shotcrete Rebound from Walls with Random Roughness

Zhaoxia Liu,[1,2] Wenhui Bian,[1] Gang Pan,[1,2] Pengcheng Li,[1] and Wenxin Li[1,2]

[1]College of Mining and Safety Engineering, Shandong University of Science and Technology, Qingdao 266590, China
[2]State Key Laboratory of Mining Disaster Prevention and Control Co-founded by Shandong Province and the Ministry of Science and Technology, Shandong University of Science and Technology, Qingdao 266590, China

Correspondence should be addressed to Gang Pan; threven@sdust.edu.cn

Academic Editor: Michael Aizenshtein

Concrete slurry can be sprayed on walls for reinforcement; however, there is a certain amount of rebound which is hazardous, lowers production quality, and wastes material. To investigate this problem, we studied single slurry droplets at the mesoscopic level. We deduced the factors influencing droplet spreading and wall adhesion to create models of shotcrete rebound. Then, a numerical simulation orthogonal experiment investigating droplet-wall impacts was performed. The relationship between the spreading coefficient and each influencing factor is discussed, and numerical models are presented. Finally, the obtained models are verified by physical experiments. The results show that the spreading coefficient can be used to better characterize the effect of slurry droplet adhesion to walls. Modeled and experimentally observed droplet-wall impacts showed good consistency. The influence of each factor on the spreading coefficient was determined in the following order of strength: droplet velocity and viscosity, wall roughness, and surface tension. The spreading coefficient increases with velocity, decreases with viscosity and roughness, and increases first and then decreases with surface tension. This study improves the fluid dynamics-based theory of multiphase flow in concrete slurry and provides a theoretical basis for mitigating shotcrete rebound.

1. Introduction

Shotcrete is a building material formed by spraying concrete slurry onto a wall. Shotcrete rebound occurs when the stream of sprayed concrete hits a solid wall and partially bounces off [1]. The rebound phenomenon not only affects the spraying process and wastes concrete but also affects the quality of the project and presents potential safety hazards [2–4]. At present, during the spraying process, many of the operating parameters are determined approximately based on operator experience. The main focus is on the shotcrete materials [5–8], material properties [9–12], and spray equipment [13, 14]. Most existing theoretical studies have investigated aggregate rebound [15–18], while research into the impact of slurry on walls is lacking.

As shown in Figure 1, according to the sprayed concrete spraying process, the jet, adhesion, and deposition process of the shotcrete are divided into three stages. After the material is ejected from the nozzle in the form of a thin stream, the stream is divided into two parts by the action of the air stream: the pure slurry liquid phase and the slurry-wrapped aggregate phase. Among them, the aggregate phase firstly occupies the mortar wrapped on the surface after the wall surface collides with the wall surface. Because there is no cement paste bonding effect, aggregate rebounding is 100%. Some of the slurry is attached to the wall after springback. The slurry phase then reaches the wall and hits the wall alone. Due to the impact force, wall condition, and slurry properties, the slurry droplets are divided into splash rebound, adhesion to the wall, and sliding/loss of adhesion. The slurry that adheres to the wall in the above two stages provides wall adhesion and impact buffering for the subsequent stream. As well, the aggregate particles are gradually deposited from small to large, the subsequent stream is

FIGURE 1: Using a high-speed camera to capture the jet flow process of shotcrete.

impacted and compacted, and the slurry is adhered to fill the aggregate void. It is evident that the cement paste adhering to the wall is the primary condition for the deposition of the sprayed concrete wall. The cohesive force and impact buffer provided by the cement slurry can significantly reduce the rebound loss of the aggregate. Therefore, it is important to study the rebound of cement slurry for the overall rebound of shotcrete, and doing so would enhance the use of sprayed concrete in engineering applications.

The impingement of slurry spray is a complex process that requires adjustment of many parameters to achieve the best slurry deposition. Due to the impact of slurry on a wall being an unstable process, many phenomena cannot be explained clearly. In the process of investigating this phenomenon, droplets of slurry can be considered as elements in a slurry mass. The behavior of the slurry as it spreads on a wall can be regarded as the sum of multiple droplet impacts. Hence, the dynamics of slurry droplets as they impact a wall need to be studied using multiphase fluid mechanics.

Research into droplets impacting walls is over 100 years old and has two main streams: (1) spreading of droplets as they strike a smooth surface and (2) the spreading of droplets striking a rough wall. The study of droplets spreading on smooth surfaces began in 1877. Worthington [19] used a small experimental platform to study the process of droplets impacting metal planes from different heights. San Marchi [20] used numerical analysis to study the deformation and deposition of single droplets impinging on flat substrates. The recent development of high-speed camera technology has allowed more detailed observations of the morphological changes that occur as a droplet strikes a wall. Zhang [21] used high-speed cameras to capture the impact of Newtonian fluids containing two different surfactants on a solid substrate and analyzed the effect of surfactants on the droplet impact and diffusion processes. Park [22] proposed a hypothesis that droplet diffusion is spherical-shaped and examined the spreading behavior of four fluids at low impact velocity on four different surfaces.

However, there are few studies on droplet spreading on rough wall surfaces. Stow [23] used roughness (Ra) values to characterize the roughness of wall surfaces. The results show that roughness influences droplet impact due to the presence of points on the substrate. The eccentricity caused by anisotropy causes different effects. To make a droplet splash on an inclined panel requires a higher impact velocity than normal. Rioboo [24] analyzed the roughness of several droplets (such as water, alcohol, etc.) in detail. In the wall collision process, droplets get "crushed", and droplet diameter and wall roughness were found to influence this process. However, roughness was characterized by a single parameter, and the experimental error was large and non-repeatable. Li [25] produced an adjustable pitch columnar array of silicon plates to study the impact of water droplets on silicon plates with regular roughness from an energetic perspective.

The slurry droplet collision process is complicated and affected by both droplet and impact properties [26–29]. In addition, because the sprayed surface is mostly random and rough in the construction process, if the surface is assumed to be an ideal plane and a regular rough surface, the results will be very inconsistent. In order to solve the above-mentioned problems, this paper first establishes a mechanical model of a deviscous sliding wall based on classical mechanics and functional relationships and establishes the connection between the spreading coefficient, wall impact, and drop shape. We model the morphological parameters of a test specimen and simulate slurry-wall impacts by an orthogonal test method. We then discuss the relationship between the spreading factor and its various influences and, finally, verify the applicability and accuracy of the spreading coefficient model via physical experiments.

2. Theoretical Model of Slurry Droplet-Wall Impacts

2.1. Main Form and Classification of the Model. Scholars from various countries have categorized the droplet collision process in order to define different droplet impact models. For example, Bai and Gosman subdivided the collision process into adherence, rebound, spreading, boiling and rupture, rebound with crushing, crushing, and splashing, according to the shapes of droplets under different conditions [30, 31]. As some of these processes are similar and interchangeable, they are further classified on the basis of predecessors to the collision process. Liu and Xu proposed that the processes are analogous to a spring-mass system; the process of droplet-wall impact can be divided into three stages: rebound, crushing, and adhesion [32]. Stanton assumed there to be a liquid film on the wall and considered only four impact modes: adhesion, spreading, rebound, and splashing [33]. In this paper, it is the standard for whether rebound of slurry droplets striking a wall during the spraying of shotcrete, and droplet-wall impacts are divided into three categories according to the tendency of droplets to move against the wall. As shown in Figure 2, these categories are as follows: splash rebound, adhesion to the wall, and sliding/loss of adhesion. Among them, adhesion to the wall is the one we hope for in shotcrete.

The rebound of cement paste after striking a wall generally has two components. Firstly, the slurry loss caused by crushing, spattering, and rebound of slurry droplets is defined as *splash rebound* and is generally influenced by collision velocity, inertial force, or the absolute dominant influence of gas vorticity [34, 35]. Secondly, slurry loss due to slurry not fully spreading on a wall is defined as sliding/loss of adhesion.

Because slurry does not get fully spread during low-velocity wall collision, the smaller the relative contact area is, the smaller the adhesion force is. The slurry seat is generally spherical, and the center of gravity is away from the wall surface, resulting in debonding and sliding of the slurry. In order to solve the above problem, a spreading coefficient D′ (the ratio of the spread diameter to the initial diameter at equilibrium) is defined to represent the droplet's form and the effect of wall adherence after slurry impact. The following will be based on the classical mechanics: study the effect of spreading coefficient on the rebound of the slurry wall and find the factors that affect the spreading coefficient.

2.2. Analysis of the Main Factors Affecting the Spreading Coefficient. According to the actual situation in the experiment, the shape of slurry droplets sprayed onto a rock wall is approximately hemispherical (Figure 3(a)). The geometric model of droplet adhesion after impact is shown in Figure 3(b). The height of the spherical crown is defined as H and the radius is R.

When the slurry is sprayed at a certain velocity, the slurry droplets adhere to the wall surface as spherical crowns. From the mechanical point of view, when the spreading coefficient is larger, the moment generated by the force of gravity and fluid tangential friction force is less likely to lower adhesion, and the droplets firmly adhere to the rock wall surface. The attachment surface is circular. When the slurry spreading coefficient reaches a critical value, the upper part of the slurry (MM′, Figure 3) first breaks away from the wall surface, the slurry rolls along the wall surface, and the lower part (NN′, Figure 3) readheres. However, due to the resultant forces at the new position, the droplet still breaks adhesion and the slurry continues to roll off until it comes off the wall.

The descent distance of the first arrival of the critical spread coefficient is ds, and the descent area is dA (the crescent portion in Figure 3). Then, the work of gravity in this process, represented by $p \cdot ds$, overcomes the stickiness. The work done by the attaching force is $W_a \cdot dA$, and the work done by viscous friction is $\mu \cdot dv/ds \cdot da$. The free enthalpy lost when a new adhesive interface is generated is negligibly small, and the work balance relationship is

$$P \cdot ds = W_a \cdot dA + 2C_k \cdot \mu \frac{dv}{ds} \cdot ds \cdot R. \tag{1}$$

In the formula, the correction coefficient of the friction force, which is a constant, is determined by the wall surface roughness.

The volume of the slurry is $V = 1/6 \cdot \pi \cdot H(3R^2 + H^2)$ and the spherical cap, $H = R \cdot \tan\theta$, is substituted into the above formula:

$$V = \frac{\pi R^3}{6} \tan\theta\left(3 + \tan^2\theta\right). \tag{2}$$

After reaching the debonding radius, the distance of each debonding and sliding action is S. The corresponding sliding area dA can be obtained by analytical methods. As shown in Figure 2(c), the orthogonal coordinate system is established with the O' center of the sliding center as the origin. Then, the analytical formulae for the circles O and O' are

$$\begin{aligned} x^2 + (y - s)^2 &= R^2, \\ x^2 + y^2 &= R^2. \end{aligned} \tag{3}$$

The coordinates of k at the intersection of the above formulae are $x = \sqrt{R^2 - (s^2/4)}$; $y = s/2$.

The area enclosed in the first quadrant is A_1, A_2, according to the definite integral knowledge:

$$\begin{aligned} A_1 &= \int_0^{\sqrt{R^2-(s^2/4)}} \left(\sqrt{R^2 - x^2} + s\right)dx, \\ A_2 &= \int_0^{\sqrt{R^2-(s^2/4)}} \sqrt{R^2 - x^2}\,dx. \end{aligned} \tag{4}$$

As shown in Figure 3(c), there is a relationship between the area of the shaded region and the sliding distance: when s tends to zero, the area tends to zero, so this part is negligible.

$$dA = 2(A_1 - A_2) = 2\int_0^{\sqrt{R^2-(s^2/4)}} s\,dx = 2\,ds \cdot \sqrt{R^2 - \left(\frac{ds^2}{4}\right)}. \tag{5}$$

In terms of magnitude, due to $R^2 \gg ds^2/4$, then $dA \approx 2ds \cdot R$.

So, the work balance equation can be written as

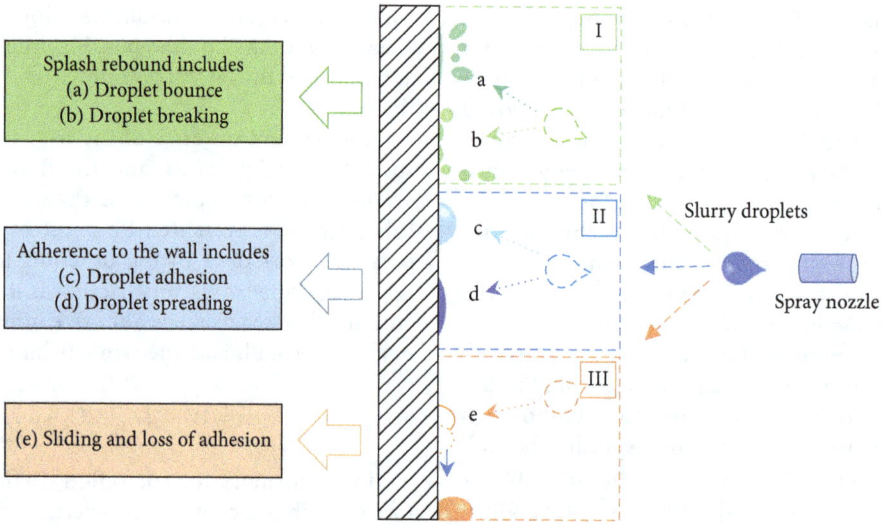

FIGURE 2: Classification of slurry droplet impacts on a wall: (I) splash rebound, (II) adhesion to the wall, and (III) sliding and loss of adhesion.

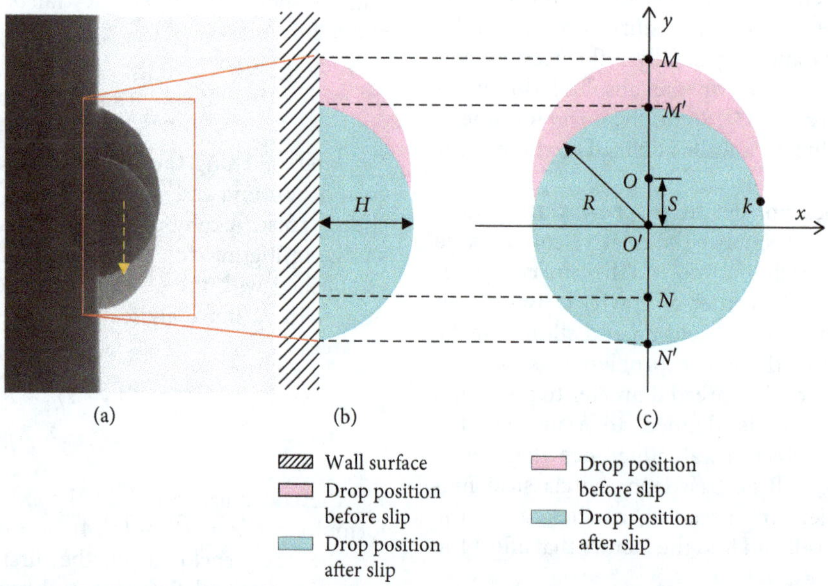

FIGURE 3: Droplet on a wall sliding/losing adhesion. (a) Real slurry sliding away from the wall and losing adhesion. (b) Slide away from wall and lose adhesion model. (c) Analytical geometry of the model.

$$\frac{\pi\varphi g R^3}{6}\tan\theta\left(3+\tan^2\theta\right)\cdot ds = e_{lv}(1+\cos\theta)\cdot 2\,ds \cdot R + 2C_k\cdot\mu\frac{dv}{ds}\cdot ds\cdot R. \quad (6)$$

The slurry droplet radius is

$$R = \left[\frac{12\sigma(1+\cos\theta)ds + C_k\cdot\mu\cdot dv}{\pi\varphi g\tan\theta(3+\tan^2\theta)ds}\right]^{1/2}. \quad (7)$$

Assuming that the initial diameter of the droplet is D_0, the relationship between the spreading coefficient D' and the droplet radius R is $D' = 2R/D0$.

The spreading coefficient D' is

$$D' = \frac{2}{D_0}\left[\frac{12e_{lv}(1+\cos\theta)ds + C_k\cdot\mu\cdot dv}{\pi\varphi g\tan\theta(3+\tan^2\theta)ds}\right]^{1/2}. \quad (8)$$

The above analysis shows that surface tension, contact angle, roughness, viscosity, velocity, and density all affect the spreading coefficient. There is a correspondence between the surface tension of the liquid and the contact angle, so the surface tension can be studied alone [36]. Secondly, the millimeter droplet density changes very little, so density has negligible influence on the spreading coefficient.

3. Slurry-Wall Impact Simulation and Orthogonal Test

In order to quantitatively study the effects of surface properties, impact properties, and droplet properties on slurry spreading during wall impact, COMSOL Multiphysics software was used for mathematical modeling and numerical simulation analysis. COMSOL Multiphysics is a numerical simulation software based on the finite element method. It can realize multiphysics coupling calculations by solving partial differential equations, which reflect the most basic scientific principles and their two-phase flow level. The set physics interface simulates the motion of a droplet impacting a solid surface.

3.1. Selection of Control Equations. The model simulates fluid mass and momentum transfer based on the Navier–Stokes equation for incompressible fluids. The effects of surface tension must also be included in the model. Therefore, the Navier–Stokes equation is

$$\rho\left(\frac{\partial u}{\partial t}+u\cdot\nabla u\right)=-\nabla p+\nabla\cdot\mu\left(\nabla u+\nabla u^{T}\right)+\rho g+F_{\mathrm{st}},$$
$$\nabla\cdot u=0. \tag{9}$$

In the formula, u is the velocity field (m/s); ρ is density (kg/m^3); p is pressure (Pa); μ is viscosity (mPa.s); ρg is gravity; and F_{st} is surface tension.

3.2. Establishment of Geometric Models and Selection of Boundary Conditions. In order to facilitate numerical simulation, the dynamic process of a single slurry droplet impacting a surface with random roughness at low velocity can be simplified as follows. A slurry droplet in the atmosphere with a certain initial velocity is located directly above the rough surface. Wall roughness is modeled according to the actual topographic parameters of the experimental substrate. The numerical simulation uses a 30 mm × 50 mm calculation area. The entire physical model is shown in Figure 4.

At the same time, due to the complex motion of droplets in the computational domain, some reasonable assumptions have been made to simplify the calculations. (1) The fluid is incompressible during movement; (2) the surface tension of the fluid is constant during the collision process; (3) there is no change in the contact angle within a surface of consistent roughness; and (4) there is no phase change at the gas-liquid interface, nor any fluid heat transfer.

The open boundary condition of the upper boundary of the model can be used to achieve the free entry and exit of air. The left and right boundaries of the model are pressure exit boundaries, and the air velocity is 0. The lower boundary of the model is the boundary condition of the wetting wall. At the same time, during the numerical simulation calculations, it was found that adding the initial velocity directly did not converge the calculation result. The reason was that there was a sudden change in the value. By defining the volume force, the collision velocity can reach the set value, and the above calculation is not solved [37]. There is a problem of convergence, expressed as

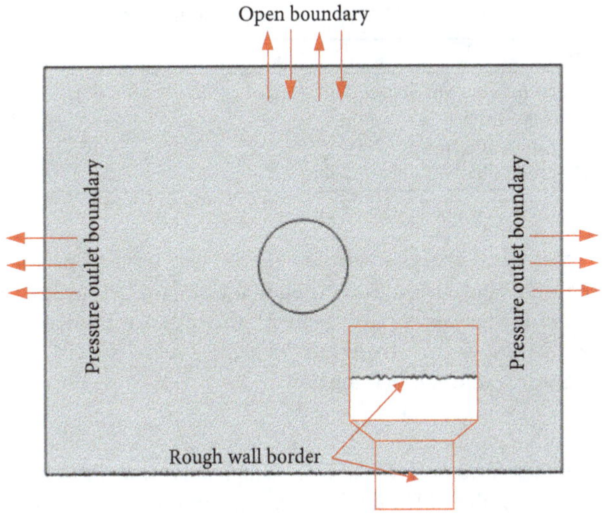

FIGURE 4: Physical model of droplet-wall interaction.

$$F_{v}=-tpf1.rho2*\left(v\_in^{2}/2/h\_in-g\_const\right)*tpf1.Vf2*(t<=t\_in). \tag{10}$$

The parameters in the formula are software-preset variables, where $tpf1.rho2$ refers to droplet density; v_in is the velocity; h_in is the drop height; g_const is gravitational acceleration; $tpf1.Vf2$ is the droplet volume fraction; t_in is the drop time; and $t<=t\_in$ means only the volume force exists in the falling process. After the droplet hits the wall, the volume force disappears.

3.3. Orthogonal Test. Several factors influence the spreading coefficient described in Equation (10). In order to further analyze their influence, the numerical simulation method described in Section 3.2 is used. The surface tension, viscosity, wall roughness, and velocity of impact are orthogonal experimental factors. The spreading coefficient after the liquid droplet impacts the wall is an orthogonal test index, and the L_9 (3^4) orthogonal table is used to perform four factors and three levels of orthogonality. Tests and orthogonal test factor levels are shown in Table 1.

Orthogonal tests and numerical simulation results are shown in Table 2 and Figure 5.

Analysis of Orthogonal test results by range analysis. Range analysis is a visual analysis method that assesses the extent of the effect by analyzing the range of range R. If X_{ij} represents i levels of j factors and Y_{ij} represents the result of X_{ij}, then the range R is calculated as follows [38]:

$$k_{ij}=\sum_{k=1}^{s}\frac{Y_{ijk}}{S}\quad(i=1,2,3,\ldots,\ j=A,B,C,\ldots),$$
$$R_{j}=\max\left\{k_{1j},k_{2j},\ldots,k_{rj}\right\}-\min\left\{k_{1j},k_{2j},\ldots,k_{rj}\right\}. \tag{11}$$

Here, R_j represents the range of j factors. The increase of R_j indicates that the influence of j factors is more important. Variable s is the number of values, and r is the level of factors.

Table 1: Orthogonal design parameters.

Factor	Level
Surface tension (mN/m)	6.3, 8, 10.5
Viscosity (mPa.s)	53, 97, 135
Roughness (Ra, μm)	0.5, 1.5, 2.5
Velocity (m/s)	1, 2, 4

Statistical analysis software (SPSS) was used to perform the range analysis on the orthogonal test data. According to the range analysis results (Table 3), the order of influence of the factors, from strong to weak, is velocity, viscosity, roughness, and surface tension.

3.3.1. Variation of Spreading Coefficient and Influencing Factors. Using SPSS, the least significant difference method (LSD) was used to conduct post hoc multiple comparison analysis to obtain the single factor marginal mean value of the estimated spread coefficient, as shown in Figure 6. Postmortem multiple analysis is a common method for obtaining information from data; its basic principle is to use a critical difference to test each pair of means. If the difference between a pair of means is greater than the critical difference, they are considered to be significantly different. The LSD method is a multiple comparison method proposed by Fisher in 1935. Using *t*-tests to conduct pairwise comparison between the groups has the advantage of high sensitivity of detection, such that small differences between means may also be tested [39–41].

From Figure 6, we can see the following relationships:

(1) The spreading coefficient increases first and then decreases with surface tension. There is a sudden change in the value. There are two stages of spreading and retraction in the stage of droplet impact. The surface tension provides resistance during the spreading stage and provides force during the retraction stage. The surface tension of 6.3–8 mN/m is relatively small, and the resistance during spreading is small, leading to a gradual increase in the spreading coefficient. When the surface coefficient is gradually increased to about 10.5 mN/m, the spreading resistance is large and the droplets are difficult to spread. The greater the power of the retraction phase, the smaller the spreading coefficient.

(2) The spreading coefficient is negatively correlated with viscosity and wall roughness. That is, the spreading coefficient decreases with increasing viscosity and wall roughness. This is because viscosity and wall roughness provide resistance over the entire process of droplet-surface collision.

(3) The spreading coefficient increases with velocity. The literature [37] also reports similar conclusions from experiments of collisions between Newtonian and non-Newtonian fluids at room temperature. The inertial force caused by velocity is significantly larger than the other influences by an order of magnitude.

(4) It can be seen from the trendline of the graph that within the range studied, the spreading coefficient has a tendency to increase with velocity and decrease with increasing surface tension, viscosity, and surface roughness.

3.3.2. Expression of Factors Influencing the Spreading Coefficient. According to Figure 6, a single factor is used to estimate the marginal mean value of the spreading coefficient. SPSS was used to carry out a one-dimensional regression analysis, and mathematical equations of the spreading factor and individual influencing factors were derived to obtain Equations (12) to (15):

$$D' = -0.0255\sigma^2 + 0.4116\sigma + 0.6788, \quad (12)$$

$$D' = -0.0069\mu + 2.9294, \quad (13)$$

$$D' = -0.095F_{\text{Ra}} + 2.4192, \quad (14)$$

$$D' = -0.0517V^2 + 0.765V + 0.8267. \quad (15)$$

In the formulae, σ is surface tension (mN/m); μ is viscosity (mPa.s); F_{Ra} is wall surface roughness (μm); and V is velocity (m/s).

From Equations (12) to (15), the spreading factor can be expressed as a linear function of viscosity and roughness and a quadratic function of surface tension and velocity. Therefore, a multivariate regression analysis was conducted and Equation (16) was used to express the relationship between the spreading factor and various influencing factors.

$$D' = k_{11}V^2 + k_{12}V + k_{21}\sigma^2 + k_{22}\sigma + k_{31}\mu + k_{41}F_{\text{Ra}} + k_{51}, \quad (16)$$

where k_{ij} is a constant.

Matlab software was used to make Equation (16) fit the orthogonal experimental results of Table 2 to obtain Equation (17):

$$\begin{aligned} D' &= -0.04833V^2 + 0.74833V - 0.02017\sigma^2 + 0.31978\sigma \\ &\quad - 0.00663\mu - 0.08167F_{\text{Ra}} + 0.3937, \\ R^2 &= 0.99139. \end{aligned} \quad (17)$$

In order to verify the accuracy of the above numerical simulation and the rationality of the influencing factors on the spreading factor, a different combination with orthogonal testing was selected, and the obtained function was verified by a physical experiment.

4. Droplet Impact Physical Experiment

4.1. Experiment Preparation

4.1.1. Experimental Substrate. Experimental substrate was taken from the Tangkou Coal Mine of the Zibo Mining

Table 2: Orthogonal tables and range analysis results.

Experiment no.	Surface tension (mN/m)	Viscosity (mPa.s)	Roughness (Ra, μm)	Velocity (m/s)	Spreading coefficient
1	6.3	53	0.5	1	1.89
2	6.3	97	1.5	2	2.2
3	6.3	135	2.5	4	2.7
4	8	53	1.5	4	3.4
5	8	97	2.5	1	1.55
6	8	135	0.5	2	2
7	10.5	53	2.5	2	2.25
8	10.5	97	0.5	4	3.1
9	10.5	135	1.5	1	1.2

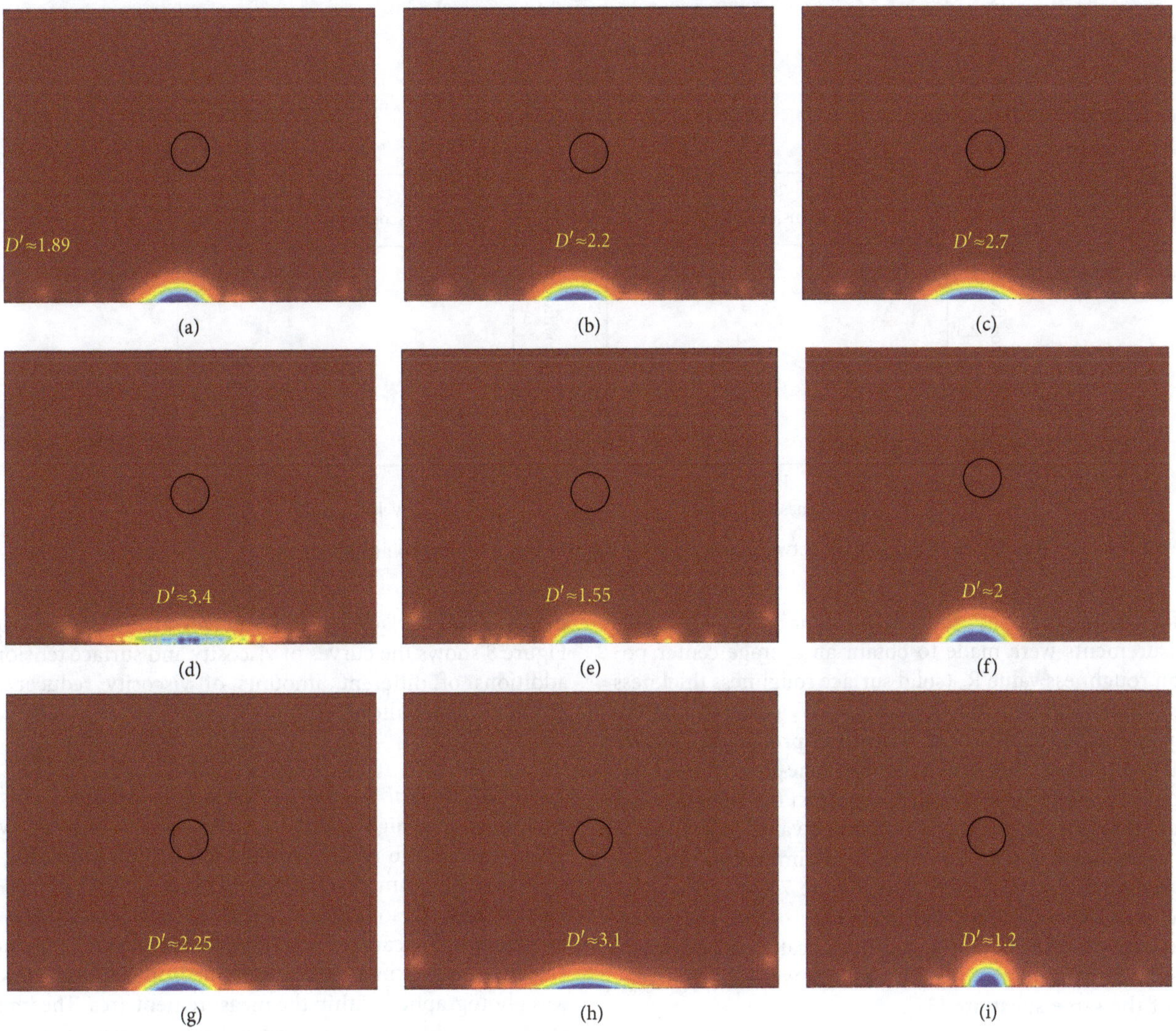

Figure 5: Orthogonal experimental simulation results.

Group. The specific location was a rock formation in the 1–2 track alley where a wet spray operation was being performed on the 630 track main roadway. In order to ensure the consistency of the raw material's composition, the experimental substrates were all taken from the same stone. First, the stones were cut into 50 mm × 50 mm × 10 mm cubes, then 500–7000 CW sandpaper was used to Polish them on a p-2 metallographic polishing machine. Afterwards, they were washed with detergent and pure water. Finally, they were put in a 99.5% mass fraction ethanol solution and subjected to ultrasonic vibration for 1 h.

Before the experiment, the surfaces of the substrates were blown dry with nitrogen, and the morphology of the substrate was measured using a Leica DCM8 (Leica DCM, Germany), as shown in Figure 7(a). A surface roughness measuring instrument (TR200, China) was used to measure

Table 3: Range analysis results.

	Surface tension (mN/m)	Viscosity (mPa.s)	Roughness (Ra, μm)	Velocity (m/s)
K1	2.263	2.513	2.330	1.547
K2	2.317	2.283	2.267	2.150
K3	2.183	1.967	2.167	3.067
R	0.134	0.546	0.163	1.520
PSF	$V > \mu > Ra > \sigma$			

Note. PSF = primary and secondary factors; K is the sum of the test results corresponding to the horizontal number i on any column; $R = \max\{K1, K2, K3\} - \min\{K1, K2, K3\}$.

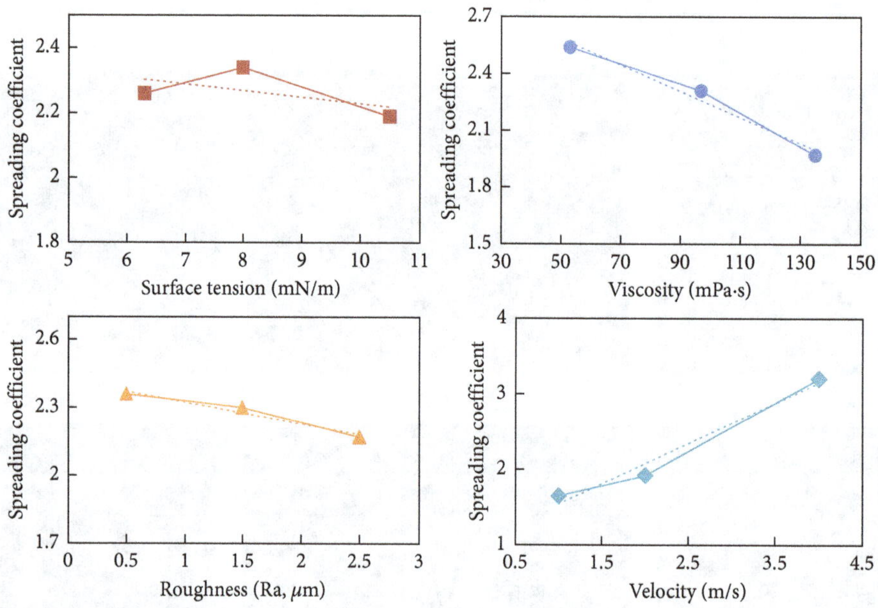

Figure 6: Relationships between marginal means and factors influencing the spreading coefficient.

surface roughness, as shown in Figure 7(b). Multiple measurements were made to obtain an average center position roughness value R_a (solid surface roughness thickness RMS value) and the roughness value W_r (the ratio of the actual contact length to the horizontal projection length). Matlab programming based on two one-dimensional topological parameters was used to construct two-dimensional parametric lines to provide a roughness value based on the real specimen's surface for numerical simulation. The reconstruction results are shown in Figure 7(c). At the same time, in order to prevent the occurrence of two roughness obscurations in the course of the numerical simulations and experiments, uniform roughness values were used to represent the same substrate [37].

4.1.2. Experimental Materials. We used standard PO42.5 cement (China United Cement Corporation) with a water to cement ratio of 1 : 2. Table 4 shows the basic properties of the cement slurry. The composite modified CR-I viscosity reducer significantly improved the viscosity of the slurry [42], using a concrete rotary rheometer (eBT-2, Germany) to measure viscosity changes at different levels. Modified lignosulfonate can vary the surface tension of the slurry without affecting its other properties greatly [43, 44], and a surface tension meter (KRUSS-K12, Germany) was used to measure the surface tension changes at different dosages. Figure 8 shows the curves of viscosity and surface tension vs additions of different amounts of viscosity reducer and modified lignosulfonate.

4.2. Experimental System. As shown in Figure 9, in the experiment, a high-speed video camera (Phantom VRI, USA) was used to record droplets falling and impacting the substrate. The camera used a frame rate of 1000 frames per second and a resolution of 1024 × 512 pixels. During the experiment, the camera was tilted and leveled. The included angle was approximately 30°, and the droplet impact process was photographed within the measurement area. The images were transmitted to a computer through a data acquisition system. A single drop generator was connected to a syringe using a syringe pump (top-5300, Japan) with a flow rate of 33.5 $ml \cdot h^{-1}$. The diameter of the droplets produced was related to the outer diameter of the injection port of the syringe, which was 2 mm. The theoretical initial diameter of the droplet was approximately 3.7 mm. The theoretical diameter was measured by weighing 200 drops and calculating the mean diameter according to mass/volume relationships. The mean droplet diameter was calculated as 4 mm, which was used as the initial diameter of the droplets used in the experiments. The mean velocity of a droplet just before it

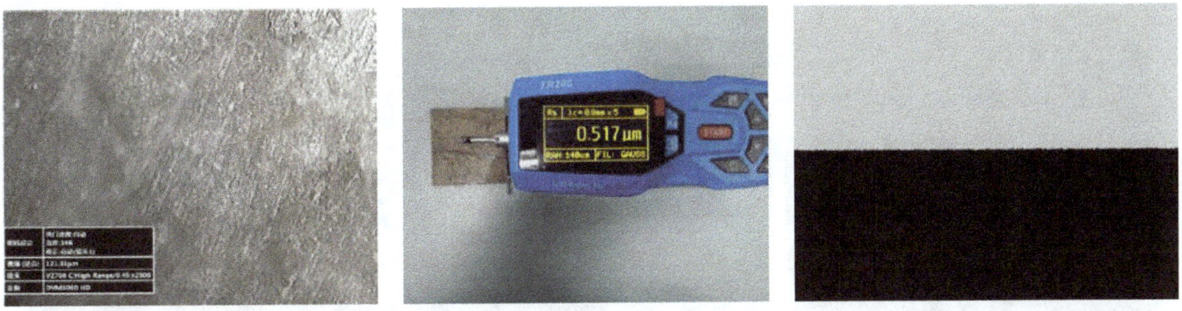

(a) (b) (c)

FIGURE 7: Experimental substrate morphology measurement and reconstruction. (a) Substrate topography measurement. (b) Substrate roughness measurement. (c) Substrate topography reconstruction results.

TABLE 4: Basic properties of the grout used.

W/C (w%)	Viscosity (mPa.s)	Surface tension (mN/m)	Density (g/cm^3)	Gelation time	
				Initial setting	Final setting
1 : 2	135	10.5	1.82	7 h 40 min	11 h 32 min

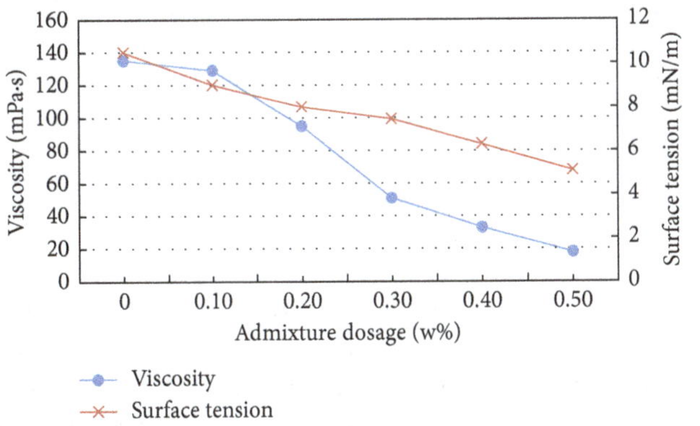

FIGURE 8: Viscosity and surface tension curves with different amounts of viscosity reducer and modified lignosulfonate.

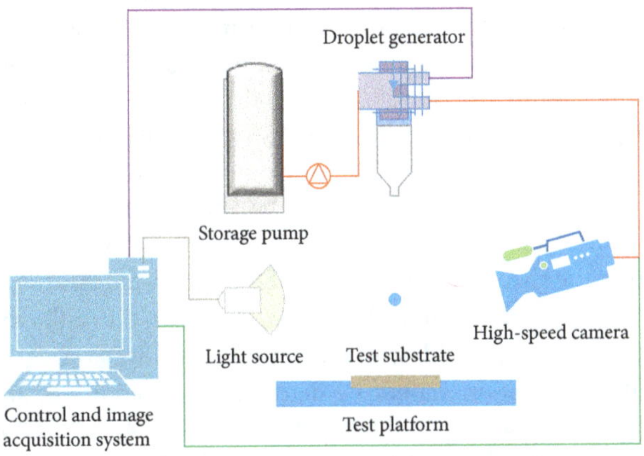

FIGURE 9: Schematic diagram of the single-drop substrate impact test system.

struck the substrate was 0.2 m/s. The impact velocity was controlled with a range of 1–4 m/s with an error of ±0.02 m/s.

4.3. Experimental Results. To verify Equation (17), according to the combinations in orthogonal Table 2, various combinations of values of surface tension, viscosity, roughness, velocity, and other factors were selected for the droplet-substrate impact tests. The tests and various spread coefficient sizes are shown in Figure 10. The spreading coefficient of each group was obtained, and the results were compared with the spreading coefficient model shown in Table 5.

Table 5 shows that the mean relative error between the experimental and theoretical data was 3.65%, with a range of 0.3%–6.3%. It shows that the influence factor expression of spreading coefficients can better quantify the effect of viscosity, surface tension, roughness, and velocity on the spreading coefficient.

5. Conclusions

(1) By means of mechanical analysis, spreading coefficients were used to characterize the effect of slurry droplet-wall adhesion, allowing the derivation of a formula for spreading coefficient based on the influential factors. The key factors influencing the impact and adhesion of slurry on a wall are droplet velocity and viscosity, wall roughness, and surface tension.

(2) Through numerical simulation and based on the results of an orthogonal experiment, a model of the spreading coefficient was derived. It quantitatively characterizes the relationship between the spreading coefficient and its influencing factors, thereby achieving the transition from qualitative theoretical research to quantitative experimental research. From the expression's constants for the influencing factors, it can be seen that velocity (constant = 0.74833) has a significant effect on the spreading coefficient. Since

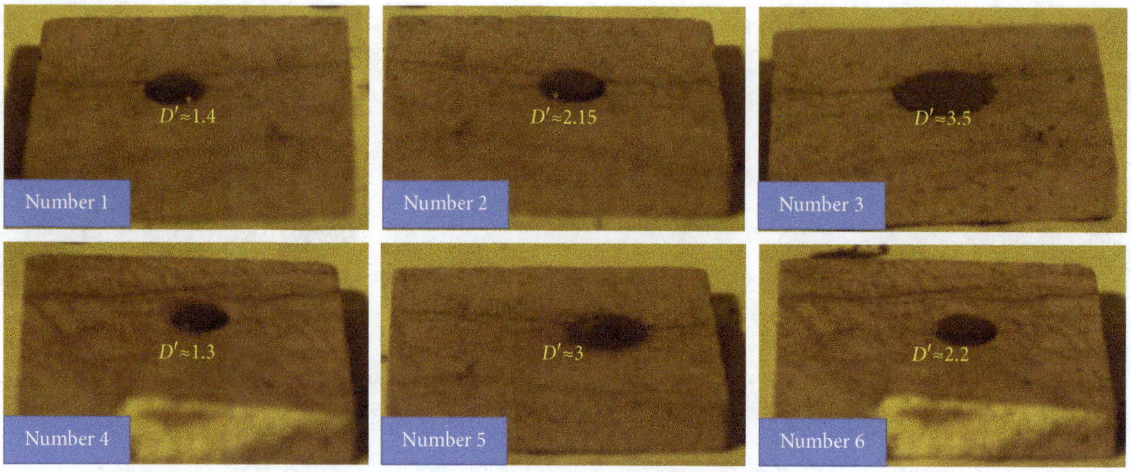

FIGURE 10: Droplet spreading after striking the substrate of the test apparatus.

TABLE 5: Experimental verification of spreading coefficients.

Test no.	Surface tension (mN/m)	Viscosity (mPa.s)	Roughness (μm)	Velocity (m/s)	Spreading coefficient		RE (%)
					EV	TV	
1	6.3	135	0.5	1	1.40	1.37	2%
2	6.3	97	2.5	2	2.15	2.06	4%
3	8	53	0.5	4	3.50	3.49	0.3%
4	8	135	1.5	1	1.30	1.34	3.4%
5	10.5	53	2.5	4	3.02	3.19	6.3%
6	10.5	97	1.5	2	2.21	2.07	5.9%

EV = experimental value; TV = theoretical value; RE = relative error.

the droplet size (5 mm) was much larger than the range of the roughness (0.5–2.5 μm), at the low velocity used in this study, roughness was also an important determinant of the spreading coefficient.

(3) Through physical droplet-wall impact tests, a numerical simulation of the factors affecting the spreading coefficient was validated. The experimental and theoretical values were in good agreement, and a further explanation was given based on the actual morphological parameters. A random wall roughness model can best represent the rough surface morphology of real walls.

At present, experimental research into the rebound of wet sprayed concrete slurry is relatively rare and its mechanisms are poorly understood. Therefore, the rebound mechanism requires further extensive theoretical and experimental research.

6. Discussion

This paper combines the stage division of shotcrete jets used in engineering practice and, from a phenomenological perspective, creatively proposes the problem of slurry rebound. Furthermore, based on the fact that slurry acts as a binder in shotcrete, it is recognized that the rebound loss of slurry plays an important role in shotcrete's overall rebound characteristics. This paper focuses on the sliding and loss-of-adhesion mechanisms of slurry rebound. The critical conditions of detachment adhesion are applied not only for vertical walls but also for horizontal top walls, which provides a theory for describing slurry rebound. At the same time, a linear relationship between slurry impact and spreading on a rough wall is obtained. This provides information for the optimization of shotcrete parameters and mixtures. It is conceivable that an admixture could be created that changes the viscosity and surface tension of shotcrete to reduce rebound.

As for future work to be done for this study, other types of slurry rebound losses should be studied first. Secondly, there are many factors influencing the behavior of slurry. From the rheological properties of the slurry, the yield stress is introduced as an input parameter, and the difference between using yield stress and interface adhesion as the critical condition should be determined. In addition, after considering all the influencing parameters, an admixture should be developed that results in significantly less rebound than existing shotcrete mixtures.

In summary, experimental research into the rebound of wet sprayed concrete slurry is relatively rare and its mechanisms are poorly understood. Therefore, the rebound mechanism requires extensive theoretical and experimental research.

Conflicts of Interest

The authors declare that they have no conflicts of interest.

Acknowledgments

This work was supported by the National Key Research and Development Plan of the 13th Five-Year Period (2017YFC0805203), the National Natural Science Foundation of China (51604163), and the Applied Research Project Foundation of Qingdao Postdoctoral Researchers (201576).

References

[1] GB 50086-2015, *Technical Code for Engineering of Ground Anchorages and Shotcrete Support*, Chinese Standard Press, Beijing, china, 2015.

[2] H. S. Armelin, N. Banthia, D. R. Morgan et al., "Rebound in dry-mix shotcrete," *Concrete International*, vol. 19, 1997.

[3] J. Kaufmann, K. Frech, P. Schuetz et al., "Rebound and orientation of fibers in wet sprayed concrete applications," *Construction and Building Materials*, vol. 49, no. 6, pp. 15–22, 2013.

[4] H. S. Armelin, *Rebound and toughening mechanisms in steel fiber reinforced dry-mix shotcrete*, The University of Columbia, New York, NY, United States, 1997.

[5] J. P. Won, U. J. Hwang, C. K. Kim et al., "Mechanical performance of shotcrete made with a high-strength cement-based mineral accelerator," *Construction and Building Materials*, vol. 49, no. 12, pp. 175–183, 2013.

[6] K. K. Yun, S. Y. Choi, and J. H. Yeon, "Effects of admixtures on the rheological properties of high-performance wet-mix shotcrete mixtures," *Construction and Building Materials*, vol. 78, pp. 194–202, 2015.

[7] J. Pickelmann and J. Plank, "A mechanistic study explaining the synergistic viscosity increase obtained from polyethylene oxide (PEO) and β-naphthalene sulfonate (BNS) in shotcrete," *Cement and Concrete Research*, vol. 42, no. 11, pp. 1409–1416, 2012.

[8] P. Choi, J. H. Yeon, and K. K. Yun, "Air-void structure, strength, and permeability of wet-mix shotcrete before and after shotcreting operation: the influences of silica fume and air-entraining agent," *Cement and Concrete Composites*, vol. 70, pp. 69–77, 2016.

[9] K. K. Yun, P. Choi, and J. H. Yeon, "Correlating rheological properties to the pumpability and shootability of wet-mix shotcrete mixtures," *Construction and Building Materials*, vol. 98, pp. 884–891, 2015.

[10] P. Choi, K. K. Yun, and J. H. Yeon, "Effects of mineral admixtures and steel fiber on rheology, strength, and chloride ion penetration resistance characteristics of wet-mix shotcrete mixtures containing crushed aggregates," *Construction and Building Materials*, vol. 142, pp. 376–384, 2017.

[11] P. Choi, K. K. Yun, and J. H. Yeon, "Rheological properties of wet-mix shotcrete mixtures made with crushed aggregate," *Journal of Materials in Civil Engineering*, vol. 29, no. 11, article 04017227, 2017.

[12] R. G. Pileggi, Y. A. Marques, D. V. Filho et al., "Aditivos para concretos de projeção WET-Shotcrete Additives," *Cerâmica*, vol. 48, no. 308, pp. 199–205, 2002.

[13] J.-D. Lemay, M. Jolin, and R. Gagné, "Ultra rapid strength development in dry-mix shotcrete for ultra rapid support in challenging mining conditions. Deep mining 2014," in *Proceedings of 7th International Conference on Deep and High Stress Mining*, Australian Centre for Geomechanics, Nedlands, Australia, September 2014.

[14] N. Ginouse, M. Jolin, and B. Bissonnette, "Effect of equipment on spray velocity distribution in shotcrete applications," *Construction and Building Materials*, vol. 70, no. 70, pp. 362–369, 2014.

[15] D. Beaupre and M. Jolin, "Effects of particle-size distribution in dry process shotcrete," *Aci Materials Journal*, vol. 101, no. 2, pp. 131–135, 2004.

[16] H. S. Armelin and N. Banthia, "Mechanics of aggregate rebound in shotcrete—(Part I)," *Materials and Structures*, vol. 31, no. 2, pp. 91–98, 1998.

[17] H. S. Armelin and N. Banthia, "Development of a general model of aggregate rebound for dry-mix shotcrete—(Part II)," *Materials and Structures*, vol. 31, no. 3, pp. 195–202, 1998.

[18] H. S. Armelin, N. Banthia, and D. R. Morgan, "Particle kinematics in dry-mix shotcrete-research in progress," *Sprayed Concrete Technology*, vol. 27, p. 243, 1996.

[19] A. M. Worthington III, "A second paper on the forms assumed by drops of liquids falling vertically on a horizontal plate," *Proceedings of the Royal Society of London*, vol. 25, no. 171–178, pp. 498–503, 1877.

[20] C. San Marchi et al., "Numerical analysis of the deformation and solidification of a single droplet impinging onto a flat substrate," *Journal of Materials Science*, vol. 28, no. 12, pp. 3313–3321, 1993.

[21] X. Zhang and A. B. Osman, "Dynamic surface tension effects in impact of a drop with a solid surface," *Journal of Colloid and Interface Science*, vol. 187, no. 1, pp. 166–178, 1997.

[22] H. Park, W.W. Carr, J. Zhu, and J. F. Morris, "Single drop impaction on a solid surface," *AIChE journal*, vol. 49, no. 10, pp. 2461–2471, 2003.

[23] C. D. Stow and M. G. Hadfield, "An experimental investigation of fluid flow resulting from the impact of a water drop with an unyielding dry surface," *Proceedings of the Royal Society A: Mathematical, Physical and Engineering Sciences*, vol. 373, no. 1755, pp. 419–441, 1981.

[24] R. Rioboo, T. Cameron, and M. Marengo, "Outcomes from a drop impact on solid surfaces," *Atomization and Sprays*, vol. 11, no. 2, 2001.

[25] X. Li, L. Mao, and X. Ma, "Dynamic behavior of water droplet impact on microtextured surfaces: the effect of geometrical parameters on anisotropic wetting and the maximum spreading diameter," *Langmuir*, vol. 29, no. 4, pp. 1129–1138, 2013.

[26] B. A. Dwiyantoro, "Capillary effects during droplet formation on the solid surface," *Advanced Materials Research*, vol. 683, pp. 889–893, 2013.

[27] K. L. Pan, K. C. Tseng, and C. H. Wang, "Breakup of a droplet at high velocity impacting a solid surface," *Experiments in Fluids*, vol. 48, no. 1, pp. 143–156, 2010.

[28] B. L. Scheller and D. W. Bousfield, "Newtonian drop impact with a solid surface," *AIChE Journal*, vol. 41, no. 6, pp. 1357–1367, 2010.

[29] R. Crooks, J. Cooper-White, and D. V. Boger, "The role of dynamic surface tension and elasticity on the dynamics of drop impact," *Chemical Engineering Science*, vol. 56, no. 19, pp. 5575–5592, 2001.

[30] C. Bai and A. D. Gosman, "Development of methodology for spray impingement simulation," SAE Technical Paper, Report no. 950283, 1995.

[31] C. X. Bai, H. Rusche, and A. D. Gosman, "Modeling of gasoline spray impingement," *Atomization and Sprays*, vol. 12, no. 1–3, pp. 1–28, 2002.

[32] H. Xu, Y. Liu, P. He et al., "The TAR model for calculation of droplet/wall impingement," *Journal of Fluids Engineering*, vol. 120, no. 3, pp. 593–597, 1998.
[33] D. W. Stanton and C. J. Rutland, "Modeling fuel film formation and wall interaction in diesel engines," SAE Technical Paper, 1996.
[34] C. Mundo, M. Sommerfeld, and C. Tropea, "Droplet-wall collisions: experimental studies of the deformation and breakup process," *International Journal of Multiphase Flow*, vol. 21, no. 2, pp. 151–173, 1995.
[35] C. Mundo, C. Tropea, and M. Sommerfeld, "Numerical and experimental investigation of spray characteristics in the vicinity of a rigid wall," *Experimental Thermal and Fluid Science*, vol. 15, no. 1, pp. 80–86, 1980.
[36] H. Y. Erbil, *Surface Chemistry: Of Solid and Liquid Interfaces, Surface chemistry of solid and liquid interfaces*, Wiley-Blackwell, Hoboken, NJ, USA, 2006.
[37] H. Zhang, "Study on spreading dynamics of droplets on random rough surface," MS thesis, Suzhou University, Jiangsu, China, 2015.
[38] X. M. Zhao, *Experimental Design Method*, Science Press, Beijing, China, 2006.
[39] G. Keppel, *Design and Analysis: A Researcher's Handbook*, Prentice-Hall, Upper Saddle River, NJ, USA, 1973.
[40] G. A. Milliken and D. E. Johnson, *Analysis of Messy Data, Analysis of Messy Data*, pp. 1206-1207, Chapman and Hall, CRC, CA, USA, 1992.
[41] U. Meier, "A note on the power of Fisher's least significant difference procedure," *Pharmaceutical Statistics*, vol. 5, no. 4, pp. 253–263, 2006.
[42] J. W. Geng, L. H. Li, and B. Tian, "Study on effect of viscosity reducer on rheological behavior and pumpability of concrete mixture," *journal of Highway and Transportation Research and Development*, vol. 8, 2017.
[43] J. F. Li and X. H. Huang, "The dispersion of lignin water reducing agent on cement raw slurry," *Journal of Guizhou University of Technology: Natural Science Edition*, vol. 30, no. 3, pp. 79–81, 2001.
[44] J. F. Li, "Research on water-reducing mechanism of modified lignosulfonate water-reducing agent for wet cement raw slurry," MS thesis, Guizhou University of Technology, Guiyang, China, 2001.

15

Flexural Behavior of RC Slabs Strengthened in Flexure with Basalt Fabric-Reinforced Cementitious Matrix

Sugyu Lee,[1] Kinam Hong,[1] Yeongmo Yeon,[1] and Kyusan Jung[2]

[1]*Department of Civil Engineering, Chungbuk National University, 1 Chungdae-ro, Seowon-Gu, Cheongju, Chungbuk 28644, Republic of Korea*
[2]*Structural Engineering Research Institute, Korea Institute of Civil Engineering and Building Technology, 283 Daehwa-dong, Goyangdae-ro, Ilsanseo-gu, Goyang-si, Gyeonggi-do 10223, Republic of Korea*

Correspondence should be addressed to Kinam Hong; hong@cbnu.ac.kr

Academic Editor: Doo-Yeol Yoo

This paper presents both experimental and analytical research results for predicting the flexural capacity of reinforced concrete (RC) slabs strengthened in flexure with basalt fabric-reinforced cementitious matrix (FRCM). A total of 13 specimens were fabricated to evaluate the flexural behavior of RC slabs strengthened with basalt FRCM composite and were tested under four-point loading. The fiber type, tensile reinforcement ratio, and the number of fabric layers were chosen as experimental variables. The maximum load of FRCM-strengthened specimens increased from 11.2% to 98.2% relative to the reference specimens. The energy ratio and ductility of the FRCM-strengthened specimens decreased with the higher amount of fabric and tensile reinforcement. The effective stress level of FRCM fabric can be accurately predicted by a bond strength of ACI 549 and Jung's model.

1. Introduction

Reinforced concrete structures experience degradation due to design errors, construction errors, aging, and increased loads caused by changes in usage [1]. These deteriorated structures can recover their load bearing capacity through strengthening [2, 3]. Strengthening methods are largely divided into enlargement of concrete section, externally bonded reinforcement (EBR), and near-surface mounted reinforcement (NSMR) [4]. Particularly, the EBR method is easy to apply because reinforcing materials such as steel plate and fiber are adhered to the concrete surface by an epoxy resin and also exhibit excellent strengthening effect [5]. Steel plates were used mainly as the EBR strengthening material 20 to 30 years ago. However, it had been known that steel plates attached to the outside of concrete structures are vulnerable to corrosion and fire and greatly increase the weight of the structure [6]. Thus, at present, sheets or plates made up of carbon fiber-reinforced polymer (CFRP), glass fiber-reinforced polymer (GFRP), and aramid fiber-reinforced polymer (AFRP) are used as strengthening materials in replace of steel plates. The FRP strengthening material has a high strength-to-weight ratio and has excellent corrosion resistance, chemical resistance, and endothelial strength [7, 8]. Despite these many advantages, FRP strengthening material attached to the exterior of the structure by the epoxy resin is prone to damage by fire and ultraviolet radiation [9]. Additionally, epoxy used as adhesive has the following limitations: (1) difficult to control quality in wet condition, difficult to harden at low temperatures in winter, and very low fire resistance due to the lower glass transition temperatures and (2) due to the epoxy resin matrix kept the moisture at interface between the concrete surface and the epoxy resin matrix, the strengthening material might be debonded from the concrete surface [3, 10].

Since inorganic cementitious matrix has a higher glass transition temperature than the organic epoxy resin, it is excellent in stability at high temperature, can be used under wet conditions due to hydration reaction, and also can prevent the fiber damage due to the ultraviolet ray [11]. In order to compensate the disadvantages of the epoxy resin, a fabric-reinforced cementitious matrix (FRCM) method, which is a strengthening method using the cementitious matrix as an adhesive, has been studied [4, 8–10, 12, 13]. Babaeidarabad et al. [12] conducted a bending test considering the number of polybenzoxazole (PBO) fabric layers and the concrete compressive strength as experimental variables. The experimental test results were compared with the flexural strengths predicted by ACI 549 [14]. The comparison results showed that the predicted flexural strengths were lower than the experimental values, but the criterion had an appropriate accuracy [12]. Ombres [7] performed a four-point bending test for FRCM-strengthened beams considering the tensile reinforcement ratio and the number of FRP fabric layers as experimental variables in order to analyze the flexural behavior, ultimate strength, deflection, ductility, and failure mode. Experimental test results showed that the ultimate strengths of strengthened specimens increased by 10% to 44% compared to the reference specimen. The specimen strengthened with one FRP fabric layer failed by flexural behavior, and the specimen with more than two FRP fabric layers failed due to FRP debonding at the midspan. In addition, the predicted value for maximum load is significantly different from the experimental test value when the specimen was destroyed by FRCM peeling [7]. D'Ambrisi et al. [11] conducted flexural tests on FRCM-strengthened beams considering the type of fiber, the number of fabric layers, and the fixing method as experimental variables. From the experimental results, it was confirmed that the PBO fabric is more efficient than the CFRP fabric in the FRCM method. Furthermore, the specimens strengthened with FRCM were destroyed due to the loss of reinforcing effect caused by peeling of the FCRM composite [11]. Azam and Soudki [15] conducted an experimental study to investigate the strengthening performance of beams strengthened in shear with the FRCM method. Three-point loading tests were carried out on FRCM shear-strengthened beams considering the type of fibers and reinforcement as experimental variables. Experimental results showed that the ultimate load increase of 19% to 105% from the type of fibers and strengthening type has no significant effect on failure mode and failure load of the specimen [15]. Jung et al. [16] conducted flexural tests on FRCM-reinforced beams by varying the type of fabric and the number of fabric layers. Experimental results showed that FRCM-reinforced specimens increased by 131.7% to 219.8% compared to the reference specimen and all of FRCM-strengthened specimens were destroyed by FRCM composite peeling. Additionally, Jung et al. [16] suggested a bonding strength model for considering the bond-slip behavior between the BFRP fabric and the cementitious matrix on FRCM.

PBO fabric, CFRP fabric, and GFRP fabric were used as reinforcing materials in most studies regarding the FRCM composites [17, 18]. However, due to decomposition of epoxy matrix, the tensile strength of CFRP and GFRP at 600°C decreased to less than 60% of the strength measured at room temperature, and the tensile strength of PBO fabric at 500°C decreased to less than 40% of the strength measured at room temperature [17, 19]. Therefore, even if the epoxy is replaced with a cementitious matrix, it should be considered that the sufficient fire resistance cannot be secured if these fabrics are used as strengthening materials. Unlike other fibers, basalt fiber is known to have more than 90% of the tensile strength measured at room temperature even at 600°C [3]. However, only few researchers have reported the application of the basalt fiber as a material of FRCM composites [3, 20, 21]. Therefore, this study reports the flexural performance of FRCM-strengthened beams with basalt fiber-reinforced polymer (BFRP) fabric. Moreover, the strength prediction model for FRCM-strengthened beams was examined for FRCM-strengthened slab structure.

2. Experimental Program

2.1. Test Specimens. A total of 13 specimens were fabricated to evaluate the flexural behavior of RC slabs strengthened with FRCM composite. The type of fibers, tensile reinforcement ratio, and the number of fabric layers were chosen as experimental variables. The specimens had a rectangular cross section of 400 mm × 200 mm, and their cover concrete depth was constantly 30 mm. The total length and net span of the specimen were 3,400 mm and 3,000 mm, respectively. Two deformed steel bars with nominal diameter of 10 mm were placed as compressive reinforcement in all specimens. Deformed steel bars with nominal diameter of 10 mm, 13 mm, and 16 mm were utilized as the variable of tensile reinforcement in the specimens.

Also, in order to prevent the occurrence of shear failure before flexural failure, U-shaped stirrups with nominal diameter of 10 mm were placed at 200 mm intervals on the specimens. Specimen details are presented in Figure 1.

2.2. Materials. The specimens were fabricated by using ready-mixed concrete with the design strength of 30 MPa. In order to confirm the mechanical properties of concrete, nine cylinders of ∅100 mm were manufactured in accordance with ASTM C39/39M [22] at the time of concrete casting. The cylinders were cured under the same condition as the specimens. The average compressive strength of specimens was 33.8 MPa at 28 days. The mixture properties of the concrete used are presented in Table 1. Symbols in Table 1 are water-cement ratio (W/C), sand percent of total aggregate by solid volume (S/a), water (W), cement (C), sand (S), gravel (G), and water-reducing admixture (Ad).

Deformed steel bars with nominal diameters of 10 mm, 13 mm, and 16 mm were used in specimen fabrication. Their mechanical properties were determined in accordance with ASTM A370 [23], and the test results are shown in Table 2.

BFRP and CFRP fabrics were used in this study. The used CFRP fabric in this research was manufactured in Switzerland, consisting of carbon fiber in the warp direction and glass fiber in the weft direction.

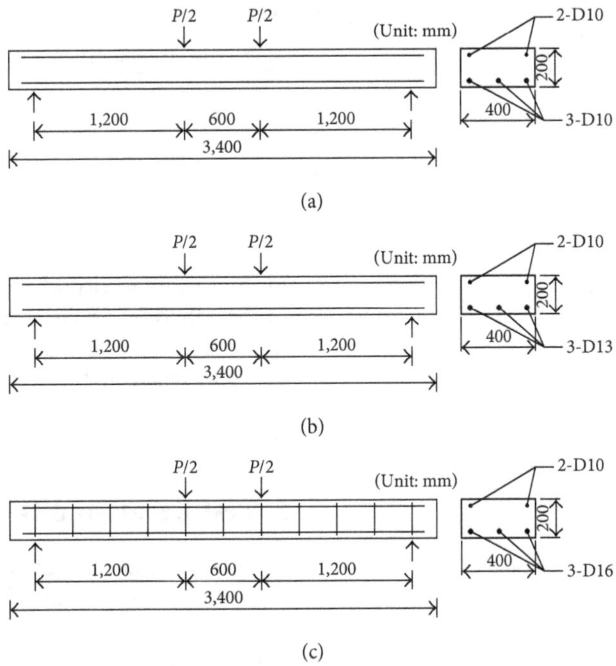

FIGURE 1: Details of test specimens. (a) R41. (b) R55. (c) R87.

TABLE 1: Mixture properties of concrete.

W/C (%)	S/a (%)	Unit weight (kg/m^3)				
		W	C	S	G	Ad
46.7	48.1	168	360	860	949	2.02

The spacing of the carbon fiber strips and the glass fiber strips was 15 mm and 30 mm, respectively. The BFRP fabric has excellent fire resistance performance compared to the CFRP fabric, while its tensile strength is 30% of CFRP strength. The fire resistance of the basalt fabric is 1.5 times greater than the fire resistance of the CFRP or GFRP fabric [24]. Thus, a direct tensile test as shown in Figure 2 was conducted as a preliminary experiment to determine the amount of BFRP fabrics having an equivalent to tensile strength of the CFRP fabric. Thus, as a preliminary test, direct tensile test as shown in Figure 2 was conducted to determine the required amount of BFRP fabrics with an equivalent to tensile strength of the CFRP fabric.

As a result, the thickness of the BFRP fabric with the same tensile strength as the CFRP fabric in the warp direction was determined to be 4.88 times the thickness of the CFRP fabric. The BFRP fabric consisted of basalt fiber in both warp and weft directions. Also, the spacing of warp and weft strips of BFRP fabric was 15 mm and 30 mm, respectively. Table 3 and Figure 3 present the mechanical properties and shapes of the fabrics used as a strengthening material in this research.

Cementitious matrix used as binder consisted of silica sand, polypropylene fiber, cement, fly ash, and polymer powder. The compression test for the cementitious matrix was performed at the curing age of 28 days in accordance with ASRM C109/C109M [25]. Its compression strength was 46.3 MPa.

2.3. Experimental Variables. The experimental variables tested were tensile reinforcement ratio (0.41%, 0.55%, and 0.87%), different strengthening materials (CFRP fabric and BFRP fabric), and different number of fabric layers (1-layer, 2-layers, and 3-layers). Test parameter details are given in Table 4. The nomenclature of the specimens follows the experimental variables used. In this research, unstrengthened specimen was fabricated as a reference specimen.

A single specimen strengthened with the CFRP fabric was tested to compare its behavior to a behavior of reinforced concrete slab with the BFRP fabric with the tensile reinforcement ratio of 0.41%. The comparison of the global behavior between the slabs reinforced with the CFRP fabric and BFRP fabric presents feasibility of the BFRP fabric as the reinforcement material.

The tensile reinforcement ratio of the specimen was labeled using a letter abbreviation and an Arabic number. The first letter, R, represents the tensile reinforcement ratio. The following Arabic number, 41, 55, or 87, represents the tensile reinforcement ratios of 0.41%, 0.55%, and 0.87%, respectively. Next to the symbol "-", the letter, N, C, and B represent the nonstrengthened CFRP fabric, and BFRP fabric, respectively. The following Arabic numbers, 1, 2, and 3, represent the 1-layer, 2-layers, and 3-layers of fabric, respectively. For example, "R55-B3" denotes a strengthen specimen with the tensile reinforcement ratio of 0.55% and 3-layers of BFRP fabric.

Table 2: Material properties of the used steel.

Nominal diameter (mm)	Yield stress (MPa)	Ultimate stress (MPa)	Modulus of elasticity (GPa)
10	487	621	204
13	469	671	199
16	449	622	194

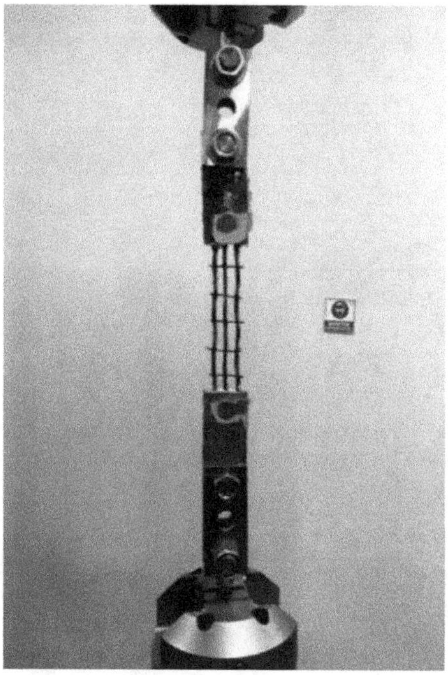

Figure 2: Direct tensile test of FRP fabric.

2.4. Specimen Preparation. Prior to strengthening the concrete slab by FRCM composite, the concrete specimens were rotated 180° so that the strengthening surface of specimens would be the upper surface. The laitance on their strengthening surface was removed using sandblaster and washed with high pressure water.

FRCM strengthening was proceeded as follows: (a) apply the cementitious matrix on the strengthening surface with a thickness of 5 mm by using a shotcrete machine; (b) place the FRP fabric on the matrix; and (c) apply the cementitious matrix with a thickness of 5 mm on FRP fabric. For the specimens with 2-layers and 3-layers of BFRP fabric, the above procedure was repeated two and three times, respectively. The flexural test was conducted after 28 days of strengthening procedure, so that the cement matrix would have enough time to develop its strength.

2.5. Test Setup. The specimens were loaded with a 2,000 kN capacity actuator at a stroke speed of 3 mm/min. The test was carried out under a four-point loading system, and the distance between loading points was 600 mm. Two linear variable differential transformers (LVDTs) were installed on the lower surface of the specimen to measure the deflection of the specimen at midspan. Additionally, the strain values of the concrete, tensile steel bar, compression steel bar, and fabric of FRCM composite were measured by the strain gages installed at the midspan section of specimens. The measured strain values were recorded at a frequency of 1 second by a data logger.

3. Test Result and Discussion

3.1. Failure Mode. Figure 4 shows the deformed shape of the representative specimens after test. Reference specimens R41-NN, R55-NN, and R87-NN were failed by the yielding of tensile rebar followed by the concrete crushing of the compression zone. All of the FRCM-strengthened specimens had fabric slippage in FRCM composite after the maximum load and were ultimately failed by the tensile fracture of the fabric followed by the concrete crushing of the compression zone. Additionally, it should be noted that the peeling failure of the FRCM composite did not occur in any specimen. Figure 5 shows the crack pattern at the midspan of R55-B1, R55-B2, and R55-B3 specimens. It can be seen from Figure 5 that the flexural cracks were developed at regular intervals in the midspan and propagated toward the compression zone. Furthermore, the FRCM composite was broken with a concrete hammer after the end of the test to investigate the BFRP fabric if the fabric was damaged. It was confirmed that about 60% of the basalt fabric was broken.

3.2. Load-Displacement Curves. Table 5 and Figure 6 show the summary of test results and the comparison of load-displacement curves for all specimens, respectively. Figure 6(a) shows the comparison of the load-displacement curves of R41-C1 and R41-B1 specimens strengthened with 1 layer of CFRP fabric and BFRP fabric, respectively. They show very similar flexural behavior before yielding of steel rebar. After yielding of steel rebar, the load of R41-C1 strengthened with CFRP fabric with relatively high elastic modulus increased more rapidly than that of R41-B1. However, the maximum loads of R41-C1 and R41-B1 were increased by 30.7% and 40.9%, respectively, compared with that of R41-NN. This is because the CFRP fabric has about 8 times higher elastic modulus than the BFRP fabric, resulting in the slip of R41-C1 earlier than R41-B1.

Figure 6(b) shows a comparison of the load-displacement curves of the specimens strengthened with the 1, 2, and 3 layers of BFRP fabric, respectively, all of which had the same tensile reinforcement ratio of 0.41%. As the number of fabric layers increased, the initial crack load

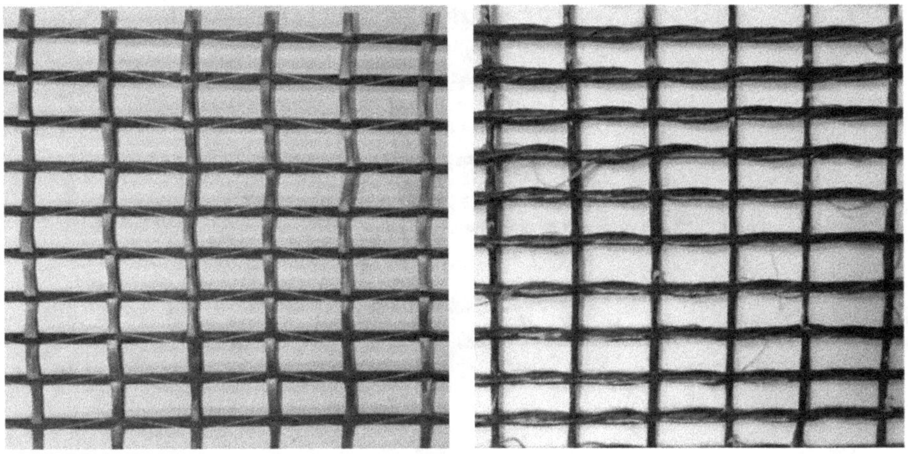

(a) (b)

FIGURE 3: FRP fabrics. (a) CFRP fabric. (b) BFRP fabric.

TABLE 3: Material properties of FRP fabric.

Fabric	Nominal thickness (mm^2/mm)	Ultimate tensile strain	Ultimate tensile stress (MPa)	Modulus of elasticity (GPa)
BFRP fabric	0.522	0.023	274.85	11.84
CFRP fabric	0.107	0.014	1314.68	94.85

TABLE 4: Test parameter details.

Specimen	Tensile reinforcement ratio (%)	Strengthening material	Number of fabric layers
R41-NN	0.41	None	0
R41-C1	0.41	CFRP	1
R41-B1	0.41	BFRP	1
R41-B2	0.41	BFRP	2
R41-B3	0.41	BFRP	3
R55-NN	0.55	None	0
R55-B1	0.55	BFRP	1
R55-B2	0.55	BFRP	2
R55-B3	0.55	BFRP	3
R87-NN	0.87	None	0
R87-B1	0.87	BFRP	1
R87-B2	0.87	BFRP	2
R87-B3	0.87	BFRP	3

increased. The yielding load of R41-B1, R41-B2, and R41-B3 increased to 27.1%, 25.3%, and 43.6%, respectively, compared with that of R41-NN. The slope of the curve increased steeply in proportion to the number of fabric layers after yielding of steel rebar. The maximum loads of R41-B1, R41-B2, and R41-B3 were increased to 40.9%, 52.0%, and 98.2%, respectively, compared with R41-NN.

Figure 6(c) shows a comparison of the load-displacement curves of the specimens strengthened with the 1, 2, and 3 layers of BFRP fabric, all of which had the same tensile reinforcement ratio of 0.55%. As with specimens with a tensile reinforcement ratio of 0.41%, the initial crack loads and yielding loads of specimens with a tensile reinforcement ratio of 0.55% increased in proportion to the number of fabric layers. In addition, the flexural rigidity of the specimens was not improved significantly before yielding of steel rebar but improved in proportion to the number of fabric layer after yielding of steel rebar. The maximum load of R51-B1, R51-B2, and R51-B3 increased to 17.1%, 40.2%, and 62.8%, respectively, compared with that of R51-NN.

(a)

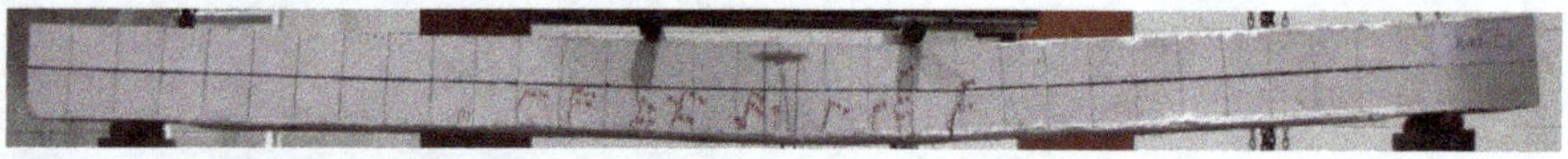

(b)

(c)

FIGURE 4: Deformed shapes of specimens after testing. (a) R41-NN. (b) R41-C1. (c) R41-B1.

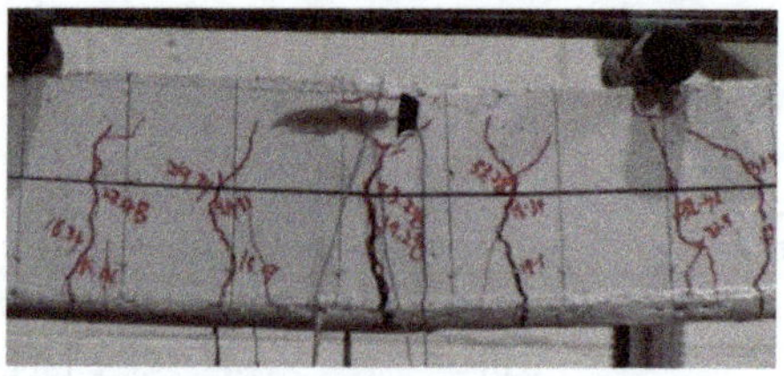

(a)

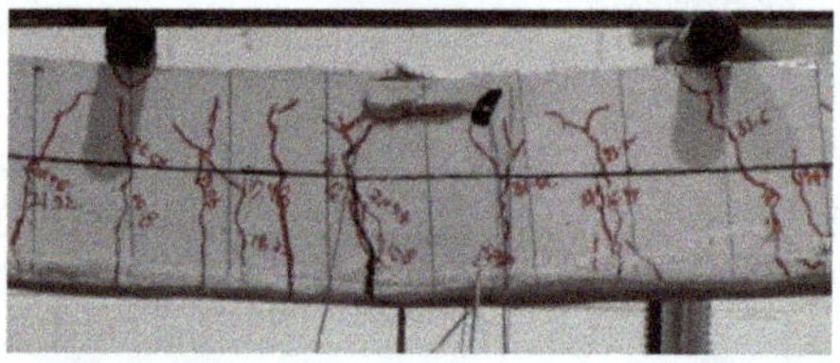

(b)

(c)

FIGURE 5: Crack pattern at midspan. (a) R55-B1. (b) R55-B2. (c) R55-B3.

Figure 6(d) shows a comparison of the load-displacement curves of the specimens strengthened with the 1, 2, and 3 layers of BFRP fabric, all of which had the same tensile reinforcement ratio of 0.87%. As with specimens with other tensile reinforcement ratios, the flexural rigidity of the specimens was not improved significantly before yielding of steel rebar, but improved in proportion to the number of fabric layer after yielding of steel rebar. The maximum load of R87-B1, R87-B2, and R87-B3 increased to 11.2%, 29.0%, and 48.8%, respectively, compared with that of R51-NN. It should also be noted that the increasing ratio of maximum load decreases as the tensile reinforcement ratio increases.

3.3. Load-Strain Curves. Figure 7 shows the strain distribution along the depth at a midspan cross section of representative specimens. The strains of concrete, tensile reinforcement, and FRP fabric were evaluated at representative load stages. It can be observed from Figure 7 that the strain distribution is linear at the low load level, while it becomes nonlinear and the slippage between FRP fabric and the cementitious matrix occurs after 60% of the maximum load. From 80% of the ultimate load stages, the strains of the FRP fabric were reduced compared to the strains of steel, which might be because the cementitious matrix was broken and slippage occurred between FRP fabric and cementitious

TABLE 5: Summary of test results.

Specimens	Yielding load (kN)	Maximum load (kN)	Load increase over reference (%)
R41-NN	28.94	34.90	—
R41-C1	39.70	45.60	30.7
R41-B1	36.82	49.16	40.9
R41-B2	36.26	53.06	52.0
R41-B3	41.36	69.16	98.2
R55-NN	43.28	50.42	—
R55-B1	47.50	59.06	17.1
R55-B2	49.14	70.70	40.2
R55-B3	54.22	82.08	62.8
R87-NN	64.84	70.72	—
R87-B1	66.08	78.66	11.2
R87-B2	74.56	91.20	29.0
R87-B3	75.54	105.24	48.8

material. Consequently, it should be noted that the strain distribution at low load level can be assumed to be linear, but it cannot be considered as linear at the high load level due to the slippage between FRP fabric and cementitious matrix. Thus, the slippage between FRP fabric and cementitious material needs to be taken into account in order to accurately predict the flexural performance of FRCM-strengthened slabs.

3.4. Ductility. Ductility is generally expressed as the ratio of the deflection at the time of steel yielding to the ultimate deflection. In this research, the fracture mode of the specimen was determined by the ductility index "μ" proposed by Jeong [26] shown in Figure 8. The ductility index is expressed in (1):

$$\mu = \frac{1}{2}\left(\frac{E_{tot}}{E_{el}} + 1\right), \tag{1}$$

$$E_{tot} = E_{in} + E_{el}, \tag{2}$$

where μ is the ductility index, E_{tot} is the total energy which is calculated as the area under the load-displacement curve up to the failure load, E_{el} is the elastic energy which is a part of the total energy, and E_{in} is the inelastic energy, as illustrated in Figure 8.

Based on the ductility index proposed by Jeong [26], Grace et al. [27] developed failure modes depending on the energy ratio. The energy ratio (E_{in}/E_{tot}) is classified into a range of 75% or more, 70% to 75%, or 70% or less and is defined as the ductile, semibrittle, or brittle failure mode, respectively. Table 6 shows the failure mode evaluation results of each specimen defined by this criterion. For nonreinforced specimens, the total energy and elastic energy increase while the energy ratio decreases as the tensile steel reinforcement ratio increases. The energy ratios of the R41-NN, R55-NN, and R87-NN were 83.5%, 76.5%, and 69.3%, which are defined as the ductile mode, ductile mode, and brittle mode, respectively.

The ductile fracture mode absorbs a large amount of energy before fracture, which was shown in R41-NN, R41-B1, and R55-NN. With increase of the tensile reinforcement ratio, the failure mode was changed from the ductile to the brittle mode. The brittle fracture was shown in the R41-C1, R41-B3, R55-B2, R55-B3, R87-B1, R87-B2, and R87-B3. All of those mentioned specimens failed by the debonding of the FRP fabric with the broken cementitious matrix.

Compared to the R41-NN, the total energy of the R41-C1 specimen with a layer of CFRP fabric was reduced by 28.2%, while the elastic energy was increased by 74.4%. The energy ratio of R41-C1 was 60.3%, confirming that the failure mode of this specimen was brittle failure. However, for the R41-B1 specimen with a layer of BFRP fabric, both the total energy and the elastic energy increased compared with those of R41-NN, and its energy ratio was 76.0%, indicating ductile failure.

For the specimens with the tensile reinforcement ratio of 0.41%, the energy ratio decreased with the increase of the number of BFRP fabric layers. The energy ratios of the R41-B1, R41-B2, and R41-B3 were 76.0%, 70.9%, and 63.6%, which are defined as the ductile mode, semiductile mode, and brittle mode, respectively.

For the specimens with the tensile reinforcement ratio of 0.55%, the energy ratio and the total energy decreased with the increase of the number of BFRP fabric layers. The energy ratios of the R55-B1, R55-B2, and R55-B3 were 74.0%, 67.0%, and 65.1%, which are defined as the ductile mode, brittle mode, and brittle mode, respectively.

For the specimens with the tensile reinforcement ratio of 0.87%, the energy ratio and the total energy decreased with the increase of the number of BFRP fabric layers. The energy ratios of the R87-B1, R87-B2, and R87-B3 were 67.4%, 61.2%, and 56.5%, which are defined as brittle mode.

Also, it can be seen that the ductility of the FRCM-strengthened specimens decreased with the higher amount of fabric and the higher tensile reinforcement ratio.

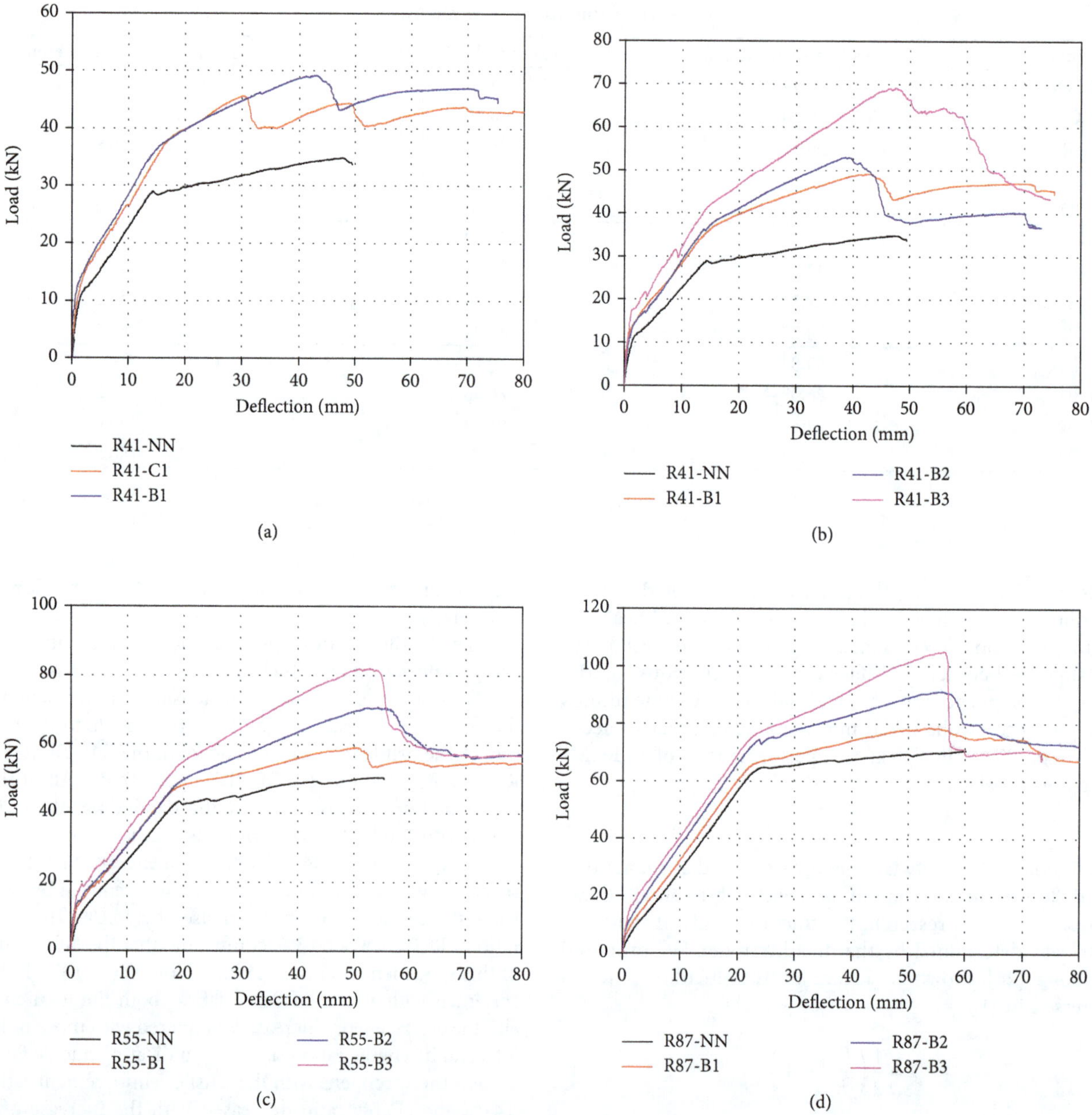

FIGURE 6: Comparison of load-deflection curves. (a) Specimens with CFRP and BFRP fabric. (b) Specimens with the tensile reinforcement ratio of 41%. (c) Specimens with the tensile reinforcement ratio of 55%. (d) Specimens with the tensile reinforcement ratio of 87%.

4. Numerical Approach

4.1. Effective Tensile Stress Level of FRCM Fabric

4.1.1. ACI 549. In ACI 549 [14], the effective tensile strain level in the FRCM fabric attained at failure, ε_{fe}, is limited to the design tensile strain of the FRCM fabric, ε_{fd}, as defined in

$$\varepsilon_{fe} = \varepsilon_{fd} \leq 0.012. \tag{3}$$

Also, the effective tensile stress level in the FRCM fabric attained at failure, f_{fe}, in the FRCM composite is calculated in accordance with

$$f_{fe} = E_f \varepsilon_{fe}, \tag{4}$$

where E_f is the elastic modulus of FRCM fabric.

4.1.2. Jung Model. Jung et al. [16] proposed the FRCM bond strength model with regard to both the slippage and the debonding phenomenon of the FRCM composite by (5), based on the FRCM research performed by D'Ambrishi et al. [18] and the model proposed by Teng et al. [28]. The effective tensile stress level in the FRCM fabric at failure can be computed by the following equations:

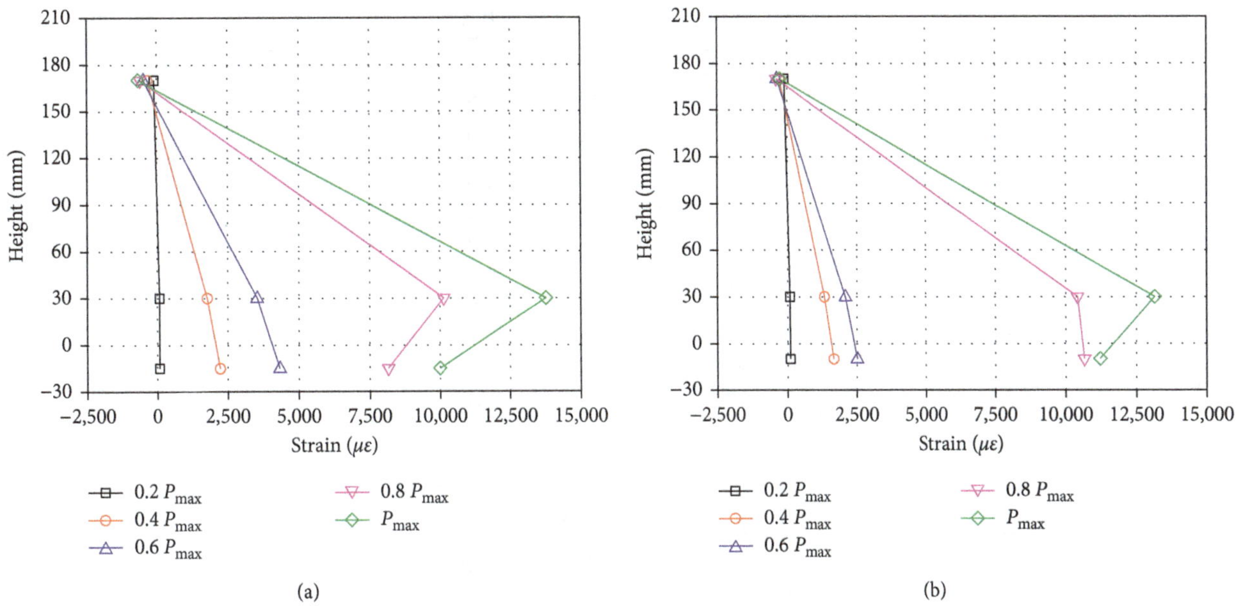

FIGURE 7: Load-strain relation. (a) R55-B1. (b) R55-B3.

$$f_{fe} = \alpha\beta_F\beta_L\sqrt{\frac{E_F\sqrt{f_c'}}{t_1\sqrt{n}}},$$

$$\beta_F = \sqrt{\frac{2 - b_F/b_c}{1 + b_F/b_c}},$$

$$\beta_L = \begin{cases} 1, & \text{if} \quad L \geq L_e \\ \sin\dfrac{\pi L}{2L_e}, & \text{if} \quad L < L_e, \end{cases} \tag{5}$$

$$L_e = \sqrt{\frac{E_F t_F}{f_c'}},$$

where b_F is the width of the FRCM fabric, b_c is the width of the specimen, L is the bond length, L_e is the effective bond length, E_F is the elastic modulus of the FRCM fabric, t_F is the total thickness of the FRCM fabric, f_c' is the cylinder compressive strength for concrete, and α is the reduction factor and given as 0.727 for FRCM by Jung et al. [16]. Additionally, t_1 is the nominal thickness of a layer of the FRCM fabric and n is the number of fabric layers.

4.2. Nominal Flexural Strength. Nominal flexural strength was calculated by using ACI 318 [13] and ACI 549 [14]. First, the neutral axis depth, c_u, satisfying the internal force equilibrium expressed by (6), (7) was determined by the trial and error method:

$$T_f + T_s = C, \tag{6}$$

where

$$T_s = A_s f_y,$$

$$T_f = nA_f f_{fe},$$

$$C = \alpha_1 f_c' \beta_1 c_u b,$$

$$\alpha_1(c_u) = \frac{3\varepsilon_c'\varepsilon_c(c_u) - [\varepsilon_c(c_u)]^2}{3\beta_1(c_u)\varepsilon_c'^2},$$

$$\beta_1(c_u) = \frac{4\varepsilon_c' - \varepsilon_c(c_u)}{6\varepsilon_c' - 2\varepsilon_c(c_u)}, \tag{7}$$

$$\varepsilon_c' = \frac{1.7f_c'}{E_c},$$

$$E_c = 4700\sqrt{f_c'},$$

$$\varepsilon_{fe} = \frac{f_{fe}}{E_f},$$

$$\varepsilon_c = \frac{c_u}{h - c_u}\varepsilon_{fe},$$

where T_s, T_f, C, A_s, A_f, f_y, n, α_1, β_1, ε_c', ε_c, E_c, f_{fe}, ε_{fe}, and E_f are the tensile force provided by tensile rebar, tensile force provided by the FRCM composite, compressive force provided by concrete, area of tensile steel rebar, area of the FRP fabric, yielding strength of steel rebar, number of FRP fabrics, concrete stress block factor, compressive strain attained in the maximum compressive strength of unconfined

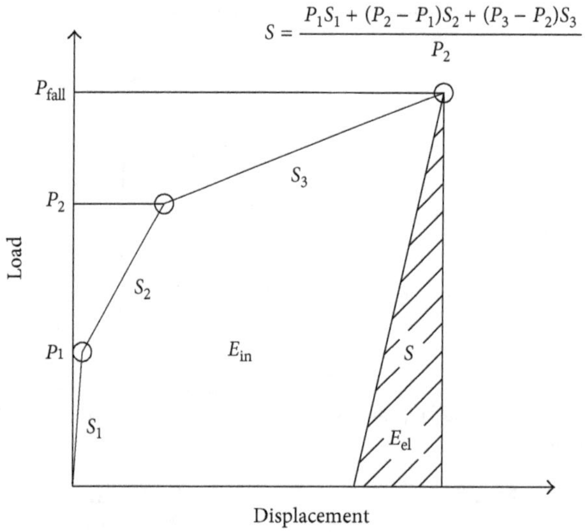

FIGURE 8: Definition of ductility index [17].

TABLE 6: Results of ductility evaluation.

Specimen	E_{el} (J)	E_{in} (J)	E_{tot} (J)	μ_{energy}	Energy ratio (%)	Failure mode
R41-NN	217	1100	1317	3.53	83.5	Ductile
R41-C1	379	575	954	1.76	60.3	Brittle
R41-B1	372	1175	1547	3.53	76.0	Ductile
R41-B2	413	1006	1419	2.22	70.9	Semiductile
R41-B3	808	1409	2217	1.87	63.6	Brittle
R55-NN	506	1641	2146	2.62	76.5	Ductile
R55-B1	581	1653	2234	2.42	74.0	Semiductile
R55-B2	880	1786	2665	2.01	67.0	Brittle
R55-B3	1027	1912	2939	1.93	65.1	Brittle
R87-NN	995	2250	3245	2.13	69.3	Semiductile
R87-B1	1005	2078	3082	2.03	67.4	Brittle
R87-B2	1414	2235	3649	1.79	61.2	Brittle
R87-B3	1754	2275	4029	1.65	56.5	Brittle

concrete, strain in the concrete, elastic modulus of concrete, strength of the FRCM composite, strain in the FRCM composite, and elastic modulus of the FRCM composite.

The nominal flexural strength, M_n, can be calculated in accordance with the following equations:

$$M_n = M_s + M_f,$$
$$M_s = A_s f_y\left(d - \frac{\beta_1 c_u}{2}\right), \tag{8}$$
$$M_f = A_f b f_{fe}\left(h - \frac{\beta_1 c_u}{2}\right),$$

where d and h are the effective depth and height of specimen cross section, respectively.

4.3. Prediction of Nominal Flexural Strength. Table 7 shows the comparisons of nominal flexural strength predicted by ACI 549 [14] and the Jung model [16]. The ratios of the test result to the predicted flexural strength by ACI 549 were ranged from 0.97 to 1.14. The mean and standard deviation of the ratios of the test result to the prediction value by ACI 549 were 1.08 and 0.05, respectively. The ratios of the test result to the predicted value by Jung model were ranged from 0.94 to 1.14. The mean and standard deviation of the ratios of the test result to the predicted value by the Jung model were 1.05 and 0.06, respectively. Consequently, the Jung model predicts the effective stress level more accurately than ACI 549. However, it can be noted that both ACI 549 and the Jung model predict the effective stress level of the FRCM fabric with the sufficient accuracy to be applied for the FRCM strengthening design of the RC slab.

5. Conclusion

The following conclusions are drawn from the results:

Table 7: Comparison of nominal flexural strength.

Specimen	Nominal flexural strength			Test/ACI 549	Test/Jung model
	Test (kN)	ACI 549 (kN)	Jung model (kN)		
R41-B1	29.50	27.33	29.18	1.08	1.01
R41-B2	31.84	32.69	33.94	0.97	0.94
R41-B3	41.50	38.00	38.08	1.09	1.09
R55-B1	35.44	32.97	34.80	1.07	1.02
R55-B2	42.42	38.25	39.48	1.11	1.07
R55-B3	49.25	43.47	43.55	1.13	1.13
R87-B1	47.20	45.39	46.94	1.04	1.01
R87-B2	54.72	50.44	51.52	1.08	1.06
R87-B3	63.14	55.37	55.43	1.14	1.14
				Mean: 1.08, SD: 0.05	Mean: 1.05, SD: 0.06

(1) All of the FRCM-strengthened specimens had fabric slippage in FRCM composite after the maximum load and were ultimately destroyed by the tensile fracture of the fabric followed by the concrete crushing of the compression zone.

(2) The initial crack loads and yielding loads of specimens with three tensile reinforcement ratio of 0.41%, 0.55%, and 0.87% showed increasing trend in proportion to the number of fabric layers. Also, the maximum loads of FRCM-strengthened specimens increase from 11.2% to 98.2% relative to the reference specimens. Additionally, it should be noted that the increasing ratio of maximum load over the reference specimen decreases as the tensile reinforcement ratio increases.

(3) The strain distribution of the FRCM-strengthened specimen is linear at the low load level, while it becomes nonlinear after 60% of maximum load, resulting from the slippage between FRP fabric and the cementitious matrix. Thus, the slippage between FRP fabric and cementitious material needs to be considered in order to accurately predict the flexural performance of the FRCM-strengthened slabs.

(4) The energy ratio and ductility of the FRCM-strengthened specimens decreased with the higher amount of fabric and the higher tensile reinforcement. With the higher amount of fabric, the fracture mode was changed from ductile failure to brittle failure.

(5) The Jung model predicts the effective stress level more accurately than ACI 549. However, it can be noted that both ACI 549 and the Jung model predict the effective stress level of the FRCM fabric with sufficient accuracy to be applied for FRCM strengthening design of the RC slab.

Conflicts of Interest

The authors declare no conflicts of interest.

Acknowledgments

This work was supported by the Korea Science and Engineering Foundation (KOSEF) grant funded by the Korea government (MOST) (no. 2015R1D1A1A01057453).

References

[1] K. Kovler and V. Chernov, "Types of damage in concrete structures," in *Failure, Distress and Repair of Concrete Structures*, pp. 32–56, Woodhead Publishing Limited, Cambridge, UK, 2009.

[2] G. Loreto, L. Leardini, D. Arboleda, and A. Nanni, "Performance of RC slab-type elements strengthened with fabric-reinforced cementitious-matrix composites," *Journal of Composites for Construction*, vol. 18, no. 3, 2014.

[3] H. M. Elsanadedy, T. H. Almusallam, S. H. Alsayed, and Y. A. Al-Salloum, "Flexural strengthening of RC beams using textile reinforced mortar—experimental and numerical study," *Composite Structures*, vol. 97, pp. 40–55, 2013.

[4] S. W. Kim, *Experimental Study on Bond Flexural Behavior of RC Members Using CFRP-Bar NSM System*, Ph.D. thesis, pp. 1–4, Yeungnam University, Gyeongsan, Republic of Korea, 2015.

[5] H. J. Kim, *Failure Behavior of Reinforced Concrete Beams with Flexural Strengthened by Steel Plates*, Ph.D. thesis, pp. 17–22, Chonnam University, Gwangju, Republic of Korea, 2005.

[6] J. Y. Park, *Behavior Characteristics and Strengthening Design of Reinforced Concrete Beam Strengthened with Carbon Fiber Reinforced Polymer Plate*, Ph.D. thesis, pp. 1–3, Chungbuk University, Cheongju, Republic of Korea, 2003.

[7] L. Ombres, "Flexural analysis of reinforced concrete beams strengthened with a cement based high strength composite material," *Composite Structures*, vol. 94, no. 1, pp. 143–155, 2011.

[8] H. H. Lee and E. S. Koo, "A study on the strength and the deformation capacity of RC beams strengthened with aramid fiber sheet," *Journal of the Korea Concrete Institute*, vol. 10, pp. 151–158, 1998.

[9] T. C. Triantafillou and C. G. Papanicolau, "Shear strengthening of reinforced concrete members with textile reinforced mortar (TRM)," *Materials and Structures*, vol. 39, no. 1, pp. 93–103, 2006.

[10] H. Ko, S. Matthys, A. Palmieria, and Y. Sato, "Development of simplified bond stress-slip model for bonded FRP-concrete interfaces," *Construction and Building Materials*, vol. 68, pp. 142–157, 2014.

[11] A. D'Ambrisi and F. Focacci, "Flexural strengthening of RC beams with cement-based composites," *Journal of Composites for Construction*, vol. 15, no. 5, pp. 707–720, 2011.

[12] S. Babaeidarabad, G. Loreto, and A. Nanni, "Flexural strengthening of RC beams with an externally bonded fabric-reinforced cementitious matrix," *Journal of Composites for Construction*, vol. 18, no. 5, 2014.

[13] American Concrete Institute (ACI), *Building Code Requirements for Reinforced Concrete (ACI 318)*, American Concrete Institute, Farmington Hills, MI, USA, 2011.

[14] American Concrete Institute (ACI), *Design and Construction Guide of Externally Bonded FRCM System for Concrete and Masonry Repair and Strengthening (ACI 549)*, American Concrete Institute, Farmington Hills, MI, USA, 2013.

[15] R. Azam and K. Soudki, "FRCM strengthening of shear-critical RC beams," *Journal of Composites for Construction*, vol. 18, no. 5, 2014.

[16] K. Jung, K. Hong, S. Han, J. Park, and J. Kim, "Prediction of flexural capacity of RC beams strengthened in flexure with FRP fabric and cementitious matrix," *International Journal of Polymer Science*, vol. 2015, Article ID 868541, 11 pages, 2015.

[17] T. Kuroki, Y. Tanaka, T. Hokudoh, and K. Yabuki, "Heat resistance properties of poly (p-phenylene-2,6-benzobisoxazole) fiber," *Journal of Applied Polymer Science*, vol. 65, no. 5, pp. 1031–1036, 1997.

[18] A. D'Ambrisi, L. Feo, and F. Focacci, "Experimental analysis on bond between PBO-FRCM strengthening materials and concrete," *Composites Part B: Engineering*, vol. 44, no. 1, pp. 524–532, 2013.

[19] G. P. Lignola, C. Caggegi, F. Ceroni et al., "Performance assessment of basalt FRCM for retrofit applications on masonry," *Composites Part B: Engineering*, vol. 128, pp. 1–18, 2017.

[20] J. Sim, C. Park, and D. Y. Moom, "Characteristics of basalt fiber as a strengthening material for concrete structures," *Composites Part B: Engineering*, vol. 36, no. 6-7, pp. 504–512, 2005.

[21] Q. Liu, M. T. Shaw, R. S. Parnas, and A. M. McDonnell, "Investigation of basalt fiber composite mechanical properties for applications in transportation," *Polymer Composites*, vol. 27, no. 1, pp. 41–48, 2006.

[22] ASTM International, *Standard Test Method for Compressive Strength of Cylindrical Concrete Specimens (ASTM C39/C39M)*, ASTM International, West Conshohocken, PA, USA, 2013.

[23] ASTM International, *Standard Test Method and Definitions for Mechanical Testing of Steel Products (ASTM A370)*, ASTM International, West Conshohocken, PA, USA, 2013.

[24] V. Dhand, G. Mittal, K. Y. Rhee, S. J. Park, and D. Hui, "A short review on basalt fiber reinforced polymer composites," *Composites Part B: Engineering*, vol. 73, pp. 166–180, 2015.

[25] ASTM International, *Standard Test Method for Compressive Strength of Hydraulic Cement Mortars (ASTM C109/C109M)*, ASTM International, West Conshohocken, PA, USA, 2013.

[26] S. M. Jeong, *Evaluation of Ductility in Prestressed Concrete Beams Using Fiber Reinforced Plastic Tendons*, Ph.D. thesis, University of Michigan, Ann Arbor, MI, USA, 1994.

[27] N. F. Grace, A. K. Soliman, G. Abdel-Sayed, and K. R. Saleh, "Behavior and ductility of simple and continuous FRP reinforced beams," *Journal of Composites for Construction*, vol. 2, no. 4, pp. 186–194, 1998.

[28] J. G. Teng, S. Y. smith, J. Yao, and J. F. Chen, "Intermediate crack-induced debonding in RC beams and slabs," *Journal of Construction and Building Materials*, vol. 17, no. 6-7, pp. 447–462, 2003.

16

Influences of Environmental Conditions on the Cracking Tendency of Dry-Mixed Plastering Mortar

Ni Tongyuan,[1,2,3] Yang Yang,[2,3] Wu Dandan,[2] and Jiang Chenhui[4]

[1]*College of Materials Science and Engineering, Zhejiang University of Technology, Hangzhou 310023, China*
[2]*College of Civil Engineering and Architecture, Zhejiang University of Technology, Hangzhou 310023, China*
[3]*Key Laboratory of Civil Engineering Structures and Disaster Prevention and Mitigation Technology of Zhejiang Province, Hangzhou 310023, China*
[4]*Zhejiang College of Construction, Hangzhou 311231, China*

Correspondence should be addressed to Yang Yang; yangyang@zjut.edu.cn

Academic Editor: Robert Cerný

Cracking tendency is one of the important performances of dry-mixed plastering mortar (DMPM). Environmental condition is a key factor to affect the cracking tendency of DMPM. For the purpose of evaluating the cracking resistance of DMPM and revealing the influence of environmental conditions on the cracking tendency of DMPM, a series of experiments were performed on restriction-induced cracking behaviors as well as free shrinkage, water loss, and mechanical properties of DMPM. The restricted shrinkage tests were based on ring tests and plate experiments. The results showed that the initial drying age exhibits significant influence on the cracking tendency of DMPM, and there was a stress balance period when the initial drying age was 2 days. But, the phenomena cannot be observed when the initial age was 3 d, 5 d, and 7 d. In order to eliminate the cracking tendency of DMPM, it should avoid water loss from the plaster layer during construction in practical engineering, especially, before initial drying ages.

1. Introduction

Compared with traditional plaster mortars, dry-mixed plaster mortar (DMPM) is a more environmentally friendly building material by reducing air pollution and waste production on construction sites. It can reduce carbon dioxide emissions, help the construction site keep clean, and allow for more flexibility in storage space for materials while reducing cement redundancy after the cement finishing is completed. Moreover, this new and advanced construction material is convenient to deploy and transport, delivers product with stable quality, helps us to facilitate construction, and saves materials [1–6]. There are many causes resulting in cracking of concrete, and shrinkage is of the upmost importance among them [7–11]. If the concrete is restrained, when the tensile stress induced by drying shrinkage exceeds the tensile strength, cracking would happen [9, 12–15]. By the same token, shrinkage cracking is also an important problem in wall plastering mortar. There are a variety of wall plastering mortar cracking causes including thermal gradients, moisture gradients, and attacks from external and internal environments [16]. Drying shrinkage, which is one of the major causes of cracking of cement mortar, is related to the water loss from mortar. The early age cracking behavior is very complex because it not only depends on the manifestation rate and magnitude of the shrinkage of the hardened cement paste with ages but also other factors such as strength development, degree of restraint, stress relaxation, and shape of the structure.

Currently, many investigators focus on revealing the mechanisms of cracking by several kinds of experiments and developing prediction models of crack development [12, 16–18]. The ring test is a commonly used method for assessing the potential of shrinkage cracking. Moon and Weiss used the ring test to assess the restrained shrinkage behavior of mortar and concrete under various conditions [5, 8, 9, 12, 17, 19]. They proposed revisions to the estimation equations developed in previous studies to consider the

TABLE 1: Mix proportions of dry-mix plaster mortar (g).

Grade	Cement	Thickening powder	Fly ash	Water	Sand
DMPM 15	160	7	23	34.2	810

FIGURE 1: Water loss rate test of the DMPM prism specimen.

changes in stress in the cross section because of differential drying shrinkage. These stress changes were caused by different levels of ambient relative humidity, and the position of the cross section of the concrete ring [7, 19].

In this paper, the ring tests were performed to qualitatively assess shrinkage and cracking tendency of DMPM under different environmental conditions, including different wind speed, types of subbase, and initial drying age. In addition, free shrinkage and tensile strength of DMPM specimens were performed with the same cross section in the same environmental condition.

2. Experiment Program

2.1. Raw Materials and Mixture Proportion. Cement type used in this study is P·O 42.5 in accordance with Chinese GB 175-2007, and bentonite was used as a thickening powder agent. Grade I fly ash in accordance with Chinese GB/T 1596-2005 was employed as mineral admixture. The fineness modulus of natural river sand was 2.3, and the rate of water and dry material was 0.18. Table 1 shows the mix proportions of dry-mix plaster mortar.

2.2. Water Loss Rate. In order to study the relationship between drying shrinkage and water loss of mortar specimens, the free shrinkage and the water loss rate of a pyramid specimen were tested, respectively. The size of specimen was 40 mm × 40 mm × 160 mm. The specimens were placed in a 20 ± 2°C, 50 ± 5% RH environment at initial drying age.

It is generally known that the wind can promote water evaporation. In order to investigate the influence of wind environment on DMPM, the water loss rate of the pyramid prism specimen was tested under wind speed at 0 m/s, 4 m/s, and 8 m/s. Figure 1 shows the method for testing the water loss rate of the pyramid prism specimen under different wind speeds.

2.3. Restrained Shrinkage Test A: Ring Test. Due to its practicability, and easy operation to measure strain or stress, the "ring test" was commonly used to assess the potential for

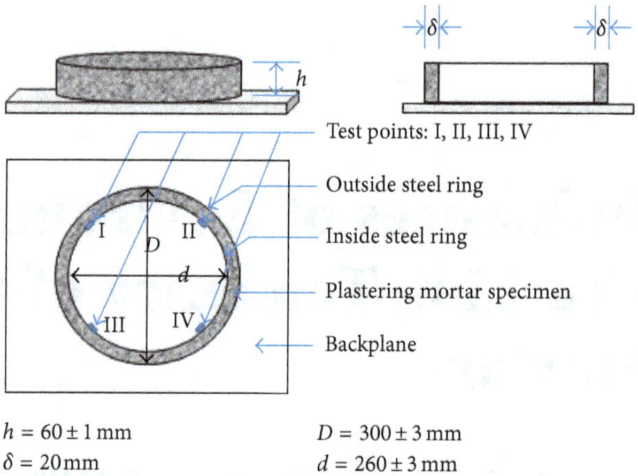

FIGURE 2: Schematic diagram of the ring restraint test device.

shrinkage cracking. The device consists of a mortar ring specimen that was cast around a steel concentric ring [7, 9, 12, 15, 17, 20, 21]. As the mortar ring dries, the shrinkage was prevented by the steel ring, which results in the development of restrained tensile stress in the mortar specimen. The simple ring specimen geometry allows it to be fabricated easily. In addition, the low cost of the system enables several tests to be conducted concurrently over long periods of time.

This paper utilized the restrained ring test to gather information as internal stress development in the mortar system. Figure 2 shows the ring restraint test device used in this study. The outer steel ring was used as mould during casting the DMPM specimen, and the inner steel ring was used as restraint. The inner ring was fixed on a subbase which was made of hydrophobic surface smooth material.

The internal ring strain induced by the DMPM shrinkage was measured by 4 strain gauges, axisymmetrically fixed at the midheight of the inner surface of the metal rings (Figure 2). On the other hand, in this paper, "a quarter bridge" strain gage was collected to test stress, and data acquisition automatic logging interval was set to 30 min. The constant temperature of test environment is at 20 ± 2°C, and the relative humidity is 50 ± 5 percent.

The ring specimen is shown in Figure 3. The ring was restrained against horizontal movement in the radial direction, and internal pressure, p, develops when the ring shrinks. Based on theory of elasticity, the stress was composited by one component σ_θ (in the circumferential direction) and the other component σ_r (in the radial directions). The stress σ_θ and σ_r can be calculated according by the following equations:

$$\sigma_\theta = \frac{(r_e/r)^2 + 1}{(r_e^2/r_i^2) - 1} p, \quad (1)$$

$$\sigma_r = \frac{(r_e/r)^2 - 1}{(r_e^2/r_i^2) - 1} p, \quad (2)$$

$$\alpha = \frac{(r_e/r)^2 + 1}{(r_e^2/r_i^2) - 1}, \quad (3)$$

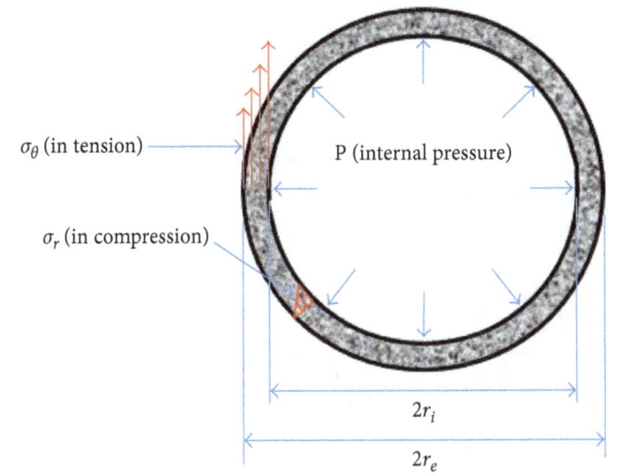

Figure 3: Distribution of internal stress and simplification for fracture analysis.

where r_e and r_i are the external radius and internal radius, respectively. Distributions of σ_θ and σ_r are indicated in Figure 3. α is a constant associated with the device. p is the internal pressure of the internal ring. The internal pressure, p, can be tested by strain gages at test points I, II, III, and IV. The value of p is average measurements of these four test points. The relationship between circumferential shrinkage and the internal pressure p can be obtained by elastic stress analysis.

$$p = \frac{E\varepsilon_\theta}{(r_e^2 + r_i^2)/(r_e^2 - r_i^2) + \mu},$$
$$\beta = \frac{E}{(r_e^2 + r_i^2)/(r_e^2 - r_i^2) + \mu}, \quad (4)$$

where ε_θ is the shrinkage strain, E is the modulus of elasticity, and μ is Poisson's ratio, and in this paper, $\mu = 0.28$. β is a constant associated with the device.

So, (1) can be rearranged to yield

$$\sigma_\theta = \alpha\beta\varepsilon_\theta. \quad (5)$$

In this experiment, two identical specimens were conducted. In order to reduce friction between the subbase and the specimens, two layers of plastic film were covered on the upper surface of the subbase. The DMPM mixture was casted into the mould with two layers. And the specimens were placed in standard conditions (the constant temperature of test environment was 20 ± 2°C, and the relative humidity was 50 ± 5%) until the initial drying age of 2 d, 3 d, 5 d, and 7 d. Before the dry experiment being started, the specimen surface must be covered with moist layer of linen and stamped with plastic film to prevent moisture to loss. The next step was to remove the outer ring and to seal the outside of the specimen with aluminum foil to ensure that water loses only through the outer surface of the specimen. Then, samples began to dry.

2.4. Restrained Shrinkage Test B: Plate Test. The method of plate test was performed following "cement mortar crack resistance test method" (Chinese technical specification JC/T 951-2005). In order to study the influence of drying conditions on cracking tendency of DMPM, the specimens were exposed to ambient condition at the age of 1 d, 2 d, 3 d, 5 d, and 7 d. During the drying period, the direction of the wind from electric fan was parallel to the surface of the plate specimen. The wind speed in the specimen transverse centerline was 0 m/s, 4 m/s, and 8 m/s. At the same time, two 1000 W power halogen lamps were lighting for 4 h. After 24 h, the width and length of crack were measured and the crack index is calculated. The schematic diagram of the plate-restrained test is shown in Figure 4.

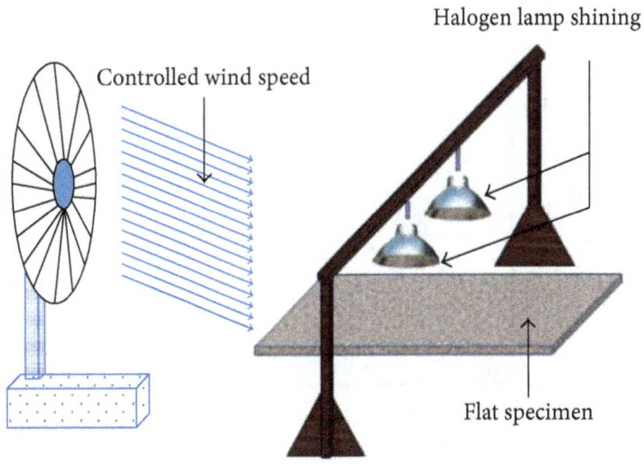

Figure 4: Schematic diagram of the plate-restrained test.

3. Results and Discussion

3.1. Influences of Initial Drying Age and Wind Speed on Flexural Strength and Compressive Strength of DMPM. Strength (compressive strength and flexural strength) characteristics of DMPM played a decisive role in its crack resistance. After an initial moist curing period of 1 d, 2 d, 3 d, 5 d, and 7 d, the development of compressive strength and flexural strength of DMPM was performed with the age increasing.

Figures 5(a)–5(c) show that the initial drying age and wind speed would have adverse effects on the flexural strength of mortar and the same as on compressive strength. On one hand, when DMPM be exposed to drying condition earlier, more water would be evaporated from it. On the other hand, greater wind speed means faster rate of moisture evaporation. The both undermine hydration of DMPM and in turn affect the strength. The influence of DMPM's initial drying age on its strength is more mild than that of different wind speeds.

3.2. Drying Shrinkage (DS) Characteristics of DMPM under Different Conditions. In order to discuss water loss rate (WLR) laws and drying shrinkage characteristics of DMPM, we analyzed the relationship between water loss rules and drying shrinkage of DMPM by experiments.

3.2.1. Influence of Wind Speed on Water Loss Rate (WLR) of DMPM. During the drying process, the free and absorbed

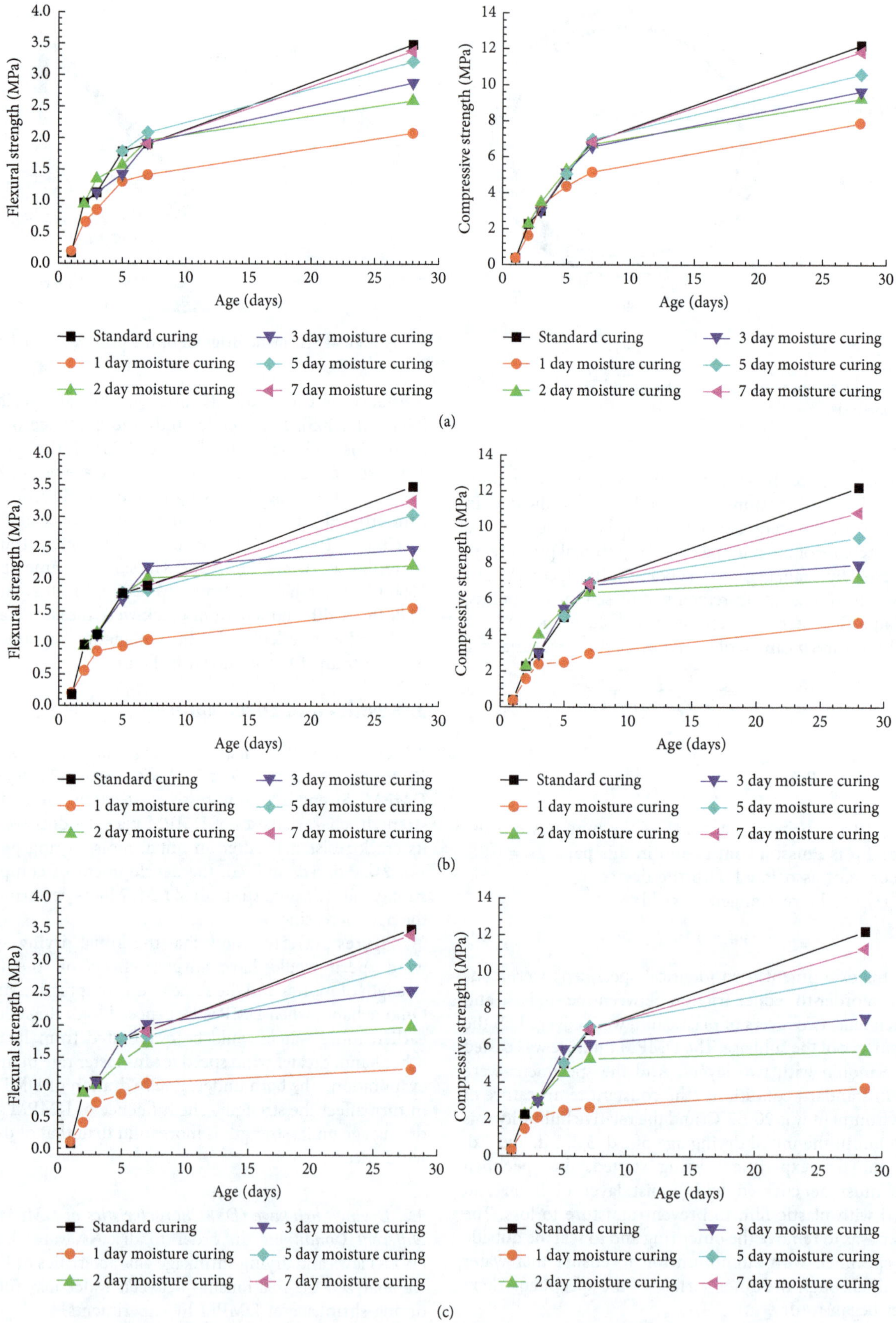

FIGURE 5: Flexural strength and compressive strength with different wind speeds. (a) 0 m/s. (b) 4 m/s. (c) 8 m/s.

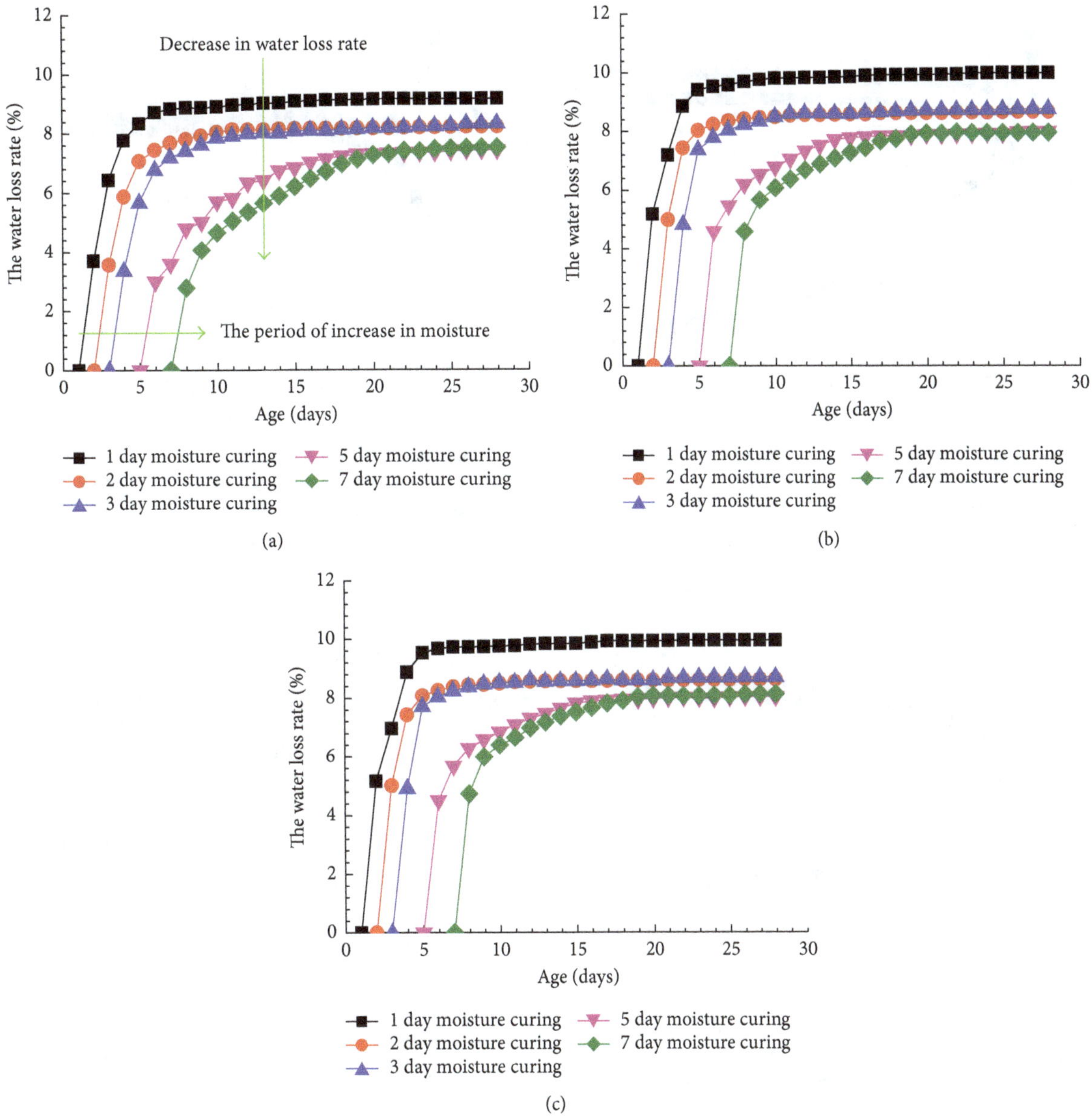

FIGURE 6: The water loss rate with different wind speeds. (a) 0 m/s. (b) 4 m/s. (c) 8 m/s.

water is lost from DMPM. It affects the performance of DMPM. We investigated WLR under different curing conditions, including wind conditions and initial drying age. Figures 6(a)–6(c) show the values of WLR of DMPM in different wind speeds of 0 m/s, 4 m/s, and 8 m/s. The WLR of DMPM was rapid at the early stage, but it decreased with the age, and WRL became slower gradually, until it comes to constant values.

3.2.2. Relationships between Water Loss Rate (WLR), Initial Drying Age, Wind Conditions, and Drying Shrinkage (DS) of DMPM. Obviously, water loss rate, initial drying age, and wind conditions affect the drying shrinkage of DMPM. The curves of Figures 7(a)–7(e) show the relations between wind speed conditions, initial drying age, and drying shrinkage. According to the ultimate shrinkage value, we can classify the rate of water loss to three grades: first, maximum WLR, for the initial drying age was 1 day; second, middle WLR, the initial drying age was 2 d and 3 d; and third, minimum WLR, the initial drying age was 5 d and 7 d. We can make different strategies to deal with the water loss of DMPM.

In the early age, drying shrinkage of DMPM increased with age, and then the development of drying shrinkage slowed down. The rate of DMPM drying shrinkage was faster at initial age and then tended to be gentle. On the other hand, in the same wind speed, the later the initial drying age, the larger the drying shrinkage value of DMPM was, which was due to much more water lose, and the volume change was larger. But the effect was remarkable at drying age from 5th day to 20th day. And it was more obvious while the wind speed was 4 m/s and 8 m/s.

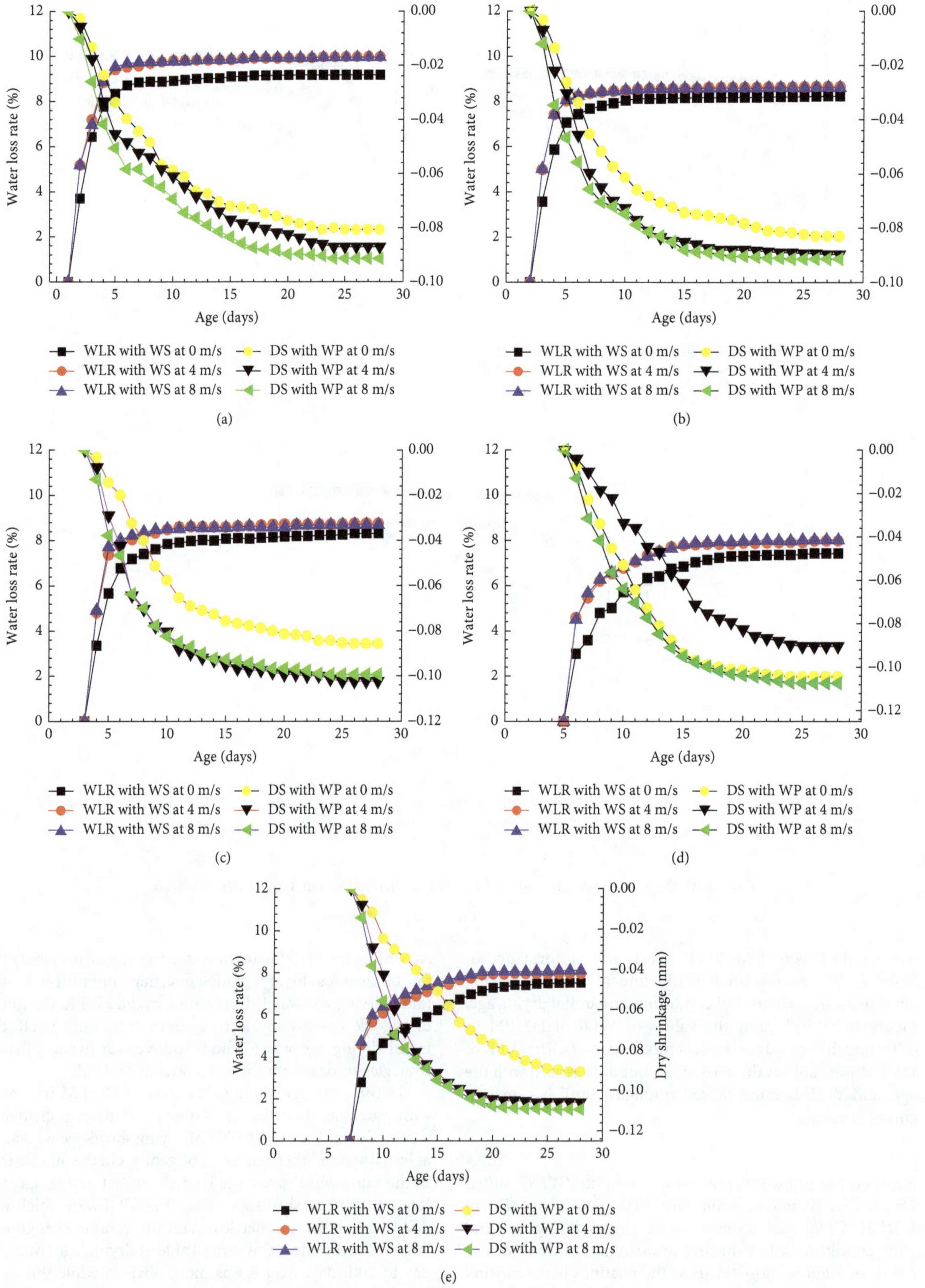

FIGURE 7: Water loss rate of DMPM with different moisture curing times. (a) 1 day. (b) 2 days. (c) 3 days. (d) 5 days. (e) 7 days.

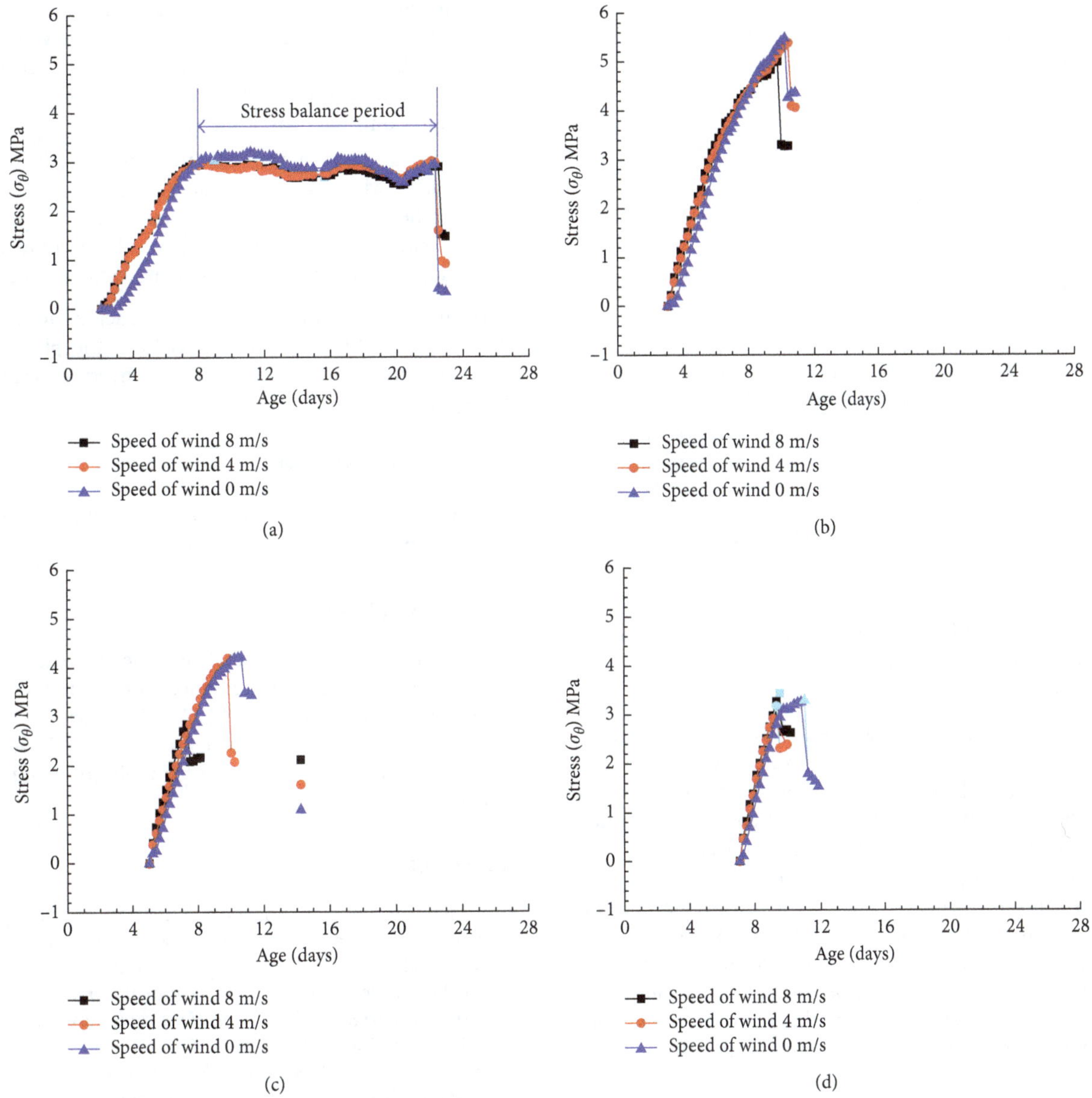

FIGURE 8: Shrinkage stress with different moisture curing ages. (a) 2 days. (b) 3 days. (c) 5 days. (d) 7 days.

3.3. *Cracking Tendency of Restrained Shrinkage Test: Ring Test and Plate Test*

3.3.1. Ring Test. Obviously, the ring test can provide quantitative information on DMPM early age stress and cracking of DMPM. The tests showed that, firstly, under the same initial drying age, the greater the wind speed, the faster the development of the DMPM ring test strain was and the greater the restraint effect of the steel ring on the DMPM ring, the larger the tensile stress induced by drying shrinkage. The reason was that the water loss occurs earlier while DMPM is exposed in the dry environment, and the elastic modulus of DMPM ring was lower. The test results also showed that the drying shrinkage value that corresponds to the cracking moment was greater as the cracking age of the DMPM ring was delayed. Conversely, if the cracking age of the DMPM ring was earlier, the cracking stress of the DMPM ring was smaller. It indicated that the DMPM drying shrinkage deformation was smaller. It means that the anticrack performance is weak. Finally, under the same wind speed, the development rate of the tensile stress was smaller while the DMPM sample was exposed to the dry environment sooner.

Figure 8(a) shows that the shrinkage stress development of the DMPM sample, which was exposed to the dry environment after 2 d moisture curing, had a stress balance period and continued for two weeks. But the phenomenon was not observed during the tests of DMPM with other initial drying age. Our interpretation of this is that, first of all, the elastic modulus of the DMPM ring specimen was smaller as the initial drying age was 2 d, and the restrained stress was smaller; secondly, the internal free water of pieces of the DMPM specimen was enough to hydrate and evaporate, and then drying shrinkage of the ring specimen

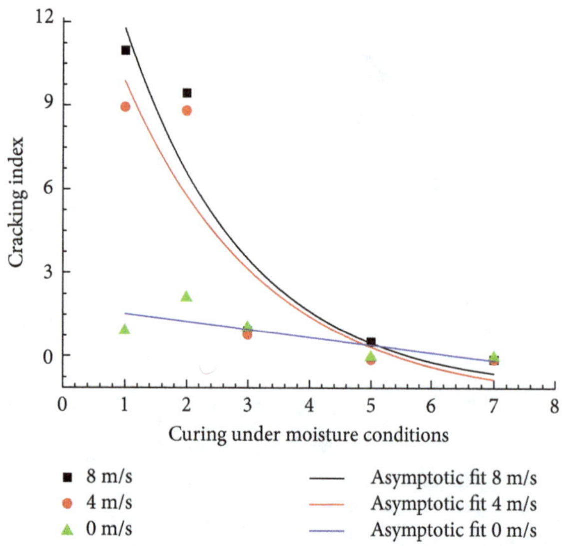

FIGURE 9: The relationship between cracking index and initial drying age.

increases slowly. Due to the abovementioned three reasons, a stress balance period occurs.

3.3.2. Plate Test. The cracking index data of flat test (Figure 9) showed that the cracking occurs more easily while the initial drying age of DMPM was earlier. More cracks would be produced when the plate sample moisture evaporated faster, and the value of cracking index test was larger as the wind speed was larger. But this phenomenon holded only for initial 3 d moisture curing. The plate specimen starts to dry after 3 d curing under moisture conditions, and the value of the crack index had a same appearance in different wind conditions. The results also showed that the cracking index presented an exponential decay with age increasing in wind environment, but appeared linear attenuation in nonwind environment.

4. Conclusions

The aim of this experimental work was to study the influences of environment conditions on cracking tendency of DMPM. Relevant results were obtained during the experimentation:

(1) The steel ring restrained test is an effective experiment method to measure restrained stress and strain of DMPM.

(2) The ring tests showed that under the same initial drying age of DMPM, the greater the wind speed, the faster the development of the ring test strain was, and the greater the restraint effect of the steel ring on the ring test piece, the larger the tensile stress caused by the shrinkage was. If cracking age of circular test pieces was earlier, the DMPM drying shrinkage deformation was smaller. It means that the anticrack performance is weak. Under the same wind speed, the sooner the DMPM sample exposed to the dry environment, the smaller the development rate of the tensile stress is. The development of internal shrinkage stress of DMPM, which pieces were stored in dry environment after 2 day age curing in moisture condition, possessed a stress balance period and continued for some time

(3) The flat tests showed that the cracks produced more easily while the initial drying age of DMPM was earlier. The cracking index presented an exponential decay with age increasing in wind environment, but appeared linear attenuation in nonwind environment.

In summary, in order to eliminate cracking tendency of DMPM, it should avoid water loss from the plaster layer during construction in practical engineering, especially, before initial drying ages.

Conflicts of Interest

The authors declare that there are no conflicts of interest regarding the publication of this paper.

Acknowledgments

The authors gratefully acknowledge the financial support from the National Natural Science Foundation of China (Project nos. 51378471 and 51778583). They also thank everyone for providing assistance for this study. In addition, thanks are extended to everyone for their contributions to the experimental work.

References

[1] C. F. Huang, R. L. Chen, and J. L. Lin, "The current practice and promotion strategy of dry-mix mortars in Taiwan—taking technology acceptance model as an approach," in *Proceedings of the 2011 International Conference on Consumer Electronics, Communications and Networks, CECNet 2011*, Xianning, China, April 2011.

[2] J. N. Eiras, J. S. Popovics, M. V. Borrachero, J. Monzó, and J. Payá, "The effects of moisture and micro-structural modifications in drying mortars on vibration-based NDT methods," *Construction and Building Materials*, vol. 94, pp. 565–571, 2015.

[3] H. Bian and J. Plank, "Effect of heat treatment on the dispersion performance of casein superplasticizer used in dry-mix mortar," *Cement and Concrete Research*, vol. 51, pp. 1–5, 2013.

[4] J. Xianyu, T. Ye, and J. Nanguo, "Early age properties and cracking control of concrete," *Journal of Building Structure*, vol. 30, no. 6, pp. 24–30, 2010.

[5] J. Nanguo, J. Xianyu, and T. Ye, "Analytical prediction and experimental research on shrinkage cracking of restrained early-age concrete," *Journal of Zhejiang University*, vol. 41, no. 9, pp. 1499–1502, 2007.

[6] T. Ohkubo, T. Inohara, and S. Matsumoto, "A study on influence the construction time of year affects the cracks of plastering mortar restrained by concrete," *Journal of Structural and Construction Engineering*, vol. 77, no. 673, pp. 333–340, 2012.

[7] S. A. Al-Saleh, "Comparison of theoretical and experimental shrinkage in concrete," *Construction and Building Materials*, vol. 72, pp. 326–332, 2014.

[8] T. H. Nguyen, A. Toumi, A. Turatsinze, and F. Tazi, "Restrained shrinkage cracking in steel fibre reinforced and rubberised cement-based mortars," *Materials and Structures*, vol. 45, no. 6, pp. 899–904, 2012.
[9] D. P. Bentz, M. R. Geiker, and K. K. Hansen, "Shrinkage-reducing admixtures and early-age desiccation in cement pastes and mortars," *Cement and Concrete Research*, vol. 31, no. 7, pp. 1075–1085, 2001.
[10] C. Zhicheng and Y. Peiyu, "Autogenous shrinkage of shrinkage-compensating concrete," *Journal of the Chinese Ceramic Society*, vol. 38, no. 4, pp. 568–573, 2010.
[11] M. Miao and Y. Peiyu, "Influences of water-binder ratio and dosage of fly ash on autogeneous shrinkage of shrinkage-compensating concrete," *Journal of the Chinese Ceramic Society*, vol. 40, no. 11, pp. 1607–1612, 2012.
[12] C. Z. Hu Hui, W. Xiaoying, C. Ke, W. Xin, and C. Ke, "Cracking tendency prediction model of early-age restrained mortar ring," *Journal of Civil, Architectural and Environmental Engineering*, vol. 3, pp. 19–25, 2015.
[13] I. B. Topcu and T. Bilir, "Experimental investigation of drying shrinkage cracking of composite mortars incorporating crushed tile fine aggregate," *Materials and Design*, vol. 31, no. 9, pp. 4088–4097, 2010.
[14] S. Xu and Y. Zhu, "Experimental determination of fracture parameters for crack propagation in hardening cement paste and mortar," *International Journal of Fracture*, vol. 157, no. 1-2, pp. 33–43, 2009.
[15] S. P. Shah, C. Ouyang, S. Marikunte et al., "A method to predict shrinkage cracking of concrete," *ACI Materials Journal*, vol. 95, no. 4, pp. 339–346, 1998.
[16] C. Jiang, Y. Yang, C. Ma et al., "Evaluation of anti-cracking performance for concrete based on temperature-stress test and analytic hierarchy process," *Journal of the Chinese Ceramic Society*, vol. 43, no. 8, pp. 1017–1023, 2015.
[17] H. Choi, M. Lim, R. Kitagaki, T. Noguchi, and G. Kim, "Restrained shrinkage behavior of expansive mortar at early ages," *Construction and Building Materials*, vol. 84, pp. 468–476, 2015.
[18] G. Rishi and B. Nemkumar, "Plastic shrinkage cracking prediction in cement-based materials using factorial design," *Journal of Materials in Civil Engineering*, vol. 27, no. 9, p. 04014244, 2015.
[19] A. B. Hossain and J. Weiss, "Assessing residual stress development and stress relaxation in restrained concrete ring specimens," *Cement and Concrete Composites*, vol. 26, no. 5, pp. 531–540, 2004.
[20] X. Ma, X. Li, C. Wang, and Z. Han, "Numerical simulation and assessment of self-induced tensile stresses in steel ring restrained concrete," *Journal of Wuhan University of Technology-Materials Science Edition*, vol. 25, no. 3, pp. 530–533, 2010.
[21] C. F. Nilsen and K. N. Subramanian, "The role of strain-rate and phase boundary geometry on the deformation behaviour of two-phase bicrystals of alpha-beta brass," *Journal of Materials Science*, vol. 19, no. 3, pp. 768–776, 1984.

17

Study of the Thermal Properties of Raffia Bamboo Vinifera L. Arecaceae

E. Foadieng,[1,2,3] P. K. Talla,[2] G. B. Nkamgang,[2] and M. Fogue[3]

[1] *Higher Technical Teachers' Training College, University of Buea, Kumba, Cameroon*
[2] *Laboratoire de Mécanique et de Modélisation des Systèmes Physiques, Faculty of Sciences, University of Dschang, Dschang, Cameroon*
[3] *Laboratoire d'Ingénierie des Systèmes Industriels et de l'Environnement, IUT-Fotso Victor, University of Dschang, Dschang, Cameroon*

Correspondence should be addressed to E. Foadieng; foidienge@yahoo.fr

Academic Editor: Fernando Lusquiños

Raffia is a kind of fast-growing palm tree, from the family of Arecaceae, encountered in marshy areas and along rivers. In this study, the "Raffia Bamboo" is the stalk of a palm, made of a fragile marrow inside a thin shell, smooth and hard to protect the latter. In our region, this material is widely used to build all the low-cost traditional houses and furniture, to make granaries storage of dry products, to build chicken coops, to make decoration. Thus, various jobs are organized around this material, with the fight against poverty. To our knowledge, information on its thermal properties is almost nonexistent. The experimental determination of the transverse thermal properties of the dry shell, the dry marrow, and the whole dry bamboo helped to find, for each, a specific heat, a thermal diffusivity, a thermal conductivity, and finally a thermal effusivity. From the analysis of results, we deduce that the thermal properties of raffia bamboo vinifera L. Arecacea make it a very good thermal insulator.

1. Introduction

The "raffia bamboo" is one of the oldest building materials used by man in our region. The reduction in timber reserves and the awareness of deforestation problems caused by crops have increased interest for raffia bamboos forest which are fast-growing resources [1]. The raffia palm, often underused, is present in the tropics. This is a multifunctional plant: it is solicited in the protection of the environment (water and soil conservation) and nutrition [2]; the petiole designated as "African bamboo" [3] and the raw leaves are used as building materials [4].

To our knowledge, unfortunately, there is no current study on the determination of thermal properties of raffia bamboo vinifera L. Arecaceae. Furthermore, the production and use of raffia bamboo are based solely on an ancestral know-how. Despite its apparent fragility, raffia bamboo offers a real opportunity to build among the poor who inhabit the tropical regions, because it is a cheap, abundant, and fast-growing material that can meet the need for broad economy housing.

Some information on raffia bamboo basic properties was discussed, but the study of its thermal properties and applications as a raw material for composite products is very limited. It is an underutilized and neglected material, so that its full potential is not used. Further studies are needed to assist and promote its application in the modern world.

The optimization of properties of raffia bamboo vinifera L. Arecaceae for the development of its products requires knowledge of not only its physical properties [5] but also its thermal characteristics. In this study, we propose to evaluate the different thermal properties of raffia bamboo vinifera L. Arecaceae.

2. Materials and Methods

2.1. Materials

2.1.1. Characteristics of the Species Studied. The raffia vinifera L. Arecaceae is in the form of a tuft is constituted of several feet, which themselves are composed of palms. A raffia palm has 4 parts: the leaflet, the spine, the petiole, and sheath basalt. The basalt sheath is the lower part of the fin characterized by more increased relative hardness. The petiole designated as "raffia bamboo" has a bright green color when fresh and gray when dry [5]. It is 5 to 10 meters high and its diameter varies from 2 to 10 cm. The petiole is smooth and the leaves are opposite needles, having a parallel nervation.

Bamboo Raffia vinifera L. Arecaceae is known locally as "dink." Our samples for the study were taken from Mbieng I quarter, in the village of Bandjoun, Poumougne Subdivision, Koung-Khi Division, Western Region Cameroon. Mbieng I is located at 5°25 north latitude, 10°25 longitude, and 1509 meters above sea level.

2.1.2. Test-Tube Preparation. On each test-tube of marrow or shell, we have prepared a smooth flat surface that will be exposed to energy flash. The probe of the thermohygrometer is inserted into the test-tube at 2 mm behind said surface for obtaining the temperature variations and to be simultaneously protected from the heat exchange by convection with the laboratory air. This arrangement allows the probe to be in agreement with the model of Parker where the specimen thickness (distance between front and back) is 2 mm.

The test-tubes used for testing have a size of 20 × 20 × 40 mm^3 for the marrow, 2 × 20 × 40 mm^3 for the shell, and $h = d/2$ for the whole bamboo, taking in account the size of our calorimeter. About 120 ml of water is weighed by a brand OHAUS balance (FLORHAM PARK) precision 1/10 and placed in a Leybold-Heraeus Calorimeter brand of German manufacture and heat capacity 66 Jk^{-1} with a thermometer precision of 1/10. The content of the latter is at the laboratory temperature close to 24°C. After a stay of 72 hours in an oven set to 100°C the test-tubes are weighed by Mettler AE160 electronic balance of 10^{-6} precision and reintroduced into the oven.

2.2. Methods

2.2.1. Specific Heat. The selected method of measuring the specific heat capacity or specific heat is known as calorimetric method [6].

After 24 hours, the test-tube is immediately taken out from the oven and added to the contents of the calorimeter [7]. Following the exchange of heat, the maximum temperature T_m of the calorimeter content is reached after a time t_1. T_m is similar to the contents temperature at thermal equilibrium when the calorimeter is perfectly adiabatic.

(1) Method Principle. A sample heated for at least 24 hours in an oven set at 100°C is immersed in water of the preceding calorimeter. One detects the maximum temperature reached by the water of the calorimeter. The heat balance between the initial time $t_0 = 0$ corresponding to the introduction of the solid sample in the calorimeter [8] and time t corresponding to the maximum temperature [9] reached by the water of the calorimeter is

$$m_S C_S \left(T_{Si} - T_{\text{emoy}}\right) = C_e \left(m_{e+}\mu_e\right)\left(T_{\text{emax}} - T_{ei}\right) + \int_0^t US\left(T_{\text{emoy}} - T_a\right) dt, \quad (1)$$

where T_{Si} the initial temperature of the specimen, T_{emax} is the maximum temperature of the water, T_{emoy} is the average water temperature, T_{ei} is the initial temperature of the water, and T_a is the ambient temperature of the experimental room.

We deduce the specific heat capacity C_S of the sample in J/kg·°C:

$$C_S = \frac{1}{m_S\left(T_{Si} - T_{\text{emax}}\right)}\left[C_e \cdot \left(m_{e+}\mu_c\right)\left(T_{\text{emax}} - T_{ei}\right) + \int_0^t US \cdot \left(T_{\text{emoy}} - T_a\right) dt\right], \quad (2)$$

where m_e is the mass of water, μ_c is the water equivalent of the calorimeter and its accessories, U is the overall heat loss coefficient, and S is the heat exchange surface.

(2) Determination of Overall Heat Loss Coefficient. One pours mass m of hot water at temperature T_{ei} in a calorimeter in thermal equilibrium with the laboratory room which is at temperature T_a. Whole water calorimeter reaches thermal equilibrium at temperature $T_{\text{eq}} < T_{ei}$. The variation of the water temperature is due to thermal losses because the calorimeter's water exchanges heat with the laboratory by convection, conduction, and radiation. If T_e is the temperature of the water at a later time, the heat balance which results can be written as follows:

$$U\left(T_e - T_a\right) = -m_e C_e \frac{dT_e}{dt}, \quad (3)$$

where U is the overall coefficient of heat loss, m_e is the mass of water, and C_e is the specific heat of water.

During the experiment, the laboratory temperature is practically constant. The integration of (3) gives time t:

$$\ln\left(\frac{T_e(t) - T_a}{T_{ei} - T_a}\right) = -\frac{U}{m_e C_e} t \quad (4)$$

Slope a_t of the straight line $\ln((T_e(t) - T_a)/(T_{ei} - T_a)) = f(t)$ is used to calculate U coefficient:

$$a_t = -\frac{U}{m_e C_e C_e}, \qquad U = m_e C_e C_e \cdot a_t. \quad (5)$$

(3) Determination of the Water Equivalent of the Calorimeter. One pours hot water of mass m_{ec} at temperature T_{ec} in a calorimeter containing cold water of mass m_{ef} at the temperature T_{ef}. If T_{emax} is the maximum temperature reached by the mixture, the heat balance gives

$$\left(m_{ef} + \mu_C\right)\left(T_{\text{emax}} + T_{ef}\right) = m_{ec}\left(T_{ec} + T_{\text{emax}}\right). \quad (6)$$

Here the heat losses are neglected because the experience is so fast that one can admit that the heat exchange between the calorimeter and the laboratory is practically negligible. We can deduce the following:

$$\mu_c = m_{ec}\frac{T_{ec} - T_{\text{emax}}}{T_{\text{emax}} - T_{ef}} - m_{ef}. \quad (7)$$

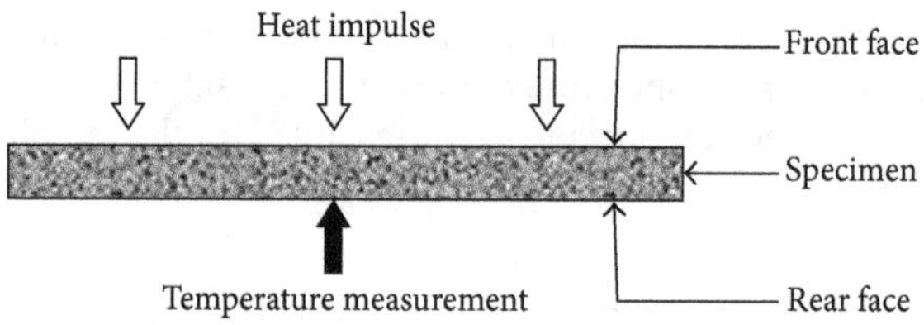

FIGURE 1: Principle of the flash methodology and implementation.

(4) Determination of the Specific Heat of a Wet Material. The specific heat of the dry test-tube is known; one can determine specific heat C of the specimen having a known moisture content. If we neglect the voluminal heat of air and water vapor compared with the liquid and solid phases, we have

$$\rho C = \rho_0 (C_S + C_e H) \tag{8}$$

knowing that

$$\rho = \rho_0 (1 + H), \quad H = \frac{m_e}{m_S} \tag{9}$$

specific heat of wet specimen is deduced by

$$C = \frac{m_S C_S + m_e C_e}{m_H}, \tag{10}$$

where ρ_0 is density of dry specimen, m_e is the mass of water contained in the specimen, m_s is the mass of anhydrous specimen, m_H is the mass of the wet specimen, C_e is the specific heat of water, and C_S is the specific heat of the dry sample.

2.2.2. Thermal Diffusivity. The process uses flash method, in transient regimen. On each specimen, a flat and smooth surface was made, covered with silicone grease, which limits the loss of energy during the heat transfer. This surface was briefly (about 30 seconds) contacting a hot plate hydraulic press EMCO brand of American manufacturing.

The temperature sensor of an electronic thermohygrometer, ETHG913R model Oregon Scientific, was introduced in the test-tube, 2 mm from the flat surface as Parker and Degiovanni models (Figure 2) to minimize heat exchange with the environment. Ten raffia bamboo samples were used for the experiments.

Principle of Plate Flash Method. The principle of the method (Figure 1) is to apply a more or less localized thermal perturbation in time and space (short flow of heat provided by a flash tube, a laser, a resistor, etc.) on the front of a test-tube and to measure, on the back, the change in temperature with respect to time. Thus, we establish the thermogram of the material. From this thermogram, the thermal diffusivity of the specimen is determined.

One of the main difficulties of the method is the consideration of heat loss by radiation and natural convection. Many techniques have been proposed to determine the thermal diffusivity of a specimen comparing the experimental thermogram with a theoretical thermogram from a model [10–18].

Thermogram of the Rear Panel. In the ideal case it is assumed that the thermal energy of the pulse is instantaneous and uniformly distributed on the front face of the sample with a surface density Q. The studied material is assumed to be homogeneous, isotropic, and opaque; these thermal characteristics are assumed to be constant with temperature, at least under the conditions of the experiment; one considers that the heat losses of different faces are void; we denote by **e** the sample thickness and by $\boldsymbol{\alpha}$ the thermal diffusivity [11].

The resolution of the heat equation, whose the solution is easily obtained in this case using the Laplace transform, leads to the expression of temperature variations on the rear face of the material and given as [10]

$$T(e,t) = T_m \left[1 + 2\sum_{n=1}^{\infty} (-1)^n \exp\left(-\frac{n^2 \pi^2 \alpha}{e^2} t \right) \right], \tag{11}$$

where T_m is the maximum value of the temperature reached by the rear face.

Pradhan et al. [9] proposed to determine the thermal diffusion from the time $t_{1/2}$ which is the time required for the temperature $T(e, t_{1/2})$ to become $T_m/2$.

Expression (11) then leads to

$$\alpha = \frac{1.39 e^2}{\pi^2 t_{1/2}}. \tag{12}$$

In this ideal case, the unique measurements of the thickness e of the sample and the time of half rise thermogram thus allow the determination of the thermal diffusivity [14]. In particular, it is not necessary to know the density per unit area Q energy absorbed by the sample. This quantity can be calculated if one knows the product value ρC where ρ is the density of the sample and C its specific heat capacity. [10] Then

$$Q = \rho C e T_m. \tag{13}$$

Conversely, if we know the amount of energy absorbed Q and T_m, the value of ρC and hence the thermal conductivity can be deduced.

Thermogram of the Front Face. The evolution of the temperature on the front face is given by

$$T(0,t) = T_m \left[1 + 2\sum_{n=1}^{\infty} \exp\left(-\frac{n^2 \pi^2 \alpha}{e^2} t \right) \right]. \tag{14}$$

For small values of t, the sample has been similar to a semi-infinite wall and the temperature is expressed as

$$T(0,t) \approx \frac{T_m \cdot e}{(\pi \alpha t)^{1/2}}. \tag{15}$$

If one knows the energy density Q absorbed, one can deduce the thermal effusivity.

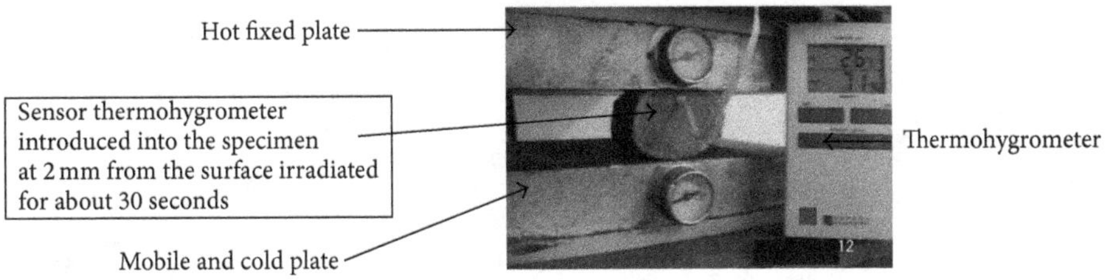

FIGURE 2: Thermal flash tester.

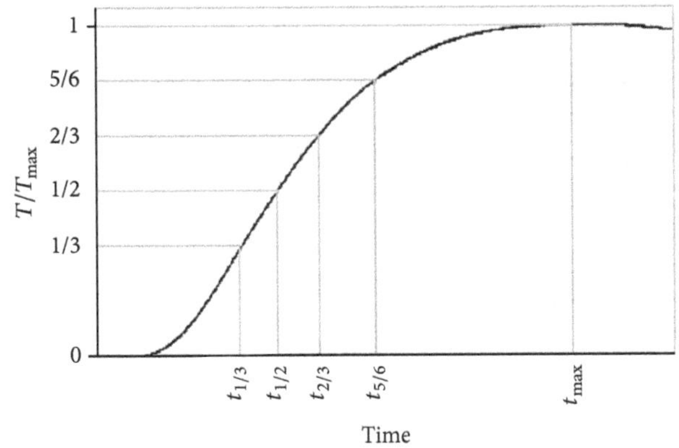

FIGURE 3: Definition of times, $t_{1/3}, t_{1/2}, t_{2/3}, t_{5/6}$ and t_{max}, used in Parker and Degiovanni models.

Analytical Models Used for the Determination of Thermal Diffusivity

(1) Parker Model. This model is to consider a perfectly isolated disc of thickness e of about 2 mm and of radius R. Initially, the sample receives an energy pulse of very short duration [19]; the heat absorption occurs on the surface uniformly [14]. Exchange coefficients are assumed to be zero on all sides and the heat flux propagates parallel to the optical axis [20] (Astm STP 885). The temperature of the rear side, expressed in reduced coordinates, is given by

$$T^* (e,t) = 1 + 2\sum_{n}^{\infty} (-1)^n \exp\left(-n^2\Omega\right), \quad (16)$$

where n is whole number, e is the thickness of the sample, t is the time, and Ω and T^* are time and reduced temperature and are, respectively, defined by the following relationships:

$$T^* (e,t) = \frac{T(e,t) - T_0}{T_{max} - T_0}, \quad (17)$$

$$\Omega = \frac{\pi^2 \cdot \alpha \cdot t}{e^2}, \quad (18)$$

where T_0 is the initial temperature and T_{max} is the maximum temperature.

The diffusivity is then obtained from the thickness of the sample and the half rise ($t_{1/2}$), that is to say the time necessary for the temperature to be equal to $T_{max}/2$ (Figure 3).

The thermal diffusivity α is obtained by

$$\alpha = 0.139 \times \frac{e^2}{t_{1/2}} \left(\mathrm{m^2 \cdot s^{-1}}\right), \quad (19)$$

where e is the thickness of the specimen and $t_{1/2}$ is the time at which the temperature of the rear face of the specimen is the half of its maximum elevation reached.

(2) Degiovanni Model. The heat losses by the various faces of the sample mean that the maximum temperature on the rear is lower than in the ideal case considered by the model of Parker. It overestimates the diffusivity of the material. Degiovanni model, on the contrary, takes into account the losses by introducing heat exchange coefficients on the upper, lower, and lateral surfaces [18]. The diffusivity is then calculated from equations (20). As part of this study, presented conductivities are those obtained by the analysis.

$$\alpha_{1/3} = \frac{e^2}{t_{5/6}} \left[0.8498 - 1.8451\frac{t_{1/3}}{t_{5/6}} + 1.0315\left(\frac{t_{1/3}}{t_{6/6}}\right)^2\right],$$

$$\alpha_{1/2} = \frac{e^2}{t_{5/6}} \left[0.968 - 1.6382\frac{t_{1/2}}{t_{5/6}} + 0.6148\left(\frac{t_{1/2}}{t_{5/6}}\right)^2\right], \quad (20)$$

$$\alpha_{2/3} = \frac{e^2}{t_{5/6}} \left[5.1365 - 11.9554\frac{t_{2/3}}{t_{5/6}} + 7.1793\left(\frac{t_{2/3}}{t_{5/6}}\right)^2\right],$$

where $t_i = t_{1/3}, t_{1/2}, t_{2/3}$, and $t_{5/6}$ (Figure 2) is the time at which the temperature of the rear rises i time its maximum elevation reached ($T_{max}/3, T_{max}/2, 2T_{max}/3, 5T_{max}/6$).

In practice we note that the value of the diffusivity is

$$\alpha = \frac{1}{3}\left(\alpha_{1/3} + \alpha_{1/2} + \alpha_{2/3}\right). \quad (21)$$

The application of this method to test pieces of wood was used to determine their thermal diffusivity and thermal conductivities in the direction perpendicular to the fiber [15].

2.2.3. Conductivity and Thermal Effusivity in the Direction Perpendicular to the Fiber. These two parameters are derived from thermal diffusivity and experimental formulas that relate to the mass density [21].

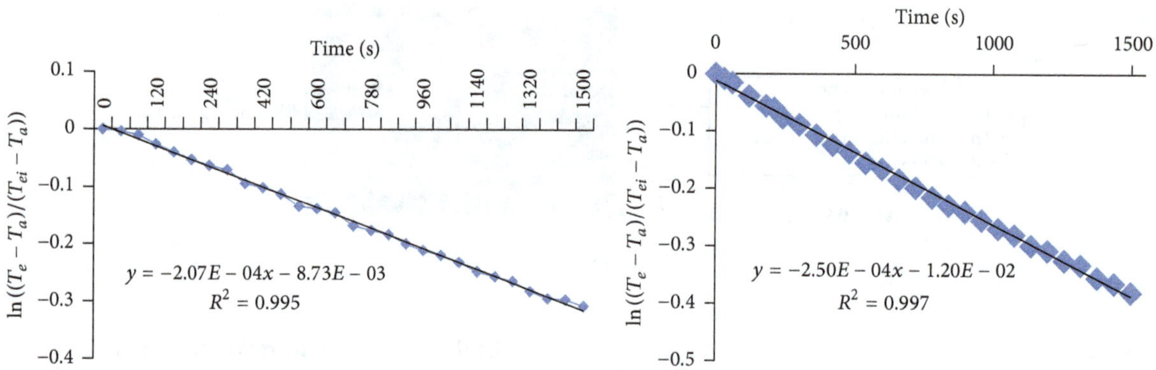

FIGURE 4: Graphic determination of the overall coefficient of heat loss calorimeter.

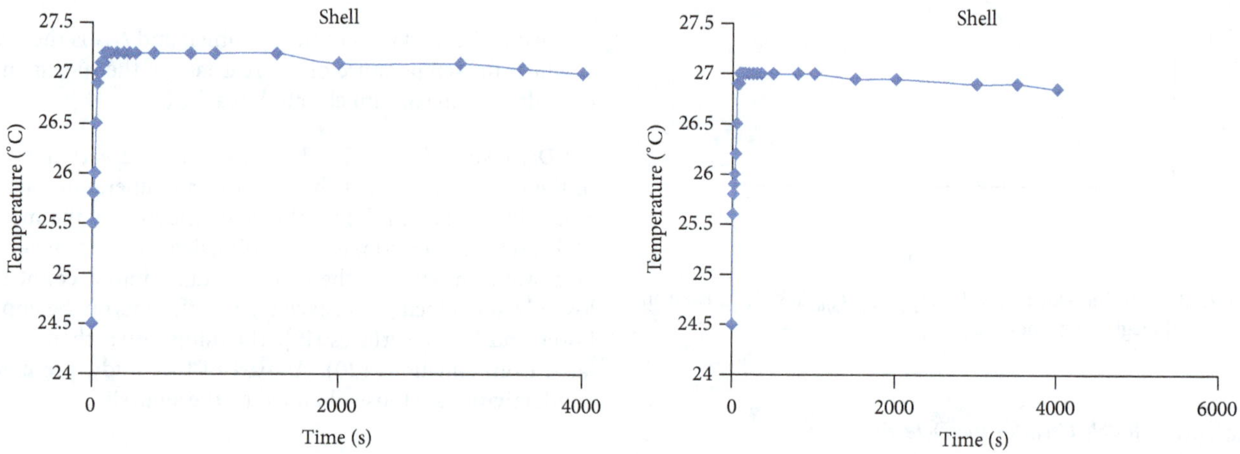

FIGURE 5: Variation of the water temperature of the calorimeter after the introduction of the hot shell.

TABLE 1: Experimental values of the overall heat loss coefficient.

Test-tube	M (g)	a_t	U (W/m^2 K)
N°1	276.32	$2.07E - 04$	0.239
N°2	211.14	$2.50E - 04$	0.221
N°3	245.23	$2.34E - 04$	0.240
N°4	255.76	$2.17E - 04$	0.232
Average			*0.233*
Standard deviation			*0.009*

3. Results

3.1. Specific Heat Capacity. A series of 10 measurements allowed us to estimate the average experimental value of the "water equivalence" of the calorimeter and accessories equal to 16.72 g, as well as the overall heat loss coefficient (Figure 4).

By linear regression analysis applied to the curves in Figure 4 we determined the slope a_t of the curve $\ln((T_e(t) - T_a)/(T_{e0} - T_a)) = f(t)$ which allowed us to calculate the overall heat loss coefficient U according to (5). Its average value is $0.233\,\mathrm{W/m^2\,K}$ knowing that the heat exchange surface S is equal to $0.03181\,\mathrm{m}^2$. The results are summarized in Table 1. The high value of the correlation coefficient $R^2 > 0.99$ allows us to believe that we are close to the good result.

After introduction into the calorimeter the anhydrous specimen was immediately removed from the oven and the temperature of the water of the calorimeter increases and reaches a maximum value. The curves (Figure 5) show changes in the temperature of the water and are used to identify the temperature values needed to determine the specific heat of each sample. The results are shown in Tables 3, 4, and 5 where each value is the average of 10 measurements.

Specific heat obtained is that for anhydrous specimen.

3.2. Thermal Diffusivity, Thermal Conductivity, and Thermal Effusivity of Marrow and Shell. During testing, we found that excitation time of 30 seconds is sufficient to achieve a significant rise in the temperature of a point in the plane located 2 mm from the irradiated side (Figure 9). Experimental data have allowed us to represent thermograms of such a plan on 10 specimens. Two of the obtained thermograms are shown in Figure 6.

Each t_i value is determined through a correlation obtained by plotting a curve of time with respect to the temperature rise. Two examples of these curves are shown in Figure 7.

The different values of time t_i deducted allowed us to represent $T/T_{\max}$ over time. The shape of the curves in Figure 8 is in agreement with the model of Parker and

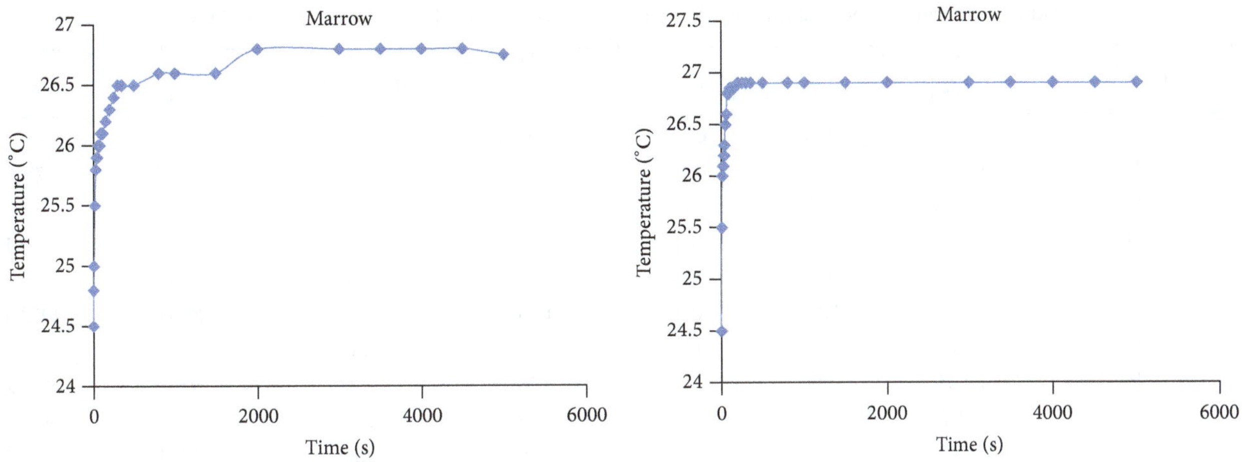

FIGURE 6: Variation of the water temperature of the calorimeter after the introduction of the hot marrow.

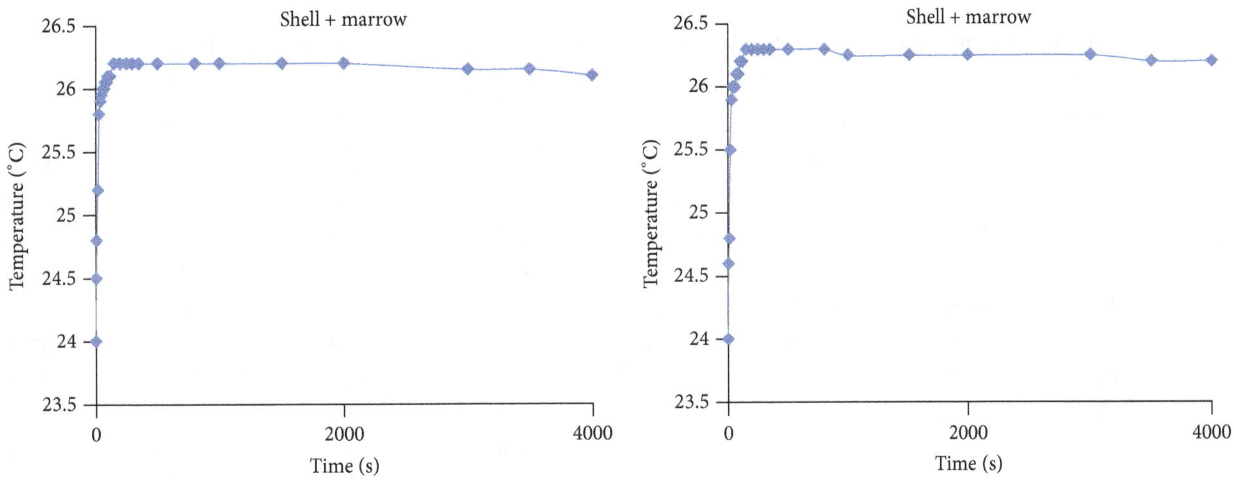

FIGURE 7: Variation of the water temperature of the calorimeter after the introduction of hot bamboo.

Degiovanni. This allows us to exploit this analytical model to determine the thermal diffusivity of raffia bamboo.

The approximate values of the thermal diffusivity α_{exp} are estimated using the formulas of Parker and Degiovanni model ((17), (18), and (19)) (Figure 10). Knowing the experimental values of the specific heat C_{exp} we deduce those of the thermal conductivity λ_{exp} and effusivity E_{exp} from (21). Formulas (18) and (21) in the literature allow us to calculate the theoretical thermal conductivity λ_{theo} in the transverse direction to the fiber in order to deduce the thermal diffusivity α_{theo} and theoretical thermal effusivity E_{theo}.

The moisture content values and densities are summarized in Table 2. The experimental average values of thermal properties, compared with the values obtained using the theoretical formulas, are grouped in Tables 3, 4, and 6.

TABLE 2: Moisture content and density of different specimens.

Specimen	Shell			Marrow		
	$H\%$	d_a	d_H	$H\%$	d_a	d_H
A	13.31	0.79	0.65	13.92	0.174	0.124
B	14.34	0.71	0.58	13.37	0.127	0.069
C	12.44	0.78	0.63	14.33	0.222	0.175
D	13.22	0.97	0.92	14.04	0.231	0.187
E	11.57	0.91	0.83	13.76	0.159	0.113
F	13.36	0.98	0.81	13.07	0.179	0.129
G	14.33	0.78	0.68	13.36	0.207	0.135
H	12.14	0.86	0.71	13.82	0.125	0.095
I	12.47	0.8	0.66	12.92	0.173	0.137
J	12.28	0.83	0.68	14.71	0.189	0.145
Average	*12.95*	*0.84*	*0.72*	*13.73*	*0.1786*	*0.1309*
Standard deviation	0.925	0.09	0.10	0.559	0.036	0.035

3.3. Thermal Properties of Transverse Bamboo Raffia Vinifera. The basics of physical properties are summarized in Table 5. Knowing the experimental values of the heat capacity and density, we evaluated the values of the transverse thermal conductivity, thermal diffusivity, and the thermal effusivity presented in the Table 3 by using of relations ((14) and (21)).

TABLE 3: Thermal capacity, diffusivity, conductivity, and effusivity of the shell of raffia bamboo.

	Specimen	C_{Sexp} (J/kg·K)	λ_{ca} (W/m·K)	$\lambda_{théo}$	$\alpha_{exp} \times 10^{-8}$ (m^2/s)	$\alpha_{theo} \times 10^{-8}$	E_{fca} (J·s$^{1/2}$/K·m^2)	E_{ftheo}
Shell	A	7833.5	0.15	0.17	2.37	2.69	963.5	930.4
	B	6964.0	0.10	0.15	2.06	2.07	703.2	778.4
	C	7178.2	0.12	0.16	2.13	2.94	819.7	850.6
	D	8822.2	0.18	0.2	2.44	2.35	1241.1	1274.1
	E	7323.8	0.14	0.19	2.17	2.81	965.9	1074.7
	F	6836.2	0.12	0.2	1.76	2.07	896.6	1052.4
	G	6286.7	0.12	0.16	2.40	2.19	767.1	827.0
	H	7642.4	0.14	0.18	2.16	2.47	959.2	988.3
	I	7580.8	0.12	0.17	2.07	2.33	853.1	922.3
	J	6557.7	0.11	0.17	1.96	2.87	773.8	870.7
Average		*7302.60*	*0.15*	*0.17*	*2.35*	*2.58*	*894.30*	*956.90*
Standard deviation		724.10	0.08	0.02	0.76	0.34	152.40	146.80

TABLE 4: Thermal capacity, diffusivity, conductivity, and effusivity of anhydrous marrow of raffia bamboo.

	Specimen	C_{Sexp} (J/kg·K)	λ_{ca}(W/m·K)	$\lambda_{théo}$	$\alpha_{exp} \times 10^{-8}$ (m^2/s)	$\alpha_{theo} \times 10^{-8}$	E_{fca} (J·s$^{1/2}$/K·m^2)	E_{ftheo}
Marrow	A	13761.0	0.042	0.05	1.72	1.71	317.1	346.0
	B	15474.2	0.049	0.041	2.47	2.09	310.3	283.9
	C	12643.1	0.051	0.059	1.83	2.09	378.3	407.0
	D	10666.3	0.043	0.06	1.74	2.43	325.5	384.5
	E	17893.7	0.047	0.047	1.66	1.67	365.7	365.7
	F	11167.5	0.043	0.051	2.17	1.89	293.2	319.3
	G	13771.7	0.049	0.056	1.72	1.57	373.7	399.6
	H	13623.1	0.038	0.041	2.23	1.99	254.4	264.2
	I	12242.4	0.038	0.05	1.80	1.82	283.7	325.4
	J	12877.0	0.055	0.053	2.27	1.79	365.9	359.2
Average		*13412.0*	*0.046*	*0.051*	*1.96*	*1.91E – 08*	*326.8*	*345.5*
Standard deviation		2093.8	0.006	0.007	0.29	0.25	42.8	47.5

TABLE 5: Moisture content and densities of raffia bamboo for the tests.

	Specimen	m_H (g)	m_a (g)	$H\%$	d_a	d_H
Raffia bamboo	A	6.25	5.52	13.22	0.30	0.34
	B	5.75	5.10	12.75	0.29	0.41
	C	8.04	7.10	13.24	0.26	0.28
	D	8.38	7.35	14.01	0.29	0.32
	E	6.01	5.26	14.26	0.23	0.27
	Average	6.89	*6.07*	*13.5*	*0.27*	*0.32*
	Standard deviation	1.23	1.07	0.62	0.03	0.06

4. Discussions

The main constants were determined, although the values are scattered like any other characteristic of the biomaterial. The specific heat of the shell and marrow is higher than that of wood, yet the product ρCP is close to that of heavy wood which allows providing a high specific heat with low thermal conductivity. Thus, the material can store heat, or it can be stored for release at the appropriate time. This is probably thanks to this property as raffia bamboo is the main material used to build the drying attic and the garret for conservation of dry products in our region. The transverse thermal conductivity of the shell is close to that of wood ranging from 0.12 to 0.23 for an average of 0.15, but that of the marrow is estimated at 0.046 less than 0.065 which is the higher limit of this property in the insulation [22]. Their transverse average thermal diffusivities are, respectively, 1.96×10^{-8} for the marrow and 2.35×10^{-8} for the shell, while the timber is approximately 4.5×10^{-7} [22].

The thermal constants vary slightly from one sample to another. The thermal conductivity of the shell in the direction perpendicular to the fiber is close to that of light wood (λ = 0.12 W/m·K), which is a good thermal insulator and

TABLE 6: Global thermal properties dry raffia bamboo.

	Specimen	C_a (J/kg·K)	λ_a (W/m·K)	$\alpha_a \times 10^{-8}$ (m^2/s)	E_{fa} ($J{\cdot}s^{1/2}/K{\cdot}m^2$)
Raffia bamboo	A	11759.5	0.07	2.08	507.5
	B	13737.0	0.07	1.78	531.8
	C	7750.5	0.07	3.28	364.78
	D	7829.4	0.07	3.13	401.5
	E	9706.7	0.06	2.69	366.0
	Average	*10156.6*	*0.07*	*2.59*	*434.3*
	Standard deviation	2588.3	0.005	6.50	79.78

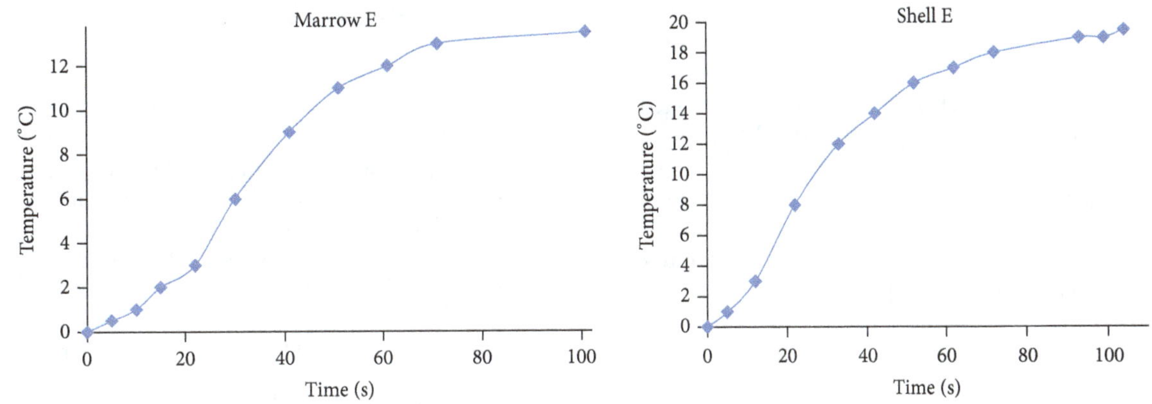

FIGURE 8: Experimental thermogram of a point in the plane located 2 mm from the surface irradiated (B).

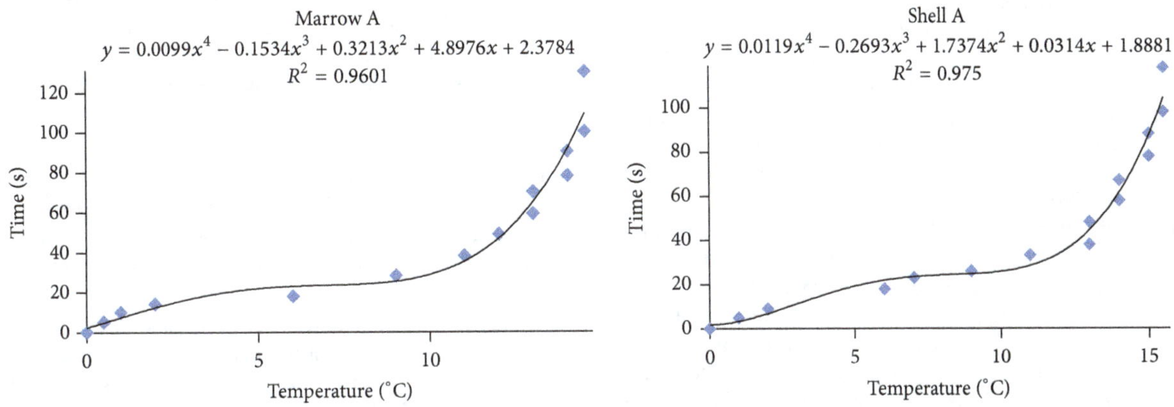

FIGURE 9: Correlation between changes in terms of temperature located 2 mm from the irradiated surface and the heat diffusion length.

that of the marrow is smaller and close to 0.04 W/m·K, characterizing wool and expanded polystyrene to be more insulating than wood [23].

An evaluation of the overall transverse thermal conductivity of anhydrous raffia bamboo gives an average of 0.07 W/m·K, smaller than that of the timber. We can justify this aspect by the fact that the raffia bamboo is more porous than wood [24]. Given the above results, we can say that the raffia bamboo is a very good thermal insulator.

5. Conclusion

The experimental study focused on the thermal properties of raffia bamboo and gives results which confirm its effectiveness as an insulation material having good heat storage capacity. It is probably this energy storage capacity that makes this bamboo building material for drying garrets, for henhouse, and for conservation of dry products in our region. From the results, raffia bamboo must be used as

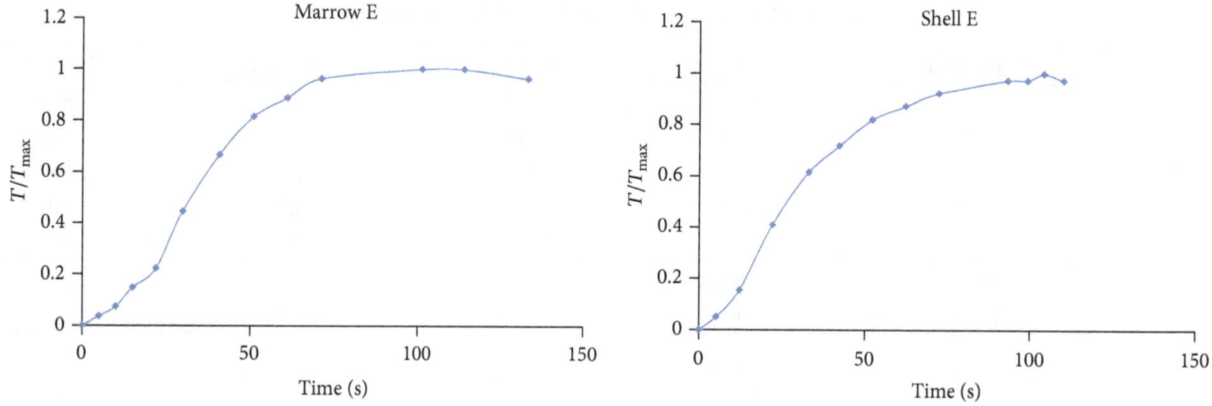

FIGURE 10: Curve T/T_{max} with respect to time of the analytical model of Parker and Degiovanni.

thermal insulation material in the structures mentioned in the introduction (houses, drying lofts, and ceilings).

Competing Interests

The authors declare that there is no conflict of interests regarding the publication of this paper.

References

[1] N. P. Shiembo, *Development and utilization of minor forest products in Cameroon with particular reference to raffia (Raphia spp.) and cane (Rattan palms) [M.Phil Thesis]*, Department of Forestry, University of Ibadan, Ibadan, Nigeria, 1986.

[2] J. K. Tangka, *Raphia palm (Raphia taedigera) as structural material in the North West province of Cameroon [Master of science seminar]*, Department of Agricultural Engineering University of Ibadan, Ibadan, Nigeria, 2001.

[3] V. Ingram, J. C. Tieguhong, E. M. Nkamgnia, J. P. Eyebe, and M. Npawe, *Bamboo Production to Consumption System, Cameroon*, CIFOR (Center for International Forestry Research), Bogor, Indonesia, 2010.

[4] J. K. Tangka, B. Ndongo, and M. Onabid, "Raffia as a structural material in the grassland region of Cameroon," *African Journal of Building Materials*, vol. 5, no. 1, pp. 10–18, 2001.

[5] E. Foadieng, P. K. Talla, F. B. Pelap, M. Fogue, S. T. Mabekou, and A. F. Et Sinju, "Contribution à l'étude de l'anatomie et des proprieties physiques du bamboo de raphia vinifera (arecaceae)," *Revue Scientifique et Technique Forêt et Environnement du Bassin du Congo (RIFFEAC)*, vol. 3, pp. 9–18, 2014.

[6] H. Urakami and M. Fukuyama, "The influence of specific gravity on thermal conductivity of wood," Bulletin of the Kyoto Prefect University for 25, 1981 (Japanese).

[7] D. L. Lang, "A quick thermal conductivity test on insulating materials," ASTM Bulletin 216, 1956.

[8] S. E. Gustafsson, E. Karawacki, and M. N. Khan, "Determination of the thermal-conductivity tensor and the heat capacity of insulating solids with the transient hot-strip method," *Journal of Applied Physics*, vol. 52, no. 4, pp. 2596–2600, 1981.

[9] P. R. Pradhan, K. Sachder, K. Bala, N. S. Saxena, and M. P. Sacksena, "Temperature dependence of thermal conductivity an thermal diffusivity of somme composites using the transient plane source technique," *International Journal of Energy Research*, vol. 13, pp. 411–417, 1991.

[10] Hladik, *Métrologie des Propriétés Thérmophysiques des Matériaux*, édition MASSON, 1990.

[11] M. Babacar Samba, *Étude du Champ Thermique dans une Plaque: Application a la Mesure de la Conductivité Thermique des Matériaux*, Projet de fin d'études en vue de l'obtention du diplôme d'ingénieur de conception, Ecole Supérieure Polytechnique, Centre de Thiès, département du Génie Mécanique, Thiès, Senegal, 2005.

[12] T. Harada, "Thermal constants of wood during the heating process measured with the laser flash method," *Journal of Wood Science*, vol. 44, no. 6, pp. 425–431, 1998.

[13] M. Mohamed Filali, *Conductivité thermique apparente des milieux granulaires soumis à des contraintes mécaniques: modélisation et mesures [Ph.D. thesis]*, Institut National Polytechnique de Toulouse, École Doctorale TYFEP, Paris, France, 2006.

[14] M. Gustavsson, E. Karawacki, and S. E. Gustafsson, "Thermal conductivity, thermal diffusivity, and specific heat of thin samples from transient measurements with hot disk sensors," *Review of Scientific Instruments*, vol. 65, no. 12, pp. 3856–3859, 1994.

[15] M. Devisme, *Conductivité et diffusivité thermiques simultanées par mesure de flux en régime non stationnaire [Ph.D. thesis]*, 1973.

[16] le. Maréchal and l. M. Devisme, *Mesure rapide et simultanée de la conductivité et de la diffusivité thermique, Entropie*, no. 64, 1975.

[17] W. J. Parker, R. J. Jenkins, C. P. Butler, and G. L. Abbott, "Flash method of determining thermal diffusivity, heat capacity, and thermal conductivity," *Journal of Applied Physics*, vol. 32, no. 9, pp. 1679–1684, 1961.

[18] A. Degiovanni, "Diffusivité et méthode Flash," Revue Générale de Thermique Française 185, 1977.

[19] C. Langlais and J. Boulant, "Use of a two heat flow transducer apparatus for transient thermal measurements on porous insulating materials," in *Proceedings of the ASTM Symposium*, Bal Harbour, Fla, USA, 1987.

[20] Astm STP 885, "*Building applications of heat flux transducers*", 1985.

[21] J. P. Cull, "Thermal conductivity probes for rapid measurements in rock," *Journal of Physics E: Scientific Instruments*, vol. 7, no. 9, pp. 771–774, 1974.

[22] P. Sallenave, *Propriétés Physiques et Mécaniques des Bois Tropicaux de l'Union Française*, Centre technique forestier tropical (CTFT), Nogent-sur-Marne, France, 1955.

[23] J. Natterer, J. L. Sandoz, M. Rey, and M. Fiaux, *Construction en Bois, Matériau, Technologie et Dimensionnement*, vol. 13 of *Traité de Génie Civil de l'Ecole Polytechnique Fédérale de Lausanne*, Deuxième édition Revue et Augmentée, Presses Polytechniques et Universitaires Romandes, CH-1015 Lausannes, 2004.

[24] CERTU/CETE de Lyon, *Confort Thermique pour Chargé d'Opération de Construction Publique*, Momento Thechnique du Bâtiment, Lyon, France, 2003.

18

Physical and Mechanical Properties of Cemented Ash-Based Lightweight Building Materials with and without Pumice

Neslihan Doğan-Sağlamtimur,[1] **Adnan Güven,**[2] **and Ahmet Bilgil**[3]

[1]*Department of Environmental Engineering, Niğde Ömer Halisdemir University, 51240 Niğde, Turkey*
[2]*ÇİMSA Cement Industry and Trade Inc. Co. Niğde Plant, Hacı Sabancı Avenue, Niğde, Turkey*
[3]*Department of Civil Engineering, Niğde Ömer Halisdemir University, 51240 Niğde, Turkey*

Correspondence should be addressed to Neslihan Doğan-Sağlamtimur; neslihandogansaglamtimur@gmail.com

Academic Editor: Cristina Leonelli

Pumice, cements (CEM I- and CEM II-type), waste fly and bottom ashes (IFA, GBA, and BBA) supplied from international companies were used to produce lightweight building materials, and physical-mechanical properties of these materials were determined. Axial compressive strength (ACS) values were found above the standards of 4 and 8 MPa (Bims Concrete (BC) 40 and 80 kgf/cm^2 class) for cemented (CEM I) pumice-based samples. On the contrary, the ACS values of the pumice-based cemented (CEM II) samples could not be reached to these standards. Best ACS results (compatible with BC80) from these cemented lightweight material samples produced with the ashes were found in 50% mixing ratio as 10.6, 13.2, and 20.5 MPa for BBA + CEM I, GBA + CEM II, and IFA + CEM I, respectively, and produced with pumice were found as 8.4 MPa (same value) for GBA + pumice + CEM II (in 25% mixing ratio), BBA + pumice + CEM I (in 100% mixing ratio), and pumice + IFA + CEM I (in 100% mixing ratio), respectively. According to the results, cemented ash-based lightweight building material produced with and without pumice could widely be used for constructive purposes. As a result of this study, an important input to the ecosystem has been provided using waste ashes, whose storage constitutes a problem.

1. Introduction

In the industry, there is the tripod of raw material-process product, and waste formation is an inevitable problem in most industrial establishments [1]. Today, at the peak of environmental problems and raw material prices, the utilization of industrial wastes as an additive material in lightweight construction technology is one of the important issues that should be considered [1, 2].

The waste ash, which poses a serious problem for their storage and disposal, is being used for different purposes throughout the world and Turkey. Utilization of millions of tonnes of new waste that emerges every year and contributes unignorably to the already existing environmental catastrophe can only be substantial in sectors that consume large amounts, such as the construction sector. For this reason, the utilization of waste as building material, thus preventing the increase of deposited waste amounts, is one of the main targets of many scientific research [3, 4]. In this study, a selection was made within the coal ash utilization alternatives, and the conversion to lightweight building material came to the forefront. In this selection, the construction sector was considered because it meets the most basic need of people that is housing, it has a leveraging role for the world and the Turkish economy with the added value and employment opportunities it has created, and it has an annual growth rate of 7.1% and a share of 6% in the gross national product (GNP); and lightweight building materials were considered because they have a high market value, they have superiority as a building material, they demonstrate high strength, and they are one of the most used materials per person after water [5].

Bottom ash (BA), which is a kind of industrial solid waste, is released after combustion of coal [1, 4, 6].

BA having a size of 0.1–10 mm and a distinctive dark grey colour is usually not spherical [7]. It is mostly composed of silicate, carbonate, aluminate, iron, and various metals and metalloids. The exact composition of BA may depend on the source of raw coal product, its size, the type of coal burner, and the operating conditions of the burner [8]. Large-sized BAs (>6 mm) can be used as coarse aggregates, while small ones are suitable for use as fine aggregates [7]. Bai et al. [9] investigated the effect of BA on concrete properties in which BA was replaced with natural sand at various rates from 30% to 100%. They investigated the strength and shrinkage properties of concrete where BA was used as a fine aggregate and reached the conclusion that BA can be used, replacing fine aggregate at a ratio of 30%. Yüksel et al. [10] reported that the use of BA as a fine aggregate improves the porosity of concrete and that the use of BA makes it possible to produce durable concrete. Nataraj et al. [11] suggested that the use of BA in concrete is an alternative to the use of aggregate in concrete. Jayaranjan et al. [8] pointed out based on the literature that BAs with relatively high specific gravity (1.9–3.0) should be used in engineering applications.

As a type of industrial solid waste, the fine-grained ash is called fly ash (FA). Its atmospheric release from the dust-holding units of pulverized coal-fired burners, especially in thermal power plants, is blocked by electrofilters [12]. The basis of FA is mainly composed of alumina silicates; alkalis, metals, and heavy metals, which can be found in small quantities with alumina silicates; and rarely soil. FA, known as the most modern material obtained as artificial pozzolan based on its chemical composition, can be used in many areas such as ceramics, plastic, wastewater treatment, cement, concrete, brick, lightweight aggregate, aerated concrete, and roads, especially in the construction sector [13]. Alkaya [14] examined the use of FAs in soil improvement. In their study, Sancak and Şamandar [12] investigated the amounts of use of blast furnace slag per FA, silica fume, and granule. They pointed out the possible contribution of waste materials to the self-compacting concrete production in terms of sustainability. Phonphuak et al. [15] demonstrated that they could achieve a lightweight building material by using FA and glass powder as the raw material with borax.

The benefits that can be achieved with the use of BAs and FAs are not limited to environmental protection and energy savings. Within the concept of industrial ecology, Basheer et al. [16] and Deja [17] had investigated the effects of this type of waste ash on concrete by using it as a mineral additive or aggregate in concrete production. Today, only 30% of the total ash produced in the world is reused, so the reuse of FAs and BAs is an important job [8].

Pumice, which is formed as a result of magma's solidification on the earth or at depths very close to the earth, is a magmatic (volcanic) rock. It has a unit weight (UW) less than 1 g/cm^3 and a porous, glassy, acidic, and basic characteristic. The pumice is in white, grey, and cream colours. This raw material does not sink when thrown into the water because it contains a large amount of gas space in the form of a foam or sponge. While it has an acid-medium composition, it contains crystal water. Because of these properties, it does not expand when heated [18]. Pumice concrete can exhibit more elastic behaviour against earthquake loads than normal concrete. Moreover, pumice concrete and building components are not affected by frost [19]. Acidic pumice, which is the most widely found on earth and the most widely used type, is white and dingy in colour, and basic pumice is blackish in colour. There are widespread uses in the construction industry due to the higher silica content in pumice with acidic character [20]. 80% of the pumice produced in Turkey is used as lightweight concrete aggregate in construction sector in the domestic market. In their study, Ünal et al. [21] investigated the effect of using pumice and Afyon Tınaztepe regional diatomite materials in the production of lightweight concrete block elements on properties of concrete. In their experimental study, they produced different series at different ratios by using pumice and diatomite of three different grain groups as lightweight aggregate, sand and crushed sand as normal aggregate. At the end of their study, they reached the conclusion that pumice and diatomite could be used instead of brick-like partition elements by utilizing the production of lightweight concrete block elements. They also concluded that besides the benefits to the economy, the issue of insulation in buildings could be solved with the use of them. Turkey, which has an important place in terms of pumice reserves in the world (with almost 40% of the world pumice reserves), has a very high market opportunity with pumice varieties of about ten different colours and texture qualities.

Lightweight building materials give the building a special character. Turkey has a geological structure in which fault and fracture systems can lead large earthquakes. There have been earthquakes in Turkey that cause big loss of life and property in short time intervals like 3 to 5 years. In developed countries such as Japan and the USA, it is no coincidence that there is little or no loss of life and property in 7.4 magnitude earthquakes. The reason for this is advanced construction technologies and the use of lightweight materials such as pumice, perlite, and aerated concrete in buildings. Lightweight materials significantly reduce building loads. The severe tremors in the horizontal and vertical directions, which are formed by an earthquake, are absorbed by porous materials, and thus buildings are shaken but not collapsed [22]. Another important point is that energy savings of up to 50% in the heating and cooling costs of buildings using lightweight building materials are ensured. Even in cold countries such as Sweden and Norway, which import and use lightweight building materials such as pumice, perlite, and aerated concrete, the heating cost of a house is less than half the heating cost of the equivalent house in Turkey. Qiao et al. [23] produced a lightweight building material (less than 1.8 g/cm^3 in density) with a compressive strength of 28 days, which was higher than 10 MPa. They found that the samples containing BA of a waste incineration plant were suitable for use as a lightweight building material based on the density and compressive strength results. Wongkeo and Chaipanich [24] conducted compressive strength, microstructural, and thermal analyses on the lightweight building material made of BA and silica fume. The compressive strength of the BA-added lightweight building material was found to increase

Table 1: Chemical analysis of waste ashes and pumice.

Element (%)	Raw material			
	Pumice	IFA	GBA	BBA
SiO_2	70.09	58.05	27.36	27.98
Al_2O_3	11.92	21.23	12.68	10.30
Fe_2O_3	2.74	6.73	8.23	—
CaO	3.84	4.76	39.31	7.97
MgO	0.88	1.71	0.74	1.64
SO_3	0.10	0.15	5.94	10.35
K_2O	2.71	1.80	0.43	0.82
Na_2O	4.70	—	1.88	0.46
Cl	0.13	—	—	—
P_2O_5	—	0.20	—	0.35
TiO_2	—	1.22	—	7.97

when silica smoke was added into the mixture. When using 5% silica fume and 20% BA mixture, they determined that a higher compressive strength was reached than the Portland cement control group. Blomqvist et al. [25] addressed the use of reusable lightweight materials in temporary buildings.

In this study, it was aimed to solve the waste ash problem emerging from coal, which is used for producing energy in the industry and thermal power plants, as well as producing lightweight building materials by evaluating the utilization efficiency of BA, FA, pumice, and cement. The study is quite detailed and different with various combinations of waste ashes, pumice, and cement of different types from different industries. It has given a new dimension to the existing knowledge of lightweight building materials.

2. Materials and Methods

2.1. Material. Cement is a calcium-based material that is used as a binder in the construction sector. Two types of cement that are CEM I 42.5 R-type Portland cement (hereinafter referred to as CEM I) and CEM II/A-L 42.5-type (Super Bims, hereinafter referred to as CEM II) Portland calcareous cement were used in this study as binders. Physical and chemical properties of these cements were not determined. Instead of this, the production values of the producer company were taken into consideration. The initial setting time and the final setting time were determined as 210 and 265 minutes and 510 and 550 minutes for CEM I and CEM II cements, respectively. The UWs of these cements were found to be 1.40 and 1.35 g/cm^3, respectively.

BA and FA used in this study as raw material were obtained from international companies that are Göknur Foodstuff Co. Nigde Factory (GBA: 700–1500 tons/year), Bor Sugar Factory (BBA: 17,000–19,000 tons/year), and İSKEN Inc. Co. Sugözü Thermal Power Plant (IFA: 300,000 tons/year). After bringing to laboratory, the raw materials were oven dried and sieved with 1 mm square sieve. During the sieving, coarse materials (1%, 3%, and 25% from IFA, GBA, and BBA, resp.) were stayed on the sieve.

The acidic pumice used in the study is supplied from Nevşehir province. The pumice beds in Nevşehir and in its vicinity have a quality that is preferred by many sectors. Grain size of this pumice generally ranges from 1 to 70 mm. The reason for preferring acidic pumice is that its UW is about 0.6 g/cm^3 and it can react better with cement. Pumice was separated into two parts including coarse and fine particles that were stayed on 1 mm square sieve (particle size is between 1 and 8 mm) and passed through this sieve (particle size is from 0 to 1 mm), respectively. The UWs of coarse and fine pumices are 0.53 and 0.91 g/cm^3. On the other hand, the UWs of GBA, BBA, and IFA were determined as 0.55, 0.78, and 0.91 g/cm^3, respectively.

According to standard test methods, chemical analysis of raw materials used in this study was done by Laboratory of ÇİMSA Cement Inc. Co. Niğde Factory and is given in Table 1.

The granulometric distribution of the pumice sieved with 8 mm square sieve, IFA, GBA, and BBA sieved with 1 mm square-mesh sieve are given in Figure 1.

2.2. Method. The production parameters of pumice-based building blocks used in construction technology should be in compliance with Turkish Standard (TS) 3234 that are based on ASTM standard [26]. According to upper and lower limits of this standard, the granulometric distribution of the materials was adjusted to produce samples having maximum compactness and is given in Figure 2. When making of mix design of GBA + pumice-, BBA + pumice-, and IFA + pumice-based sample, the materials adjusted according to Figure 2 were used.

According to TS 3234, it is recommended to do mixing calculation on the weight basis. In this standard, when considering axial compressive strength (ACS) values, pumice-based samples are separated into three groups as Bims Concrete (BC) 40, BC80, and BC120 (28-day ACS values of these concrete types are 4 MPa (40 kgf/cm^2), 8 MPa (80 kgf/cm^2), and 12 MPa (120 kgf/cm^2), resp.). Mixing design differs for each group. In this study, mixing calculations were used to obtain samples in accordance with BC40 and BC80. However, since the UW values of the materials in the study were quite different from each other, the volume basis was taken into consideration for determination of mixture parameters. In the study, two types of cement were used, and the samples were produced from five groups of different materials.

Coarse (1–8 mm) and fine (0–1 mm) materials were used in forming the groups and in the mixing calculations. Pumice was used as coarse material in the samples consisting of waste ash + pumice mixing. Waste ashes were used as fine material as a substituent of pumice.

In group 1, samples that include mixing of CEM I + pumice and CEM II + pumice were produced. This group is a comparison group to waste ash-added samples. BC40 and BC80 samples were produced according to mixing design that conforms to the TS 3234 standard [26].

In group 2, samples that include mixing of CEM I + pumice + IFA and CEM II + pumice + IFA were produced. IFA (as fine material) was substituted by pumice in 25, 50, 75, and 100% ratios. In this group, all coarse materials (1–8 mm) are pumice. BC40 and BC80 samples were produced according to mixing design that conforms to the TS 3234 standard [26].

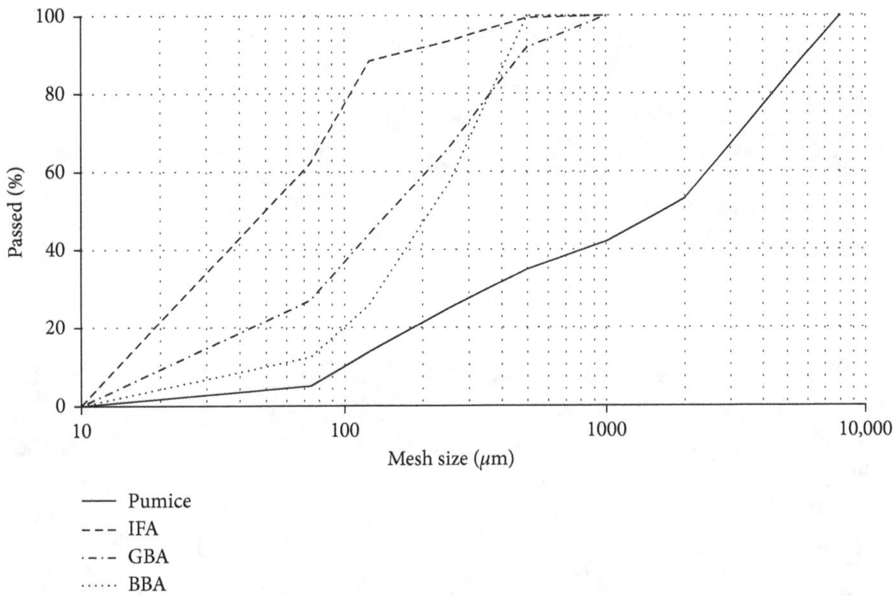

FIGURE 1: Granulometric distribution of pumice, IFA, GBA, and BBA.

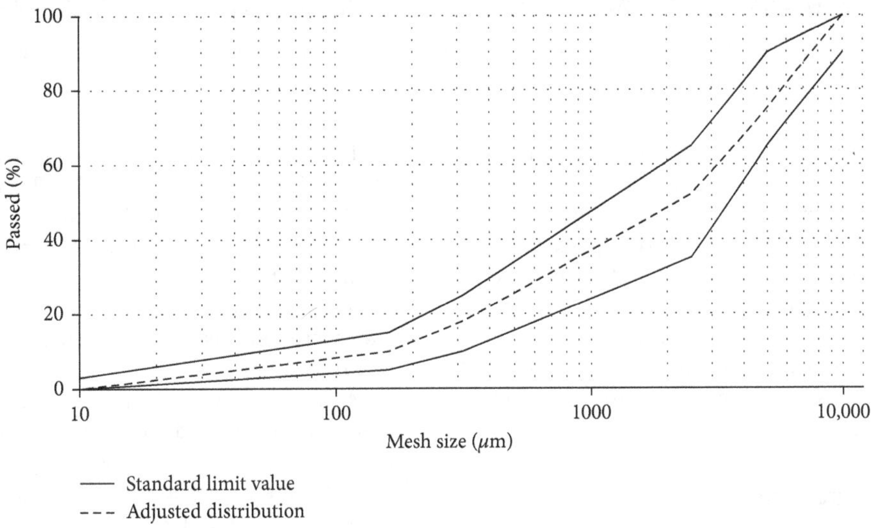

FIGURE 2: Grain-size curve of the pumice [26].

In group 3, BC40 and BC80 samples that conform to the TS 3234 standard [26] were produced from CEM I + pumice + GBA and CEM II + pumice + GBA mixing. GBA (as fine material) was substituted by pumice in 25, 50, 75, and 100% ratios. In this group, all coarse materials are pumice.

In group 4, samples that include mixing of CEM I + pumice + BBA and CEM II + pumice + BBA were produced. BBA (as fine material) was replaced by pumice in 25, 50, 75, and 100% ratios. In this group, all coarse materials are pumice. According to mixing design that conforms to the TS 3234 standard [26], BC40 and BC80 samples were produced.

In group 5, samples including only waste ashes were produced. The mixing calculation was not done in this group. In the sample production, waste ashes were bound with cement in 25, 50, 75, 100% ratios (in volume basis). Considering CEM I + IFA, CEM II + IFA, CEM I + GBA, CEM II + GBA, CEM I + BBA, and CEM II + BBA mixing, the experiments were designed.

3. Experimental

Samples produced were kept in molds ($40 \times 40 \times 160$ mm in size) and placed with the help of vibration (Figure 3). 12 replicates were produced from each sample (Figure 4).

The consistency test—because water ratios may influence the sample properties—was carried out during the sample preparation, and the consistency intervals of all samples were kept constant. The percentage of combustible substances in FA and BA at 750°C was found to be 7, 23, and 24% for IFA, GBA, and BBA, respectively. The test samples,

FIGURE 3: Some molded samples in this study.

FIGURE 4: After water absorption percentage (WAP) experiment, photographs of some samples produced in this study. Note: the acronym of BK in this figure is in Turkish, indicating BBA.

which were separated into five different groups, were produced in the same environmental conditions as much as possible and were prepared for the experiment. Samples were kept in molds for 24 h after they were placed. These samples were taken out from the molds and water cure in $20 \pm 2^{\circ}C$ temperature was applied for them. After curing/staying in water (Figure 4), 7- and 28-day ACS, UW, porosity, and WAP were determined. All the mentioned tests were made and assessed according to the relevant standard (TS 3234) [26].

4. Results and Discussion

Experiments conforming to the relevant standards were carried out on different proportions of pumice-, IFA-, BBA-, and GBA-based cemented (CEM I and CEM II) sample. These experiments included 7- and 28-day ACS, UW, WAP, porosity, and ACS after freeze-thaw. Findings of the results (as arithmetic mean values of each 3 samples) obtained from the experiments are given in Tables 2 and 3 altogether.

In this section, important parameters of cemented low-cost, durable lightweight building materials that are produced from pumice and waste ashes of two industrial plants and a thermal power plant that have different waste ash problems and different ash compositions are visualized in Figures 5–10 to better interpret their performance assessment.

According to the increase of IFA fine material ratio, the ACS values are given in Table 2 and Figure 5 to samples produced with CEM I and CEM II and IFA. The ACS values were found in the range of 3.45–5.20 MPa (mean: 4.51 MPa) and 3.00–3.45 MPa (mean: 3.20 MPa) for BC40 and 6.70–8.40 MPa (mean: 7.75 MPa) and 4.20–5.20 MPa (mean: 4.65 MPa) for BC80 in the samples produced with CEM I- and CEM II-type cement, respectively. For BC40 and BC80 Bims concrete classes, unlike the CEM I samples, the ACS values were decreased as the ash content increased in the CEM II samples.

As the ash content increased, the ACS values were increased and decreased in the CEM I + GBA and CEM II + GBA mixing samples, respectively, for BC40 and BC80. The ACS values were found in the range of 3.10–3.90 MPa (mean: 3.44 MPa) and 2.75–3.30 MPa (mean: 3.00 MPa) for BC40 and 6.80–8.40 MPa (mean: 7.78 MPa) and 7.10–8.40 MPa (mean: 7.73 MPa) for BC80 in the samples produced with CEM I- and CEM II-type cement, respectively. However, the ACS values reached the standard [26] values (BC80) in the CEM I samples mixed with 75 and 100% GBA ratios and in the CEM II samples mixed with 25 and 50% GBA ratios (Table 2, Figure 5).

For BC40, the ACS values of the CEM I + BBA and CEM II + BBA mixed samples were increased in the range of 4.30–5.20 MPa (average: 4.75 MPa) and decreased from 4.50 to 3.30 MPa (average: 3.93 MPa), respectively, as a function of the percentage of BBA. On the other hand, for BC80, these values were increased from 6.70 to 8.40 MPa (mean: 7.85 MPa) and decreased from 7.90 to 6.70 MPa (mean: 7.25 MPa), in parallel with the increase of the BBA ratio. The CEM I and BBA samples having 50, 75, and 100% mixing ratios exceeded the standard [26] (BC80) values (Table 2, Figure 5).

TABLE 2: Experimental results of pumice-based lightweight materials.

Sample*	7-day ACS (MPa)	28-day ACS (MPa)	UW (g/cm^3)	WAP (%)	Porosity	ACS after freeze-thaw (MPa)	Loss of ACS after freeze-thaw (%)
				Pumice + IFA			
BC40 CEM I + 0	2.25	5.24	0.77	45.60	38.40	3.80	27
BC40 CEM I + 25	1.35	3.45	0.90	53.50	38.10	2.80	18
BC40 CEM I + 50	2.00	4.60	0.92	52.70	37.60	3.70	20
BC40 CEM I + 75	1.90	4.80	0.95	52.80	36.80	3.80	21
BC40 CEM I + 100	2.45	5.20	0.97	52.60	36.70	4.00	24
BC40 CEM II + 0	2.10	4.17	0.75	42.60	39.80	3.20	23
BC40 CEM II + 25	0.80	3.45	0.91	41.60	38.70	2.80	19
BC40 CEM II + 50	0.95	3.30	0.94	40.90	38.40	2.60	21
BC40 CEM II + 75	1.30	3.05	1.07	40.30	39.20	2.30	25
BC40 CEM II + 100	1.60	3.00	1.16	40.00	39.10	2.20	26
BC80 CEM I + 0	3.82	9.25	0.87	41.70	40.90	7.80	16
BC80 CEM I +25	2.00	6.70	0.98	46.00	31.30	5.70	15
BC80 CEM I + 50	3.85	7.80	1.00	47.30	31.50	6.50	17
BC80 CEM I + 75	3.35	8.10	1.02	51.70	33.50	6.50	20
BC80 CEM I + 100	4.25	8.40	1.09	53.80	34.80	6.70	20
BC80 CEM II + 0	4.15	8.36	0.85	39.80	44.50	7.20	14
BC80 CEM II + 25	2.50	5.20	0.94	53.30	38.30	4.30	17
BC80 CEM II + 50	2.40	4.70	0.96	53.40	38.80	3.80	19
BC80 CEM II + 75	2.25	4.50	1.04	42.80	38.00	3.60	21
BC80 CEM II + 100	1.70	4.20	1.18	41.10	37.20	3.20	23
				Pumice + GBA			
BC40 CEM I + 0	2.25	5.24	0.77	45.60	38.40	3.80	27
BC40 CEM I + 25	1.50	3.10	0.95	42.90	40.20	2.60	16
BC40 CEM I + 50	1.70	3.25	0.94	43.80	40.20	2.60	20
BC40 CEM I + 75	1.70	3.50	0.91	44.80	40.20	2.80	20
BC40 CEM I + 100	1.80	3.90	0.89	46.90	43.00	3.10	21
BC40 CEM II + 0	2.10	4.17	0.75	42.60	39.80	3.20	23
BC40 CEM II + 25	0.90	3.30	0.97	43.00	40.70	2.70	18
BC40 CEM II + 50	1.25	2.95	0.95	45.60	40.00	2.40	19
BC40 CEM II + 75	1.70	3.00	0.92	51.40	44.00	2.30	22
BC40 CEM II + 100	0.90	2.75	0.90	53.20	44.10	2.10	23

Table 2: Continued.

Sample*	7-day ACS (MPa)	28-day ACS (MPa)	UW (g/cm^3)	WAP (%)	Porosity	ACS after freeze-thaw (MPa)	Loss of ACS after freeze-thaw (%)
BC80 CEM I + 0	3.82	9.25	0.87	41.70	40.90	7.80	16
BC80 CEM I + 25	3.40	6.80	1.05	35.20	35.30	5.70	16
BC80 CEM I + 50	2.80	7.80	1.04	35.60	35.60	6.40	18
BC80 CEM I + 75	2.70	8.10	1.02	34.10	33.20	6.60	18
BC80 CEM I + 100	3.70	8.40	1.00	34.40	33.50	6.80	19
BC80 CEM II + 0	4.15	8.36	0.85	39.80	44.50	7.20	14
BC80 CEM II + 25	2.90	8.40	1.09	38.70	38.70	6.70	20
BC80 CEM II + 50	3.00	8.10	1.06	37.60	39.70	6.50	20
BC80 CEM II + 75	2.70	7.30	1.03	34.40	34.40	5.50	24
BC80 CEM II + 100	2.40	7.10	1.00	34.20	34.10	5.50	23
				Pumice + BBA			
BC40 CEM I + 0	2.25	5.24	0.77	45.60	38.40	3.80	27
BC40 CEM I + 25	1.45	4.30	1.05	44.40	40.40	3.40	20
BC40 CEM I + 50	0.90	4.70	0.97	44.20	40.30	3.80	20
BC40 CEM I + 75	1.00	4.80	0.93	43.30	41.60	3.70	22
BC40 CEM I + 100	0.75	5.20	0.92	44.60	42.90	3.90	25
BC40 CEM II + 0	2.10	4.17	0.75	42.60	39.80	3.20	23
BC40 CEM II + 25	1.25	4.50	1.08	45.20	41.10	3.60	21
BC40 CEM II + 50	0.80	4.30	1.06	44.30	40.80	3.40	22
BC40 CEM II + 75	1.30	3.60	1.01	38.50	38.60	2.80	23
BC40 CEM II + 100	1.40	3.30	1.00	38.30	37.80	2.40	26
BC80 CEM I + 0	3.82	9.25	0.87	41.70	40.90	7.80	16
BC80 CEM I + 25	3.80	6.70	1.05	34.50	37.80	5.60	17
BC80 CEM I + 50	3.20	8.10	0.98	38.90	38.10	6.60	19
BC80 CEM I + 75	2.20	8.20	0.93	41.20	39.70	6.60	20
BC80 CEM I + 100	1.80	8.40	0.92	45.10	40.70	6.70	20
BC80 CEM II + 0	4.15	8.36	0.85	39.80	44.50	7.20	14
BC80 CEM II + 25	3.50	7.90	1.10	43.90	40.20	6.50	18
BC80 CEM II + 50	2.75	7.50	1.04	39.60	41.50	6.00	20
BC80 CEM II + 75	1.80	6.90	1.01	39.40	41.00	5.40	22
BC80 CEM II + 100	0.60	6.70	0.98	38.90	39.10	5.00	25

*The values like +25 and +50 indicate fine material (waste ash) mixing ratios.

Table 3: Experimental results of ash-based lightweight materials.

Cement amount (in volume) (%)	7-day ACS (MPa)		28-day ACS (MPa)		UW (g/cm^3)		WAP (%)		Porosity		ACS after freeze-thaw (MPa)		Loss of ACS after freeze-thaw (%)	
	CEM I	CEM II	CEM I	CEM II	CEM I	CEM II	CEM I	CEM II	CEM I	CEM II	CEM I	CEM II	CEM I	CEM II
IFA														
10	2.7	1.9	6.8	6.2	1.15	1.16	30.8	31.7	37.3	38.0	4.4	3.8	36.0	38.0
20	3.3	2.5	9.0	10.1	1.22	1.25	30.8	30.8	35.8	37.3	5.9	6.7	34.0	34.0
30	9.4	4.1	14.9	13.7	1.30	1.30	30.2	29.2	34.3	36.2	10.6	9.6	29.0	30.0
40	10.2	5.3	18.9	13.8	1.36	1.39	29.4	28.7	33.1	35.2	14.4	10.2	24.0	26.0
50	11.1	5.6	20.5	14.3	1.40	1.43	27.9	27.5	32.0	31.9	15.6	11.2	24.0	22.0
GBA														
10	0.5	0.5	4.7	1.6	0.67	0.68	46.3	45.9	46.3	47.5	3.2	1.1	32.0	31.0
20	1.6	0.8	7.0	3.9	0.77	0.78	45.8	45.2	45.1	46.1	5.0	2.8	28.0	28.0
30	3.0	3.1	7.4	6.6	0.90	0.91	45.2	43.9	42.2	43.7	5.6	5.1	25.0	23.0
40	5.5	5.4	9.4	7.8	1.02	1.01	44.1	42.6	40.8	41.7	7.4	6.1	21.0	22.0
50	7.5	6.0	13.2	12.1	1.09	1.10	43.1	41.4	39.3	39.1	10.7	9.7	19.0	20.0
BBA (0-1 mm)														
10	0.5	0.6	2.4	1.6	0.93	0.94	47.2	47.5	48.3	46.2	1.9	1.2	21.0	24.0
20	0.9	1.0	3.8	2.9	0.99	0.98	48.8	46.8	47.4	44.2	3.1	2.2	19.0	23.0
30	2.1	2.3	5.1	4.5	1.06	1.17	45.0	44.8	44.3	42.4	4.1	3.6	19.0	20.0
40	3.6	2.5	6.8	5.3	1.11	1.27	41.4	41.6	42.7	39.7	5.6	4.3	17.0	18.0
50	5.3	2.9	9.3	6.7	1.18	1.31	40.5	39.1	41.1	38.1	7.9	5.6	15.0	16.0
BBA (0–8 mm)														
10	0.5	0.1	1.2	0.8	0.95	0.98	45.0	46.1	45.7	44.3	1.0	0.6	19.0	22.0
20	1.1	0.7	3.3	1.4	0.98	1.07	38.9	41.5	42.9	43.3	2.7	1.1	17.0	20.0
30	4.0	2.5	8.5	3.8	1.15	1.22	36.3	38.6	41.1	42.5	7.2	3.2	15.0	17.0
40	4.8	4.1	9.4	7.2	1.20	1.24	34.6	37.3	40.0	41.5	8.2	6.0	13.0	16.0
50	7.3	5.7	10.6	10.3	1.22	1.26	31.0	35.0	39.6	40.1	9.4	9.0	11.0	13.0

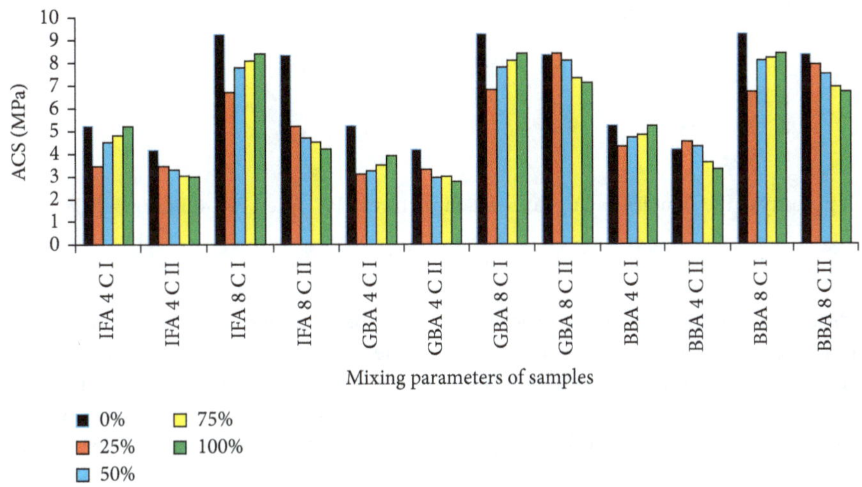

Figure 5: Relationship of ACS of the samples having different admixtures. IFA: İsken FA; 4: BC40; 8: BC80; C: CEM; GBA: Göknur BA; BBA: Bor BA.

During the substitution of pumice with fine material (IFA, GBA, and BBA), the reasons of decreasing the ACS values in the samples produced with CEM II are (a) the excess amount of calcareous found in CEM II cement and (b) the reaction between cement oxides and ashes, which influence mechanical properties.

As seen in Table 2 and Figure 6, the UW values of the samples that were produced with 100% pumice according to BC40 and BC80 lightweight building material mixing calculations were approximately 20% lower than those that were produced with ashes. As increasing fine material in the samples produced with CEM I, CEM II, and IFA, the UW values showed an increasing trend. The reason for this is the UWs of IFA and BAs are higher than those of pumice. In the samples produced with GBA and BBA, the UW decreased as the ash ratio increased. However, this decrease is in

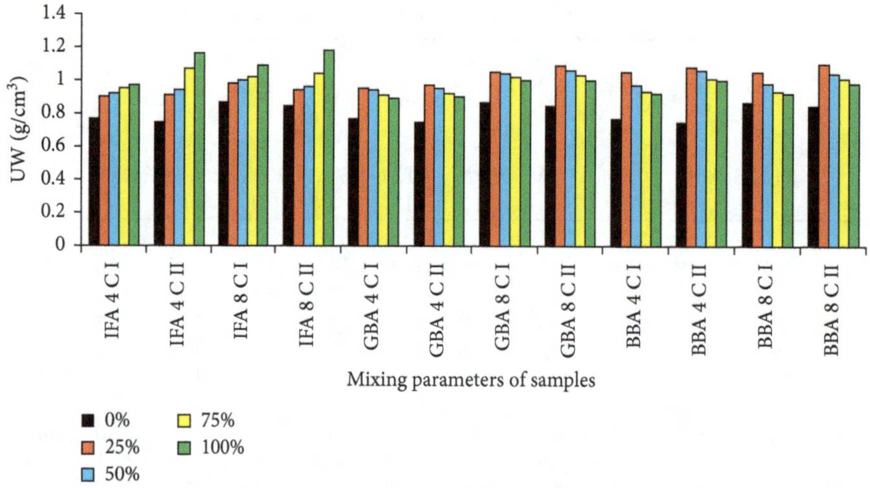

FIGURE 6: Relationship of UW of the samples having different admixtures. IFA: İsken FA; 4: BC40; 8: BC80; C: CEM; GBA: Göknur BA; BBA: Bor BA.

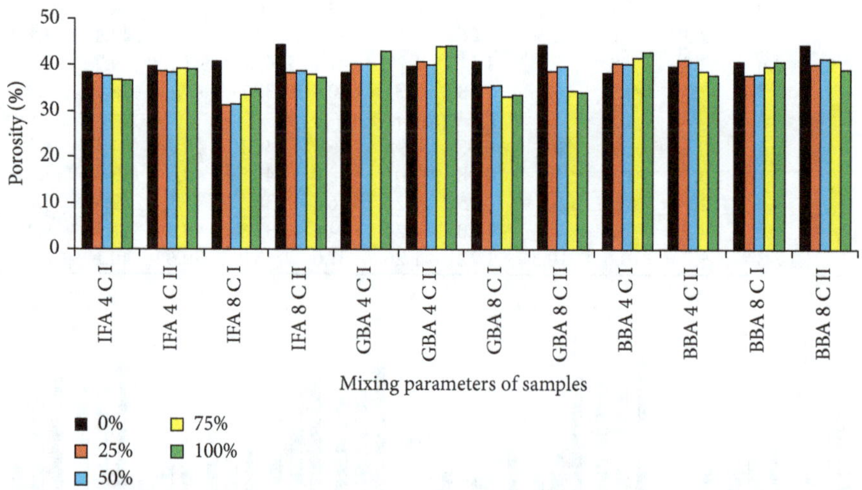

FIGURE 7: Relationship of porosity of the samples having different admixtures. IFA: İsken FA; 4: BC40; 8: BC80; C: CEM; GBA: Göknur BA; BBA: Bor BA.

negligible level. In addition, although all samples produced with ash show approximately 20% increase for the UW values, the products are in the class of lightweight building material.

In Table 2 and Figure 7, the porosities of the samples produced with CEM I, CEM II, and fine materials (IFA, GBA, and BBA) according to BC40 and BC80 did not change very much, and the results were close to each other. However, the porosities of the samples produced with pumice were higher. This can be said to be due to the fact that very fine FA and BA particles fill the pores of the pumice.

The amount of cement (Table 3) used in the 5th group and in the production of samples including 100% IFA, GBA, and BBA was determined according to the volume ratios. The BC40 standard could be achieved even with the IFA-based samples including 10% CEM I and CEM II ratios. Similar result was also obtained from the GBA-based cemented (CEM I) samples. However, this result could not be obtained from cemented (CEM II) samples. This is due to the reaction of ashes and CEM II cement and originates from the inherent properties of the cement.

The samples including 100% BBA were produced in two groups according to grain distribution. In the first group (BBA1), samples with grain distribution of 0-1 mm and in the second group (BBA2), those of 0–8 mm were produced. The ACS values of BBA1, CEM I, and CEM II mixed samples could be reached to BC40 when they include more than 30% cement. The ACS values that comply with BC40 for BBA2 could be reached in samples including more than 30% CEM I and 40% CEM II (Table 3, Figure 8).

The porosities of samples produced with 100% IFA, GBA, and BBA showed similar results with those produced with 10, 20, 30, 40, and 50% (in volume) CEM I and CEM II (Table 3, Figure 9). However, parallel to the increase in volumetric amounts of cement, the porosity of the produced samples decreased in a similar way.

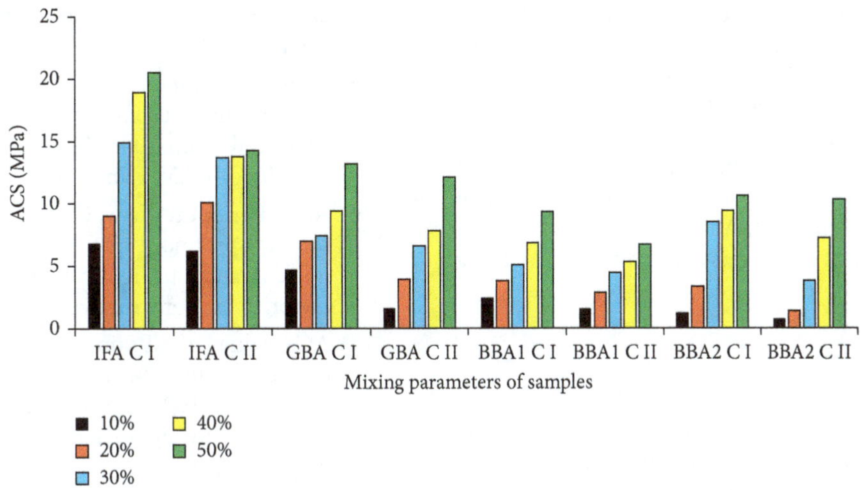

FIGURE 8: Relationship of ACS for FA- and BA-based sample. IFA: İsken FA; C: CEM; GBA: Göknur BA; BBA: Bor BA.

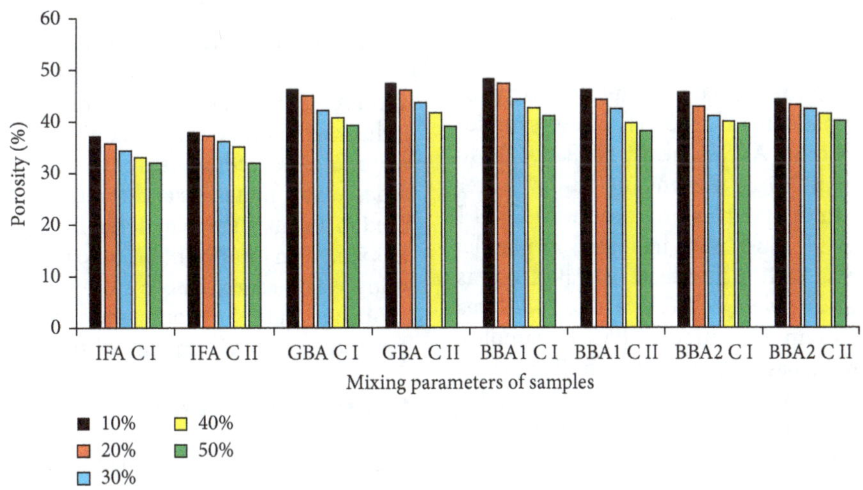

FIGURE 9: Relationship of porosity for FA- and BA-based sample. IFA: İsken FA; C: CEM; GBA: Göknur BA; BBA: Bor BA.

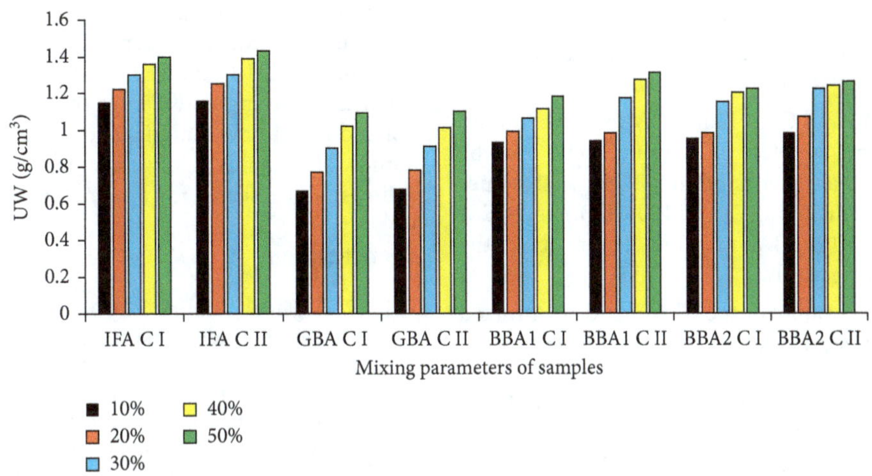

FIGURE 10: Relationship of UW for FA- and BA-based sample. IFA: İsken FA; C: CEM; GBA: Göknur BA; BBA: Bor BA.

In terms of the UWs, similar results were obtained for all IFA-, GBA-, and BBA-based samples that were produced with 10, 20, 30, 40, and 50% CEM I- and CEM II-type cement. As the volumetric cement ratio for each mixture increased, their UWs increased. The lowest UW was obtained in samples including GBA. The highest UW was found in samples produced with IFA due to higher value of IFA's UW (Table 3, Figure 10). Although the UWs of each sample varied, each was in the class of lightweight building materials.

5. Conclusions

This work has been carried out with the aim of finding solutions for the waste ash problem with known technologies in the building industry. For this purpose, waste ash is added to the mixture of constructive building block produced by pumice. 100% waste ash-based cemented samples were also produced. The physical and mechanical properties of the produced samples were determined and the following results were obtained:

(1) 100% pumice-based standard samples were produced for comparison to TS 3234. These samples must reach minimum ACS values of 4 MPa for BC40 and 8 MPa for BC80 at the end of 28-day water curing period. In the samples produced with CEM I and IFA, the ACS values were increased in parallel with the increase in IFA fine material ratio. It was concluded that an average ACS of 4.51 MPa was found for BC40 in cemented (CEM I) samples. Therefore, pumice-based lightweight building blocks with IFA additive can be produced for constructive purposes. The targeted results for the samples produced with CEM II could not be obtained.

(2) For the samples produced with CEM I and IFA, the ACS values increased with increasing IFA fine material ratio. The average ACS values of 8.12 MPa for BC80 were obtained for 50, 75, and 100% IFA-based samples. It was found that the IFA-additive pumice-based lightweight building blocks for constructive purposes can be produced. By using CEM II, the targeted results could not be obtained.

(3) For the CEM I and CEM II and GBA mixed samples for BC40, the targeted ACS value could not be achieved. On the other hand, for BC80, ACS values of the samples including 75 and 100% GBA and CEM I and 25 and 50% GBA and CEM II were above the standards, respectively.

(4) For BC40, ACS values of the CEM I and BBA mixed samples increased with increasing percentage of waste ash and found to be above the standard value with a minimum of 4.30 MPa. For the CEM II and BBA mixed samples, standard ACS values were found at 25 and 50% ratios of BBA. For BC80, the ACS values of samples including CEM I and BBA in 50, 75, and 100% mixture ratios were found to be above the standard values. ACS values for CEM II and BBA mixed samples were not reached the standard value.

(5) It is determined that lightweight concrete and building material for constructive purposes can be produced from 100% IFA-, GBA-, and BBA-based cemented (CEM I and CEM II) material.

(6) ACS after freeze-thaw, UW, WAP, and porosity of all samples produced in this study are obtained in required values to lightweight concrete and building material for constructive purposes.

(7) ACS values of all samples after freeze-thaw are decreased about 20%. It is similar with the obtained results from 100% pumice-based samples. Moreover, there are no significant adverse effects on visual inspection, and samples resistant to external influences can be produced.

(8) It was found that IFA is the most suitable fine material for producing pumice-ash-based cemented lightweight building materials.

The waste formation constitutes an inevitable problem in the majority of industrial establishments. In this study, BA and FA were used in order to solve the waste ash problem of industrial and thermal power plant. The lightweight building material obtained by using the effectiveness of pumice and ashes as raw materials and cement as a binder can be widely used as a constructive material in the construction sector. In this study, an important input to the ecosystem was provided using waste ashes, whose storage constitutes a problem. A contribution to the economy was also provided as well as the protection of environment with natural materials.

Conflicts of Interest

The authors declare that they have no conflicts of interest.

Acknowledgments

This study was supported by the Turkish Scientific and Technical Research Council (TUBITAK) 2241/B Programme, Project Grant no. 1139B411402723. Authors acknowledge ÇİMSA Cement Industry and Trade Inc. for providing CEM I and CEM II cements and Department of Civil Engineering and Center of Application and Research for Industrial Raw Material and Building Materials of Niğde Ömer Halisdemir University for laboratory conditions, editors, and referees for comments.

References

[1] N. Doğan-Sağlamtimur, A. Bilgil, M. Demir et al., "A reuse study from Niğde, Turkey: the conversion of industrial ash to geopolymer," *Desalination and Water Treatment*, vol. 57, no. 6, pp. 2604–2615, 2016.

[2] B. Postacıoğlu, "Beton Bağlayıcı Maddeler," cilt 1, Teknik Kitaplar Yayınevi, İstanbul, 1986.

[3] H. Yazıcı, *Termik Santral Atiği Yapay Alçi-Uçucu Kül-Taban Külü Esasli Yapi Malzemesi Geliştirilmesi*, Dokuz Eylül Üniversitesi, Fen Bilimleri Enstitüsü Doktora Tezi, İzmir, Turkey, 2004.

[4] N. Doğan-Sağlamtimur, A. Bilgil, T. Cila et al., "Niğde fabrikası kazan altı küllerinden çimento bağlayıcılı hafif yapı elemanı üretimi," *Çevre Bilim ve Teknoloji*, vol. 1, pp. 50–57, 2016.

[5] INTES, *Türkiye İnşaat Sanayicileri İşveren Sendikası İnşaat Sektörü Raporu Mart*, 2014, http://www.intes.org.tr/content/MArt-2014.pdf.

[6] İ. Yüksel, Ö. Özkan, and T. Bilir, "Kazanaltı külü ile briket üretimi," *Gazi Üniversitesi Mühendislik-Mimarlık Fakültesi Dergisi*, vol. 21, no. 3, pp. 527–532, 2006.

[7] R. Siddique, "Utilization of coal combustion by-products in sustainable construction materials," *Resources, Conservation and Recycling*, vol. 54, no. 12, pp. 1060–1066, 2010.

[8] M. L. D. Jayaranjan, E. D. van Hullebusch, and A. P. Annachhatre, "Reuse options for coal fired power plant bottom ash and fly ash," *Reviews in Environmental Science and Bio/Technology*, vol. 13, no. 4, pp. 467–486, 2014.

[9] Y. Bai, F. Darcy, and P. A. M. Basheer, "Strength and drying shrinkage properties of concrete containing furnace bottom ash as fine aggregate," *Construction and Building Materials*, vol. 19, no. 9, pp. 691–697, 2005.

[10] İ. Yüksel, T. Bilir, and Ö. Özkan, "Durability of concrete incorporating non-ground blast furnace slag and bottom ash as fine aggregate," *Building and Environment*, vol. 42, no. 7, pp. 2651–2659, 2007.

[11] K. Nataraj, S. M. Midun, K. M. Hari, G. M. Ganesh, and A. S. Santhi, "Bottom ash concrete," *International Journal of Advanced Research in Mechanical and Production Engineering and Development*, vol. 1, no. 1, pp. 6–11, 2013.

[12] E. Sancak and A. Şamandar, "The role of usage of self-consolidating concrete with regard to sustainability," in *Proceedings of International Sustainable Building Symposium*, pp. 134–140, Ankara, Turkey, May 2010.

[13] İ. Bentli, A. O. Uyanık, U. Demir, O. Şahbaz, and M. S. Çelik, "Seyitömer termik santrali uçucu küllerinin tuğla katki hammaddesi olarak kullanimi," in *Türkiye 19, Uluslararası Madencilik Kongresi ve Fuarı*, pp. 385–389, İzmir, Turkey, Haziran 2005.

[14] D. Alkaya, "Uçucu küllerin zemin iyileştirmesinde kullanılmasının incelenmesi," *Yapı Teknolojileri Elektronik Dergisi*, vol. 5, no. 81, pp. 61–72, 2009.

[15] N. Phonphuak, S. Kanyakam, and P. Chindaprasirt, "Utilization of waste glass to enhance physical and mechanical properties of fired clay brick," *Journal of Cleaner Production*, vol. 112, pp. 3057–3062, 2016.

[16] P. A. M. Basheer, P. R. V. Gilleece, A. E. Long, and W. J. Mc Carter, "Monitoring electrical resistance of concretes containing alternative cementitious materials to assess their resistance to chloride penetration," *Cement and Concrete Composites*, vol. 24, no. 5, pp. 437–449, 2002.

[17] J. Deja, "Freezing and de-icing salt resistance of blast furnace slag concretes," *Cement and Concrete Composites*, vol. 25, no. 3, pp. 357–361, 2003.

[18] O. Ünal, A. Ergün, T. Uygunoğlu, and G. Kürklü, *Hafif Agregalı Elemanların Fiziko-Mekanik Özelliklerin Araştırılması ve Modellenmesi*, TÜBİTAK, Project No: 109M391, Ankara, Turkey, 2008.

[19] Ş. G. Özkan and G. Tuncer, "Pomza madenciliğine genel bir bakiş," in *4. Endüstriyel Hammaddeler Sempozyumu*, pp. 200–207, İzmir, Turkey, Ekim 2001.

[20] İ. Demir, M. S. Başpınar, G. Görhan, E. Kahraman, and O. Akyol, "Pomza agregali hafif beton isil özelliklerine polistiren köpük ve uçucu kül katkisini etkileri, x. ulusal tesisat mühendisliği kongresi," in *Binalarda Enerji Performansı Sempozyumu*, pp. 911–916, İzmir, Turkey, Nisan 2011.

[21] O. Ünal, İ. Demir, and T. Uygunoğlu, "Pomza ve diyatomitin hafif blok eleman üretiminde kullanilmasinin araştirilması," in *III. Ulusal Kırmataş Sempozyumu*, pp. 107–113, İstanbul, Turkey, Aralık 2003.

[22] M. Eriç, *Yapı Fiziği ve Malzemesi*, Literatür Yayıncılık, İstanbul, Turkey, 1994.

[23] X. C. Qiao, B. R. Ng, M. Tyrer, C. S. Poon, and C. R. Cheeseman, "Production of lightweight concrete using incinerator bottom ash," *Construction and Building Materials*, vol. 22, no. 4, pp. 473–480, 2008.

[24] W. Wongkeo and A. Chaipanich, "Compressive strength, microstructure and thermal analysis of autoclaved and air cured structural lightweight concrete made with coal bottom ash and silica fume," *Materials Science and Engineering A*, vol. 527, no. 16-17, pp. 3676–3684, 2010.

[25] L. Blomqvist, A. Pfriem, and J. Johansson, "Temporary buildings in reusable lightweight material design," in *Proceedings of World Conference on Timber Engineering*, Vienna, Austria, August 2016.

[26] TS 3234, *Bimsbeton Yapım Kuralları, Karışım Hesabı ve Deney Metotları, Türk Standartları Enstitüsü (Mining, placing, curing, mix design and methods of testing of pumice concrete)*, Turkish Standards Institution (TSE), Ankara, Turkey, 2005.

19

Influences of Specimen Geometry and Loading Rate on Compressive Strength of Unstabilized Compacted Earth Block

Guan-qi Lan, Yi-hong Wang, and Si-si Chao

School of Civil Engineering, Chang'an University, Middle-section of Nan'er Huan Road, Xi'an, Shaanxi 710064, China

Correspondence should be addressed to Yi-hong Wang; 2016028001@chd.edu.cn

Academic Editor: Carlo Santulli

Loading rate and specimen geometry are critical parameters used in simple compressive tests to determine the strength of earth materials. In this study, cubic and cylindrical unstabilized compacted earth block of different sizes were manufactured and tested in unconfined compression to investigate the size and shape effects on the compressive strength of earth materials under different loading rates. The correlation between the compressive strength of earth materials and the geometric parameters of the specimen was determined. The recommended loading rate and sample for testing the compressive strength of earth materials were given. Finally, a constitutive model for describing the stress-strain response of earth materials was proposed. This model modifies the defects of the constitutive equations suggested in previous studies and may be applied to the numerical analysis of earth structure.

1. Introduction

Earth as a natural building material has been widely used all over the world [1], and it can be attributed to the many advantages such as low cost, excellent thermal performance, and easy to use local materials [2]. However, compared with industrial building materials like concrete, earth materials have the apparent deficiency in physical property, mechanical property, and durability [3, 4]. It hindered the development and application of earth materials in the field of construction. Due to the improvement in people's attention to the environment, the problem of sustainable development has attracted much attention in the field of civil engineering. Meanwhile, the application of low-cost and sustainable materials in construction has become the focus of research [5, 6]. Consequently, earth materials have attracted people's attention again attributing to its ecological benefits and widespread regional adaptability [7–9]. Compressive strength is the most important mechanical property of earth materials and one of the fundamental mechanical parameters required for earth structure design. Therefore, accurate measurement of the compressive strength of earth materials was of high research and engineering significance. However, in most of the existing study, the size and shape of the specimen for determining the uniaxial compressive strength of earth materials were different, and the loading rates used in the simple compressive test were not uniform [10–13]. It has been noted that the experimental conditions, such as specimen shape, size, and loading rate, significantly influence testing results, and the differences in test conditions make it difficult to compare the previous results.

Platen restraint is one of the reasons that cause the specimen geometry effect on compressive strength. The treatment method of geometrical effects used in both New Zealand [14] and Australia [15] was to reflect the platen restraint effects through a single aspect ratio correction factor. The advantage of this method was that it can be applied to a variety of different specimen sizes, but it also depended on the accuracy of correction factors. According to the Standards New Zealand "NZS 4298:1998" [14] and Australian handbook [15], the compressive strength of the specimen with an aspect ratio of 2 was 0.88 times the compressive strength of the cube specimen. Morel et al. [16] indicated that another reason for geometrical effects on the compressive strength of earth specimens was the friction during specimen manufacture. They stated that for the samples were produced using single compaction, the density increases with the height close to the compression surface due to friction along the mould sides.

Several studies reported in the previous literature suggested that the compressive strength of earth materials with comparability can be obtained by standardizing the size and shape of the specimens. Silveira et al. [17] established the relationship between the compressive strength of cubic specimens and the compressive strength of the cylinder, with a correction factor of about 0.94 for converting the cube strength to cylinder strength. They also concluded that the slight difference in the compressive strength of the specimens with different shapes could be minimized by the regularization mortar at the end of specimens. However, the testing specimens were taken from the existing adobe building in Aveiro district, Portugal, which caused the high variability of results. Ciancio and Gibbings [11] measured the compressive strength of cement-stabilized rammed earth by simple compression test of the specimen with slenderness ratio equal to 2. Due to the limited number of specimens, the results failed to reflect the effect of the specimen size on compressive strength. Meanwhile, they found that the cylinder specimens with loading end contact with the steel plate show an evident slenderness effect, but the slenderness effect was substantially eliminated after the steel plate was replaced by the Teflon sheet. Aubert et al. [18] measured the compressive strength of small cube specimens dry sawn from earth bricks. Because the practical implementation of the manufacturing process was rather difficult, he advised against using this method to obtain the compressive strength of earth brick. However, this problem is easily solved by making test specimens with the same process as the brick. Although the test results with comparability can be obtained by using a standard test specimen, the shape and size of the standard specimen still need further study.

Loading rate effects on the mechanical properties of earth materials play an essential part in the measurement of material compressive strength. However, a small amount of test data has now been acquired for earth materials under a wide range of loading rates. According to results by unconfined compression test for frozen soil, Chen et al. [19] found that compressive strength increased 100% in average when loading rates rose from 5×10^{-5} to 1×10^{-3} and proposed (1) to correlate the compressive strength f_c with loading rates:

$$f_c = A\left(\frac{\dot{\nu}}{\dot{\nu}_0}\right)^n, \tag{1}$$

where $\dot{\nu}$ represents loading rate, $\dot{\nu}_0$ represents benchmark loading rate, A is a temperature-related parameter, and n is an index. After the analysis by Chen et al., Du et al. [20] suggested that the elasticity modulus of the frozen soil increased with the loading rate in a specific range, but when the loading rate exceeded this range, the modulus of elasticity decreased with the loading rate. The above studies were focused on frozen soil, and the analysis of the reasons for loading rate effect was not given. For concrete-like materials, a significant number of studies indicated that the enhancement of compressive strength at high loading rate was mainly caused by a combination of fundamental material behavior and inertial confinement effects [21–23]. These analyses provided a reference for the study of the loading rate effect on earth materials.

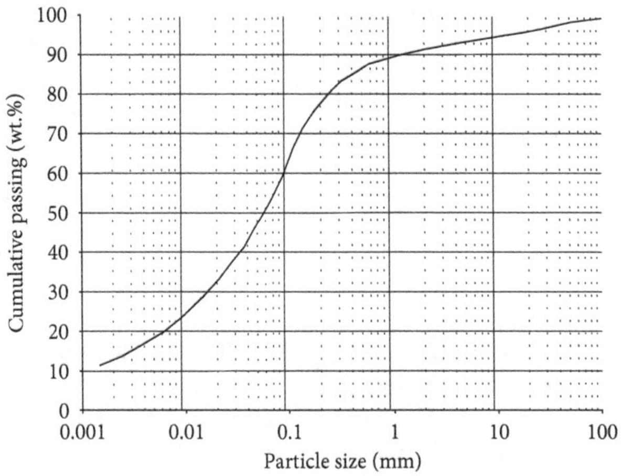

Figure 1: Particle size distribution of soil.

At present, only a few countries have given the recommended loading rate for the simple compression test of earth materials. The loading rate used in most of the simple compression tests of earth materials was carried out according to the test method for the strength of concrete and stone [24]. Consequently, the study on loading rate effect of earth materials and determination of standard loading rate used in uniaxial compression test has emerged as an urgent issue.

This study performed simple compression tests on cubic and cylindrical compressed earth blocks (CEBs) to investigate the influences of specimen size, specimen shape, and loading rate on the compressive properties of earth materials. The empirical relations between geometric parameters of the specimen and compressive strength were established. The failure patterns and failure mechanism of different specimens in simple compression tests were analyzed. Besides, the constitutive model for describing CEB's stress-strain behavior under compression was proposed.

2. Materials and Methods

2.1. Materials. The CEBs tested were produced using soil collected from the region of Xi'an in China. The particle analysis test and limit water content test were carried out following Specification of Soil Test (SL237-1999) [25] to obtain composition and Atterberg limits of soil particles. Figure 1 shows the grain size distribution of the tested soil. The composition of the soil was as follows: 17.6% clay (<5 μm), 38.5% silt (5–75 μm), 36.1% sand (75–2000 μm), and 7.8% gravel (>2000 μm). Atterberg limits of soil were obtained as follows: $W_L = 33.6\%$ (liquid limit), $W_P = 17.2\%$ (plastic limit), and $I_P = 16.4\%$ (plasticity index). It can be classified as low liquid limit clay (CL) according to the Standard for Engineering Classification of Soil (GB T 50145-2007) [26]. The optimum moisture content (OMC) and maximum dry density (MDD) of soil were 18.2% and 1.9 g/cm^3, respectively, and it was also carried out in compliance with SL237-1999 [25].

Table 1: Number of specimens.

Label	Shape	Size (mm)	V_c (mm/min) 1	3	5	7
Cu-100	Cube	100	10	10	10	10
Cu-150		150	10	10	10	10
Cu-200		200	10	10	10	10
Cy-100	Cylinder	100	10	10	10	10

The cross-sectional diameter of cylindrical specimens is equal to the height of it. V_c, loading rate.

2.2. Specimen Preparation. Cubic CEBs with different sizes (70.7 mm, 100 mm, 150 mm, and 200 mm) were used for investigating the size effect on compressive strength of earth materials. Meanwhile, ten cylindrical CEBs were produced as the contrast group. The number of specimens, per size, shape, and type of test are indicated in Table 1.

Test specimens were compacted by a hydraulic Jack. Details of the preparation of these specimens were stated below:

(i) The soil was dried in an oven and sieved through a 5 mm screen. The optimum moisture content of soil (18.2%) was chosen as the compaction moisture content to make the density of specimen reach the maximum dry density of soil. Water was evenly spread to the soil until the optimum moisture content was reached.

(ii) The wet soil was filled with the steel mould and compacted in two layers. The mass of the damp soil in each layer was controlled to ensure that the maximum dry density was obtained. The close contact between the two layers was secured by rough surface treatment on the first layer of compacted soil.

(iii) The specimen was removed from the mould after the compaction process. The curing of CEB was realized in a homothermal chamber at a temperature of 26 ± 2°C and relative humidity higher than 60%. After 28-day curing, the weight of the CEB must be kept constant before starting the test. The average dry density of the specimen was 1.88 g/cm^3. The average water content of the specimen at the time of the test was 5.62%.

2.3. Test Device and Procedure. An electrohydraulic servo testing machine with the extern digital controller was used to measure the compressive strength of CEB. The max load of the testing machine was 300 kN. The force-displacement curve of specimens can be automatically drawn by the computer system.

The testing procedures, in this case, were referenced in the Test Method of Mechanical Properties on Ordinary Concrete (GB/T 50081-2002) [27]. Because the CEB has strong deformability but low strength, the displacement control was adopted in loading. Four loading rates (1 mm/min, 3 mm/min, 5 mm/min, and 7 mm/min) were selected to investigate the influence of loading rate on CEB. A total of 160 specimens were examined.

Table 2: Mechanical properties of CEB.

Label	V_c (mm/min)	f_{mean} (MPa)	CV	$\varepsilon_{bc}\cdot 10^{-2}$
Cu-100	1	3.24	0.37	2.64
	3	3.64	0.25	2.86
	5	2.90	0.54	2.74
	7	2.81	0.37	2.62
Cu-150	1	2.46	0.32	2.54
	3	2.30	0.19	2.68
	5	1.82	0.33	2.90
	7	2.03	0.22	2.94
Cu-200	1	1.54	0.19	3.68
	3	1.44	0.22	3.32
	5	1.27	0.28	5.40
	7	1.11	0.17	6.05
Cy-100	1	2.71	0.14	2.71
	3	2.73	0.15	2.51
	5	3.13	0.18	2.47
	7	2.96	0.22	2.73

f_{mean} is the average of each group after the abnormal values were eliminated. CV is coefficient of variation of the compressive strength. ε_{bc} is the strain corresponding to the peak load.

The unconfined compressive strength (f_{cu}/MPa, (2)) was defined as the ratio of the maximum axial force (F_{max}/N) to the compressed area (A/mm^2):

$$f_{cu} = \frac{F_{max}}{A}. \tag{2}$$

The Grubbs statistics theory was used to eliminate the abnormal values in each group of unconfined compressive strength. For example, the compressive strength (MPa) of ten cubic specimens with the size of 150 mm was 2.47, 2.62, 2.49, 2.16, 2.06, 1.60, 2.18, 2.40, 2.12, and 2.23, respectively. The maximum, minimum, mean, and standard deviation of the above data were 2.62, 1.60, 2.53, and 0.27, respectively. The Grubbs test statistic was defined as the ratio of the residual to the standard deviation and calculated from the following equation:

$$G'_n = \frac{(\sigma_{max} - \sigma_{mean})}{S_\sigma} = 0.33,$$
$$G_n = \frac{(\sigma_{mean} - \sigma_{min})}{S_\sigma} = 3.07, \tag{3}$$

where σ_{min} is the minimum compressive strength, σ_{max} is the maximum compressive strength, σ_{mean} is mean of compressive strength, S_σ is the standard deviation of compressive strength, n is the number of specimens, and G_n and G'_n are Grubbs test statistics. The reject level (α) was selected as 0.05, and the corresponding discriminant value was $G_{0.95} = 2.176$. The σ_{min} (1.60) was judged to be an abnormal value and was eliminated due to $G_n > G'_n$ and $G_n > G_{0.95}$. The above steps were used to judge the remaining data until the abnormal values in the group were eliminated. Table 2 lists the mean and characteristic values of CEB's compressive strength.

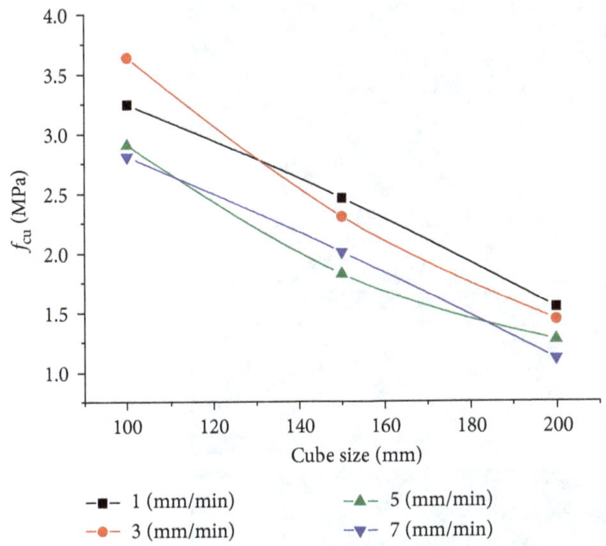

FIGURE 2: Compression strengths of differently sized cubes.

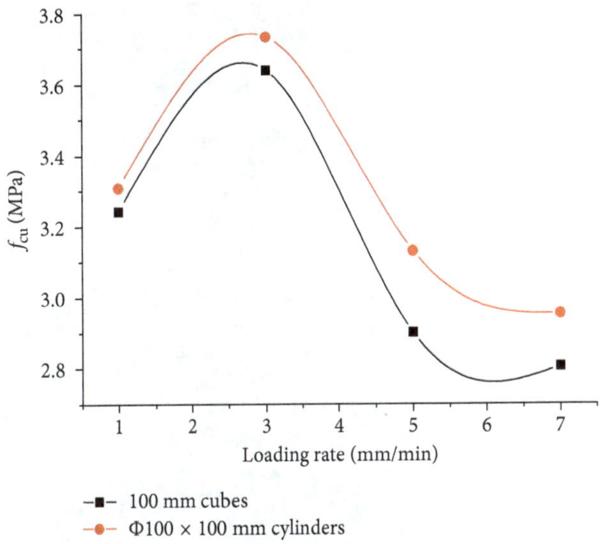

FIGURE 3: Compression strengths of 100 mm cubes to Φ100 × 100 mm cylinders.

3. Results and Discussion

3.1. The Influence of Specimen Size on Compressive Strength. The mean compressive strengths of different size cube specimens were obtained after the abnormal values were eliminated. Figure 2 shows the results of the unconfined compression tests with loading rates ranging from 1 mm/min to 7 mm/min on differently sized cubes. The x-axis indicates the size of the cubic specimen, and the y-axis indicates the mean of compressive strength. As expected, this figure shows the significant influence of specimen size on the compressive strength of CEB. The compressive strength decreased as the specimen size increased at four different loading rates. For the cube with small size, the compressive strength was greater than 3 MPa. In contrast, for the jumbo size, the compressive strength was low, and it decreased to a value of 1.11 MPa for the cubic specimen with a size of 200 mm. Two reasons may cause this. On the one hand, the deformation of the specimen in unconfined compression test is not only along the direction of load but also in the horizontal direction. As the flat deformation of the steel plate is less than the deformation of the specimen, friction is produced between the specimen and the plate. The conflict restricts the horizontal deformation of the specimen, and the strength of CEB is improved. However, this constraint is gradually reduced along the height of specimen. The large size specimen leads to the weakening of the constraint. On the other hand, the probability of defects such as pores, cracks, and partial damages in large specimens is substantial, and it also reduces the strength of CEB.

3.2. The Influence of Specimen Shape on Compressive Strength. The uniaxial compression test results of cubic and cylindrical specimens with the same sectional dimension and height, namely, 100 mm, are shown in Figure 3. The x-axis means the loading rate, and the y-axis denotes the mean of the compressive strength. This figure shows the slight influence of the specimen size on the compressive strength, and the cylindrical specimens gave slightly higher compressive strength than the cubic at different loading rates. One possible reason is that the stress concentration appears in the corner of the cubic specimen. Nonetheless, the compressive strength difference between specimens with cylindrical and cubic shapes was trivial under axial loading with loading rate ranging from 1 mm/min to 3 mm/min. The maximum difference in compressive strength was less than 8.4% at the loading rate of 3 mm/min, indicating that the effect of specimen shape on compressive strength is negligible.

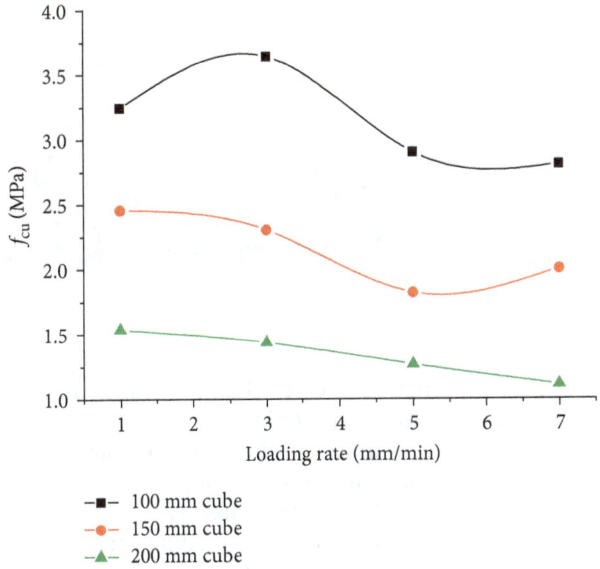

FIGURE 4: Compression strengths of cubes under different loading rates.

3.3. The Influence of Loading Rate on the Compressive Strength. The variation of mean compressive strength with loading rate is shown in Figure 4. As can be seen, the

Figure 5: The failure patterns of specimens.

compressive strength decreased slightly with the increase of loading rate from 1 mm/min to 7 mm/min, but the loading rate effect is minor in this tests compared to the size effect. The maximum difference of compressive strength between both loading rates is less than 25%. These results were opposite to the loading rate effect on concrete. Existing research results show that the high loading rate caused the internal microcracks in the mortar not fully expanded, which led to the destruction of the concrete aggregate. Higher loading rate resulted in the increase of aggregate destruction and the improvement in concrete strength [22]. However, the failure of soil is due to the appearance of the continuous sliding surface. The high loading rate accelerates the generation and slip of the sliding surface, which leads to the reduction in the CEB strength. It can also be noted that there was a definite difference in the relationship between loading rate and compressive strength at different size cubes. For example, the compressive strength of 100 mm cubes increased first and then decreased with the increase in loading rate but those of 200 mm cubes decreased with the increase in loading rate. This phenomenon may be due to the limitation of the range of loading rate, and the rule of compressive strength with loading rate cannot be fully reflected. Moreover, the heterogeneity of earth materials also leads to a significant variability of results.

3.4. Size and Shape Effects on Failure Patterns. Figure 5 shows the failure patterns of specimens with different sizes and shapes after uniaxial compression tests. As it can be noticed, the failure pattern was sensitive to the specimen shape but insensitive to the specimen size. The damage of cubes was mainly due to the cracks in the corner of cubes and the exfoliation of the lateral sides as shown in Figure 5. One cause may be the friction between steel plate and specimen that formed stress concentration on every corner of cubic specimens. In contrast, the stress concentration was relieved on the cylinder specimen because there was no mutation edge in the circular section. For the cylindrical specimens, it is observed that the failures crack showed columnar distribution. Meanwhile, similar to the failure pattern of cubes, the lateral sides of the cylinder specimen were also spalling. This cause of the phenomenon could be that the friction limits the end deformation of specimen followed by expansion producing on the lateral sides of the specimen. The remaining hourglass core is shown in Figure 6 also reflected the influence of boundary confinement.

FIGURE 6: The remaining hourglass core of cube and cylinder.

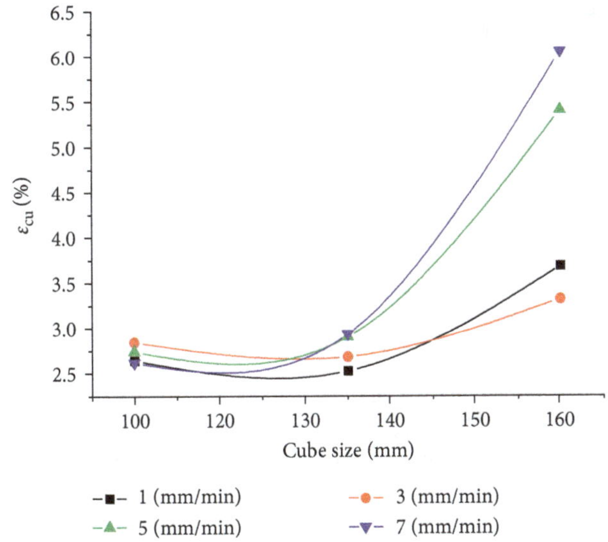

FIGURE 7: Peak strains of cubes under different loading rates.

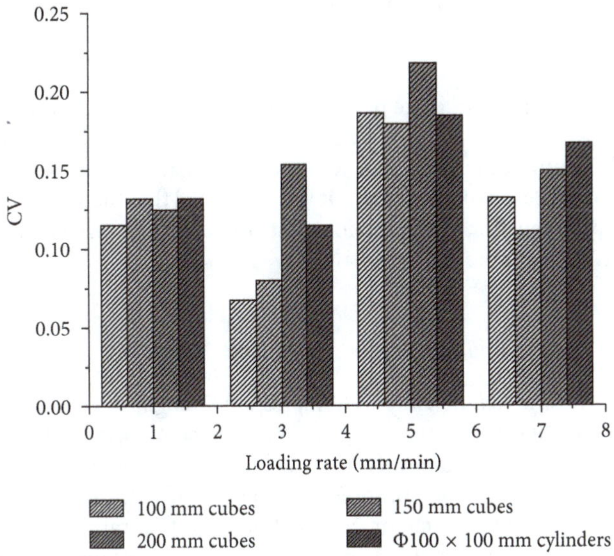

FIGURE 8: CV of compressive strength.

3.5. Size and Shape Effects on Peak Strain. The peak strain is defined as the strain corresponding to the peak stress. The mean peak strains of the specimens were obtained in the uniaxial compression tests conducted on cubic specimens. The variation of mean peak strain with loading rate on differently sized cubes is shown in Figure 7. It can be seen that the mean peak strain of small size specimen (100 mm, 150 mm) was less than that of large size specimen (200 mm). For 100 mm cubes, the mean peak strain ranged between 2.62% (7 mm/min) and 2.85% (3 mm/min) with a global mean value of 2.71%. By comparison, the mean peak strain of 200 mm cubes ranged between 3.32% (3 mm/min) and 6.05% (7 mm/min) with a global mean value of 4.61%. This phenomenon is because the large size specimens contained more pores and cracks than small size specimens. These pores and cracks are fully compressed at the peak load, and the increase in compression increased peak strain. Moreover, it can also be noted that the effect of loading rate on the peak strain of large size specimen was more significant than that of small size specimen. The mean peak strain of 150-mm cubes increased only by 0.38% as the loading rate increased from 1 mm/min to 7 mm/min. However, the mean peak strain of the 200 mm cubes increased by 2.73% within the range of loading rate. This phenomenon may be because the high loading rate accelerates the compression of the pores and cracks and increases the compression of the specimen before the formation of the continuous sliding surface.

3.6. Recommended Standard Loading Rate and Standard Specimen. Due to the inhomogeneity of the spatial distribution of the earth materials, the discreteness of the experimental data was notable. The coefficient of variation (CV) of each group of compressive strength was calculated and used to analyze the stability of experimental data. The variation of CV as specimen size and loading rate is shown in Figure 8.

It can be observed that the strength CV of all types of specimens reached the minimum at the loading rate of 3 mm/min. Consequently, 3 mm/min was selected as the

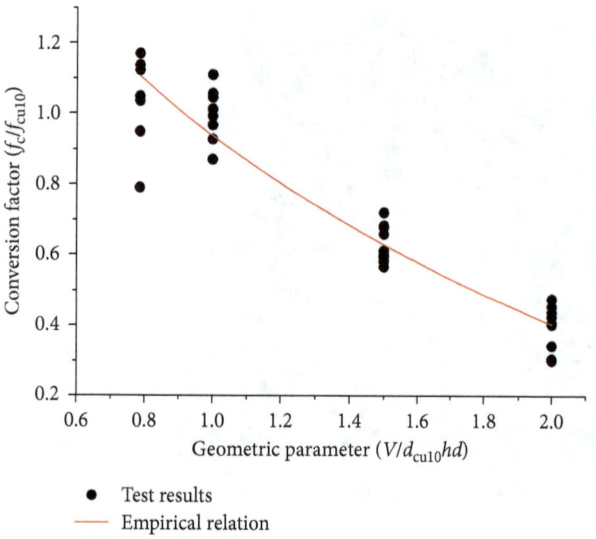

FIGURE 9: Empirical relation of conversion factors for compressive strength.

TABLE 3: The elastic modulus of CEB.

Label	Cu-100	Cu-150	Cu-200
Elastic modulus (MPa)	1.688	1.863	0.711
	2.177	1.301	0.518
	2.000	1.363	0.726
	1.755	1.564	0.703
	1.750	1.317	0.570
	2.061	0.955	0.534
	2.084	1.223	0.923
	1.678	1.733	0.556
	2.263	1.533	0.550
	1.794	1.317	0.770
Mean (MPa)	1.924	1.417	0.656

standard loading rate of compression test of earth materials. Meanwhile, it can also be noted that the strength CV of 100 mm cubic specimen was less than that of other specimens. Based on the results, it was suggested that the standard specimen for compression test of earth materials was 100 mm cubic specimen.

3.7. Correlations

3.7.1. The Relationship of Compressive Strength of Specimens with Different Shape and Size. A conversion coefficient was introduced to quantify the effects of shape and size of specimens on the compressive strength of earth materials. The conversion factor was defined as the ratio of the compressive strength f_c of specimens with various shapes and sizes to f_{cu10}, the strength of 100 mm cubic specimen. As the compressive strength was dependent on both the size and shape of the specimens, the empirical relation, in this case, considers the conversion factor as a function of lateral dimension d, height h, volume V, and h/d (ratio of height to lateral dimension) and is proposed in (4):

$$\frac{f_c}{f_{cu10}} = F\left[\frac{V/((h/d)\times d^2)}{V_{cu10}/((h_{cu10}/d_{cu10})\times d_{cu10}{}^2)}\right] = F\left(\frac{V}{d_{cu10}hd}\right). \tag{4}$$

The variation of the conversion factor (f_c/f_{cu10}) as a function of the geometric parameter ($V/d_{cu10}hd$) is shown in Figure 9. The best-fit regression line for test results is given by (5). The corresponding R^2 value of 0.9025 shows a good correlation between the conversion factor and the parameter.

$$\frac{f_c}{f_{cu10}} = 2.13e^{-0.83(V/d_{cu10}hd)}. \tag{5}$$

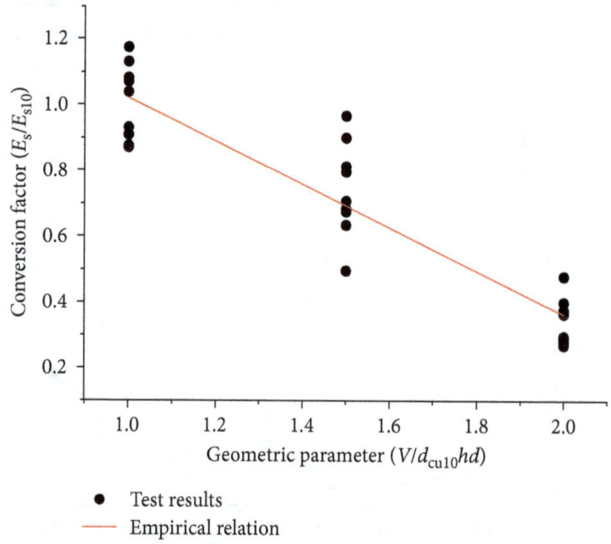

FIGURE 10: Empirical relation of conversion factors for elastic modulus.

3.7.2. The Relationship of the Elastic Modulus of Specimens with a Different Size. Elastic modulus was defined as the slope of the initial linear portion of compressive stress-strain curves of CBD. The elastic modulus of different size cubes at the loading rate of 3 mm/min and the mean elastic modulus are shown in Table 3. This Table shows the significant influence of specimen size on the elastic modulus. The mean elastic modulus decreased with the increase in specimen size. It ranged from 0.51 MPa for 200 mm cube up to 2.26 MPa for 100 mm cube. It can be explained by the fact that the increase in specimen size is accompanied by an increase in defects, such as pores, cracks, and partial damages, which can weaken the CEB.

The variation of the conversion factor (E_s/E_{s10}) as a function of the geometric parameter ($V/d_{cu10}hd$) is shown in Figure 10. The best-fit regression line for test results is given by (6):

$$\frac{E_s}{E_{s10}} = -0.66\left(\frac{V}{d_{cu10}hd}\right) + 1.68. \tag{6}$$

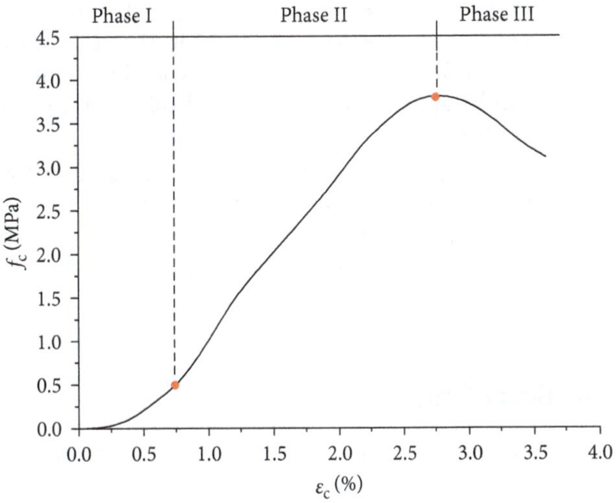

FIGURE 11: Stress-strain diagram of CEB to compression.

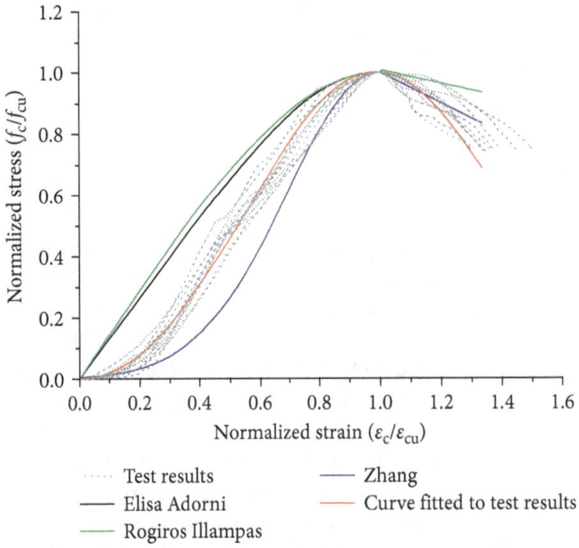

FIGURE 12: Normalized stress-strain test results obtained from compression tests on ten cubic specimens. The curves fitted to these data are compared to other three models.

3.8. Stress-Strain Curve. Figure 11 shows an example of stress-strain curves for 100 mm cubic specimens at the loading rate of 3 mm/min. It can be found that the stress-strain curve can be divided into the following three phases. Initially, a compaction phase occurred, during which, the preexisting protuberance on the surface of the specimen was eliminated. Meanwhile, the specimen underwent recompacted at the initial stage of loading. The cracks and pores existing in the CEBs microstructure were filled with redistributed soil particles. It was noted that the compaction of the specimen occurs in a broad strain range at Phase I. In the second phase, the recompaction process of specimens led to linear stress increase with the growth of strain until the peak stress was reached. After the peak stress was exceeded, the compaction softening effect appears in the earth materials, up to the specimen failure. The reason may be that the continuous sliding surface was produced inside the material.

The results, which were obtained from tests on 100 mm cubic specimens that loaded at the loading rate of 3 mm/min, were used to establish the stress-strain relations describing the response of materials to compressive loading. The stress and strain value of CEBs was normalized to propose the constitutive equation of earth materials. The normalized test data are presented in Figure 12. The x-axis represents the ratio of axial strains ε_c to the peak stress, ε_{cu} ($\overline{\varepsilon}_c = \varepsilon_c/\varepsilon_{cu}$), and the y-axis represents the ratio of normal stresses f_c to the peak stress, f_{cu} ($\overline{f}_c = f_c/f_{cu}$).

The Weibull equation was adopted for capturing the nonlinear compressive behavior of earth materials. The best Mimic Curve of the test data was determined by selecting the equation corresponding to the maximum of the coefficient of determination R^2. Moreover, in order to ensure the effectiveness of the equation, the following constraints were proposed: (a) the fitting curve passes through the origin, (b) the fitting curve passes through the point with $\overline{\varepsilon}_c = 1$ and $\overline{f}_c = 1$, and (c) the continuity of equation and its first derivatives in the variation range of $\overline{\varepsilon}_c$.

The fitting curve that satisfies the above conditions was obtained using standard optimization routines of MATLAB, and the corresponding equation is shown below:

$$\overline{f}_c = 2 \times \overline{\varepsilon}_c^{2} \times 0.5^{\overline{\varepsilon}_c^{3}}. \quad (7)$$

The corresponding satisfactory overall R^2 value of 0.917 indicated that stresses of earth materials can be well predicted by (7), as shown in Figure 12.

Figure 12 shows the comparison of the stress-strain model derived in this study with the models proposed by Zhang et al. [28], Illampas et al. [29], and Adorni et al. [30], respectively. Equation (8) reported in [30] was calibrated by experimental data from uniaxial compression test on earth blocks with the ratio of height to thickness $\beta > 1$. It was noted that (8) was only fitted to the ascending branch of the stress-strain curve, and the equation neglected the compaction phase of earth materials at the initial stage of loading.

$$\overline{f}_c = \frac{4.11 \times \overline{\varepsilon}_c}{3.11 + \overline{\varepsilon}_c^{4.11}}. \quad (8)$$

Further from Adorni's model, Rogiros Illampas used a piecewise function to fit the stress-strain curve of earth blocks and proposed (9). Although (9) contained the descending branch of the stress-strain curve, it still did not reflect the compaction phase. Furthermore, the continuity of (9) and its first derivative cannot be guaranteed at the piecewise point:

$$\overline{f}_c = \overline{\varepsilon}_c \begin{cases} 1.441\overline{\varepsilon}_c + 0.119\overline{\varepsilon}_c^2 - 0.559\overline{\varepsilon}_c^3 & \text{for } 0 \le \overline{\varepsilon}_c \le 1.07 \\ 1.266 - 0.285\overline{\varepsilon}_c + 0.028\overline{\varepsilon}_c^2 - 0.0001\overline{\varepsilon}_c^3 & \text{for } 1.07 \le \overline{\varepsilon}_c \le 4. \end{cases} \quad (9)$$

Table 4: R^2 of different constitutive modes.

Constitutive model	R^2
Elisa Adorni	0.807
Rogiros Illampas	0.823
Youchao Zhang	0.836
Equation (1)	0.915

After that, Youchao Zhang recommended a constitutive model reflecting the compaction phase of earth materials. For (10), exponential function and linear function were used to fit the ascending and descending branches of the stress-strain curve, respectively. However, not only this piecewise function did not pass through the origin but also its first derivative was discontinuous at the piecewise point. The equation proposed in this study overcomes the defects of above constitutive equations. The satisfactory overall R^2 values of different constitutive equations were listed in Table 4. The parameters R^2 of (7) was the largest, and it also indicated that (7) was preferable to the other three model:

$$\overline{f}_c = \overline{\varepsilon}_c \begin{cases} e^{-5.3(\overline{\varepsilon}_c - 1)^2} & \text{for } \overline{\varepsilon}_c \le 1 \\ 1.5 - 0.5\overline{\varepsilon}_c & \text{for } \overline{\varepsilon}_c > 1. \end{cases} \tag{10}$$

4. Conclusions

This study aimed to evaluate the influences of specimen geometry and loading rate on the mechanical property of compressed earth block used in simple compression tests. Based on the analysis of results, the following main conclusions can be drawn:

(i) The effect of specimen size on the compressive strength was evident in simple compression tests. The unconfined compressive strength decreases with the increase in specimen size. This phenomenon was due to the compound influence of plate constraint effect and distribution inhomogeneity of earth materials defects.

(ii) The effect of loading rate on the compressive strength was insignificant. It may be due to the narrow range of loading rates chosen in this study. Further research work was required to investigate the loading rate effects on the compressive strength under a wide range of loading rate.

(iii) The effect of specimen shape was trivial in unconfined compressive tests. The cylindrical specimens are given slightly higher compressive strength than the cubic specimens at different loading rates. The maximum difference in the compressive strength was less than 8.4%, indicating that the shape effect on compressive strength was negligible.

(iv) According to the variability of test data, 100 mm cube was recommended as a standard specimen, and 3 mm/min was selected as the standard loading rate in the simple compression test.

(v) Based on the experimental results, a constitutive model for describing CEBs' stress-strain behavior under compression was proposed. Defects in the previous studies were modified in the constitutive model. It was expected that the constitutive model can be applied in the numerical calculation of the earth structure.

(vi) Besides the geometric effect and loading rate effect, the influence of manufacture method and test method were significant in the compression test. The next paper will be systematically studied for these two points.

Conflicts of Interest

The authors declare that they have no conflicts of interest.

Acknowledgments

The work described in this article is part of a research project on the "Study on the Standard Test Method of Materials and Masonry Based on Raw-soil" funded by the "Science Fund for Creative Research Groups of the National Natural Science Foundation of China" under Grant no. 51478043.

References

[1] L. Widder, "Earth eco-building: textile-reinforced earth block construction," *Energy Procedia*, vol. 122, pp. 757–762, 2017.

[2] H. V. Damme and H. Houben, "Earth concrete. Stabilization revisited," *Cement and Concrete Research*, 2017, In press.

[3] M. B. Mansour, A. Jelidi, A. S. Cherif et al., "Optimizing thermal and mechanical performance of compressed earth blocks (CEB)," *Construction and Building Materials*, vol. 104, pp. 44–51, 2016.

[4] N. M. Miller and F. M. Tehrani, "Mechanical properties of rubberized lightweight aggregate concrete," *Construction and Building Materials*, vol. 147, pp. 264–271, 2017.

[5] T. Sturm, L. F. Ramos, and P. B. Lourenço, "Characterization of dry-stack interlocking compressed earth blocks," *Materials and Structures*, vol. 48, no. 9, pp. 3059–3074, 2015.

[6] L. F. Cabeza, C. Barreneche, L. MirÓ et al., "Low carbon and low embodied energy materials in buildings: a review," *Renewable and Sustainable Energy Reviews*, vol. 23, pp. 536–542, 2013.

[7] M. Achenza and L. Fenu, "On earth Stabilization with natural polymers for earth masonry construction," *Materials and Structures*, vol. 39, no. 1, pp. 21–27, 2006.

[8] L. Miccoli, D. V. Oliveira, A. S. Rui et al., "Static behaviour of rammed earth: experimental testing and finite element modelling," *Materials and Structures*, vol. 48, no. 10, pp. 3443–3456, 2015.

[9] F. Pacheco-Torgal and S. Jalali, "Earth construction: lessons from the past for future eco-efficient construction," *Construction and Building Materials*, vol. 29, no. 4, pp. 512–519, 2012.

[10] P. Zak, T. Ashour, A. Korjenic et al., "The influence of natural reinforcement fibers, gypsum and cement on compressive strength of earth bricks materials," *Construction and Building Materials*, vol. 106, pp. 179–188, 2016.

[11] D. Ciancio and J. Gibbings, "Experimental investigation on the compressive strength of cored and molded cement-stabilized rammed earth samples," *Construction and Building Materials*, vol. 28, no. 1, pp. 294–304, 2012.

[12] F. Wu, G. Li, H. N. Li et al., "Strength and stress-strain characteristics of traditional adobe block and masonry," *Materials and Structures*, vol. 46, no. 9, pp. 1449–1457, 2013.

[13] B. V. V. Reddy and A. Gupta, "Characteristics of soil-cement blocks using highly sandy soils," *Materials and Structures*, vol. 38, no. 6, pp. 651–658, 2005.

[14] Standards Australia Handbook 194, *The Australian Earth Building Handbook*, Standards Australia, Sydney, Australia, 2002.

[15] New Zealand Standard 4298, *Materials and Workmanship for Earth Buildings*, Standards New Zealand, New Zealand, 1998.

[16] J. C Morel, A. Pkla, and P. Walker, "Compressive strength testing of compressed earth blocks," *Construction and Building Materials*, vol. 21, no. 2, pp. 303–309, 2007.

[17] D. Silveira, H. Varum, and A. Costa, "Influence of the testing procedures in the mechanical characterization of adobe bricks," *Construction and Building Materials*, vol. 40, no. 3, pp. 719–728, 2013.

[18] J. E. Aubert, P. Maillard, J. C. Morel et al., "Towards a simple compressive strength test for earth bricks?," *Materials and Structures*, vol. 49, no. 5, pp. 1641–1654, 2016.

[19] Y. L. Chen, L. Q. Chang, and S. Xu, "Experimental study on uniaxial compressive strength," *Journal of Shanghai University (Natural Science)*, vol. 15, no. 3, pp. 310–315, 2009.

[20] H. M. Du, W. Ma, and S. J. Zhang, "Effects of strain rate and water content on uniaxial compressive characteristics of frozen soil," *Rock and Soil Mechanics*, vol. 37, no. 5, pp. 1373–1379, 2016.

[21] X. X. Zhang, G. Ruiz, and A. M. A. Elazim, "Loading rate effect on crack velocities in steel fiber-reinforced concrete," *International Journal of Impact Engineering*, vol. 76, pp. 60–66, 2015.

[22] H. Xu and H. M. Wen, "Semi-empirical equations for the dynamic strength enhancement of concrete-like materials," *International Journal of Impact Engineering*, vol. 60, no. 60, pp. 76–81, 2013.

[23] X.-h. Wang, S.-r. Zhang, C. Wang, R. Song, C. Shang, and X. Fang, "Experimental investigation of the size effect of layered roller compacted concrete (RCC) under high-strain-rate loading," *Construction and Building Materials*, vol. 165, pp. 45–57, 2018.

[24] F. Parisi, D. Asprone, L. Fenu et al., "Experimental characterization of Italian composite adobe bricks reinforced with straw fibers," *Composite Structures*, vol. 122, pp. 300–307, 2015.

[25] Standards Chinese Standard SL237-1999, *Specification of Soil Test*, Standards Chinese, Beijing, China, 1999.

[26] Standards Chinese Standard GB T 50145-2007, *Standard for Engineering Classification of Soil*, Standards Chinese, Beijing, China, 2007.

[27] Standards Chinese Standard GB/T 50081-2002, *Test Method of Mechanical Properties on Ordinary Concrete*, Standards Chinese, Beijing, China, 2002.

[28] Y. Zhang, Y. Wang, N. Zhao, and T. Wang, "Experimental and stress-strain equation investigation on compressive strength of raw and modified soil in loess plateau," *Advances in Materials Science and Engineering*, vol. 2016, no. 203, pp. 1–10, 2016.

[29] R. Illampas, I. Ioannou, and D. C. Charmpis, "Adobe bricks under compression: experimental investigation and derivation of stress-strain equation," *Construction and Building Materials*, vol. 53, no. 4, pp. 83–90, 2014.

[30] E. Adorni, E. Coïsson, and D. Ferretti, "In situ characterization of archaeological adobe bricks," *Construction and Building Materials*, vol. 40, no. 7, pp. 1–9, 2013.

20

Physicochemical Properties of Handere Clays and Their Use as a Building Material

Yasin Erdoğan

Department of Petroleum & Natural Gas Engineering, Faculty of Engineering, Mustafa Kemal University, İskenderun, 31200 Hatay, Turkey

Correspondence should be addressed to Yasin Erdoğan; erdogan@mku.edu.tr

Academic Editor: Fa-Nian Shi

Handere clay deposits were discovered at Adana in Turkey. These clay units primarily consist of uncoloured claystone, pebbly sandstone, sandstone, siltstone, and mudstone marl and include gypsum lenses and clay levels of various thicknesses in places. The physicochemical properties of these clays have been investigated by different techniques including Scanning Electron and Elemental Analysis (SEM and EDS), mineralogical analyses, chemical and physical analyses, X-ray diffraction (XRD), thermogravimetric differential thermal analysis (TG-DTA), and Atterberg (Consistency) Limits Test. The mineralogical composition deduced from XRD is wide (smectite + palygorskite + illite ± feldspar ± chlorite ± quartz ± calcite ± serpentine) due to the high smectite contents (≈85%). SEM studies reveal that smectite minerals are composed of irregular platy leaves and show honeycomb pattern in the form of wavy leaves in places. The leaves presenting an array with surface edge contact are usually concentrated in the dissolution voids and fractures of volcanic glass. Organic matter content and loss on ignition analysis of raw materials are good for all the studied samples. In summary, Handere clays can be used as building materials in bricks, roof tiles, and cement and as a binder.

1. Introduction

Human beings found various applications of layered clay minerals since prehistoric civilization due to their widespread distribution and a great diversity of reactions in nature. Depending on the layer structure and specific properties, such as high specific surface area, ion exchange capacity, or hydration property, clay minerals were widely used in ceramics and building materials, paper industries, oil drilling, foundry moulds, and pharmaceuticals and were also used as adsorbents, catalysts or catalyst supports, ion exchangers, and decolourizing agents [1–5].

Clay minerals are a class of phyllosilicates which usually form as a result of chemical weathering of other silicate minerals at the surface of the earth. This mineral mostly contains limestone, silica, mica, and iron oxide and has a yellowish, reddish, or brown colour. Clay is a valid concept for decomposition products generated by hydrothermal activity and for particles deposited by sedimentation, and chemical classifications were made according to the minerals comprising such small particles [6–10].

Clay minerals are included in kaolinite, halloysite, illite, chlorite, smectite, and attapulgite groups according to their chemical composition and formation. Clay minerals which used to be called montmorillonite group are referred to as smectite. Smectite minerals are composed of two silica tetrahedral layers and a single Al octahedral layer. Important smectite group minerals include montmorillonite, beidellite, nontronite, hectorite, saponite, and sauconite. The minerals in this group have a density of 2-3 gr/cm^3 and hardness between 1 and 2 [11–13].

Depending on the amount of water mixed into them, clays exhibit a variety of properties, including plasticity, fluidity, and colloidal and thixotropic properties. These various properties affect industrial and engineering use of clay minerals [14–18]. Chemical composition of various clay minerals was presented in Table 1.

In this study, Handere clays were investigated in order to characterise the raw material available at Adana in Turkey. The study covered over approximately 20 km including Handere formation (Figure 1).

TABLE 1: Chemical composition of various clay minerals [17].

Clay	SiO_2	Al_2O_3	Fe_2O_3	MgO	CaO	K_2O	Na_2O	TiO_2	H_2O	Total
Kaolinite	46.90	37.40	0.66	0.27	0.29	0.84	0.44	0.18	12.95	99.92
Nacrite	44.75	39.48	0.53	0.19	0.13	0.34	0.22	—	14.40	100
Stalagmites	46.86	37.12	1.43	0.09	0.22	0.60	0.07	0.51	12.99	99.89
Halloysite	44.75	36.94	0.31	—	0.11	0.60	—	—	17.42	100.01
Anauxite	54.32	29.96	2.00	0.14	0.32	—	0.37	—	12.64	99.75
Nontronite	40.54	5.19	31.63	0.06	1.92	0.24	0.14	—	20.75	100.47
Chlorite	31.44	17.62	—	37.64	—	—	—	—	13.19	99.89
Prochlorite	23.69	21.26	26.52	17.60	3.22	—	—	—	7.63	99.92
Sepiolite	52.50	0.60	2.99	21.31	0.47	—	—	—	21.27	99.14
Attapulgite	57.85	7.89	2.82	13.44	0.30	0.08	0.53	—	16.95	99.86

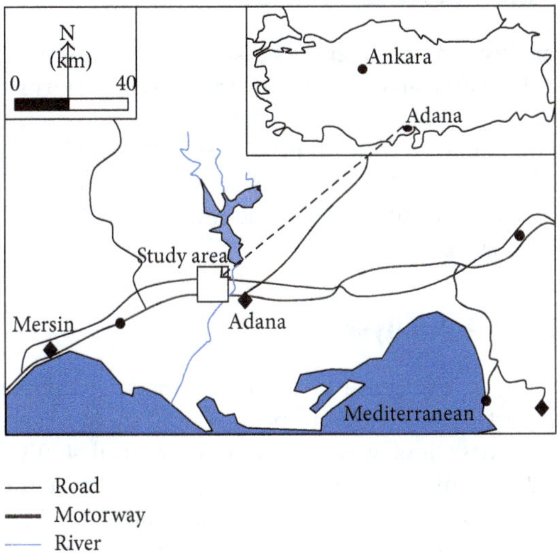

FIGURE 1: Location of studied area at Handere in Turkey.

Their best uses require an exact knowledge of their physicochemical properties. This paper describes studies of crude clay samples with a set of techniques comprising mineralogical analyses, Scanning Electron Microscope (SEM), Elemental Analysis (EDS), chemical and physical analyses, X-ray diffraction (XRD), thermogravimetric-differential thermal analysis (TG-DTA), and Atterberg (Consistency) Limits Test. In addition, organic matter content and loss on ignition analysis of raw materials test were performed for determination of applicability of the material as a building material. The work had three distinct aims:

(1) To identify Handere clay present in different samples and its purity and associated minerals.

(2) To characterise the physicochemical properties of the clay to enable assessment of its suitability for different uses.

(3) To determine applicability of the material as a building material.

2. Geological Description of Handere Clays

Clay deposits investigated in the present study are from a cluster at Adana in Turkey. Handere geological formation was named by Schmidt in 1961 [19]. The unit primarily consists of uncoloured claystone, pebbly sandstone, sandstone, siltstone, and mudstone marl and includes gypsum lenses and clay levels of various thicknesses in places. Trough cross-stratification in pebbles and parallel lamination in those with tiny particles are observed. The thickness of the formation is in the range of 120 to 700 m [20]. The bottom unit has a transitional contact on Kuzgun formation, and the top unit covered by young alluvial deposits in places has a widespread terrace set formation of Adana basin. According to the fossil groups discovered in the formation, the formation is assigned a Messinian-Pliocene age. The study samples representative of the area were collected, classified by sieving them (0.2 to 2 mm) at the laboratory, and made ready for tests [21–24].

3. Chemical Composition

Samples representative of the area were collected from Handere formation and chemical analyses were performed in order to determine their chemical composition. Chemical analysis of the samples was performed using XRF device (Siemens SRS 300 X-Ray Fluorescence Spectrometer) after they were dried at 105°C. The results obtained from the analyses are given in Table 2. When the results were interpreted, it was identified that the clay content (%) is largely composed of SiO_2 (mean: 61.71 ± 0.21) and Al_2O_3 (mean: 16.42% ± 0.58) and the clay group contains smectite mineral, and these results were consistent with previous studies in the literature.

4. Petrographic Studies

Thin sections with a thickness of 0.02 mm were prepared from clays. These sections were examined by polarizing microscope. As a result of their microscopic examinations, it was determined that the clay is smectite, a type of clay mineral rich in silica (Si) and aluminum (Al). The smectite mineral was found to contain amorphous silica, calcite, chlorite, quartz, and illite minerals.

Table 2: The results of chemical analysis of Handere clay (%).

Chemical composition	Mean	Standard deviation
SiO_2	61.71	0.21
Al_2O_3	16.42	0.58
Na_2O	2.26	0.19
K_2O	4.58	0.24
Fe_2O_3	7	1.76
CaO + MgO	3.92	0.33
Others	4.12	1.83

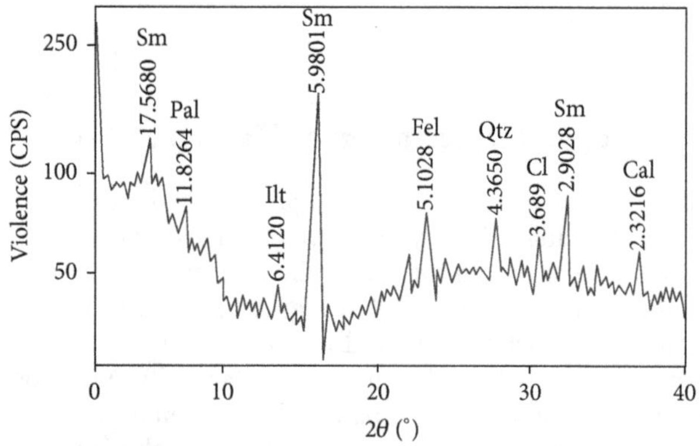

Figure 2: X-ray diffraction patterns for a Handere clay sample (glycol solvated).

5. Mineralogical Composition (XRD)

Mineralogical analysis was conducted on the samples in powder form by Cu X-ray tube Rigaku Geigerflex XRD apparatus. As a result of phase (mineralogical) analysis, it was revealed that the clay samples are largely comprised of smectite 14 (85%), followed by palygorskite, illite, feldspar, chlorite, quartz, calcite, and trace amounts of serpentine in decreasing order of abundance (Figure 2).

6. Scanning Electron Microscope (SEM)

Clay samples were enlarged in the range of 105–170 μm and were examined by image analysis system and their photographs were taken (Figure 3). It is clear that clay samples are smectite clay minerals. Smectite minerals are composed of irregular platy flakes and show honeycomb pattern in the form of wavy leaves in places. The flakes presenting an array with surface edge contact are usually concentrated in the dissolution voids and fractures of volcanic glass. In addition, smectite flakes show growth towards the pores.

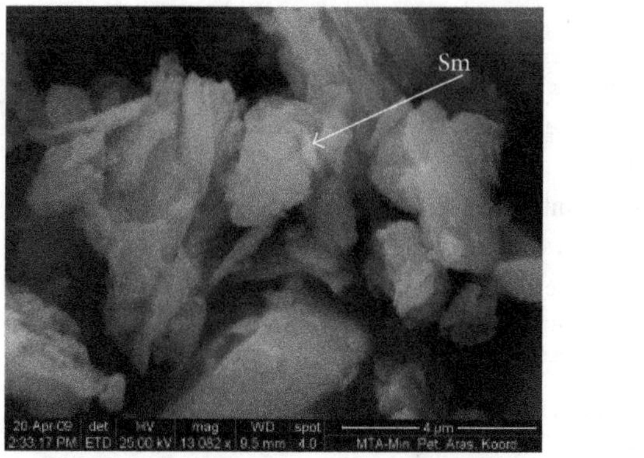

Figure 3: SEM image of Handere clay (smectite).

7. Elemental Analysis (EDS)

SEM analysis and EDS (energy dispersive spectrometer) and point (elemental) analyses of the samples were performed and distribution percentages of the elements were determined. EDS point analysis was conducted using EDAX Genesis XM 4i model detector. The clay samples were identified to have oxygen (46.95%), silica (27.3%), aluminum (7.11%), and iron (2.53%) (Figure 4).

8. TG-DTA Analysis

According to TG curve of the clay (smectite) sample, the sample lost its moisture at about 50–100°C. In DTA analysis, the mass loss was not significant until 480°C, and the endothermic peak at 713.4°C supported that thermal degradation took place between 480 and 800°C. In TG curve, it was determined that there was no mass loss after 790°C and the total mass loss was about 5.5%. Sintering started at around 440°C and was complete at 700°C. The reaction decomposition started at 810°C, made an endothermic peak at about 834.6°C, and was complete at 855°C (Figure 5). In a literature review, it has been found to show similar characteristic of clay in the Mediterranean zone [12, 21, 25–27].

9. Physical Properties

In order to determine the physical properties (loose unit weight, specific gravity, porosity, water absorption, and compactness ratio) of Handere clays, the samples were ground and sieved through sieves with 1 mm of sieve opening and their impurities were removed. Their loose unit weight analyses were conducted in accordance with TS 3529 and TS 1114 standards [28, 29]. The results obtained from the analyses are given in Table 3.

10. Atterberg (Consistency) Limits Test of Handere Clays

Water content values defined by Atterberg, a Swedish chemist, are used to determine consistency limits of soils

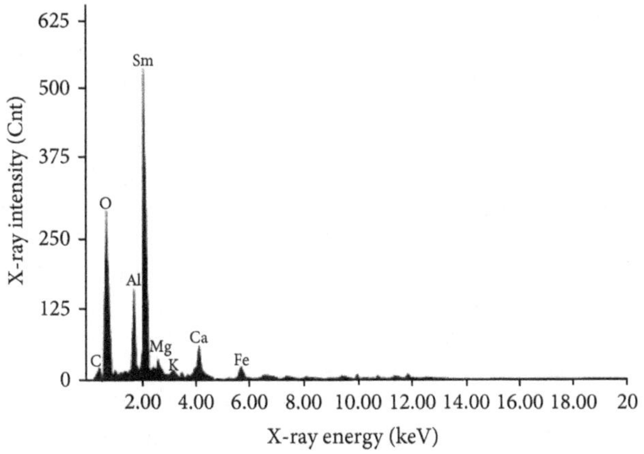

Figure 4: EDS analysis of Handere clay.

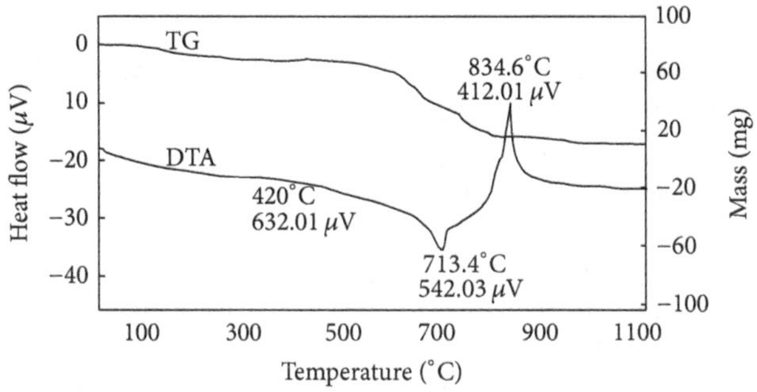

Figure 5: Clay example 20°C/min. heating rate conversion TG-DTA curves.

Table 3: The results of the physical analysis of Handere clay.

Tests	Mean	Standard deviation
Unit weight (gr/cm^3)	1.32	0.06
Specific gravity (gr/cm^3)	2.73	0.02
Porosity (%)	12.80	1.60
Water absorption (%)	52.64	2.65
Compactness ratio (%)	0.46	0.03

[30]. In the present study, Casagrande method was used to determine the consistency limits of Handere clays [31].

As a result of the test, liquid limit and plastic limit were 52% and 23%, respectively. According to these results, calculated plasticity index, PI, was 29%, and the soil class was identified as highly plastic clay.

11. Tests for Determination of Applicability of the Material as a Building Material

Limit values of a material to be used in the building and construction sector should be determined by examining the general structural characteristics of that material. For this purpose, whether the clay mineral is suitable to be used as a building material was investigated by organic matter content and loss on ignition analyses.

12. Organic Matter Content

Presence of organic matters in a raw material is inconvenient and undesirable. Organic compounds particularly reduce the strength of a building material, cause corrosion and softening in the material over time, and, most importantly, make natural and chemical materials, which are used as binders, lose their binding properties. As a result of the test conducted according to the test criteria indicated in TS EN 1744-1 and TS 4790 [32–34], Handere clays were found to contain no organic matter. The results of chemical analysis conducted on the clay are consistent with the results obtained from this test.

13. Loss on Ignition Analysis of Raw Materials

At high temperatures, as in the case of a fire, the composition of aggregate and natural binder materials, used in the manufacture of building materials in the construction industry, deteriorates, and such materials undergo mass loss. Therefore, whether a raw material intended for use in the construction industry undergoes any mass loss as a result of the effect of temperature should be determined.

According to the results obtained from loss on ignition analyses, loss on ignition of Handere clays was determined as 4.85%. As stipulated in the standard, TS 1114, the maximum loss on ignition of a raw material can be 5% by weight. TS 1114 stipulates that if the loss by weight of a raw material at maximum temperature is ≤5%, it can be used as a raw material in bricks, cement, and roof tiles.

According to the analysis results, it was determined that the clays of the region can be used as building materials in bricks, roof tiles, and cement and as a binder; however, they cannot be used in the production of ceramics, porcelain, glass, and tiles.

14. Conclusion

In this study, material properties of the clays from Handere formation were investigated in order to determine their applicability as a building material. As a result of petrographic and chemical analyses, it was determined that the clays were smectite, a clay mineral rich in Si and Al. The smectite mineral was found to contain amorphous silica, calcite, chlorite, quartz, and illite. According to the results of qualitative mineralogical (XRD) analyses, the samples were found to be largely comprised of smectite 14 (85%), followed by palygorskite, illite, feldspar, chlorite, quartz, calcite, and trace amounts of serpentine in decreasing order of abundance.

In SEM examinations, it was seen that the clay samples are smectite clay minerals. Smectite minerals are composed of irregular platy leaves and show honeycomb pattern in the form of wavy leaves in places. In elemental (EDS) analyses, the clay samples were identified to have oxygen (46.95%), silica (27.3%), aluminum (7.11%), and iron (2.53%).

In TG-DTA analysis, the mass loss was not significant until 480°C, and the endothermic peak at 713.4°C supported the fact that thermal degradation took place between 480 and 800°C. In TG curve, it was determined that there was no mass loss after 790°C and the total mass loss was about 5.5%.

Consistency limits of Handere clays were established, and, at the end of the test, a liquid limit of 52% and a plastic limit of 23% were found. According to these results, calculated plasticity index was 29% and the soil class was identified as highly plastic clay.

As a result of the studies performed according to test criteria specified in EN 1744-1, no organic matter was found in Handere clays. According to the results of loss on ignition analyses conducted according to TS 1114 standard, loss on ignition of Handere clays was 4.85%. When the results were examined, loss on ignition of the clays was less than 5%, so it was determined that the clays can be used as aggregate raw materials in the production of bricks and lightweight building materials.

In conclusion, when the studies and the analysis results were evaluated together, it was established that the clays from the region can be used as building materials, including bricks, roof tiles, and fillers; however, they cannot be used in the production of ceramics, porcelain, glass, and tiles.

Conflict of Interests

The author declares that there is no conflict of interests regarding the publication of this paper.

References

[1] A. Vaccari, "Preparation and catalytic properties of cationic and anionic clays," *Catalysis Today*, vol. 41, no. 1–3, pp. 53–71, 1998.

[2] M. I. Carretero and G. Lagaly, "Clays and health: an introduction," *Applied Clay Science*, vol. 36, no. 1–3, pp. 1–3, 2007.

[3] C. Nkoumbou, F. Villieras, D. Njopwouo et al., "Physicochemical properties of talc ore from three deposits of Lamal Pougue area (Yaounde Pan-African Belt, Cameroon), in relation to industrial uses," *Applied Clay Science*, vol. 41, no. 3-4, pp. 113–132, 2008.

[4] D. Zhang, C.-H. Zhou, C.-X. Lin, D.-S. Tong, and W.-H. Yu, "Synthesis of clay minerals," *Applied Clay Science*, vol. 50, no. 1, pp. 1–11, 2010.

[5] R. Srinivasan, "Advances in application of natural clay and its composites in removal of biological, organic, and inorganic contaminants from drinking water," *Advances in Materials Science and Engineering*, vol. 2011, Article ID 872531, 17 pages, 2011.

[6] R. B. Grim, *Clay Mineralogy*, McGraw-Hill, New York, NY, USA, 2nd edition, 1953.

[7] G. S. Odin, "Observations on the structure of glcuconite vermicular pellets," *Sedimentalogy*, vol. 19, pp. 204–285, 1972.

[8] H. Tanişan and Z. Mete, "Ceramic technology and application," in *Proceedings of the Ceramic Congress*, pp. 23–34, Istanbul, Turkey, 1985.

[9] R. L. Vitra, *U.S. Geological Survey, Mineral Commodity Summaries*, Clays, San Antonio, Tex, USA, 2009.

[10] A. Nzeugang Nzeukou, N. Fagel, A. Njoya, V. Beyala Kamgang, R. Eko Medjo, and U. Chinje Melo, "Mineralogy and physicochemical properties of alluvial clays from Sanaga valley (Center, Cameroon): suitability for ceramic application," *Applied Clay Science*, vol. 83-84, pp. 238–243, 2013.

[11] Y. Erdoğan, "Engineering properties of Turkish travertines," *Scientific Research and Essays*, vol. 6, no. 21, pp. 4551–4566, 2011.

[12] A. G. Çelik and A. M. Kiliç, "Characteristic properties of Ankara (imrahor) clays and its use as a construction," in *Proceedings of the 8th International Industrial Minerals Symposium*, pp. 243–252, Istanbul, Turkey, 2012.

[13] A. G. Çelik, A. M. Kiliç, and G. Ö. Çakal, "Expanded perlite aggregate characterization for use as a lightweight construction raw material," *Physicochemical Problems of Mineral Processing*, vol. 49, no. 2, pp. 689–700, 2013.

[14] G. Rajasekaran, K. Murali, and R. Srinivasaraghavan, "Microfabric, chemical and mineralogical study of Indian marine clays," *Ocean Engineering*, vol. 26, no. 5, pp. 463–483, 1998.

[15] D. Ogle, "Making lightweight refractory ceramic from perlite and clay," *Aprovecho Research Center*, pp. 2–7, 2003.

[16] C. E. Weaver, "The clay petrology of sediments," in *Clays and Clay Minerals: Proceedings of the 6th Conference*, vol. 566, pp. 154–187, Pergamon Press, New York, NY, USA, 1959.

[17] R. L. Vitra, "U.S. geological survey, mineral commodity summaries," *Clays*, 2008.

[18] Y. Erdogan, *Investigation of engineering properties of building materials made with acidic and alkaline pumice [Ph.D. thesis]*, Cukurova University, Institute of Science, Adana, Turkey, 2007.

[19] G. C. Schmidt, "Stratigraphic nomenclature for the Adana region petroleum district VII," *Petroleum Administration Bulletin*, vol. 6, pp. 47s–63s, 1961.

[20] C. Yetis and C. Demirkol, "Detailed study of the western part of the Adana basin," M.T.A. Compilation Report 8037, 1986.

[21] A. Palali, *The effect of changıng of water content and compaction energy to strength in Handere clay [M.S. thesis]*, Cukurova University, Institute of Science, Adana, Turkey, 2006.

[22] T. Rusen, *Uniaxial strength of the effect of sample size on clayey soils [M.S. thesis]*, Geological Engineering Department, Science institute, Cukurova University, Adana, Turkey, 2011.

[23] S. Özkan, E. Yaşar, Y. Erdogan et al., "Salbas (Adana) sandstones of usability evaluation of industrial area," in *Proceedings of the 20th International Mining Congress and Exhibition of Turkey*, pp. 259–268, Ankara, Turkey, 2007.

[24] E. Yaşar and Y. Erdoğan, "Strength and thermal conductivity in lightweight building materials," *Bulletin of Engineering Geology and the Environment*, vol. 67, no. 4, pp. 513–519, 2008.

[25] M. A. Rahman and M. Muneer, "Photocatalysed degradation of two selected pesticide derivatives, dichlorvos and phosphamidon, in aqueous suspensions of titanium dioxide," *Desalination*, vol. 181, no. 1–3, pp. 161–172, 2005.

[26] M. C. Wu, S. L. Kuo, J. C. Lin, C. M. Ma, G. B. Hong, and C. T. Chang, "Studies on certain physical properties of modified smectite nanocatalysts," *Applied Surface Science*, vol. 257, no. 13, pp. 5641–5646, 2011.

[27] A. Sandler, "Clay distribution over the landscape of Israel: from the hyper-arid to the Mediterranean climate regimes," *Catena*, vol. 110, pp. 119–132, 2013.

[28] TS 3529, Specific Weight and Absorption Determination of Concrete Aggregates, 1980.

[29] TS 1114 EN 13055-1, Lightweight Aggregates—Part 1: for Use in Concrete, Mortar and The Juice, 2004.

[30] T. W. Lambe, *Soil Testing for Engineers*, John Wiley & Sons, New York, NY, USA, 1951.

[31] A. Casagrande, "Classification and identification of soils," *Transactions of the American Society of Civil Engineers*, vol. 113, pp. 901–930, 1948.

[32] TS EN 1744-1, "Tests for Chemical Properties of Aggregates, Part 1: Chemical Analysis," 2000.

[33] TS 4790, Experimental Methods of Soil Bricks and Tiles, 1986.

[34] W. Hamza, C. Chtara, and M. Benzina, "Retention of organic matter contained in industrial phosphoric acid solution by raw Tunisian clays: kinetic equilibrium study," *Journal of Chemistry*, vol. 2013, Article ID 218786, 9 pages, 2013.

21

Impact Resistance of Concrete Structures

R. Drathi

IIT Kanpur, Kanpur, Uttar Pradesh 208016, India

Correspondence should be addressed to R. Drathi; 478923@admin.iitd.ac.in

Academic Editor: Xiaoying Zhuang

We present simulations on the resistance of concrete structures due to impact loading. Therefore, a mesh-free method is exploited in combination with a viscous damage model that accounts for strain rate effects and high pressures that commonly occur in those events. In contrast to many other studies in the literature, we account for uncertainties in the material parameters by subjecting them to a probability distribution function. Furthermore, we also consider the geometric correlation inside the concrete and show that this geometric correlation has a minor influence on the predicted results. All numerical results are compared to the impact experiments performed by Hanchak and coworkers as well as our own impact experiments.

1. Introduction

Predicting the impact resistance of concrete and reinforced concrete structures has been of interest for many years. Constitutive models that account for strain rate effects, high pressure, and the associated dynamic strength increase are the key to successful predictions. Constitutive models accounting for strain rate effects can be classified into damage models, plasticity models, and damage-plasticity models [1–6]. Damage models for concrete accounting for high strain rates were developed, for instance, by [7–10]. A popular plasticity model for concrete was proposed by [11] as an extension of the Johnson-Cook [12] model originally developed for metals. In this model, the dynamic strength increase was related to the current strain rate. However, memory effects were not accounted for. Furthermore, discontinuities in the stress-strain response are allowed which are physically not meaningful. In [13], the dynamic buckling strength was therefore dependent on the strain rate history and an average strain rate. A viscous damage-plasticity model accounting for high strain rates and high pressures was proposed by [14]. Therefore, a dynamic damage variable was introduced that decayed the static damage evolution. The decaying function depends on previous damage-rates or damage increments. The scalar damage was later on extended to a damage tensor accounting for the anisotropy for concrete under tensile loading [15] while a scalar damage was maintained in compression.

Most of the studies on the impact resistance of concrete were done with the finite element method [16–24]. Commonly, element-deletion techniques [25, 26] were applied to allow for large deformations and perforations. A strong competitor to finite elements is mesh-free methods [27–37] that allow for a more robust and accurate prediction of dynamic fracture and fragmentation; see [38] for a recent overview of mesh-free methods and their applications. Reference [39], for instance, was able to predict the impact resistance of the experiments carried out by [40] with mesh-free methods quite accurately, while the FE simulations done in [40] led always to an underestimation of the impact resistance. Also, penetration was not predicted correctly in those simulations. Dynamic fracture in early mesh-free methods was treated by simple separation of particles [41–47]. However, [30, 48] showed that care needs to be taken as the fracture often is caused by numerical artefacts. More advanced methods such as the extended finite element method [49, 50], extended mesh-free methods [34, 51–59], the phantom node method [60–64], or efficient remeshing techniques [65–70] have been applied to dynamic fracture of concrete as well [71, 72]. However, they do not seem suitable to model the impact resistance of concrete with large deformations and an enormous number of cracks. A simpler and more efficient pathway is the Cracking Particles Method (CPM) [73, 74] which models the crack as set of crack segments. Similar methods were used by [75–80]. Other researchers try to combine the advantages of mesh-free methods and finite elements. Therefore, only the area around the perforation was modelled by a mesh-free method whereas a finite element discretization was employed elsewhere reducing computational cost [31, 80]. Most simulations were based on pure deterministic models that can lead to unrealistic crack patterns as the ones shown in [39]. In these simulations, the radial main cracks appear too close to each other. In order to achieve unsymmetrical crack patterns, [81] introduced randomness in the strength to avoid the unrealistic symmetric patterns they modelled before in [82–84]. However, all other parameters remained deterministic.

Several benchmark experiments have also been conducted to test the influence of the thickness of the specimen or the speed of the impactor on the impact resistance of concrete or reinforced concrete structures [85–91]. One classical experiment was done by [92] who subjected concrete structures to impact with different speeds of the impactor. They measured the residual velocity of the impactor. The recent experiments by [40] also report penetration depth.

This paper presents one of the first studies on stochastic modelling of dynamic fracture of concrete. Our constitutive model is based on the model described in [14] and combined with a Lagrangian mesh-free method for the computational studies. Fracture will be modelled by a simple particle splitting as described in [77]. This methodology will be described in the first sections before we present computational results that will be compared to the experiments of Hanchak et al. [92]. We also carried out our own experiments that will serve as validation.

2. Governing Equations and Discretization

The governing equation to be solved is the balance of linear momentum that is given in weak form of the updated description of motion by

$$\delta W_{\text{int}} - \delta W_{\text{ext}} = \delta W_{\text{kin}}, \quad \delta W_{\text{int}} = \int_{\Omega} \delta\epsilon_{ij} : \sigma_{ij} d\Omega, \ \delta W_{\text{ext}} = \int_{\Gamma_t} \delta u_i t_i d\Gamma + \int_{\Omega} \delta u_i b_i d\Omega, \ \delta W_{\text{kin}} = \int_{\Omega} \varrho u_i \ddot{u}_i d\Omega, \tag{1}$$

where δW_{int}, δW_{ext}, and δW_{kin} denote the internal, external, and kinetic energy, respectively. The domain is denoted by Ω whereas $\Gamma_t \cup \Gamma_u = \Gamma$ with $\Gamma_t \cap \Gamma_u = 0$ denotes the Lipschitz continuous boundary; ϵ_{ij} and σ_{ij} indicate the strain tensor and Cauchy stress tensor, b_i and t_i are the body forces and tractions, respectively, u_i is the displacement field, and ϱ is the density while δ denotes the first variation and the superimposed dote material time derivatives which are identical to the partial derivatives in an updated Lagrangian framework.

We have employed a stabilised [48] nodally regularised [30] element-free Galerkin method [93]. Therefore, the discretization of the displacement field is given by

$$\mathbf{u}(\mathbf{x}, t) = \sum_{I \in \mathcal{S}} N_I(\mathbf{x}) \mathbf{u}_I(t), \tag{2}$$

where $\mathbf{u}_I$ denote the nodal parameters, which are not equal to the physical displacement value at that point, that is, $\mathbf{u}(\mathbf{x}_I) \neq \mathbf{u}_I$, and $N_I(\mathbf{x})$ indicate the shape functions. They are derived from minimising a discrete weighted norm as explained in detail, for example, in [93]:

$$\mathbf{N}(\mathbf{x}) = \mathbf{p}^T(\mathbf{x}) \mathbf{A}^{-1}(\mathbf{x}) \mathbf{B}(\mathbf{x}), \tag{3}$$

where $\mathbf{A}(\mathbf{x}) = \sum_{I=1}^{n} w(\mathbf{X} - \mathbf{X}_I)\mathbf{p}(\mathbf{X}_I)\mathbf{p}^T(\mathbf{X}_I)$ is called the moment-matrix and the matrix $\mathbf{B}(\mathbf{x})$ is obtained by

$$\mathbf{B}(\mathbf{x}) = [w(\mathbf{x}-\mathbf{x}_1)\mathbf{p}(\mathbf{x}_1) \ \ w(\mathbf{x}-\mathbf{x}_2)\mathbf{p}(\mathbf{x}_2) \ \cdots \ w(\mathbf{x}-\mathbf{x}_n)\mathbf{p}(\mathbf{x}_n)] \tag{4}$$

$w(\mathbf{X} - \mathbf{X}_I)$ denoting the kernel function and $\mathbf{p}^T(\mathbf{X})$ the polynomial basis. To ensure Galilean invariance, we opted to choose a linear basis $\mathbf{p}^T(\mathbf{x}) = [1 \ x \ y]$. The exponential weighting function has been employed in all our simulations.

Substituting the discretization of the displacement field and virtual displacement field into the weak form yields the well-known equation of motion that is solved for the nodal parameter $\mathbf{u}_I(t)$ that is stored in the global vector $\mathbf{D}$:

$$\mathbf{M}\ddot{\mathbf{D}} = \mathbf{F}_{\text{ext}} - \mathbf{F}_{\text{int}} \tag{5}$$

with the mass matrix $\mathbf{M} = \int_{\Omega} \varrho \mathbf{N}^T \mathbf{N} \, d\Omega$ and the vector of external and internal forces

$$\mathbf{F}_{\text{ext}} = \int_{\Omega} \varrho \mathbf{N}^T \mathbf{b} \, d\Omega + \int_{\Gamma_t} \mathbf{N}^T \bar{\mathbf{t}} \, d\Gamma$$
$$\mathbf{F}_{\text{int}} = \int_{\Omega} \tilde{\mathbf{B}}^T \sigma \, d\Omega, \tag{6}$$

where $\mathbf{N}$ and $\tilde{\mathbf{B}}$ are matrices containing the mesh-free shape functions and their spatial derivatives, respectively. The integrals are evaluated by stabilised nodal integration which ensures bending exactness in a Galerkin method. The use of an updated Lagrangian kernel formulation as suggested in [39] ensures the stability of the method while maintaining the applicability to extremely large deformations needed for dynamic fracture and fragmentation.

3. Constitutive Model

We have used a modified version of the constitutive model presented in [14]. Instead of using a viscous damage-plasticity model, we removed the plasticity part of this model as it adds material parameters that are difficult to calibrate. Furthermore, preliminary studies indicated that the plastic

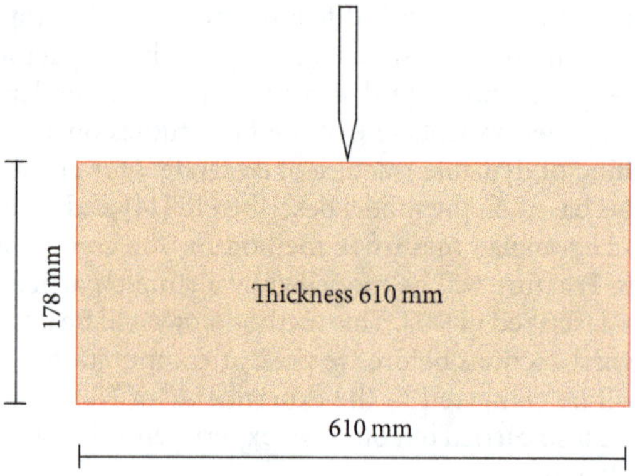

FIGURE 1: Hanchak impact experiment.

TABLE 1: Parameters of the constitutive model.

Young's modulus E [GPa]	Poisson ratio	Density [g/mm^3]		
52	0.22	2700		
e_0	e_d	g_d	c_{t1}	c_{t2}
0.0001	0.0003	2.3	0.032	0.55
c_1	c_2	c_3	c_4	
0.01234	0.0252	0.7821	0.3464	
r_t	r_c			
1.2	18.8			
a_v	b_v	e_v	$e_{v,th}$	
0.7	3.4	0.02	0.007	

response is of minor importance as the results in the next section will show. The basic idea of the damage model is given subsequently. For more detailed information, the reader is referred to [14].

4. Results

Two types of experiments are employed in order to validate the proposed computational framework:

(i) the impact experiments carried out by Hanchak [92],

(ii) contact explosion experiments done at the laboratory of our university.

Hanchak tested the impact resistance of concrete structures of different concrete strength. The main goal of the experiments of Hanchak studies was to test their high performance concrete compared to standard concrete. The high performance concrete had a compressive strength of 140 MPA while the standard concrete's compressive strength was nearly 50 MPA. Therefore, concrete specimens were subjected to impacts of different projectile velocities. The experimental set-up is illustrated in Figure 1. As our focus is not on high performance concrete but rather on validation of our framework, we present only the results of the impact simulations on standard concrete. In the results, all plates were perforated and the residual impact velocity was measured. We will focus on experiments with penetration later.

A full three-dimensional discretization has been utilized. Preliminary studies showed substantial differences in results obtained from 2D simulations based on plane strain assumptions. Around 2 million nodes were needed in order to get convergence in the residual impact velocity and simultaneously in the damage pattern. The (deterministic) parameters of the constitutive model are listed in Table 1. They are classified in different categories as indicated by the different lines: (1) bulk parameters, that is, Young's modulus, Poisson's ratio, and density; (2) damage parameters e_0, e_d, and g_d determining the "static" damage evolution: the key parameters are e_d and g_d controlling the ductility and maximum strength of the concrete and the parameters determining the dynamic part of the damage responses c_{t1} and c_{t2}: they determine how the dynamic damage is decaying over time; (3) the parameters of the damage surface c_1 to c_4; (4) the reduction factors r_c and r_t controlling how much compressive damage can be done by tensile loading and vice versa; and (5) the parameters of the Hugoniot-curve accounting for the pressure dependence of concrete under high strain rates and high volumetric strains. All parameters, except the ones associated with the damage surface, are assumed to be stochastic in our computations and are subjected to a uniform probability distribution function with a scatter of 5% and 10%, respectively. In order to identify the influence of each physical mechanism, we modified only parameters associated with this physical mechanism according to the probability distribution function while keeping all other parameters constant. We are aware that a global sensitivity analysis as proposed, for example, by [94–97] in the context of studies on polymeric composites or by [98] in the context of optimization will provide more quantitative results. However, such studies are out of the scope of this paper and left for the future.

Before discussing the results presented in Table 2, let us discuss the different physical phenomena and the related input parameters:

(1) The parameters related to the first line of 1 govern the initial elastic response of the material in the bulk.

(2) The parameters e_0, e_d, and g_d determine the stress-strain response under static conditions. The first parameter determines the transition to nonlinear material behaviour while the other two parameters determine the tensile/compressive strength and failure strain.

(3) The parameters c_{t1}, c_{t2} govern the dynamic damage evolution and are responsible for the strength increase under high strain rates.

(4) The parameters r_t, r_c are responsible for the amount of compressive damage caused under tensile loading and vice versa.

(5) The last set of parameters in Table 1 account for the pressure dependence of the concrete.

The results in Table 2 show the lower and upper bounds of the residual impact velocity for the different input parameters

TABLE 2: Residual velocity ([m/s]) of the impactor in dependence of the impact velocity ([m/s]) and the scatter in the input parameters: the first line of the same impact velocity refers to a 5% scatter while the second line refers to the 10%.

Impact speed	Determ.	Set 1	Set 2	Set 3	Set 4	Set 5	Experiment
360	75	74-75	73–76	74–76	75-76	74–76	67
360		74-75	73–77	72–77	75-76	73–76	
430	223	222-223	220–229	219–231	221–224	221–230	214
430		222-224	215–230	213–234	221–225	215–232	
750	628	625–631	618–634	615–635	623–632	619–631	615
750		624–632	610–639	601–645	621–633	611–641	
1060	957	952–961	942–971	938–974	950–962	943–973	947
1060		951–961	934–980	928–989	949–963	932–984	

TABLE 3: Penetration depth and damage diameter for the explosion experiment: the first line refers to a scatter of 5% while the second line refers to the 10% scatter in the input.

	Deterministic	Set 1	Set 2	Set 3	Set 4	Set 5	Experiment
Diameter [cm]	57	56.7–57	56–57.2	55.6–57.5	56.5–57	56.4–57.3	56
Diameter [cm]		56.7–57	55.5–57.3	55–57.8	56.1–57.3	56.3–57.3	
Penetr. depth [cm]	28	27.8–28	27.3–28.2	27–28.4	27.5–28.1	27.7–28	26
Penetr. depth [cm]		27.8–28	27.0–28.5	26.3–28.8	27.0–28.3	27.6–28	

for different impact velocities. The values of the deterministic simulation are depicted in Table 2 as well. As can be seen, uncertainties in the input substantially influence the residual velocity of the impactor. While parameter sets 1 (parameters related to the initial elastic response) and 4 barely have an effect on the residual velocity (the scatter in the output is much smaller than the scatter in the input), the static damage evolution (parameter set 2) and pressure dependence (parameter set 5) have a substantial influence in the order of the scatter in the input. This is observed consistently for both 5% and 10% scatter. The biggest influence is achieved by the parameters related to the dynamic strength increase which is more pronounced for increasing scatter. Finally, we remark that we also varied two sets of input parameters while keeping the other parameters constant. Since parameter sets 3 and 5 had the biggest influence on the output, we varied those parameters. However, the upper and lower bounds of the residual velocity barely changed. The scatter in the output is slightly higher than the scatter in the input. Note that the effect gets more pronounced with increasing impact velocity. Overall, one can see an excellent agreement between computational and experimental results.

The second test example is a concrete slab under contact explosion as depicted in Figure 2. The explosion was modelled by a (JWL) John-Wilkonson-Lee equation of state. The explosive is also modelled with a mesh-free Lagrangian method, which is one of the advantages of the proposed method: the fluid-structure interaction remains fairly simple compared to a coupled Lagrangian-Eulerian coupling such as (ALE) which is required for finite elements, for instance.

In the experiment, a penetration was observed and the penetration depth as well as the diameter of the crater at the top surface was measured. We used the same concrete as in the experiments performed by Hanchak and therefore the same material parameters of the constitutive model were used. Figure 3 shows the damage shape towards the end of the simulation; the damage is indicated by the red color. The full deformation has not been reached, yet. However, the damage evolution is fully completed already. Table 3 summarizes the key results of the simulation. We tested again the influence of uncertain input parameters for the same probability distribution function and scatter as discussed previously. The observations are somewhat similar compared to the observations for the Hanchak experiments. However, the uncertainties in the input parameters seem to be less pronounced here. The largest effect is attributed to parameter set 3 related to the dynamic damage evolution followed by parameter sets 2 and 5. Parameter sets 1 and 4 have nearly a negligible influence on the penetration depth and damage diameter.

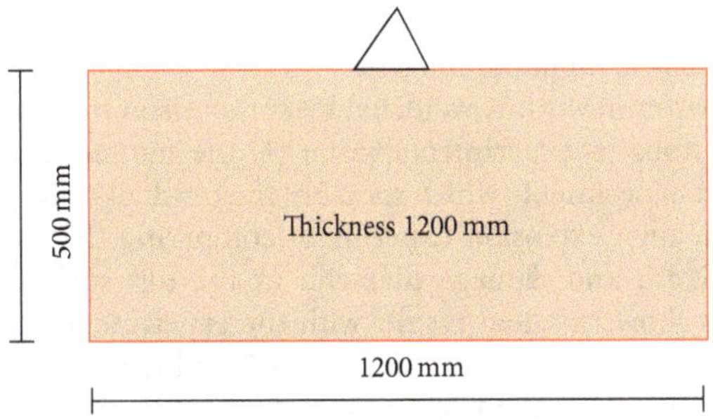

FIGURE 2: Explosion experiment.

5. Discussions

We presented a computational framework for predicting the impact resistance of reinforced concrete structure. This framework is based on a mesh-free Lagrangian method

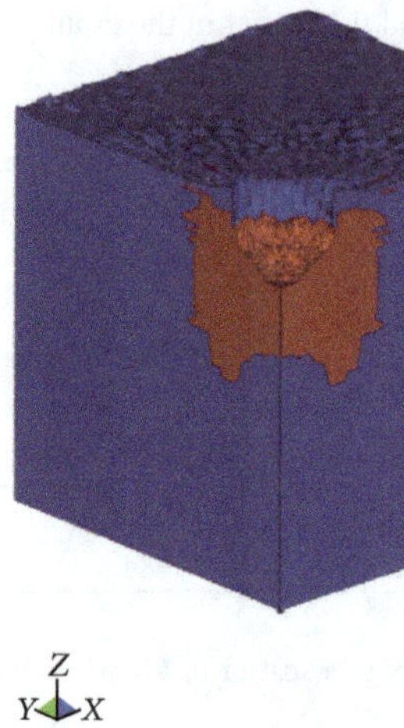

FIGURE 3: Concrete under explosive loading.

which guarantees the applicability to large deformations as they occur under high velocity impact, explosion, dynamic fracture, and fragmentation. A viscous dynamic damage model is exploited which accounts for the strength increase of concrete at high strain rates and the pressure dependent response of concrete. Furthermore, we consider the influence of stochastic input parameters.

The framework is validated by comparison of numerical predictions to experimental data: (1) the famous Hanchak impact experiments which measure the residual velocity and (2) our own explosion experiment comparing the penetration depth and damage diameter at the top surface. Our results show excellent results with the experimental results. Furthermore, we account for stochastic influence. Therefore, we classify the parameters of the constitutive model into 5 categories related to (1) the initial mechanical properties in the bulk, (2) static damage evolution, (3) dynamic damage evolution, (4) reduction factor of compressive damage at tensile load and vice versa, and (5) pressure dependence. By varying each set of parameters while keeping the others constant, we show that the key factor influencing our output is related to parameter set 3. We also reveal that the scatter in the output is not small.

Conflict of Interests

The author declares that there is no conflict of interests regarding the publication of this paper.

References

[1] Z. P. Bazant and G. Pijaudier-Cabot, "Nonlocal continuum damage, localization instabilities and convergence," *Journal of Applied Mechanics*, vol. 55, no. 2, pp. 287–293, 1988.

[2] Z. P. Bažant, "Why continuum damage is nonlocal: micromechanics arguments," *Journal of Engineering Mechanics*, vol. 117, no. 5, pp. 1070–1087, 1991.

[3] Z. P. Bažant and M. Jirásek, "Non-local integral formulations of plasticity and damage: survey of progress," *Journal of Engineering Mechanics*, vol. 128, no. 11, pp. 1119–1149, 2002.

[4] W. F. Chen, *Constitutive Equations for Engineering Materials*, vol. 2 of *Plasticity and Modeling*, Elsevier, New York, NY, USA, 1994.

[5] I. Carol and Z. P. Bažant, "Damage and plasticity in microplane theory," *International Journal of Solids and Structures*, vol. 34, no. 29, pp. 3807–3835, 1997.

[6] W. Han and B. D. Reddy, *Plasticity. Mathematical Theory and Numerical Analysis*, vol. 9 of *Interdisciplinary Applied Mathematics*, Springer-Verlag, New York, NY, USA, 1999.

[7] J. W. Tedesco, J. C. Powell, C. A. Ross, and M. L. Hughes, "A strain-rate-dependent concrete material model for ADINA," *Computers & Structures*, vol. 64, no. 5-6, pp. 1053–1067, 1997.

[8] J. F. Georgin and J. M. Reynouard, "Modeling of structures subjected to impact: concrete behaviour under high strain rate," *Cement and Concrete Composites*, vol. 25, no. 1, pp. 131–143, 2003.

[9] L. J. Malvar and C. A. Ross, "Review of strain rate effects for concrete in tension," *Materials Journal*, vol. 95, pp. 735–739, 1995.

[10] J.-F. Dubé, G. Pijaudier-Cabot, and C. La Borderie, "Rate dependent damage model for concrete in dynamics," *Journal of Engineering Mechanics*, vol. 122, no. 10, pp. 939–947, 1996.

[11] G. R. Johnson and T. J. Holmquist, "A computational constitutive model for brittle materials subjected to large strains," in *Shock Wave and High-Strain-Rate Phenomena in Materials*, M. A. Meyers, L. E. Murr, and K. P. Staudhammer, Eds., pp. 1075–1081, Marcel Dekker, New York, NY, USA, 1992.

[12] G. R. Johnson and W. H. Cook, "Fracture characteristics of three metals subjected to various strains, strain rates, temperatures and pressures," *Engineering Fracture Mechanics*, vol. 21, no. 1, pp. 31–48, 1985.

[13] T. Rabczuk, J. Y. Kim, E. Samaniego, and T. Belytschko, "Homogenization of sandwich structures," *International Journal for Numerical Methods in Engineering*, vol. 61, no. 7, pp. 1009–1027, 2004.

[14] T. Rabczuk and J. Eibl, "Simulation of high velocity concrete fragmentation using SPH/MLSPH," *International Journal for Numerical Methods in Engineering*, vol. 56, no. 10, pp. 1421–1444, 2003.

[15] T. Rabczuk and J. Eibl, "Modelling dynamic failure of concrete with meshfree methods," *International Journal of Impact Engineering*, vol. 32, no. 11, pp. 1878–1897, 2006.

[16] G. T. Camacho and M. Ortiz, "Computational modelling of impact damage in brittle materials," *International Journal of Solids and Structures*, vol. 33, no. 20–22, pp. 2899–2938, 1996.

[17] X. Teng, T. Wierzbicki, S. Hiermaier, and I. Rohr, "Numerical prediction of fracture in the Taylor test," *International Journal of Solids and Structures*, vol. 42, no. 9-10, pp. 2929–2948, 2005.

[18] A. Plotzitza, T. Rabczuk, and J. Eibl, "Techniques for numerical simulations of concrete slabs for demolishing by blasting," *Journal of Engineering Mechanics*, vol. 133, no. 5, pp. 523–533, 2007.

[19] T. Børvik, A. H. Clausen, M. Eriksson, T. Berstad, O. S. Hopperstad, and M. Langseth, "Experimental and numerical study on the perforation of AA6005-T6 panels," *International Journal of Impact Engineering*, vol. 32, no. 1–4, pp. 35–64, 2005.

[20] G. Ruiz, A. Pandolfi, and M. Ortiz, "Three-dimensional cohesive modeling of dynamic mixed-mode fracture," *International Journal for Numerical Methods in Engineering*, vol. 52, no. 1-2, pp. 97–120, 2001.

[21] G. Ruiz, M. Ortiz, and A. Pandolfi, "Three-dimensional finite element simulation of the dynamic Brazilian tests on concrete cylinders," *International Journal for Numerical Methods in Engineering*, vol. 48, no. 7, pp. 963–994, 2000.

[22] M. Polanco-Loria, O. S. Hopperstad, T. Børvik, and T. Berstad, "Numerical predictions of ballistic limits for concrete slabs using a modified version of the HJC concrete model," *International Journal of Impact Engineering*, vol. 35, no. 5, pp. 290–303, 2008.

[23] T. Børvik, A. G. Hanssen, M. Langseth, and L. Olovsson, "Response of structures to planar blast loads—a finite element engineering approach," *Computers and Structures*, vol. 87, no. 9-10, pp. 507–520, 2009.

[24] W. Riedel, M. Nöldgen, E. Straßburger, K. Thoma, and E. Fehling, "Local damage to ultra high performance concrete structures caused by an impact of aircraft engine missiles," *Nuclear Engineering and Design*, vol. 240, no. 10, pp. 2633–2642, 2010.

[25] S. R. Beissel, G. R. Johnson, and C. H. Popelar, "An element-failure algorithm for dynamic crack propagation in general directions," *Engineering Fracture Mechanics*, vol. 61, no. 3-4, pp. 407–425, 1998.

[26] S. R. Beissel, G. R. Johnson, and C. H. Popelar, "An element-failure algorithm for dynamic crack propagation in general directions," *Engineering Fracture Mechanics*, vol. 61, no. 3-4, pp. 407–425, 1998.

[27] S. Beissel and T. Belytschko, "Nodal integration of the element-free galerkin method," *Computer Methods in Applied Mechanics and Engineering*, vol. 139, no. 1–4, pp. 49–74, 1996.

[28] L. D. Libersky, P. W. Randles, T. C. Carney, and D. L. Dickinson, "Recent improvements in SPH modeling of hypervelocity impact," *International Journal of Impact Engineering*, vol. 20, no. 6–10, pp. 525–532, 1997.

[29] G. A. Dilts, "Moving-least-squares-particle hydrodynamics. I. Consistency and stability," *International Journal for Numerical Methods in Engineering*, vol. 44, no. 8, pp. 1115–1155, 1999.

[30] R. Vignjevic, J. R. Reveles, and J. Campbell, "Sph in a total Lagrangian formalism," *Computer Modeling in Engineering and Sciences*, vol. 14, no. 3, pp. 181–198, 2006.

[31] T. Rabczuk and T. Belytschko, "Adaptivity for structured meshfree particle methods in 2D and 3D," *International Journal for Numerical Methods in Engineering*, vol. 63, no. 11, pp. 1559–1582, 2005.

[32] M. A. Puso, J. S. Chen, E. Zywicz, and W. Elmer, "Meshfree and finite element nodal integration methods," *International Journal for Numerical Methods in Engineering*, vol. 74, no. 3, pp. 416–446, 2008.

[33] P. C. Guan, S. W. Chi, J. S. Chen, T. R. Slawson, and M. J. Roth, "Semi-Lagrangian reproducing kernel particle method for fragment-impact problems," *International Journal of Impact Engineering*, vol. 38, no. 12, pp. 1033–1047, 2011.

[34] F. Amiri, D. Millán, Y. Shen, T. Rabczuk, and M. Arroyo, "Phase-field modeling of fracture in linear thin shells," *Theoretical and Applied Fracture Mechanics*, vol. 69, pp. 102–109, 2014.

[35] Y. Wu, D. D. Wang, and C.-T. Wu, "Three dimensional fragmentation simulation of concrete structures with a nodally regularized meshfree method," *Theoretical and Applied Fracture Mechanics*, vol. 72, no. 1, pp. 89–99, 2014.

[36] M. Hillman, J. S. Chen, and S. W. Chi, "Stabilized and variationally consistent nodal integration for meshfree modeling of impact problems," *Computational Particle Mechanics*, vol. 1, no. 3, pp. 245–256, 2014.

[37] S.-W. Chi, C.-H. Lee, J.-S. Chen, and P.-C. Guan, "A level set enhanced natural kernel contact algorithm for impact and penetration modeling," *International Journal for Numerical Methods in Engineering*, vol. 102, no. 3-4, pp. 839–866, 2015.

[38] V. P. Nguyen, T. Rabczuk, S. Bordas, and M. Duflot, "Meshless methods: a review and computer implementation aspects," *Mathematics and Computers in Simulation*, vol. 79, no. 3, pp. 763–813, 2008.

[39] T. Rabczuk and T. Belytschko, "A three-dimensional large deformation meshfree method for arbitrary evolving cracks," *Computer Methods in Applied Mechanics and Engineering*, vol. 196, no. 29-30, pp. 2777–2799, 2007.

[40] M. Unosson and L. Nilsson, "Projectile penetration and perforation of high performance concrete: experimental results and macroscopic modelling," *International Journal of Impact Engineering*, vol. 32, no. 7, pp. 1068–1085, 2006.

[41] A. G. Petschek and L. D. Libersky, "Cylindrical smoothed particle hydrodynamics," *Journal of Computational Physics*, vol. 109, no. 1, pp. 76–83, 1993.

[42] W. Benz and E. Asphaug, "Simulations of brittle solids using smooth particle hydrodynamics," *Computer Physics Communications*, vol. 87, no. 1-2, pp. 253–265, 1995.

[43] P. W. Randles and L. D. Libersky, "Smoothed particle hydrodynamics: some recent improvements and applications," *Computer Methods in Applied Mechanics and Engineering*, vol. 139, no. 1–4, pp. 375–408, 1996.

[44] L. D. Libersky, A. G. Petschek, T. C. Carney, J. R. Hipp, and F. A. Allahdadi, "High strain lagrangian hydrodynamics a three-dimensional SPH code for dynamic material response," *Journal of Computational Physics*, vol. 109, no. 1, pp. 67–75, 1993.

[45] G. R. Johnson and S. R. Beissel, "Normalized smoothing functions for sph impact computations," *International Journal for Numerical Methods in Engineering*, vol. 39, no. 16, pp. 2725–2741, 1996.

[46] G. R. Johnson, S. R. Beissel, and R. A. Stryk, "A generalized particle algorithm for high velocity impact computations," *Computational Mechanics*, vol. 25, no. 2, pp. 245–256, 2000.

[47] T. Rabczuk, J. Eibl, and L. Stempniewski, "Numerical analysis of high speed concrete fragmentation using a meshfree lagrangian method," *Engineering Fracture Mechanics*, vol. 71, no. 4–6, pp. 547–556, 2004.

[48] T. Rabczuk, T. Belytschko, and S. P. Xiao, "Stable particle methods based on Lagrangian kernels," *Computer Methods in Applied Mechanics and Engineering*, vol. 193, no. 12–14, pp. 1035–1063, 2004.

[49] N. Moës, J. Dolbow, and T. Belytschko, "A finite element method for crack growth without remeshing," *International Journal for Numerical Methods in Engineering*, vol. 46, no. 1, pp. 131–150, 1999.

[50] T. Belytschko and T. Black, "Elastic crack growth in finite elements with minimal remeshing," *International Journal for Numerical Methods in Engineering*, vol. 45, no. 5, pp. 601–620, 1999.

[51] T. Rabczuk and P. Areias, "A meshfree thin shell for arbitrary evolving cracks based on an extrinsic basis," *Computer Modeling in Engineering & Sciences*, vol. 16, no. 2, pp. 115–130, 2006.

[52] T. Rabczuk and G. Zi, "A meshfree method based on the local partition of unity for cohesive cracks," *Computational Mechanics*, vol. 39, no. 6, pp. 743–760, 2007.

[53] T. Rabczuk, P. M. A. Areias, and T. Belytschko, "A simplified mesh-free method for shear bands with cohesive surfaces,"

International Journal for Numerical Methods in Engineering, vol. 69, no. 5, pp. 993–1021, 2007.

[54] T. Rabczuk, P. M. A. Areias, and T. Belytschko, "A meshfree thin shell method for non-linear dynamic fracture," *International Journal for Numerical Methods in Engineering*, vol. 72, no. 5, pp. 524–548, 2007.

[55] S. Bordas, T. Rabczuk, and G. Zi, "Three-dimensional crack initiation, propagation, branching and junction in non-linear materials by an extended meshfree method without asymptotic enrichment," *Engineering Fracture Mechanics*, vol. 75, no. 5, pp. 943–960, 2008.

[56] G. Zi, T. Rabczuk, and W. Wall, "Extended meshfree methods without branch enrichment for cohesive cracks," *Computational Mechanics*, vol. 40, no. 2, pp. 367–382, 2007.

[57] T. Rabczuk, S. Bordas, and G. Zi, "On three-dimensional modelling of crack growth using partition of unity methods," *Computers and Structures*, vol. 88, no. 23-24, pp. 1391–1411, 2010.

[58] T. Rabczuk and P. M. A. Areias, "A new approach for modelling slip lines in geological materials with cohesive models," *International Journal for Numerical and Analytical Methods in Geomechanics*, vol. 30, no. 11, pp. 1159–1172, 2006.

[59] H. Talebi, C. Samaniego, E. Samaniego, and T. Rabczuk, "On the numerical stability and mass-lumping schemes for explicit enriched meshfree methods," *International Journal for Numerical Methods in Engineering*, vol. 89, no. 8, pp. 1009–1027, 2012.

[60] A. Hansbo and P. Hansbo, "A finite element method for the simulation of strong and weak discontinuities in solid mechanics," *Computer Methods in Applied Mechanics and Engineering*, vol. 193, no. 33–35, pp. 3523–3540, 2004.

[61] J.-H. Song, P. M. A. Areias, and T. Belytschko, "A method for dynamic crack and shear band propagation with phantom nodes," *International Journal for Numerical Methods in Engineering*, vol. 67, no. 6, pp. 868–893, 2006.

[62] T. Rabczuk, G. Zi, A. Gerstenberger, and W. A. Wall, "A new crack tip element for the phantom-node method with arbitrary cohesive cracks," *International Journal for Numerical Methods in Engineering*, vol. 75, no. 5, pp. 577–599, 2008.

[63] T. Chau-Dinh, G. Zi, P.-S. Lee, J.-H. Song, and T. Rabczuk, "Phantom-node method for shell models with arbitrary cracks," *Computers & Structures*, vol. 92-93, pp. 242–256, 2012.

[64] N. Vu-Bac, H. Nguyen-Xuan, L. Chen et al., "A phantom-node method with edge-based strain smoothing for linear elastic fracture mechanics," *Journal of Applied Mathematics*, vol. 2013, Article ID 978026, 12 pages, 2013.

[65] P. Areias and T. Rabczuk, "Finite strain fracture of plates and shells with configurational forces and edge rotation," *International Journal for Numerical Methods in Engineering*, vol. 94, no. 12, pp. 1099–1122, 2013.

[66] P. Areias, T. Rabczuk, and D. Dias da Costa, "Assumed-metric spherically-interpolated quadrilateral shell element," *Finite Elements in Analysis and Design*, vol. 66, pp. 53–67, 2013.

[67] P. Areias, D. Dias-Da-Costa, J. M. Sargado, and T. Rabczuk, "Element-wise algorithm for modeling ductile fracture with the rousselier yield function," *Computational Mechanics*, vol. 52, no. 6, pp. 1429–1443, 2013.

[68] P. Areias, T. Rabczuk, and D. Dias-da-Costa, "Element-wise fracture algorithm based on rotation of edges," *Engineering Fracture Mechanics*, vol. 110, pp. 113–137, 2013.

[69] P. Areias, T. Rabczuk, and P. P. Camanho, "Initially rigid cohesive laws and fracture based on edge rotations," *Computational Mechanics*, vol. 52, no. 4, pp. 931–947, 2013.

[70] P. Areias, T. Rabczuk, and P. P. Camanho, "Finite strain fracture of 2D problems with injected anisotropic softening elements," *Theoretical and Applied Fracture Mechanics*, vol. 72, pp. 50–63, 2014.

[71] G. Liu, Y. Hu, Q. Li, and Z. Zuo, "XFEM for thermal crack of massive concrete," *Mathematical Problems in Engineering*, vol. 2013, Article ID 343842, 9 pages, 2013.

[72] T. Rabczuk, S. Bordas, and G. Zi, "A three-dimensional meshfree method for continuous multiple-crack initiation, propagation and junction in statics and dynamics," *Computational Mechanics*, vol. 40, no. 3, pp. 473–495, 2007.

[73] T. Rabczuk and T. Belytschko, "Cracking particles: a simplified meshfree method for arbitrary evolving cracks," *International Journal for Numerical Methods in Engineering*, vol. 61, no. 13, pp. 2316–2343, 2004.

[74] T. Rabczuk, J.-H. Song, and T. Belytschko, "Simulations of instability in dynamic fracture by the cracking particles method," *Engineering Fracture Mechanics*, vol. 76, no. 6, pp. 730–741, 2009.

[75] H. X. Wang and S. X. Wang, "Analysis of dynamic fracture with cohesive crack segment method," *Computer Modeling in Engineering & Sciences*, vol. 35, no. 3, pp. 253–274, 2008.

[76] N. Sageresan and R. Drathi, "Crack propagation in concrete using meshless method," *Computer Modeling in Engineering & Sciences*, vol. 32, no. 2, pp. 103–112, 2008.

[77] T. Rabczuk, G. Zi, S. Bordas, and H. Nguyen-Xuan, "A simple and robust three-dimensional cracking-particle method without enrichment," *Computer Methods in Applied Mechanics and Engineering*, vol. 199, no. 37–40, pp. 2437–2455, 2010.

[78] T. Rabczuk, R. Gracie, J.-H. Song, and T. Belytschko, "Immersed particle method for fluid-structure interaction," *International Journal for Numerical Methods in Engineering*, vol. 81, no. 1, pp. 48–71, 2010.

[79] F. Caleyron, A. Combescure, V. Faucher, and S. Potapov, "Dynamic simulation of damage-fracture transition in smoothed particles hydrodynamics shells," *International Journal for Numerical Methods in Engineering*, vol. 90, no. 6, pp. 707–738, 2012.

[80] Y. Chuzel-Marmot, R. Ortiz, and A. Combescure, "Three dimensional SPH-FEM gluing for simulation of fast impacts on concrete slabs," *Computers and Structures*, vol. 89, no. 23-24, pp. 2484–2494, 2011.

[81] T. Rabczuk, G. Zi, S. Bordas, and H. Nguyen-Xuan, "A geometrically non-linear three-dimensional cohesive crack method for reinforced concrete structures," *Engineering Fracture Mechanics*, vol. 75, no. 16, pp. 4740–4758, 2008.

[82] T. Rabczuk and J. Eibl, "Numerical analysis of prestressed concrete beams using a coupled element free Galerkin/finite element approach," *International Journal of Solids and Structures*, vol. 41, no. 3-4, pp. 1061–1080, 2004.

[83] T. Rabczuk, J. Akkermann, and J. Eibl, "A numerical model for reinforced concrete structures," *International Journal of Solids and Structures*, vol. 42, no. 5-6, pp. 1327–1354, 2005.

[84] T. Rabczuk and T. Belytschko, "Application of particle methods to static fracture of reinforced concrete structures," *International Journal of Fracture*, vol. 137, no. 1–4, pp. 19–49, 2006.

[85] R. John and S. P. Shah, "Mixed mode fracture of concrete subjected to impact loading," *ASCE Journal of Structural Engineering*, vol. 116, no. 3, pp. 585–602, 1990.

[86] P. Rossi, "A physical phenomenon which can explain the mechanical behaviour of concrete under high strain rates," *Materials and Structures*, vol. 24, no. 6, pp. 422–424, 1991.

[87] M. J. Forrestal, D. J. Frew, S. J. Hanchak, and N. S. Brar, "Penetration of grout and concrete targets with ogive-nose steel projectiles," *International Journal of Impact Engineering*, vol. 18, no. 5, pp. 465–476, 1996.

[88] D. J. Frew, S. J. Hanchak, M. L. Green, and M. J. Forrestal, "Penetration of concrete targets with ogive-nose steel rods," *International Journal of Impact Engineering*, vol. 21, no. 6, pp. 489–497, 1998.

[89] T. Børvik, A. H. Clausen, O. S. Hopperstad, and M. Langseth, "Perforation of AA5083-H116 aluminium plates with conical-nose steel projectiles—experimental study," *International Journal of Impact Engineering*, vol. 30, no. 4, pp. 367–384, 2004.

[90] W. Riedel, M. Wicklein, and K. Thoma, "Shock properties of conventional and high strength concrete: experimental and mesomechanical analysis," *International Journal of Impact Engineering*, vol. 35, no. 3, pp. 155–171, 2008.

[91] P. Forquin, W. Riedel, and J. Weerheijmh, "Dynamic test devices for analyzing the tensile properties of concrete," in *Understanding the Tensile Properties of Concrete*, pp. 137–180, 181e, Woodhead Publishing, 2013.

[92] S. J. Hanchak, M. J. Forrestal, E. R. Young, and J. Q. Ehrgott, "Perforation of concrete slabs with 48 MPa (7 ksi) and 140 MPa (20 ksi) unconfined compressive strengths," *International Journal of Impact Engineering*, vol. 12, no. 1, pp. 1–7, 1992.

[93] T. Belytschko, Y. Y. Lu, and L. Gu, "Element-free galerkin methods," *International Journal for Numerical Methods in Engineering*, vol. 37, no. 2, pp. 229–256, 1994.

[94] N. Vu-Bac, T. Lahmer, Y. Zhang, X. Zhuang, and T. Rabczuk, "Stochastic predictions of interfacial characteristic of carbon nanotube polyethylene composites," *Composite Part B: Engineering*, vol. 59, pp. 80–95, 2014.

[95] N. Vu-Bac, T. Lahmer, H. Keitel, J. Zhao, X. Zhuang, and T. Rabczuk, "Stochastic predictions of bulk properties of amorphous polyethylene based on molecular dynamics simulations," *Mechanics of Materials*, vol. 68, pp. 70–84, 2014.

[96] N. Vu-Bac, M. Silani, T. Lahmer, X. Zhuang, and T. Rabczuk, "A unified framework for stochastic predictions of mechanical properties of polymeric nanocomposites," *Computational Materials Science*, vol. 96, pp. 520–535, 2015.

[97] M. Silani, H. Talebi, S. Ziaei-Rad, P. Kerfriden, S. P. Bordas, and T. Rabczuk, "Stochastic modelling of clay/epoxy nanocomposites," *Composite Structures*, vol. 118, pp. 241–249, 2014.

[98] H. Ghasemi, R. Rafiee, X. Zhuang, J. Muthu, and T. Rabczuk, "Uncertainties propagation in metamodel-based probabilistic optimization of CNT/polymer composite structure using stochastic multi-scale modeling," *Computational Materials Science*, vol. 85, pp. 295–305, 2014.

Permissions

All chapters in this book were first published in AMSE, by Hindawi Publishing Corporation; hereby published with permission under the Creative Commons Attribution License or equivalent. Every chapter published in this book has been scrutinized by our experts. Their significance has been extensively debated. The topics covered herein carry significant findings which will fuel the growth of the discipline. They may even be implemented as practical applications or may be referred to as a beginning point for another development.

The contributors of this book come from diverse backgrounds, making this book a truly international effort. This book will bring forth new frontiers with its revolutionizing research information and detailed analysis of the nascent developments around the world.

We would like to thank all the contributing authors for lending their expertise to make the book truly unique. They have played a crucial role in the development of this book. Without their invaluable contributions this book wouldn't have been possible. They have made vital efforts to compile up to date information on the varied aspects of this subject to make this book a valuable addition to the collection of many professionals and students.

This book was conceptualized with the vision of imparting up-to-date information and advanced data in this field. To ensure the same, a matchless editorial board was set up. Every individual on the board went through rigorous rounds of assessment to prove their worth. After which they invested a large part of their time researching and compiling the most relevant data for our readers.

The editorial board has been involved in producing this book since its inception. They have spent rigorous hours researching and exploring the diverse topics which have resulted in the successful publishing of this book. They have passed on their knowledge of decades through this book. To expedite this challenging task, the publisher supported the team at every step. A small team of assistant editors was also appointed to further simplify the editing procedure and attain best results for the readers.

Apart from the editorial board, the designing team has also invested a significant amount of their time in understanding the subject and creating the most relevant covers. They scrutinized every image to scout for the most suitable representation of the subject and create an appropriate cover for the book.

The publishing team has been an ardent support to the editorial, designing and production team. Their endless efforts to recruit the best for this project, has resulted in the accomplishment of this book. They are a veteran in the field of academics and their pool of knowledge is as vast as their experience in printing. Their expertise and guidance has proved useful at every step. Their uncompromising quality standards have made this book an exceptional effort. Their encouragement from time to time has been an inspiration for everyone.

The publisher and the editorial board hope that this book will prove to be a valuable piece of knowledge for researchers, students, practitioners and scholars across the globe.

List of Contributors

Peng Zhang
Center for Durability & Sustainability Studies, Qingdao University of Technology, Qingdao 266033, China
Institute of Concrete Structures and Building Materials, Karlsruhe Institute of Technology, Karlsruhe 76131, Germany

Peibing Li, Hong Fan, Huaishuai Shang, Siyao Guo and Tiejun Zhao
Center for Durability & Sustainability Studies, Qingdao University of Technology, Qingdao 266033, China

Qing Qin, Shansuo Zheng, Lei Li, Liguo Dong, Yixin Zhang and Sha Ding
Department of Civil Engineering, Xi'an University of Architecture and Technology, Xi'an 710055, China

Seung-Yup Jang
Korea Railroad Research Institute, Suwon, Republic of Korea

Subbiah Karthick
Innovative Durable Building and Infrastructure Research Center, Hanyang University, 55 Hanyangdaehak-ro, Sangrok-gu, Ansan-si, Gyeonggi-do, Republic of Korea

Seung-Jun Kwon
Department of Civil and Environmental Engineering, Hannam University, 133 Ojeong-ding, Daejeon 306-791, Republic of Korea

Alena Sičáková, Matej Špak, Mária Kozlovská and Marek Kováč
Faculty of Civil Engineering, Technical University of Kŏsice, Vysokŏskolská 4, 042 00 Kŏsice, Slovakia

Ludovica Casnedi, Ombretta Cocco, Paola Meloni and Giorgio Pia
Dipartimento di Ingegneria Meccanica, Chimica e dei Materiali, Università degli Studi di Cagliari, Piazza d'Armi, 09123 Cagliari, Italy

Alim Feizrakhmanovich Kemalov, Ruslan Alimovich Kemalov, Ilmira Maratovna Abdrafikova, Pavel Sagitovich Fakhretdinov and Dinar Zinnurovich Valiev
Department of High-Viscosity Oil and Natural Bitumen, Kazan (Volga Region) Federal University, 18 Kremlyovskaya Street, Kazan 420008, Russia

Rui Sun and Bo Xie
Key Laboratory of New Technology for Construction of Cities in Mountain Area, Ministry of Education, School of Civil Engineering, Chongqing University, Chongqing 400045, China

Ricardo Perera
Department of Mechanical Engineering, Technical University of Madrid, José Gutiérrez Abascal 2, 28006 Madrid, Spain

Yongjun Pan
School of Automotive Engineering, Chongqing University, Chongqing 400044, China

Juan Wang, Junxiao He, Na Yang and Qingshan Yang
Beijing's Key Laboratory of Structural Wind Engineering and Urban Wind Environment, School of Civil Engineering, Beijing Jiaotong University, Beijing 100044, China

Ahmed A. M. AL-Shaar
Civil Engineering Department, University of Gaziantep, 27310 Gaziantep, Turkey
Civil Engineering Department, AL-Nahrain University, 10072 Baghdad, Iraq

Mehmet Tolga Göğüş
Civil Engineering Department, University of Gaziantep, 27310 Gaziantep, Turkey

Sung-Yong Song
Yoosong Engineering, Dongan-gu 14054, Anyang, Republic of Korea

Hwa-Sung Ryu and Sang-Heon Shin
Hanyang Experiment and Consulting, Hanyang University, Ansan 15588, Republic of Korea

Deuck-Mo Kim
Hwarok Construction Industry, Hanyang University, Ansan 15588, Republic of Korea

Won-Jun Park
Department of Architectural Engineering, Kangwon National University, Samcheok 25913, Republic of Korea

Zhenzhen Jiao, Ying Wang, Wenzhong Zheng and Wenxuan Huang
Key Lab of Structures Dynamic Behavior and Control of the Ministry of Education, Harbin Institute of Technology, Harbin 150090, China
Key Lab of Smart Prevention and Mitigation of Civil Engineering Disasters of the Ministry of Industry and Information Technology, Harbin Institute of Technology, Harbin 150090, China
School of Civil Engineering, Harbin Institute of Technology, Harbin 150090, China

Ki-Bong Park
Department of Architectural Engineering, College of Engineering, Kangwon National University, Chuncheon-si, Republic of Korea

Takafumi Noguchi
Department of Architecture, Graduate School of Engineering, The University of Tokyo, Tokyo, Japan

Wojciech Drozd, Agnieszka Lésniak and Sebastian Zaworski
Institute of Construction Management, Tadeusz Kósciuszko Cracow University of Technology, Warszawska 24 St., 31-155 Kraków, Poland

Wenhui Bian and Pengcheng Li
College of Mining and Safety Engineering, Shandong University of Science and Technology, Qingdao 266590, China

Zhaoxia Liu, Gang Pan and Wenxin Li
College of Mining and Safety Engineering, Shandong University of Science and Technology, Qingdao 266590, China
State Key Laboratory of Mining Disaster Prevention and Control Co-founded by Shandong Province and the Ministry of Science and Technology, Shandong University of Science and Technology, Qingdao 266590, China

Sugyu Lee, Kinam Hong and Yeongmo Yeon
Department of Civil Engineering, Chungbuk National University, 1 Chungdae-ro, Seowon-Gu, Cheongju, Chungbuk 28644, Republic of Korea

Kyusan Jung
Structural Engineering Research Institute, Korea Institute of Civil Engineering and Building Technology, 283 Daehwa-dong, Goyangdae-ro, Ilsanseo-gu, Goyang-si, Gyeonggi-do 10223, Republic of Korea

Ni Tongyuan
College of Materials Science and Engineering, Zhejiang University of Technology, Hangzhou 310023, China
College of Civil Engineering and Architecture, Zhejiang University of Technology, Hangzhou 310023, China
Key Laboratory of Civil Engineering Structures and Disaster Prevention and Mitigation Technology of Zhejiang Province, Hangzhou 310023, China

Yang Yang
College of Civil Engineering and Architecture, Zhejiang University of Technology, Hangzhou 310023, China
Key Laboratory of Civil Engineering Structures and Disaster Prevention and Mitigation Technology of Zhejiang Province, Hangzhou 310023, China

Jiang Chenhui
Zhejiang College of Construction, Hangzhou 311231, China

E. Foadieng
Higher Technical Teachers' Training College, University of Buea, Kumba, Cameroon
Laboratoire de Mécanique et de Modélisation des Systèmes Physiques, Faculty of Sciences, University of Dschang, Dschang, Cameroon
Laboratoire d'Ingénierie des Systèmes Industriels et de l'Environnement, IUT-Fotso Victor, University of Dschang, Dschang, Cameroon

P. K. Talla and G. B. Nkamgang
Laboratoire de Mécanique et de Modélisation des Systèmes Physiques, Faculty of Sciences, University of Dschang, Dschang, Cameroon

M. Fogue
Laboratoire d'Ingénierie des Systèmes Industriels et de l'Environnement, IUT-Fotso Victor, University of Dschang, Dschang, Cameroon

Neslihan Doğan-Sağlamtimur
Department of Environmental Engineering, Niğde Ömer Halisdemir University, 51240 Niğde, Turkey

Adnan Güven
ÇİMSA Cement Industry and Trade Inc. Co. Niğde Plant, Hacı Sabancı Avenue, Niğde, Turkey

Ahmet Bilgil
Department of Civil Engineering, Niğde Ömer Halisdemir University, 51240 Niğde, Turkey

Guan-qi Lan, Yi-hong Wang and Si-si Chao
School of Civil Engineering, Chang'an University, Middle-section of Nan'er Huan Road, Xi'an, Shaanxi 710064, China

R. Drathi
IIT Kanpur, Kanpur,Uttar Pradesh 208016, India

Index

www.ingramcontent.com/pod-product-compliance
Lightning Source LLC
LaVergne TN
LVHW060528210926
848485LV00005B/1165